Lecture Notes in Artificial Intelligence   9302

Subseries of Lecture Notes in Computer Science

LNAI Series Editors

Randy Goebel
  University of Alberta, Edmonton, Canada
Yuzuru Tanaka
  Hokkaido University, Sapporo, Japan
Wolfgang Wahlster
  DFKI and Saarland University, Saarbrücken, Germany

LNAI Founding Series Editor

Joerg Siekmann
  DFKI and Saarland University, Saarbrücken, Germany

More information about this series at http://www.springer.com/series/1244

Preface

The annual conference on Text, Speech and Dialogue (TSD), which originated in 1998, constitutes a recognized platform for the presentation and discussion of state-of-the-art technology and recent achievements in natural language processing. It has become a broad interdisciplinary forum, interweaving the topics of speech technology and language processing. The conference attracts researchers not only from Central and Eastern Europe but also from other parts of the world. Indeed, one of its goals has always been to bring together NLP researchers with various interests from different parts of the world and to promote their mutual cooperation. One of the ambitions of the conference is, as its title says, not only to deal with dialogue systems but also to improve dialogue between researchers in the two areas of NLP, i.e., between 'dialogue' and 'speech' people. Moreover, the TSD 2015 conference was partially organized at the new learning and research center of the University of West Bohemia in Plzeň (Pilsen), which was opened last year. This center, named NTIS (New Technologies for the Information Society), is a European project under the Operational Program 'Research and Development for Innovations,' aiming at establishing a European Center of Excellence at the Faculty of Applied Sciences of the University of West Bohemia. A new building was constructed for use by NTIS, housing offices and laboratories with total area of 12 thousand sq. meters. However, top-level research must always be coupled with education. Thus, new lecture rooms and laboratories for students' use were built there and the project goal is to improve the quality and to increase the number of graduates choosing to continue their careers in research and development.

The TSD 2015 conference was the 18th event in the series of the International Conferences on Text, Speech and Dialogue supported by the International Speech Communication Association (ISCA) and the Czech Society for Cybernetics and Informatics (ČSKI). The conference was held partially in the Parkhotel and Congress Center Pilsen, the largest hotel center in West Bohemia, with a capacity for 2000 people, and partially directly in the above-mentioned new building of the Faculty of Applied Sciences of the University of West Bohemia in Pilsen on September 14–17, 2015, immediately after the famous conference Interspeech 2015 in Dresden. Like its predecessors, TSD 2015 highlighted to both the academic and scientific world the importance of text and speech processing and its most recent breakthroughs in current applications. Both experienced researchers and professionals, as well as newcomers to text and speech processing, interested in designing or evaluating interactive software, developing new interaction technologies, or investigating overarching theories of text and speech processing, found in the TSD conference a forum to communicate with people sharing similar interests.

This volume contains a collection of submitted papers presented at the conference. Each of them was thoroughly reviewed by three members of the conference reviewing team consisting of more than 60 top specialists in the conference topic areas. A total of 67 accepted papers out of 138 submitted, altogether contributed by 152 authors and co-authors, were selected by the Program Committee for presentation at the conference and for publication in this book. Theoretical and more general contributions were presented in common (plenary) sessions. Problem-oriented sessions as well as panel discussions then brought together specialists in narrower problem areas with the aim of exchanging knowledge and skills resulting from research projects of all kinds.

Last but not least, we would like to express our gratitude to the authors for providing their papers on time, to the members of the conference reviewing team and the Program Committee for their careful reviews and paper selection, and to the editors for their hard work preparing this volume. Special thanks are due to the members of the Local Organizing Committee for their tireless effort and enthusiasm during the conference organization. We hope that you benefited from the event and that you also enjoyed the social program prepared by members of the Local Organizing Committee.

July 2015   Václav Matoušek

Organization

TSD 2015 was organized by the Faculty of Applied Sciences, University of West Bohemia in Plzeň (Pilsen), in cooperation with the Faculty of Informatics, Masaryk University in Brno, Czech Republic. The conference website is located at: http://www.kiv.zcu.cz/tsd2015/ or http://www.tsdconference.org.

Program Committee

Elmar Nöth, Germany (General Chair)
Eneko Agirre, Spain
Geneviève Baudoin, France
Vladimir Benko, Slovakia
Paul Cook, Australia
Jan Černocký, Czech Republic
Simon Dobrišek, Slovenia
Kamil Ekštein, Czech Republic
Karina Evgrafova, Russia
Darja Fišer, Slovenia
Eleni Galiotou, Greece
Radovan Garabik, Slovakia
Alexander Gelbukh, Mexico
Louise Guthrie, UK
Jan Hajič, Czech Republic
Eva Hajičová, Czech Republic
Yannis Haralambous, France
Hynek Heřmanský, USA
Jaroslava Hlaváčová, Czech Republic
Aleš Horák, Czech Republic
Eduard Hovy, USA
Maria Khokhlova, Russia
Daniil Kocharov, Russia
Miloslav Konopík, Czech Republic
Ivan Kopeček, Czech Republic
Valia Kordoni, Germany
Siegfried Kunzmann, Germany
Natalija Loukachevitch, Russia

Václav Matoušek, Czech Republic
France Mihelić, Slovenia
Roman Mouček, Czech Republic
Hermann Ney, Germany
Karel Oliva, Czech Republic
Karel Pala, Czech Republic
Nikola Pavešić, Slovenia
Maciej Piasecki, Poland
Josef Psutka, Czech Republic
James Pustejovsky, USA
German Rigau, Spain
Leon Rothkrantz, The Netherlands
Anna Rumshisky, USA
Milan Rusko, Slovakia
Mykola Sazhok, Ukraine
Pavel Skrelin, Russia
Pavel Smrž, Czech Republic
Petr Sojka, Czech Republic
Stefan Steidl, Germany
Georg Stemmer, Germany
Marko Tadić, Croatia
Tamás Varadi, Hungary
Zygmunt Vetulani, Poland
Pascal Wiggers, The Netherlands
Yorick Wilks, UK
Marcin Woliński, Poland
Victor Zakharov, Russia

Local Organizing Committee

Václav Matoušek (Chair)
Tomáš Brychcín
Kamil Ekštein
Tomáš Herzig
Michal Konkol

Miloslav Konopik
Pavel Král
Roman Mouček
Anna Habernalová (Secretary)

Acknowledgements

We would like to thank all the reviewers who have participated in the conference review process:

German Rigau
Tilman Becker
Ales Horák
Kamil Ekštein
Milan Rusko
Dietrich Klakow
Karl Weilhammer
Pavel Smrž
Genevieve Baudoin
Roman Mouček
Ulrich Kordon
Eleni Galiotou
Miloslav Konopík
Volker Fischer
Jan Černocký
Vladimir Petkevič
Björn Schuller
Pavel Skrelin
Marcin Woliński
Siegfried Kunzmann
Karina Evgrafova
Géza Németh
Karel Oliva
Dalibor Fiala
Eneko Agirre
Maciej Piasecki
Ivan Kopeček
Vera Evdokimova
Jozef Ivanecký
Daniil Kocharov
Pavel Rychlý
Karel Ježek

Josef Psutka
France Mihelič
Karel Pala
Pascal Wiggers
Hynek Heřmanský
Natalia Loukachevitch
Paul Cook
Maria Khokhlova
Yevgen Fedorov
Georg Stemmer
Pavel Král
Tamas Varadi
Nikola Ljubesic
Boris Lobanov
Jaroslava Hlaváčová
Vladimír Benko
Eva Hajičová
Simon Dobrišek
Ondřej Glembek
Stefan Steidl
Mykola Sazhok
Yannis Haralambous
Dirk Schnelle-Walka
Oleksandr Marchenko
Marco Lui
Christian Hacker
Agnieszka Mykowiecka
Yorick Wilks
Christophe Cerisara
Radovan Garabik
Tomáš Hercig
Oldřich Plchot

Contents

Invited Talks

Speech Analysis in the Big Data Era

Björn W. Schuller[1,2,3,4,5,6](✉)

[1] University of Passau, Chair of Complex and Intelligent Systems, Passau, Germany
schuller@ieee.org
http://www.schuller.it
[2] Deparment of Computing, Imperial College London, London, UK
[3] audEERING UG, Gilching, Germany
[4] Joanneum Research, Graz, Austria
[5] CISA, University of Geneva, Geneva, Switzerland
[6] Harbin Institute of Technology, Harbin, People's Republic of China

Abstract. In spoken language analysis tasks, one is often faced with comparably small available corpora of only one up to a few hours of speech material mostly annotated with a single phenomenon such as a particular speaker state at a time. In stark contrast to this, engines such as for the recognition of speakers' emotions, sentiment, personality, or pathologies, are often expected to run independent of the speaker, the spoken content, and the acoustic conditions. This lack of large and richly annotated material likely explains to a large degree the headroom left for improvement in accuracy by todays engines. Yet, in the big data era, and with the increasing availability of crowd-sourcing services, and recent advances in weakly supervised learning, new opportunities arise to ease this fact. In this light, this contribution first shows the de-facto standard in terms of data-availability in a broad range of speaker analysis tasks. It then introduces highly efficient 'cooperative' learning strategies basing on the combination of active and semi-supervised alongside transfer learning to best exploit available data in combination with data synthesis. Further, approaches to estimate meaningful confidence measures in this domain are suggested, as they form (part of) the basis of the weakly supervised learning algorithms. In addition, first successful approaches towards holistic speech analysis are presented using deep recurrent rich multi-target learning with partially missing label information. Finally, steps towards needed distribution of processing for big data handling are demonstrated.

Keywords: Speech analysis · Paralinguistics · Big data · Self-learning

1 Introduction

Speech recognition has more and more found its way into our every day lives – be it when searching on small hand-held devices, controlling home-entertainment or entering, e. g., an address into a navigation system. This is yet to come for many other speech analysis tasks – in particular the 'paralinguistic' ones. There, the most

P. Král and V. Matoušek (Eds.): TSD 2015, LNAI 9302, pp. 3–11, 2015.
DOI: 10.1007/978-3-319-24033-6_1

frequently encountered usage in day-to-day live is being the identification of the speaker per se, such as in some telephone banking settings. Next come, most likely – usually unnoticed, gender and age-group, e. g., in dialogue systems or simply to adapt the speech recogniser. A few applications, e. g., in video games such "Truth or Lies" promise to recognise decveptive speech or emotion. However, the plethora of other opportunities such as recognition of a speaker's personality, physical and mental load, health condition, eating condition, degree of nativeness, intoxication, or sleepiness have hardly found their way into applications noticed by the general public. While certainly of high usefulness if running properly, this raises the question on the cause which is likely the still too low reliability. While this is of course a matter of diverse factors such as the right pre-processing including de-noising and de-reverberation, optimal feature representation, optimal classification or regression and optimisation of models, the main bottleneck can likely be attributed to the sparseness of learning data for such systems. In comparison, a speech recogniser is partially being trained on more data than a human is exposed to throughout lifetime. For computational paralinguistic tasks, data often remains at the level of one up to a few hours and a handful to some hundred speakers. This data material is mostly annotated with a single phenomenon such as a particular speaker state at a time. In stark contrast to this, engines such as for the recognition of speakers' emotions, sentiment, personality, or pathologies, are often expected to run independent of the speaker, the spoken content, and the acoustic conditions. While one may argue that still, a human might not need as much data to learn certain paralinguistic characteristics as are needed to learn a whole language, clearly, more data are desired than are given at present – also as one may wish to aim at super-human abilities in some tasks. Three factors are mainly responsible for this sparseness of speech data and suited labels: the data are often 1) sparse per se, such as in the case of a sparsely occurring speaker state or trait, 2) considerably more ambiguous and thus challenging to annotate than, e. g., orthographic transcription of speech usually is, and 3) of highly private nature such as highly emotional or intoxicated data or such of speech disorders. Yet, in the big data era, it is becoming less and less the actual speech data that is lacking, as diverse resources such as the internet, broadcast, voice communication, and increased usage of speech-services including self-monitoring provide access to 'big' amounts. Instead, it is rather the labels that are missing. Luckily, with the increasing availability of crowd-sourcing services, and recent advances in weakly supervised, contextual, and reinforced learning, new opportunities arise to ease this fact.

In this light, this contribution first shows the de-facto standard in terms of data-availability in a broader range of speaker analysis tasks (Section 2). It then presents highly efficient 'cooperative' learning strategies basing on the combination of active and semi-supervised alongside transfer learning to best exploit available data (Section 3). Further, approaches to estimate meaningful confidence measures in this domain are suggested, as they form (part of) the basis of the weakly supervised learning algorithms (Section 4). In addition, first successful approaches towards holistic speech analysis are presented using deep recurrent rich multi-target learning with partially missing label information (Section 5).

Then, steps towards needed distribution of processing for big data handling are demonstrated (Section 6). Finally, some remaining aspects are discussed and conclusions are drawn (Seciton 8). Overall, a system architecture and methodology is thus discussed that holds the promise to lead to a major breakthrough in performance and generalization ability of tomorrow's speech analysis systems.

2 Data: The Availability-Shock

Few speech analysis tasks are lucky enough to have a day of labelled speech material available for training and testing of models. Taking the Interspeech 2009 – 2015 series of Computational Paralinguistics Challenges as a reference [1], one can see that, in fact, mostly around one or 'some' hours is all one is left with as a starting point to train a model for a new speech analysis task such as recognising Altzheimer, Autism, or Parkinson's Condition of a speaker. Obviously, one can hardly expect to train models independent of the speaker, language, cultural background, and co-influencing factors from such little data. Actually, some attempts at cross-corpus studies show the very weak generalisation observed for most systems trained in such a way (e. g., [2]).

3 On Efficiency: Learning Cooperatively

This reality of little labelled speech data, but availability of large(r) amounts of unlabelled such has led to a number of recent approaches in this field to most efficiently exploit both of these with little human labour involved.

3.1 Transfer Learning

Often, one has labelled data from a 'similar' domain or task available, such as recognising emotion of adult speakers, but little to no (labelled) data for the current situation of interest – let's say recognising emotion of children. In such a case, one can train a model that best learns how to 'transfer' the knowledge to the new domain, even if no labels are available at all in the new target domain [3]. An interesting further example has shown that this way, one can even train a model for the recognition of emotion in speech on music and then transfer this knowledge – in [4] this was reached by use of a sparse autoencoder that learns a compact representation of one of the domains (out of speech and music) to 'transfer' features to the respective other one. In [5] a more efficient approach was shown by training several autoencoders and to learn the differences with an additional neural network. Further, usage of related data for the initialisation of models such as in deep learning has been shown useful in general speech processing, e. g., in [6], but is less exploited in paralinguistics as of now.

3.2   (Dynamic) Active Learning

Better models can usually non-the-less be reached if one does label at least some data in the new domain or for the new task. To keep human efforts to a minimum, the computer can first decide which data points are of interest, such as by identifying 'sparse' instances such as emotional data (leaving it to the human to tell which emotion it is) versus 'non-emotional' data (which usually appears in much higher frequency and is thus 'less interesting' after some such data points have already been seen) [7,8]. Accordingly, rather than recognising different emotions, where data for each class may be to sparse, a coarser model is first chosen which is simply neutral versus non-neutral speech. In this sense, one can initialise an active learning system basically by collecting only emotionally neutral speech and then execute a novelty detection or alike for unlabelled data. As neutral emotional speech is available in large amounts or can even be synthesised [9], one can easily train such a 'one class' model by loads of data. Then, when newly seen speech is deviant in some form, a human can be asked for labelling aid. Other aspects can include the likely change of model parameters, i. e., the learning algorithm decides if the data would change its parameters significantly at all before asking for human aid on 'what it is'. Such approaches were also extended for actively learning regression tasks rather than discrete classes [10]. An interesting more recent option for fast labelling is crowd sourcing [11], as it offers to quickly reach a large amount of labellers (in fact often even in real-time which may become necessary when dealing with 'big' and growing amounts of data). However, as often laymen rather than experts in phonetics, linguistics, psychology, medicine or other related disciplines may be of relevance to the speech analysis task of interest form the majority of the crowd, one often needs a factor higher a number of labellers and has to cope with noisy labels. 'Learning' the labellers and dynamically deciding on how many labellers and 'whom to ask when' allows to source the crowd more efficiently [12].

3.3   Semi-supervised Learning

More efficiently, the computer can label data itself once it was trained supervised on some first data [13]. Obviously, this comes at a risk of labelling data erroneously and then re-training the system on partially noisy labels. Accordingly, one usually needs to make a decision based on some form of 'confidence measure' (cf. below) on whether to add a computer-labelled data instance to the learning material for (re-)training or not. In addition, one can use multiple 'views' in 'co-training' to decide on the labels of the data [14,15]. In [16] it was shown for a range of speech analysis tasks that this way, it is indeed possible to have a speech analysis system self-improve by giving it new (unlabelled) speech data observations.

3.4   Cooperative Learning

Putting the above two (i. e., active and semi-supervised learning) together leads to 'cooperative learning' [17]. The principle can best be described as follows:

For a new data instance, have the computer first decide if it can label it itself – if not, make a decision if it is worth or not to ask for human aid. [17] shows that this can be more efficient than any of the above to forms.

4   On Decision-Making: Learning Confidence Measures

As both, active, and semi-supervised learning mostly base on some confidence measure, it seems crucial to find ways of reliably estimating such. In stark contrast to speech recognition [18], there does not yet exist much literature on this topic for the field of paralinguistic speech analysis that would go beyond using a learning's algorithm inherent confidence such as the distance to the separating hyperplane of the winning class as compared to the next best class. As it is, however, this exact learning algorithm that makes the decision on a class and often does so wrongly, it seems more reliable to enable additional ways of measuring the confidence one has in a recognition result. Two ways have been shown recently in the field of Computational Paralinguistics partially exploiting the characteristics of this field.

4.1   Agreement-Based Confidence Measures

The first approach aims at estimating the agreement humans would likely have in judging the paralinguistic phenomenon of interest [19]. Thus, for a subjective task requiring several labellers such as emotion or likability of a speaker, one does not train the emotion class or degree of likability as target, but rather the percentage of human raters that agreed upon the label. Then, one can automatically estimate this percentage also for new speech data which serves as a measure on how difficult it is likely to assess a unique 'correct' label/opinion. Obviously, this can be interpreted as an indirect measure of confidence.

4.2   Learning Errors

Alternatively, one can train additional recognition engines alongside the paralinguistic engine 'in charge' using whether or not it made errors as learning target. If several such engines are trained on different data, their estimates can be used as confidence measure. In fact this measure's reliability can even be improved by semi-supervised learning [20].

5   On Seeing the Larger Picture: Learning Multiple Targets

As all our personal speaker traits as well as our multi-faceted state have an impact on the same voice production mechanism, it seems wise to attempt to see the 'larger picture'. Up to now, most work in the field of Computational Paralinguistics is pre-concerned with one phenomenon at a time such as recognition of exclusively emotion or exclusively health state or exclusively personality. Obviously, however, the voice sounds different not only because one is angry, but also as one has a flu and depending on whether one is of open or less open personality. Most attempts to learn multiple targets in parallel such that the knowledge of each other target positively influences the overall recognition accuracy have so far been focused on learning several emotion primitives commonly, cf., e. g., [21]. However, more recent attempts aim at learning a richer variety of states and traits as multiple targets [22]. This introduces the challenge to find data that are labelled in such manifold states and traits – something hardly met these days. One can easily imagine how this raises the demand in the above described efficient ways of quickly labelling data by the aid of the crowd in intelligent ways. In addition, it requires learning algorithms not only able to learn with multiple targets, but likely also with partially missing such given that one will not always be able to obtain a broad range of attributes for any voice sample 'found' or newly observed by the computer.

6   On Big Data: Distribution

Referring to 'big data' usually goes along with the amount of data being so large that 'conventional' approaches of processing cannot be applied [23]. This may require partitioning of the data and distribution of efforts [24]. While a vast body of literature exists in the field of 'core' Machine Learning on how to best distribute processing, it will remain to tailor these approaches to the needs of speech analysis. Distributed processing has been targeted considerably for speech recognition, but hardly for paralinguistic tasks where only very first experiences are reported, e. g., on optimal compression of feature vectors [25].

7   Conclusion

The next major leap forward for the field of Computational Paralinguistics and the broader field of Speech Analysis can likely be expected to be made by overcoming the ever-present sparseness of learning data by making efficient use of the big amounts of available speech by adding rich amounts of labels to these likely with help from the crowd. Such resources will have a partially noisy gold standard thus requiring potentially larger amounts of labelled speech than if labelled by experts, but it will be easier to reach large amounts of data. In particular, these may be labelled by a multitude of information rather than targeting a single phenomenon such as emotion or sleepiness of a speaker at a time. The labelling effort will likely become manageable by pre-processing by a machine that makes first decisions on the interest of the data, but that also learns how many and which raters to ask in which situation, i. e., that is not only learning about the phenomena of interest but also about the crowd that helps it to learn about these. One can probably best depict this by the metaphor of a child that not only learns about its world, but also whom to best ask about which parts of it and sometimes to better inquire several opinions. If one does not need to know "what is inside" the speech data, but simply makes further use of it, e.g., in a spoken dialogue system without attaching a human-interpretable label, unsupervised learning, i.e., clustering may be another suited variant [26,27] allowing for exploitation of big speech data.

Concerning technical necessities, efficient big data speech analysis will require reliable confidence measure estimation to decide which data to label by the computer and which by humans. Further, distributed processing may become necessary if data becomes 'too' large. Processing of big data handling comes with further new challenges such as sharing of the data and trained models, and ethical aspects such as privacy, transparency, and responsibility for what has been learnt by the machine once decisions are made [28,29].

On the other end of the 'big' scale, some speech analysis tasks will remain sparse in terms of data, e.g., for some pathological speech analysis tasks. Here, zero resource [30] or sparse resource approaches are an alternative to circumvent the data sparseness. Such approaches are known from speech recognition and keyword spotting and spoken term detection [31]. The usual application scenario there is to recognise words in 'new' spoken languages where only very sparse resources exist. For the recognition of paralinguistic tasks, an opportunity arises once at least something is known about the phenomenon of interest so to be able to implement rules such as "IF the speech is faster and the pitch is higher THAN the speaker is more aroused" etc.

As a final statement, one can easily imagine that these conclusions may hold in similar ways to a broader range of audio analysis tasks and in fact many other fields – the era of big data and increasingly autonomous machines exploring it has just begun.

Acknowledgments. The research leading to these results has received funding from the European Union's Seventh Framework and Horizon 2020 Programmes under grant agreements No. 338164 (ERC StG iHEARu), No. 645094 (SEWA), No. 644632 (Mixed-Emotions), and No. 645378 (ARIA VALUSPA).

References

1. Schuller, B., Steidl, S., Batliner, A., Hantke, S., Hönig, F., Orozco-Arroyave, J.R., Nöth, E., Zhang, Y., Weninger, F.: The INTERSPEECH 2015 computational paralinguistics challenge: degree of nativeness, Parkinson's & eating condition. In: Proc. INTERSPEECH, Dresden, Germany, p. 5. ISCA (2015)
2. Devillers, L., Vaudable, C., Chastagnol, C.: Real-life emotion-related states detection in call centers: a cross-corpora study. In: Proc. INTERSPEECH, Makuhari, Japan, pp. 2350–2353. ISCA (2010)
3. Deng, J., Zhang, Z., Eyben, F., Schuller, B.: Autoencoder-based Unsupervised Domain Adaptation for Speech Emotion Recognition. IEEE Signal Processing Letters **21**(9), 1068–1072 (2014)
4. Coutinho, E., Deng, J., Schuller, B.: Transfer learning emotion manifestation across music and speech. In: Proc. IJCNN, Beijing, China, pp. 3592–3598. IEEE (2014)
5. Deng, J., Zhang, Z., Schuller, B.: Linked source and target domain subspace feature transfer learning - exemplified by speech emotion recognition. In: Proc. ICPR, Stockholm, Sweden, pp. 761–766. IAPR (2014)

6. Swietojanski, P., Ghoshal, A., Renals, S.: Unsupervised cross-lingual knowledge transfer in DNN-based LVCSR. In: Proc. Spoken Language Technology Workshop (SLT), Miama, FL, pp. 246–251. IEEE (2012)

7. Bondu, A., Lemaire, V., Poulain, B.: Active learning strategies: a case study for detection of emotions in speech. In: Perner, P. (ed.) ICDM 2007. LNCS (LNAI), vol. 4597, pp. 228–241. Springer, Heidelberg (2007)

8. Zhang, Z., Deng, J., Marchi, E., Schuller, B.: Active learning by label uncertainty for acoustic emotion recognition. In: Proc. INTERSPEECH, Lyon, France. ISCA, pp. 2841–2845 (2013)

9. Lotfian, R., Busso, C.: Emotion recognition using synthetic speech as neutral reference. In: Proc. ICASSP, Brisbane, Australia, pp. 4759–4763. IEEE (2015)

10. Han, W., Li, H., Ruan, H., Ma, L., Sun, J., Schuller, B.: Active learning for dimensional speech emotion recognition. In: Proc. INTERSPEECH, Lyon, France, pp. 2856–2859. ISCA (2013)

11. Callison-Burch, C., Dredze, M.: Creating speech and language data with Amazon's mechanical turk. In: Proc. NAACL HLT 2010 Workshop on Creating Speech and Language Data with Amazon's Mechanical Turk, Los Angeles, CA, pp. 1–12. ACL (2010)

12. Zhang, Y., Coutinho, E., Zhang, Z., Quan, C., Schuller, B.: Agreement-based dynamic active learning with least and medium certainty query strategy. In: Proc. Advances in Active Learning: Bridging Theory and Practice Workshop held in conjunction with ICML, Lille, France, p. 5. IMLS (2015)

13. Yamada, M., Sugiyama, M., Matsui, T.: Semi-supervised speaker identification under covariate shift. Signal Processing **90**(8), 2353–2361 (2010)

14. Liu, J., Chen, C., Bu, J., You, M., Tao, J.: Speech emotion recognition using an enhanced co-training algorithm. In: Proc. ICME, Beijing, P.R. China, pp. 999–1002. IEEE (2007)

15. Jeon, J.H., Liu, Y.: Semi-supervised learning for automatic prosodic event detection using co-training algorithm. In: Proc. Joint Conference of the 47th Annual Meeting of the ACL and the 4th International Joint Conference on Natural Language Processing of the AFNLP, vol. 2, pp. 540–548. ACL, Stroudsburg (2009)

16. Zhang, Z., Deng, J., Schuller, B.: Co-training succeeds in computational paralinguistics. In: Proc. ICASSP, Vancouver, Canada, pp. 8505–8509. IEEE (2013)

17. Zhang, Z., Coutinho, E., Deng, J., Schuller, B.: Cooperative Learning and its Application to Emotion Recognition from Speech. IEEE/ACM Transactions on Audio, Speech and Language Processing **23**(1), 115–126 (2015)

18. Jiang, H.: Confidence measures for speech recognition: A survey. Speech communication **45**(4), 455–470 (2005)

19. Deng, J., Han, W., Schuller, B.: Confidence measures for speech emotion recognition: a start. In: Fingscheidt, T., Kellermann, W. (eds.) Proc. Speech Communication. 10. ITG Symposium, Braunschweig, Germany, pp. 1–4. IEEE (2012)

20. Deng, J., Schuller, B.: Confidence measures in speech emotion recognition based on semi-supervised learning. In: Proc. INTERSPEECH, Portland, OR. ISCA (2012)

21. Eyben, F., Wöllmer, M., Schuller, B.: A Multi-Task Approach to Continuous Five-Dimensional Affect Sensing in Natural Speech. ACM Transactions on Interactive Intelligent Systems **2**(1), 29 (2012)

22. Schuller, B., Zhang, Y., Eyben, F., Weninger, F.: Intelligent systems' Holistic evolving analysis of real-life universal speaker characteristics. In: Proc. 5th Int. Workshop on Emotion Social Signals, Sentiment & Linked Open Data (ES^{3}LOD 2014), satellite of LREC, Reykjavik, Iceland, pp. 14–20. ELRA (2014)

23. Madden, S.: From databases to big data. IEEE Internet Computing **3**, 4–6 (2012)
24. Chen, M., Mao, S., Liu, Y.: Big data: A survey. Mobile Networks and Applications **19**(2), 171–209 (2014)
25. Zhang, Z., Coutinho, E., Deng, J., Schuller, B.: Distributing Recognition in Computational Paralinguistics. IEEE Transactions on Affective Computing **5**(4), 406–417 (2014)
26. Wöllmer, M., Eyben, F., Reiter, S., Schuller, B., Cox, C., Douglas-Cowie, E., Cowie, R.: Abandoning emotion classes - towards continuous emotion recognition with modelling of long-range dependencies. In: Proc. INTERSPEECH, Brisbane, Australia, pp. 597–600. ISCA (2008)
27. Zhang, Y., Glass, J.R.: Towards multi-speaker unsupervised speech pattern discovery. In: Proc. ICASSP, Dallas, TX, pp. 4366–4369. IEEE (2010)
28. Richards, N.M., King, J.H.: Big data ethics. Wake Forest L. Rev. **49**, 393 (2014)
29. Wu, X., Zhu, X., Wu, G.Q., Ding, W.: Data mining with big data. IEEE Transactions on Knowledge and Data Engineering **26**(1), 97–107 (2014)
30. Harwath, D.F., Hazen, T.J., Glass, J.R.: Zero resource spoken audio corpus analysis. In: Proc. ICASSP, Vancouver, BC, pp. 8555–8559. IEEE (2013)
31. Jansen, A., Dupoux, E., Goldwater, S., Johnson, M., Khudanpur, S., Church, K., Feldman, N., Hermansky, H., Metze, F., Rose, R., et al.: A summary of the 2012 JHU CLSP workshop on zero resource speech technologies and models of early language acquisition, pp. 8111–8115 (2013)

Conference Papers

A Multi-criteria Text Selection Approach for Building a Speech Corpus

Chiragkumar Patel$^{(\boxtimes)}$ and Sunil Kumar Kopparapu

TCS Innovation Labs - Mumbai, Thane (West) 400601, Maharastra, India
{patel.chiragkumar,sunilkumar.kopparapu}@tcs.com
http://www.tcs.com

Abstract. Speech corpus is an important and primary requirement for several speech tasks. Building a speech corpora is a lengthy, time consuming and expensive process, it typically involves collection of a large set of textual utterances and then selective distribution of these text utterances among a set of speakers, called speaker sheets. These speaker sheets are articulated by speakers to generate the speech corpora. Depending on the task at hand the speech corpora needs to satisfy certain criteria; For example, a phonetically balanced speech corpora is essential for building an automatic speech recognition (ASR) engine, while for a text dependent speaker recognition engine there is a need for several spoken repetition of the same text by several speakers. In this paper, we formulate a method that enables creation of speaker sheets from a predetermined set of text utterances such that the speech corpora satisfies the desired requirement.

Keywords: Speech corpora · Speaker sheet generation · Optimization

1 Introduction

A speech corpus is a collection of speech audio files and their text transcripts. Speech corpora find use in building speech based solutions; the most common use being to build acoustic models for automatic speech recognition (ASR) purpose. The traditional approach to build a speech corpus (example SPEECON [1]) is to construct text speaker sheets which satisfy some desired criteria; recruited speaker in turn speak the text utterance to generate the speech audio data. The process of creating speaker sheets, generally picked up from a repository of textual utterances, satisfying a certain criteria is one of the important steps in building a speech corpora. In this paper, we address this problem of building speaker sheets so that the speech corpus developed satisfies multiple criteria.

Specifically, the problem that we are addressing can be stated as, given a set $\mathcal{U} = \{U_1, U_2, U_3, \cdots, U_N\}$ of N utterances in a language $\mathcal{L}$ having K phonemes denoted by $P = \{P_1, P_2, P_3, \cdots, P_K\}$, create m sets $S_1, S_2, S_3, \cdots, S_m$, each having p utterances, such that $S_i = \{S_{i1}, S_{i2}, \cdots, S_{ip}\}$ and $S_{ij} \in \mathcal{U}$ and $S_i \subset \mathcal{U}$. Note that $\mathcal{S} = \bigcup_{i=1}^{m} S_i$ is the generated speech corpus. Both $\mathcal{S}$ and $\{S_i\}_{i=1}^{m}$ need

© Springer International Publishing Switzerland 2015
P. Král and V. Matoušek (Eds.): TSD 2015, LNAI 9302, pp. 15–22, 2015.
DOI: 10.1007/978-3-319-24033-6_2

to satisfy some criteria jointly or individually depending on the requirement. For example, the criterion could be that all the K phonemes occur in $\bigcup_{i=1}^{m} S_i$ the same number of times (phonetically balanced) as is required to build a speech corpus for building an ASR or $\{S_i\}_{i=1}^{m}$ be such that it can be used for text dependent speaker identification, namely, $S_1 = S_2 = \cdots = S_m$.

In this paper, we propose a novel optimization approach which allows us to construct S given $\mathcal{U}$. The approach is based on construction of multiple cost functions which when minimized generates a S with desired requirement. The rest of the paper is organized as follows, in Section 2 we review related literature; we identify criteria that is required to build speech corpora for a particular kind of speech analysis in Section 3 and discuss our approach in detail. Experimental validation of the proposed approach is discussed in Section 4 and we conclude in Section 5.

2   Literature Review

Speech corpora development is by and far restricted to that of building a phonetically balanced corpora for ASR applications (for example, [2–6]). The general rule of thumb is that the more distributed the available training textual data, the better the utility of the data to enable building automatic speech recognition (ASR) systems. For example, in [7] a method for selecting training data from text databases is discussed for the task of syllabification. A proposal to choose data uniformly according to the distribution of some target speech unit (phoneme, word or character etc.) is discussed in [8]. They show that it is possible to select a highly informative subset of data that produces recognition performance comparable to a system that makes use of a much larger amount of data. Their experiments negate the common belief that there is no data like more data.

Optimal selection of speech data for ASR systems is proposed in [9]. They propose a method for selecting a limited set of maximally information rich speech data from a larger speech database for ASR training. It uses principal component analysis (PCA) to map the variance of speech database into a low-dimensional space, followed by clustering and a selection technique. A rapid method for optimal text selection is discussed in [10] and propose an implementation of a a faster version of an iterative greedy algorithm. Using diphone as the basic unit their selection criteria is to maximize the diphone coverage. In [11], with the aim of developing a Bengali speech corpus for a phone recognizer, they use an optimal text selection technique. They maximize the less frequent phones and minimize more frequent phones with minimum text. As can be observed, the criteria for building a speech corpora is majorly defined by the phonetic balance to automatic speech recognition (ASR). In this paper, we propose an approach which enables creation of a speech corpora by generating speaker sheets which can be used for different speech application, including ASR.

3   Proposed Approach

As mentioned earlier, assume that we are given a set $\mathcal{U} = \{U_1, U_2, U_3, \cdots, U_N\}$ of N utterances in a language $\mathcal{L}$ having K phonemes denoted by $P = \{P_1, P_2, P_3, \cdots, P_K\}$. Let α_{ij} denote the total number of occurrences of the phoneme P_j in the utterance U_i. Observe that

$$\#P_j(\mathcal{U}) = \sum_{i=1}^{N} \alpha_{ij} \tag{1}$$

denotes the total number of phoneme P_j in the set $\mathcal{U}$. Note that $\sum_{i=1}^{N} \#P_j(U_i) = \#P_j(\mathcal{U})$.

Say we are required to create m sets $S_1, S_2, S_3, \cdots, S_m$ (speaker sheets) such that each speaker sheet S_i contains p utterances, namely, $S_i = \{S_{i1}, S_{i2}, \cdots, S_{ip}\}$.

Additionally, both $\bigcup_{i=1}^{m} S_i$ ($\overset{def}{=} \mathcal{S}$) and $\{S_i\}_{i=1}^{m}$ satisfy the criteria for a speech recognition application; then $\mathcal{S}$ should be phonetically balanced, namely,

$$\#P_1(\mathcal{S}) = \#P_2(\mathcal{S}) = \#P_3(\mathcal{S}) = \cdots = \#P_K(\mathcal{S})$$

which implies that all the K phones in the corpora $\mathcal{S}$ occur equal number of times. One of the known methods adopted is to construct

$$f_i = \sum_{j=1}^{K} \overbrace{\frac{1}{\#P_j(\mathcal{U})}}^{w_j} \#P_j(U_i) \tag{2}$$

for each utterance $i = 1, 2, 3, \cdots, N$. Note that w_j is inversely proportional to $\#P_j(\mathcal{U})$ implying that if a phoneme j occurs more frequently in $\mathcal{U}$ compared to a phoneme l, then $w_l > w_j$. Subsequently, an utterance with higher number of rare phonemes will result in a higher value of f_i score. It is immediately clear that the utterance with higher f_i score must occur more number of times in $\mathcal{S}$ so as to enable phonetic balance of $\mathcal{S}$.

One of the approaches to build a phonetically balanced speaker sheet set $\mathcal{S}$ is to first sort the N utterances ($\in \mathcal{U}$) in the descending order of their f_i scores and select a value k (where $1 < k < N$) and partition the sorted N utterances into two sets; the top k utterances ($\mathcal{U}_t$) and the bottom $(N - k)$ utterances ($\mathcal{U}_b$). Note that $\mathcal{U} = \mathcal{U}_t \cup \mathcal{U}_b$; note that the set $\mathcal{U}_t$ will have most of the rare phonemes.

If every speaker sheet S_i contains p utterances, a percentage $\gamma_p\% = \left(\frac{\gamma}{p}\right) \times 100$ of utterances can be chosen from the set $\mathcal{U}_t$ and and the rest, namely, $(100 - \gamma_p\%)\%$ can be selected from the utterances set $\mathcal{U}_b$. A good choice of $\mathcal{U}_t$ and $\gamma_p\%$ will ensure that $\mathcal{S}$, represented by $\mathcal{S}(\mathcal{U}_t, \gamma_p\%)$ has the desired property (say, phonetically balanced). We now formulate the desired criteria that $\mathcal{S}$ needs to satisfy,

C_0 A measure of phonetically balanced corpus would be to compute
$\mathcal{P} = \{\#P_k(\mathcal{S}(\mathcal{U}_t, \gamma_p\%))\}_{k=1}^{K}$ and find

$$C_0(\mathcal{U}_t, \gamma_p\%) = \frac{1}{K} \sum_{k=1}^{K} (\#P_k(\mathcal{S}(\mathcal{U}_t, \gamma_p\%)) - \bar{\mathcal{P}})^2 \tag{3}$$

where $\bar{\mathcal{P}} = \frac{1}{K}\sum_{k=1}^{K}(\#P_k(\mathcal{S}(\mathcal{U}_t,\gamma_p\%)))$ is the mean. Note that $C_0(\mathcal{U}_t,\gamma_p\%)$ is the variance of $\mathcal{P}$. The configuration $(\mathcal{U}_t,\gamma_p\%)$ for which (3) is minimum is desired and gives the best phonetically balanced $\mathcal{S}$.

However, the phonetically balanced corpus is not the only desired criteria on $\{S_i\}_{i=1}^{m}$ or $\mathcal{S}$. We now elaborate on criteria which can allow for $\{S_i\}_{i=1}^{m}$ or $\mathcal{S}$ to have certain requirements imposed on them.

C_1 The minimum occurrence of every phoneme in $\mathcal{S}$ should be maximized, namely,

$$C_1(\mathcal{U}_t,\gamma_p\%) = \min_k \{\#P_k(\mathcal{S}(\mathcal{U}_t,\gamma_p\%))\} \tag{4}$$

Subsequently maximizing (4) ensures that even the phoneme that occur the least number of times in $\mathcal{S}(\mathcal{U}_t,\gamma_p\%)$ is maximized.

C_2 Let $\#U_n(\mathcal{S})$ denote the count of utterance U_n in the corpus $\mathcal{S}$. A measure of equal distribution of utterances in the corpus would be

$$C_2(\mathcal{U}_t,\gamma_p\%) = \frac{1}{N}\sum_{n=1}^{N}(\#U_n(\mathcal{S}(\mathcal{U}_t,\gamma_p\%))) - \bar{\mathcal{U}})^2 \tag{5}$$

where $\bar{\mathcal{U}} = \frac{1}{N}\sum_{n=1}^{N}(\#U_n(\mathcal{S}(\mathcal{U}_t,\gamma_p\%)))$ is the mean. Note that $C_2(\mathcal{U}_t,\gamma_p\%)$ captures the distribution of the utterances in $\mathcal{S}(\mathcal{U}_t,\gamma_p\%)$. The configuration $(\mathcal{U}_t,\gamma_p\%)$ for which (5) is minimum is desired so that all utterances occur uniformly in $\mathcal{S}$.

C_3 Common utterances between any two speaker sets, namely,

$$C_3(\mathcal{U}_t,\gamma_p\%) = \sum_{i,j=i+1}^{m} |S_i(\mathcal{U}_t,\gamma_p\%) \cap S_j(\mathcal{U}_t,\gamma_p\%)| \tag{6}$$

where $S_i(\mathcal{U}_t,\gamma_p\%) \cap S_j(\mathcal{U}_t,\gamma_p\%)$ captures the utterances that are common to both S_i and S_j and $|S_i \cap S_j|$ gives the count of common utterances. The configuration for which (6) is minimum is desired so that there is a rich utterance variability in corpus $\mathcal{S}$.

We hypothesize that the combination of these criteria jointly (7) produces the best possible dataset for a given speech application rather than the dataset which is based on individual criteria. Namely,

$$(\mathcal{U}_t^*,\gamma_p\%^*) = \arg\min_{(\mathcal{U}_t,\gamma_p\%)} \left\{ \begin{array}{l} w_1 C_0(\mathcal{U}_t,\gamma_p\%) + w_2\left(\frac{1}{C_1(\mathcal{U}_t,\gamma_p\%)}\right) + \\ w_3 C_2(\mathcal{U}_t,\gamma_p\%) + w_4 C_3(\mathcal{U}_t,\gamma_p\%) \end{array} \right\} \tag{7}$$

where w_i are the weights and $\sum_{i=1}^{4} w_i = 1$. Algorithm (1) describes this in more detail.

Note that in literature C_0 is the only criteria that is used to build a phonetically balanced speech corpus. The main contribution of this paper is to identify criteria that make the speech corpus usable. For example, $w_1 = 1$ and $w_2 = w_3 = w_4 = 0$ would reduce to what is done in the literature.

Algorithm 1. Multi Criteria approach for constructing $\mathcal{S}$.

---

for $(\mathcal{U}_t, \gamma_p\%)$ **do**
    Generate $\mathcal{S}(\mathcal{U}_t, \gamma_p\%)$
    for $k = 1, 2, \cdots K$ **do**
        Compute $\#P_k(\mathcal{S}(\mathcal{U}_t, \gamma_p\%))$
    end for
    for $n = 1, 2, \cdots N$ **do**
        Compute $\#U_n(\mathcal{S}(\mathcal{U}_t, \gamma_p\%))$
    end for

    Find $\bar{\mathcal{P}} = \frac{1}{K} \sum_{k=1}^{K}(\#P_k(\mathcal{S}(\mathcal{U}_t, \gamma_p\%)))$;
    $\bar{\mathcal{U}} = \frac{1}{N} \sum_{n=1}^{N}(\#U_n(\mathcal{S}(\mathcal{U}_t, \gamma_p\%)))$

    $C_0(\mathcal{U}_t, \gamma_p\%) = \frac{1}{K} \sum_{k=1}^{K}(\#P_k(\mathcal{S}(\mathcal{U}_t, \gamma_p\%)) - \bar{\mathcal{P}})^2$

    $C_1(\mathcal{U}_t, \gamma_p\%) = \min_k\{\#P_k(\mathcal{S}(\mathcal{U}_t, \gamma_p\%))\}$

    $C_2(\mathcal{U}_t, \gamma_p\%) = \frac{1}{N} \sum_{n=1}^{N}(\#U_n(\mathcal{S}(\mathcal{U}_t, \gamma_p\%)) - \bar{\mathcal{U}})^2$

    $C_3(\mathcal{U}_t, \gamma_p\%) = \sum_{i,j=i+1} |S_i(\mathcal{U}_t, \gamma_p\%) \cap S_j(\mathcal{U}_t, \gamma_p\%)|$
end for
Normalize using (8)
$C_0(\mathcal{U}_t, \gamma_p\%), C_1(\mathcal{U}_t, \gamma_p\%), C_2(\mathcal{U}_t, \gamma_p\%), C_3(\mathcal{U}_t, \gamma_p\%)$
to produce $nC_0(\mathcal{U}_t, \gamma_p\%), nC_1(\mathcal{U}_t, \gamma_p\%), nC_2(\mathcal{U}_t, \gamma_p\%), nC_3(\mathcal{U}_t, \gamma_p\%)$
$(\mathcal{U}_t^*, \gamma_p\%^*) = \arg\min_{(\mathcal{U}_t, \gamma_p\%)} \left\{ \begin{array}{l} w_1 nC_0(\mathcal{U}_t, \gamma_p\%) + w_2\left(\frac{1}{nC_1(\mathcal{U}_t, \gamma_p\%)}\right) + \\ w_3 nC_2(\mathcal{U}_t, \gamma_p\%) + w_4 nC_3(\mathcal{U}_t, \gamma_p\%) \end{array} \right\}$

---

4   Experimental Results

For the purpose of analysis we collected $N = 1493$ unique English utterances, namely, $\mathcal{U} = \{U_1, U_2, \cdots, U_{1493}\}$ [12]. The number of phonemes is $K = 39$. The distribution of the phonemes in $\mathcal{U}$ is shown in Figure 1. It can be observed that the phoneme 'AH' occurs the most number of times (11.55%) in $\mathcal{U}$ while the 'OY' occurs the least number of times (0.05%). All our experimental results are based on this set of utterances.

    Using (2) we arranged all the 1493 utterances in the descending order of their f_i score. The sorted utterances were partitioned into two sets. First set $(\mathcal{U}_t(k))$ contained the first $k = 1, 2, \cdots 1493$ utterances while the second set $(\mathcal{U}_b(k))$ contained $(1493 - k)$ utterances. The task was to build $m = 500$ sets with each S_i containing $p = 10$ utterances, such that $|\mathcal{S}| = 5000$. Each S_i gets $\gamma_p\% = 10, 20, \cdots, 90$ utterances from $\mathcal{U}_t(k)$ while the remaining $(100 - \gamma_p\%)$ utterances came from $\mathcal{U}_b(k)$. In all we constructed $1493 \times 9(= 13437)$ different sets of speaker sheets, namely, $\mathcal{S}(\mathcal{U}_t, \gamma_p\%)$ for $\mathcal{U}_t = 1, 2, \cdots 1493$, $\gamma_p\% = 10, 20, \cdots, 90$.

    The first set of experiments were based on Algorithm 1 with $w_1 = 1$ and $w_2, w_3, w_4 = 0$ which is the generic approach adopted to build a phonetically balanced corpus in literature. For each of these speaker sheet sets we computed

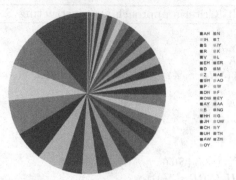

Fig. 1. Distribution of phoneme in $\mathcal{U}$.

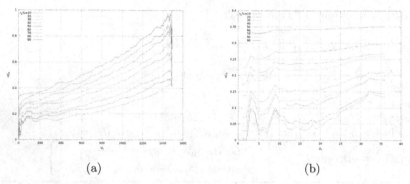

(a)   (b)

Fig. 2. (a) $C_0(\mathcal{U}_t, \gamma_p\%)$ and (b) for the first $U_t = 36$. The minimum occurs for $U_t = 2$ and $\gamma_p\% = 80$.

Table 1. $(\mathcal{U}_t, \gamma_p\%)$ determined for different criteria.

w_1	w_2	w_3	w_4	$(\mathcal{U}_t^*, \gamma_p\%^*)$
1	0	0	0	$(2, 80)$
0	1	0	0	$(126, 90)$
0	0	1	0	$(49, 10)$
0	0	0	1	$(125, 10)$
$\frac{1}{4}$	$\frac{1}{4}$	$\frac{1}{4}$	$\frac{1}{4}$	$(367, 90)$

$C_0(\mathcal{U}_t, \gamma_p\%)$ (3) and normalized it

$$nC_0 = \frac{(C_0 - \min(C_0))}{(\max(C_0) - \min(C_0))} \tag{8}$$

Figure 2 shows the plot of nC_0 for different values of $(\mathcal{U}_t, \gamma_p\%)$. The speaker sheet set (among the 13437 speaker sheet sets) with the least $C_0(\mathcal{U}_t, \gamma_p\%)$ is the set that is phonetically best balanced. As can be seen $\mathcal{U}_t(k = 2), \gamma_p\% = 80$ (see Figure 2(b)) produces the best phonetically balanced dataset suitable for ASR

Table 2. Criteria cost for different speaker sheet set.

$(\mathcal{U}_t, \gamma_p\%)$	nC_0	$1/nC_1$	nC_2	nC_3
(2, 80)	0	0.00042	0.556	0.888
(126, 90)	0.104	0	0.00714	0.0645
(49, 10)	0.260	0.00002	0	0.00098
(125, 10)	0.268	0.00007	0.00047	0
(367, 90)	0.097	9.0363e-06	0.0077	.017312

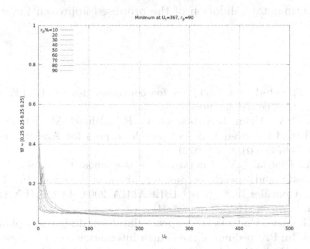

Fig. 3. $(\mathcal{U}_t^*, \gamma_p\%^*) = (367, 90)$ for $w_{1,2,3,4} = 1/4$.

type applications. Clearly, one can observe that using only the C_0 criteria does not produce the best data set (even though it is best in the sense of phonetically being balanced) because the majority of the utterances, namely $\gamma_p\% = 80$ of the dataset consists of just $\mathcal{U}_t = 2$ utterances. This motivates the need for other criteria to construct a speech corpora.

Table 1 gives the speaker sheet set (denoted by $(\mathcal{U}_t, \gamma_p\%)$) that best produces $\mathcal{S}$ if we consider different combination of the proposed criteria. Clearly the choice of speaker sheet sets depends on the emphasis given to a criteria.

When all the four criteria are given equal weightage, speaker sheet set denoted by $(\mathcal{U}_t^*, \gamma_p\%^*) = (367, 90)$ is the best (see last row in Table 1). It is clear from Table 2 that $(\mathcal{U}_t^*, \gamma_p\%^*) = (367, 90)$ is the best in terms of individual criteria $(nC_0, 1/nC_1, nC_2, nC_3)$ being minimum together.

5   Conclusion

In this paper, we proposed a multi criteria approach to generate speaker sheets which satisfy the desired requirements on the speech corpora. We believe the formulation can be used to generate speaker sheets which will assist in building a speech corpora that might be required for different speech applications and research. For example, a researcher who is doing in-depth analysis on phones may want to maximize the occurance of the phone that occurs the least number of times (using C_2). We believe that satisfying all the proposed criteria jointly will produce the best speech corpora in terms of its being useful for different speech research and development. The main contribution of this paper is (a) identification of several criteria which need to be satisfied to generate a speech corpora, (b) formulation of a multi-criteria approach by combining the criteria and (c) experimental validation of the proposed approach for speaker sheet generation.

References

1. SPEECON, Speech-driven interfaces for consumer devices (2014). http://www.speechdat.org/speecon/index.html
2. Abushariah, M.A., Ainon, R.N., Zainuddin, R., Elshafei, M., Khalifa, O.O.: Phonetically rich and balanced text and speech corpora for Arabic language. Lang. Resour. Eval. **46**(4), 601–634 (2012)
3. Pineda, L.A., Pineda, L.V., Cuétara, J., Castellanos, H., López, I.: DIMEx100: a new phonetic and speech corpus for Mexican Spanish. In: Lemaître, C., Reyes, C.A., González, J.A. (eds.) IBERAMIA 2004. LNCS (LNAI), vol. 3315, pp. 974–983. Springer, Heidelberg (2004)
4. Uraga, E., Gamboa, C.: VOXMEX speech database: design of a phonetically balanced corpus. In: Proceedings of the Fourth International Conference on Language Resources and Evaluation. LREC 2004, Lisbon, Portugal, May 26–28. European Language Resources Association (2004)
5. Asinovsky, A., Bogdanova, N., Rusakova, M., Ryko, A., Stepanova, S., Sherstinova, T.: The ORD speech corpus of Russian everyday communication "One Speaker's Day": creation principles and annotation. In: Matoušek, V., Mautner, P. (eds.) TSD 2009. LNCS, vol. 5729, pp. 250–257. Springer, Heidelberg (2009)
6. van Heerden, C., Davel, M.H., Barnard, E.: The semi-automated creation of stratified speech corpora (2013). http://www.nwu.ac.za/sites/www.nwu.ac.za/files/files/v-must/Publications/prasa2013-17.pdf
7. Tian, J., Nurminen, J., Kiss, I.: Optimal subset selection from text databases. In: Proceedings of IEEE International Conference on Acoustics, Speech, and Signal Processing, (ICASSP 2005), vol. 1, pp. 305–308, March 2005
8. Wu, Y., Zhang, R., Rudnicky, A.: Data selection for speech recognition. In: IEEE Workshop on Automatic Speech Recognition Understanding, ASRU, pp. 562–565, December 2007. http://www.cs.cmu.edu/~yiwu/paper/asru07.pdf
9. Nagroski, A. Boves, L., Steeneken, H.: Optimal selection of speech data for automatic speech recognition systems. In: ICSLP, pp. 2473–2476 (2002)
10. Chitturi, R., Mariam, S.H., Kumar, R.: Rapid methods for optimal text selection. In: Recent Advances in Natural Language Processing, September 2005
11. Mandal, S., Das, B., Mitra, P., Basu, A.: Developing Bengali speech corpus for phone recognizer using optimum text selection technique. In: 2011 International Conference on Asian Language Processing (IALP), pp. 268–271, November 2011
12. Awaz, Y.P.: Data: Speaker sheet generation for building speech corpora (2015). https://sites.google.com/site/awazyp/data/speaker

Experiment with GMM-Based Artefact Localization in Czech Synthetic Speech

Jiří Přibil[1,2]($\boxtimes$), Anna Přibilová[3], and Jindřich Matoušek[1]

[1] Department of Cybernetics, Faculty of Applied Sciences,
University of West Bohemia, Univerzitní 8, 306 14 Plzeň, Czech Republic
Jiri.Pribil@savba.sk, jmatouse@kky.zcu.cz
[2] SAS, Institute of Measurement Science, Dúbravská cesta 9,
841 04 Bratislava, Slovakia
[3] Faculty of Electrical Engineering and Information Technology,
Institute of Electronics and Photonics, Slovak University of Technology,
Ilkovičova 3, 812 19 Bratislava, Slovakia
Anna.Pribilova@stuba.sk

Abstract. The paper describes an experiment with using the statistical approach based on the Gaussian mixture models (GMM) for localization of artefacts in the synthetic speech produced by the Czech text-to-speech system employing the unit selection principle. In addition, the paper analyzes influence of different number of used GMM mixtures, and the influence of setting of the frame shift during the spectral feature analysis on the resulting artefact position accuracy. Obtained results of performed experiments confirm proper function of the chosen concept and the presented artefact position localizer can be used as an alternative to the standardly applied manual localization method.

Keywords: Quality of synthetic speech · Text-to-speech system · GMM classification · Statistical analysis

1 Introduction

At present, the synthetic speech produced by the text-to-speech (TTS) systems is often used to make human-machine interaction more effective and also to enable easy selection of a suitable methodology of dialogue management. Different speech synthesis methods are implemented in the TTS systems. The unit selection (USEL) method [1] is one of the main strategies used in the corpus-based speech synthesis. Being a concatenative method, any concatenation point can be a source of an audible artefact. During the synthesis process artefacts of different origin can occur in the finally generated speech. Apart from the wrong

The work has been supported by the Technology Agency of the Czech Republic, project No. TA 01030476, the Grant Agency of the Slovak Academy of Sciences (VEGA 2/0013/14), and the Ministry of Education of the Slovak Republic (KEGA 022STU-4/2014).

© Springer International Publishing Switzerland 2015
P. Král and V. Matoušek (Eds.): TSD 2015, LNAI 9302, pp. 23–31, 2015.
DOI: 10.1007/978-3-319-24033-6_3

description of the natural source speech (such as wrong annotation and/or segmentation [2]), the most prevalent causes of the artefacts are related mainly to F0 discontinuities [3]. Nevertheless, other causes like temporal inconsistencies or spectral mismatches at concatenation points can also result in serious artefacts [4]. The artefacts as perceived by human listeners should ideally correspond to high values of the target or concatenation costs (however, this assumption is often not met as shown in [5]). Therefore, we want to find out some other evaluation methods, which can work without any human interaction. Among the objective methods, the automatic speech recognition system yielding the final evaluation in the form of the recognition score can be used [6]. These recognition systems are often based on hidden Markov models [7] or Gaussian mixture models (GMM) [8]. The GMM of a speaker, providing a probabilistic model of the underlying sounds of a person's voice, is useful for text-independent speaker identification. Often, a fusion of different recognition methods performs best, e.g. combination of GMM with support vector machines (SVM) used for speaker recognition in the same way as for language recognition [9]. Motivation of our work was to design, test, and verify functionality of the localization of artefacts in the synthetic speech produced by the USEL method. We expect that these steps will support our main goal – application of this GMM artefact localizer for automatic evaluation of synthetic speech. The paper next analyzes and compares the influence of the number of used GMM mixture components, and the influence of different frame shift during spectral feature analysis on the final localization accuracy.

2   Method

The artefact localization experiment starts with the manual classification of the artefact type and visual identification of the beginning and ending frames of the artefact by comparing the clean original speech signal to the same sentence with the artefact. After this visual localization of the artefact position, the nearest region of interest (ROI) is determined for further detailed analysis – see an example in Fig. 1. In the next step, the parts of the artefacted synthetic speech inside the ROI are described by the spectral and prosodic parameters as the input feature vectors for creation and training of the GMMs. In general, the synthetic speech produced by the TTS system based on USEL method can generate five basic types of artefacts defined as:

1. local decrease/increase of the signal RMS (energy),
2. local increase/decrease of the pitch frequency (F0),
3. combination (superposition) of local energy and F0 decrease,
4. correlated occurrence of local increase of F0 and energy,
5. wrongly chosen speech unit from the database (with respect to the left-right context).

The GMMs represent a linear combination of multiple Gaussian probability distribution functions of the input data vector. For GMM creation, it is necessary

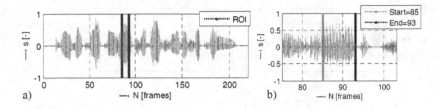

Fig. 1. Demonstration of visual localization of artefacts: sentence *"sent02d"* (male voice) with an artefact in the selected ROI (a), detailed part in the artefact neighbourhood with the determined start/end location in frames (b).

to determine the covariance matrix, the vector of mean values, and the weighting parameters from the input training data. Using the expectation-maximization (EM) iteration algorithm, the maximum likelihood function of GMM is found [8]. The EM algorithm is controlled by the number of used mixtures (N_{gmix}) and the number of iteration steps (N_{iter}). The GMM classifier returns the probability (so called *score*) that the tested utterance belongs to the GMM model. In the standard realization of the GMM classifier, the resulting class is given by the maximum overall probability of all obtained scores corresponding to K output classes using the feature vector T from the tested sentences [8]. For our purpose, only one output class is defined and the GMM classifier processes N input feature vectors T_1 to T_N corresponding to N frames of the tested sentence. The main idea of the proposed localization method is based on the assumption of correlation between the positions of determined score maxima and the location of the analyzed speech frame (maximum of probability with the trained GMM model) – see demonstration example in Fig. 2. Therefore, from the vector of obtained values of the normalized score $\{sc_1, .., sc_N\}$, the first, the second, and the third maxima are determined for the next evaluation. In our experiment, three types of the GMM models are created and trained for each of the voices (male/female):

a) starting part of the artefact – speech signal in the left border frame of the artefact and $\pm i$ frames in its neighbourhood,
b) ending part of the artefact – speech signal in the right border frame of the artefact and $\pm i$ frames in its neighbourhood,
c) body of the artefact – speech signal spanning between the starting and the ending frames.

In the classification phase, the input feature vectors from the tested sentence are compared in parallel with the three trained GMM models, so we obtain three output vectors of the normalized score (probability). Applying logical matching by predefined rules, the final localization of the artefact position is obtained – see the block diagram in Fig. 3. The rules enable to solve the basic possible situations when the localization algorithm might fail – the starting frame position must precede the ending one, the artefact body must lie within the interval from

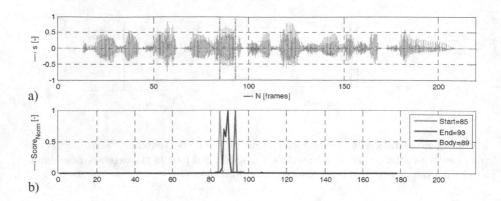

a)

b)

Fig. 2. Demonstration of artefact localization using three GMM models for starting, ending, and body parts: sentence "*sent02d*" with the manually determined artefact (a), values of the normalized GMM scores for all three models (b).

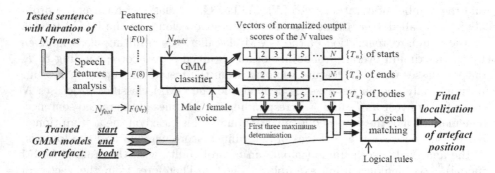

Fig. 3. Block diagram of the developed GMM classifier for artefact localization.

start to end, etc. If one of these conditions is not fulfilled, the second or the third determined score maximum (position) will be applied. By this approach, only one artefact within the tested sentence can be found, and the artefact presence must be confirmed (detected) by another method [10].

Spectral features like MFCC together with energy and prosodic parameters are most commonly used in GMM-based speaker [11] and emotional voice classification [12]. However, the other spectral properties can be used as the main indicator for speech classification. The basic properties determined from the spectral envelope comprise the spectral decrease (tilt), the spread (SPR), and the centroid (SC). Among the supplementary spectral features, the harmonic-to-noise ratio (HNR), the spectral flatness (SFM), and the spectral entropy (SE) were determined. As regards the supra-segmental properties including the speech signal energy (calculated from the first cepstral coefficient En_{c0}, or with the help of autocorrelation function – En_{R0}), the prosodic parameters like the differential contour $F0_{DIFF}$, microintonation, jitter, shimmer, etc., were calculated. Obtained values in the form of the feature vectors with the length N_{feat} are used in dependence on the voice type (male/female) for further processing.

3   Material, Experiments, and Results

The synthetic speech produced by the Czech TTS system based on the USEL method was used in the performed experiments. For the creation and training of the GMM models the speech corpus consisting of declarative sentences with one "real" artefact per sentence for male + female voices was used. This corpus includes 20 + 20 sentences with duration from 2.5 to 5 seconds, sampled at frequency of 16 kHz. Due to a limited number of usable sentences with "real" artefacts occurring during the TTS synthesis, the classical cross-validation data selection approach [8] could not be applied. Therefore, we prepared a special set of another 20 + 20 artefacted sentences for the testing phase of the localization experiment. These sentences were derived from the original ones using different manipulations (cutting or adding a part of a sentence, changing a part of it using the signal from other sentences, etc.) to change the time position of the artefact inside the original sentence. In all cases, the determination of positions of the artefacts was performed manually and these values were used for further comparison of the localization accuracy. The basic frame length w_L for speech feature analysis depends on the mean pitch period of the processed signal: 24-ms/20-ms frames were used for the male/female voices. The performed comparison experiments were aimed at analysis of:

Table 1. Comparison of the mean artefact position relative error in [frames] and its standard deviation (in parentheses) depending on the used number of GMM mixtures N_{gmix}; $w_O = w_L/4$.

Voice /N_{GMIX}	16	32	48	64	128
APE$_{\text{REL}}$ Male	19.8 (47)	2.2 (5)	9.6 (23)	9.9 (23)	11.1 (26)
APE$_{\text{REL}}$ Female	8.6 (19)	2.8 (7)	3.3 (9)	4.1 (11)	7.5 (15)
APE$_{\text{REL}}$ Total	14.2	**2.5**	6.5	7	9.3

Table 2. Mean values of the artefact position relative error in [frames] and its standard deviation (in parentheses) depending on the used frame shift w_O for speech feature determination; $N_{gmix} = 32$.

Voice / w_O	$w_L/2$ ($i=3$)	$w_L/3$ ($i=5$)	$w_L/4$ ($i=7$)	$w_L/5$ ($i=9$)	$w_L/6$ ($i=11$)
APE$_{\text{REL}}$ Male	19.2 (36)	7.6 (18)	2.2 (5)	6.6 (16)	7.2 (14)
APE$_{\text{REL}}$ Female	12.7 (24)	6.4 (10)	2.8 (7)	4.3 (10)	7.5 (15)
APE$_{\text{REL}}$ Total	15.9	7	**2.5**	5.5	7.4

- influence of the initial parameter during the GMM creation on the resulting artefact localization error: the number of applied mixtures $N_{gmix} = \{16, 32, 48, 64, \text{ and } 128\}$ – see the summarized mean values in Table 1,
- influence of the chosen frame shift w_O for computing of the input feature vectors on precision of the artefact localization; the evaluation has been done for $w_O = \{1/2, 1/3, 1/4, 1/5, \text{ and } 1/6\}$ of the frame length w_L – see the summarized mean values of relative errors in Table 2,
- comparison of computational complexity: CPU times for the GMM creation, training, and artefact localization for different number of used mixtures and frame shifts summarized for both voices presented by Table 3.

The artefact neighbourhood of $\pm i$ frames for the speech feature analysis was set to $i = \{3, 5, 7, 9, \text{ and } 11\}$ in correspondence with the chosen frame shift w_O. The artefact position relative error APE_{REL} in frames was calculated as the mean value from the absolute position error of the starting and the ending parts in every sentence as $APE_{REL} = mean(APE_{ABSstart}, APE_{ABSend}) * w_O$ – see an example of a detailed graph comparison in Fig. 4. According to the research results published in [13] the length of the input feature vector was set to $N_{feat} = 16$. The input data vector for GMM training and testing contains a mix of the basic spectral properties {SPR, SC, and tilt} and the supplementary spectral features {HNR, SFM, SE} together with the supra-segmental parameters {$En_{c0}, En_{R0}, F0_{DIFF}$, jitter, and shimmer}. The statistical types – median values, range of values, standard deviations (std), and/or relative maximum and minimum were used for implementation in the feature vectors. To determine the spectral features and the prosodic parameters, the elementary functions from Matlab ver. 2010b environment with the help of "Signal Processing Toolbox" and "Statistics Toolbox" were applied. The computational complexity was tested on the PC with the following configuration: processor Intel(R) i3-2120 at 3.30 GHz, 8 GB RAM, and Windows 7 professional OS. The basic functions from the Ian T. Nabney "Netlab" pattern analysis toolbox [14] were used. For creation of the GMM models, data training, and testing the simple diagonal covariance matrix of the mixture models was applied in this localization experiment.

Table 3. Comparison of the computational complexity (CPU time in [s]) for different number of used mixtures and frame shifts; summarized for both genders.

Phase /	N_{GMIX} ($w_O = w_L/4$)					w_O ($N_{GMIX} = 32$)				
	16	32	48	64	128	$w_L/2$	$w_L/3$	$w_L/4$	$w_L/5$	$w_L/6$
GMM creation	1.516	2.453	3.812	4.766	10.453	1.0	1.98	2.453	4.66	6.95
Localization *)	4.15	4.25	4.32	4.45	4.90	0.96	2.46	4.25	6.78	9.58
	(1.2)	(1.3)	(1.2)	(1.3)	(1.4)	(0.3)	(0.7)	(1.3)	(2.0)	(2.9)
Total time	5.66	6.71	8.13	9.22	15.35	1.96	4.44	6.71	11.44	16.53

*) Mean values per sentence including the standard deviation values in [ms] (in parentheses).

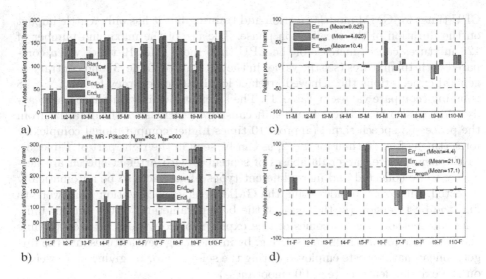

Fig. 4. Detailed comparison of the defined/identified artefact starting and ending parts positions (a-b), and the position error (c-d): for male sentences Mt1-Mt10 (a-c) and female sentences Ft1-Ft10 (b-d); $N_{gmix} = 32$; $w_O = w_L/2$.

4 Discussion and Conclusion

The obtained experimental results practically confirm functionality of the developed GMM-based artefacts localizer. From the performed experiments follows that there exists a principal influence of the used number of mixtures and the frame shift for computing of the input feature vectors on the precision of the artefact localization. The (sub)optimal setting of these parameters was $N_{gmix} = 32$ and $w_O = w_L/4$ as documented by the achieved minima of the artefact position relative error in Tables 1 and 2. Therefore, these settings were chosen for next use in our experiments. Obtained mean values of APE_{REL} (as well as their std values) are approximately the same for male and female sentences only in the case of the "optimal" duration tolerance of the starting/ending positions (uncertainty of the beginning or the end of the region of the artefact). Accumulating the starting and the ending duration tolerances gives about 2.5 frames in total, i.e. 7.5 milliseconds for $w_O = w_L/4$. For other parameter setting, the results for the female sentences are about twice better than for the male ones. It can be caused by the fact that, generally, the range of F0 contour as well as the range of energy and time duration changes are greater for the female voice, and this effect was also reflected in the production of artefacts. Therefore, these greater changes are better trained in the GMM models and then they can be better detected in the classification phase. In addition, the used frame length w_L was shorter in the case of the female voice (160 samples in comparison to 192 samples for the male voice at 16 kHz sampling frequency). The applied number of mixtures has a great influence on the computational complexity (the measured

CPU time) for GMM models creation and training but it has only a little impact on the duration of the localization phase. The use of the maximum number of 128 mixtures causes increase of the CPU time more than 10 times when compared with 16 mixtures (see the values in the left part of Table 3) and contrary to general expectations the achieved artefact localization accuracy for $N_{gmix} \geq 48$ is falling (compare values in Table 1). The artefact localization phase was heavily affected also by different frame shift causing an increase of the data size from the processed speech signal (approx. 10 times higher computational complexity for settings of $w_L/6$ than for $w_L/2$, as can be seen in the right part of Table 3). In the near future, larger databases of sentences with artefact speech are necessary to be collected and more different types of artefacts to be automatically detected. Moreover, besides using the GMM artefact localization framework to find the best version (in the sense of the best parameters) of the current unit-selection method, the outcomes of this experiment could also be used to tune the unit-selection mechanism itself, e.g. by introducing new features into the target/concatenation costs employed during the selection, or by giving more weight on the existing features (e.g. F0 smoothness).

References

1. Tihelka, D., Kala, J., Matoušek, J.: Enhancements of viterbi search for fast unit selection synthesis. In: Proceedings of Interspeech 2010, Makuhari, Japan, pp. 174–177 (2010)
2. Matoušek, J., Tihelka, D.: Annotation errors detection in TTS corpora. In: Proceeding of Interspeech 2013, Lyon, France, pp. 1511–1515 (2013)
3. Legát, M., Matoušek, J.: Identifying concatenation discontinuities by hierarchical divisive clustering of pitch contours. In: Habernal, I., Matoušek, V. (eds.) TSD 2011. LNCS, vol. 6836, pp. 171–178. Springer, Heidelberg (2011)
4. Tihelka, D., Matoušek, J., Kala, J.: Quality deterioration factors in unit selection speech synthesis. In: Matoušek, V., Mautner, P. (eds.) TSD 2007. LNCS (LNAI), vol. 4629, pp. 508–515. Springer, Heidelberg (2007)
5. Legát, M., Tihelka, D., Matoušek, J.: Configuring TTS evaluation method based on unit cost outlier detection. In: Habernal, I. (ed.) TSD 2013. LNCS, vol. 8082, pp. 177–184. Springer, Heidelberg (2013)
6. Bello, C., Ribas, D., Calvo, J.R., Ferrer, C.A.: From speech quality measures to speaker recognition performance. In: Bayro-Corrochano, E., Hancock, E. (eds.) CIARP 2014. LNCS, vol. 8827, pp. 199–206. Springer, Heidelberg (2014)
7. Juang, B.H., Rabiner, L.R.: Hidden Markov Models for Speech Recognition. Technometrics 33(3), 251–272 (1991)
8. Reynolds, D.A., Rose, R.C.: Robust Text-Independent Speaker Identification Using Gaussian Mixture Speaker Models. IEEE Transactions on Speech and Audio Processing 3, 72–83 (1995)
9. Togneri, R., Pullella, D.: An Overview of Speaker Identification: Accuracy and Robustness Issues. IEEE Circuits and Systems Magazine 11(2), 23–61 (2011)
10. Přibil, J., Přibilová, A., Matoušek, J.: Detection of artefacts in Czech synthetic speech based on ANOVA statistics. In: Proc. of the 37th International Conference on Telecommunications and Signal Processing TSP 2014, Berlin, Germany, pp. 414–418 (2014)

11. Venturini, A., Zao, L., Coelho, R.: On Speech Features Fusion, α-Integration Gaussian Modeling and Multi-Style Training for Noise Robust Speaker Classification. IEEE/ACM Transactions on Audio, Speech, and Language Processing **22**(12), 1951–1964 (2014)
12. Shah, M., Chakrabarti, C., Spanias, A.: Within and Cross-Corpus Speech Emotion Recognition Using Latent Topic Model-Based Features. EURASIP Journal on Audio, Speech, and Music Processing **2015**(4), 1–17 (2015)
13. Přibil, J., Přibilová, A.: Evaluation of Influence of Spectral and Prosodic Features on GMM Classification of Czech and Slovak Emotional Speech. EURASIP Journal on Audio, Speech, and Music Processing **2013**(8), 1–22 (2013)
14. Nabney, I.T.: Netlab Pattern Analysis Toolbox (retrieved October 2, 2013). http://www.mathworks.com/matlabcentral/fileexchange/2654-netlab

Tuned and GPU-Accelerated Parallel Data Mining from Comparable Corpora

Krzysztof Wołk[(✉)] and Krzysztof Marasek

Department of Multimedia, Polish-Japanese Academy of Information Technology,
Koszykowa 86, Warsaw, Poland
{kwolk,kmarasek}@pja.edu.pl

Abstract. The multilingual nature of the world makes translation a crucial requirement today. Parallel dictionaries constructed by humans are a widely available resource, but they are limited and do not provide enough coverage for good quality translation purposes, due to out-of-vocabulary words and neologisms. This motivates the use of statistical translation systems, which are unfortunately dependent on the quantity and quality of training data. Such has a very limited availability especially for some languages and very narrow text domains. Is this research we present our improvements to Yalign's mining methodology by reimplementing the comparison algorithm, introducing a tuning scripts and by improving performance using GPU computing acceleration. The experiments are conducted on various text domains and bi-data is extracted from the Wikipedia dumps.

Keywords: Machine translation · Comparable corpora · Machine learning · NLP · Knowledge-free learning

1 Introduction

The aim of this study is the preparation of parallel and comparable corpora. This work improves SMT quality through the processing and filtering of parallel corpora and through extraction of additional data from the resulting comparable corpora. To en-rich the language resources of SMT systems, various adaptation and interpolation techniques will be applied to the prepared data. Evaluation of SMT systems was per-formed on random samples of parallel data using automated algorithms to evaluate the quality and potential usability of the SMT systems' output [1].

As far as experiments are concerned, the Moses Statistical Machine Translation Toolkit software, as well as related tools and unique implementations of processing scripts for the Polish language, are used. Moreover, the multi-threaded implementation of the GIZA++ tool is employed to train models on parallel data and to perform their symmetrization at the phrase level. The SMT system is tuned using the Minimum Error Rate Training (MERT) tool, which through parallel data specifies the optimum weights for the trained models, improving the resulting

P. Král and V. Matoušek (Eds.): TSD 2015, LNAI 9302, pp. 32–40, 2015.
DOI: 10.1007/978-3-319-24033-6_4

translations. The statistical language models from single-language data are trained and smoothed using the SRI Language Modeling toolkit (SRILM). In addition, data from outside the the-matic domain is adapted. In the case of parallel models, Moore-Levis Filtering is used, while single-language models are linearly interpolated [1].

Lastly, the Yalign parallel data-mining tool is enhanced. Its speed is increased by reimplementing it in a multi-threaded manner and by employing graphics processing unit (GPU) horsepower for its calculations. Yalign quality is improved by using the Needleman-Wunch algorithm for sequence comparison and by developing a tuning script that adjusts mining parameters to specific domain requirements [2].

2 State of the Art

An interesting idea for mining parallel data from Wikipedia was described in [3]. The authors propose two separate approaches. The first idea is to use an online machine translation (MT) system to translate Dutch Wikipedia pages into English, and they try to compare original EN pages with translated ones. The idea, although interesting, seems computationally infeasible, and it presents a chicken-or-egg problem. Their second approach uses a dictionary generated from Wikipedia titles and hyperlinks shared between documents. Unfortunately, the second method was reported to return numerous, noisy sentence pairs.

Yasuda and Sumita [4] proposed a MT bootstrapping framework based on statistics that generate a sentence-aligned corpus. Sentence alignment is achieved using a bilingual lexicon that is automatically updated by the aligned sentences. Their solution uses a corpus that has already been aligned for initial training. They showed that 10% of Japanese Wikipedia sentences have an English equivalent.

Interwiki links were leveraged by the approach of Tyers and Pienaar in [5]. Based on Wikipedia link structure, a bilingual dictionary is extracted. In their work they measured the average mismatch between linked Wikipedia pages for different languages. They found that their precision is about 69–92%.

In [6] the authors attempt to advance the state of art in parallel data mining by modeling document-level alignment using the observation that parallel sentences can most likely be found in close proximity. They also use annotation available on Wikipedia and an automatically induced lexicon model. The authors report recall and precision of 90% and 80%, respectively.

The author of [7] introduces an automatic alignment method for parallel text fragments that uses a textual entailment technique and a phrase-based SMT system. The author states that significant improvements in SMT quality were obtained (BLEU increased by 1.73) by using this aligned data.

The authors in [8] propose obtaining only title and some meta-information, such as publication date and time for each document, instead of its full contents to reduce the cost of building the comparable corpora. The cosine similarity of the titles' term frequency vectors was used to match titles and the contents of matched pairs.

In the present research, the Yalign tool is used. The solution was far from perfect, but after improvements that were made during this research, it supplied the SMT systems with bi-sentences of good quality in a reasonable amount of time.

3   Parallel Data Mining

For the experiments in data mining, the TED lectures domain prepared for IWSLT 2014[1] evaluation campaign by the FBK[2], was chosen. This domain is very wide and covers many unrelated subject areas [9]. Narrower domains were selected as well. The first corpus is composed of documents from the European Medicines Agency (EMEA) [10]. The second corpus was extracted from the proceedings of the European Parliament (EUP) by Philipp Koehn (University of Edinburgh) [11]. In addition, experiments on the Basic Travel Expression Corpus (BTEC), tourism-related sentences, were also conducted [12]. Lastly, a big corpus obtained from the OpenSubtitles.org web page was used as an example of human dialogs. Table 1 provides details on the number of unique words (WORDS) and their forms, as well as the number of bilingual sentence pairs (PAIRS).

Table 1. Corpora specification

CORPORA	PL WORDS	EN WORDS	PAIRS
BTEC	50,782	24,662	220,730
TED	218,426	104,117	151,288
EMEA	148,230	109,361	1,046,764
EUP	311,654	136,597	632,565
OPEN	1,236,088	749,300	33,570,553

4   Yalign and Improvements

The Yalign tool was designed to automate the parallel text mining process by finding sentences that are close translation matches from comparable corpora. This presents opportunities for harvesting parallel corpora from sources like translated documents and the web. In addition, Yalign is not limited to a particular language pair. However, alignment models for two selected languages must first be created [2].

Unfortunately, the Yalign tool is not computationally feasible for large-scale parallel data mining. The standard implementation accepts plain text or web links, which need to be accepted, as input, and the classifier is loaded into memory for each pair alignment. In addition, the Yalign software is single-threaded. To improve performance, a solution that supplies the Yalign tool with articles from the database within one session, with no need to reload the classifier each

---

[1] http://www.iwslt.org
[2] http://www.fbk.eu

time, was developed. The developed solution also facilitated multi-threading and decreased the mining time by a factor of 6.1 (using a 4-core, 8-thread i7 CPU). The alignment algorithm was also reimplemented (Needleman-Wunch is used instead of A* Search) for better accuracy and to leverage the power of GPUs for additional computing requirements. The tuning algorithm was also implemented. The two NW algorithms, with and without GPU optimization, are conceptually identical, but the second has an advantage in efficiency, depending on the hardware, up to $\max(n, m)$ times. However, the results of the A* algorithm, if the similarity calculation and the gap penalty are defined as in the NW algorithm, will be the same only if there is an additional constraint on paths: Paths cannot go upward or leftward in the M matrix. Yalign does not impose these additional conditions, so in some scenarios, repetitions of the same phrase may appear. In fact, every time the algorithm decides to move up or left, it is coming back into the second and first sequence, respectively.

The quality of alignments in Yalign is defined by a tradeoff between precision and recall. The Yalign has two configurable variables:

- Threshold: the confidence threshold to accept an alignment as "good." A lower value means more precision and less recall. The "confidence" is a probability estimated from a support vector machine classifying "is a translation" or "is not a translation."
- Penalty: controls the amount of "skipping ahead" allowed in the alignment. Both of these parameters are selected automatically during training, but they can be adjusted if necessary. The solution implemented in this research also introduces a tuning algorithm for those parameters, which allows better adjustment of them [2].

5 Evaluation of Obtained Comparable Corpora

To evaluate the corpora, each corpus was divided into 200 segments, and 10 sentences were randomly selected from each segment. This methodology ensured that the test sets covered the entire corpus. The selected sentences were removed from the corpora. The testing system was trained with the baseline settings. In addition, a system was trained with extended data from the Wikipedia corpora. Lastly, Modified Moore-Levis Filtering was used for the Wikipedia corpora domain adaptation. The monolingual part of the corpora was used as language model and was adapted for each corpus by using linear interpolation [13].

The evaluation was conducted using test sets built from 2,000 randomly selected bi-sentences taken from each domain. For scoring purposes, four well-known metrics that show high correlation with human judgments were used. Among the commonly used SMT metrics are: Bilingual Evaluation Understudy (BLEU), the U.S. National Institute of Standards & Technology (NIST) metric, the Metric for Evaluation of Translation with Explicit Ordering (METEOR) and Translation Error Rate (TER) [13].

First, speed improvements were made by introducing multi-threading to the algorithm, using a database instead of plain text files or Internet links, and using

GPU acceleration in sequence comparison. More importantly, two improvements were made to the quality and quantity of the mined data. The A* search algorithm [14] was modified to use Needleman-Wunch [15], and a tuning script of mining parameters was developed. During this empirical research, it was realized that Yalign suffers from a problem that produces different results and quality measures, depending on whether the system was trained from a foreign to a native language or vice versa. To cover as much parallel data as possible during the mining, it is necessary to train the classifiers bidirectionally for the language pairs of interest. By doing so, additional bi-sentences can be found. The data mining approaches used were: directional (PL → EN classifier) mining (MONO), bi-directional (additional EN → PL classifier) mining (BI), bi-directional mining with Yalign using a GPU-accelerated version of the Needleman-Wunch [16] algorithm (NW), and mining using a NW version of Yalign that was tuned (NWT). The results of such mining are shown in Table 2.

Table 2. Number of obtained Bi-Sentences

Mining Method	Number of Bi-Sentences	Uniq PL Tokens	Uniq EN Tokens
MONO	510,128	362071	361039
BI	530,480	380771	380008
NW	1,729,061	595064	574542
NWT	2,984,880	794478	764542

As presented in Table 2, each of the improvements increased the number of parallel sentences discovered. However, there is no indication of the quality of the obtained data, SMT improvements, or information regarding the computation time of the NW version of Yalign.

Table 3. Computation Time of Different Yalign Version

Mining Method	Computation Time [s]
YALIGN	89,67
M YALIGN	14,7
NW YALIGN	17,3
GNW YALIGN	15,2

To address these aspects, two additional experiments were conducted. First, in Table 3 a speed comparison is made using different versions of the Yalign Tool. A total of 1,000 comparable articles were randomly selected from Wikipedia and aligned using the native Yalign implementation (Yalign), multi-threaded implementation (M Yalign), Yalign with the Needleman-Wunch algorithm (NW Yalign), and Yalign with a GPU-accelerated Needleman-Wunch algorithm (GNW Yalign). Second, MT experiments were conducted to verify potential gains in translation quality on the data that was tuned and aligned using a different heuristic. The TED, EUP, EMEA and OPEN domains were used for this purpose. For each of the domains, the system was trained using baseline settings.

Table 4. Results of SMT Enhanced Comparable Corpora for PL to EN and PL to EN transltion

		PL to EN				EN to PL			
		BLEU	NIST	TER	METEOR	BLEU	NIST	TER	METEOR
TED	BASE	16,96	5,26	67,10	49,42	10,99	3,95	74,87	33,64
	MONO	16,97	5,39	65,83	50,42	11,24	4,06	73,97	34,28
	BI	17,34	5,37	66,57	50,54	11,54	4,02	73,75	34,43
	NW	17,45	5,37	65,36	50,56	11,59	3,98	74,49	33,97
	TNW	17,50	5,41	64,36	90,62	11,98	4,05	73,65	34,56
EUP	BASE	36,73	8,38	47,10	70,94	25,74	6,56	58,08	48,46
	MONO	36,89	8,34	47,12	70,81	24,71	6,37	59,41	47,45
	BI	36,56	8,34	47,33	70,56	24,63	6,35	59,73	46,98
	NW	35,69	8,22	48,26	69,92	24,13	6,29	60,23	46,78
	TNW	35,26	8,15	48,76	69,58	24,32	6,33	60,01	47,17
EMEA	BASE	62,60	10,19	36,06	77,48	56,39	9,41	40,88	70,38
	MONO	62,48	10,20	36,29	77,48	55,38	9,25	42,37	69,35
	BI	62,62	10,22	35,89	77,61	56,09	9,32	41,62	69,89
	NW	62,69	10,27	35,49	77,85	55,80	9,30	42,10	69,54
	TNW	62,88	10,26	35,42	77,96	55,63	9,31	41,91	69,65
OPEN	BASE	64,58	9,47	33,74	76,71	31,55	5,46	62,24	47,47
	MONO	65,77	9,72	32,86	77,14	31,27	5,45	62,43	47,28
	BI	65,87	9,71	33,11	76,88	31,23	5,40	62,70	47,03
	NW	65,79	9,73	33,07	77,31	31,47	5,46	62,32	47,39
	TNW	65,91	9,78	32,22	77,36	31,80	5,48	62,27	47,47

The additional corpora were used in the experiments by adding parallel data to the training set using Modified Moore-Levis Filtering and by adding a monolingual language model with linear interpolation. The results are shown in Table 4, where BASE represents the baseline system; MONO, the system enhanced with a mono-directional classifier; BI, a system with bi-directional mining; NW, a system mined bi-directionally using the Needleman-Wunch algorithm; and TNW, a system with additionally tuned parameters.

The results indicate that multi-threading significantly improved speed, which is very important for large-scale mining. As anticipated, the Needleman-Wunch algorithm decreases speed (that is why authors of the Yalign did not use it, in the first place). However, GPU acceleration makes it possible to obtain performance almost as fast as that of the multi-threaded A* version. It must be noted that the mining time may significantly differ when the alignment matrix is big (text is long). The experiments were conducted on a hyper-threaded Intel Core i7 CPU and a GeForce GTX 660 GPU. The quality of the data obtained with the NW algorithm version, as well as the TNW version, seems promising. Slight improvements in the translation quality were observed, but more importantly, much more parallel data was obtained.

It was decided to train an SMT system using only data extracted from comparable corpora (not using the original in domain data) to verify the results. The mined data was used also as a language model. The evaluation was conducted using the same test sets shown in Table 4. The results are presented in Table 5, where BASE indicates the results for the baseline system trained on the original in-domain data; MONO, a system trained only on mined data in one direction; BI, a system trained on data mined in two directions with duplicate segments removed; NW, a system using bi-directionally mined data with the Needleman-Wunch algorithm; and TNW, a system with additionally tuned parameters.

Table 5. SMT Results Using only Comparable Corpora for PL to EN translation

		PL to EN				EN to PL			
		BLEU	NIST	TER	METEOR	BLEU	NIST	TER	METEOR
TED	BASE	16,96	5,26	67,10	49,42	10,99	3,95	74,87	33,64
	MONO	12,91	4,57	71,50	44,01	7,89	3,22	83,90	29,20
	BI	12,90	4,58	71,13	43,99	7,98	3,27	84,22	29,09
	NW	13,28	4,62	71,96	44,47	8,50	3,28	83,02	29,88
	TNW	13,94	4,68	71,50	45,07	9,15	3,38	78,75	30,08
EUP	BASE	36,73	8,38	47,10	70,94	25,74	6,56	58,08	48,46
	MONO	21,82	6,09	62,85	56,40	14,48	4,76	70,64	36,62
	BI	21,24	6,03	63,27	55,88	13,91	4,67	71,32	35,83
	NW	20,20	5,88	64,24	54,38	13,13	4,54	72,14	35,03
	TNW	20,42	5,88	63,95	54,65	13,41	4,58	71,83	35,34
EMEA	BASE	62,60	10,19	36,06	77,48	56,39	9,41	40,88	70,38
	MONO	21,71	5,09	74,30	44,22	19,11	4,73	74,77	37,34
	BI	21,45	5,06	73,74	44,01	18,65	4,60	75,16	36,91
	NW	21,47	5,06	73,81	44,14	18,60	4,53	76,19	36,30
	TNW	22,64	5,30	72,98	45,52	18,58	4,48	76,60	36,28
OPEN	BASE	64,58	9,47	33,74	76,71	31,55	5,46	62,24	47,47
	MONO	11,53	3,34	78,06	34,71	7,95	2,40	88,55	24,37
	BI	11,64	3,25	82,38	33,88	8,20	2,40	89,51	24,49
	NW	11,64	3,32	81,48	34,62	9,02	2,52	86,14	25,01

6   Conclusions

The results for SMT systems based only on mined data are not very surprising. First, they confirm the quality and high level of parallelism of the corpora. This can be concluded from the translation quality, especially for the TED data set. Only two BLEU scoring anomalies were observed when comparing systems strictly trained on in-domain (TED) data and mined data for EN to PL translation. It also seems reasonable that the best SMT scores were obtained on TED data. This data set is the most similar to the Wikipedia articles, overlapping with it on many topics. In addition, the Yalign classifier trained on the TED data set recognized most of the parallel sentences. The results show that the METEOR metric, in some cases, increases when the other metrics decrease. The most likely explanation for this is that other metrics suffer, in comparison to METEOR, from the lack of a scoring mechanism for synonyms. Wikipedia is a very wide domain, not only in terms of its topics, but also its vocabulary. This leads to a conclusion that mined corpora is good source for extending sparse text do-mains. It is also the reason why test sets originating from wide domains outscore those of narrow domains and also why training on a larger mined data set sometimes slightly decreases the results from very specific domains. Nonetheless, in many cases after manual analysis was conducted, the translations were good, but the automatic metrics were lower due to the usage of synonyms.

In addition, it was proven that mining data using two classifiers trained from a foreign to a native language and vice versa can significantly improve data quantity, even though some repetition is possible. Such bi-directional mining, which is logical, found additional data mostly for wide domains. In narrow text domains, the potential gain is small. From a practical point of view, the method requires neither expensive training nor language-specific grammatical resources, but it produces satisfying results. It is possible to replicate such mining for any language pair or text domain, or for any reasonably comparable input data. Such experiments are planned in future.

The results presented in Table 5 show a slight improvement in translation quality, which verifies that the improvements to Yalign positively impact the overall mining process. It must be noted that the above mining experiments were conducted using a classifier trained only on the TED data. This is why the improvements are very visible on this corpus and less visible on other corpora. What is more improvements obtained by enriching training set were observed mostly on wide-domain text, while for narrow domains the effects were negative. The improvements were mostly observed on the Ted data set, because the classifier was only trained on text samples. Nonetheless, regardless the text domain, tuning algorithm proved to always improve the transaction quality.

Acknowledgments. The work was supported by CLARIN-PL project.

References

1. Wolk, K., Marasek, K.: Alignment of the polish-english parallel text for a statistical machine translation. Computer Technology and Application **4**, 575–583 (2013). ISSN:1934–7332 (Print), ISSN: 1934–7340 (Online), David Publishing
2. Berrotarán, G., Carrascosa, R., Vine, A.: Yalign documentation (2015). Accessed 01/2015
3. Adafre, S., De Rijke, M.: Finding similar sentences across multiple languages in wikipedia. In: Proceedingsofthe 11th Conference of the European Chapter of the Association for Computational Linguistics, pp. 62–69 (2006)
4. Yasuda, K., Sumita, E.: Method for building sentence-aligned corpus from wikipedia. In: 2008 AAAI Workshop on Wikipedia and Artificial Intelligence (WikiAI08) (2008)
5. Tyers, F.M., Pienaar, J.A.: Extracting bilingual word pairs from wikipedia. In: Collaboration: interoperability between people in the creation of language resources for less-resourced languages, p. 19 (2008)
6. Smith, J.R., Quirk, C., Toutanova, K.: Extracting parallel sentences from comparable corporausing document level alignment. In: Human Language Technologies: The 2010 Annual Conference of the North American Chapter of the Association for Computational Linguistics, Association for Computational Linguistics, pp. 403–411 (2010)

7. Pal, S., Pakray, P., Naskar, S.K.: Automatic building and using parallel resourcesfor smt from comparable corpora. In: Proceedings of the 3rd Workshop on Hybrid Approaches to Translation (HyTra)@ EACL, pp. 48–57 (2014)

8. Aker, A., Kanoulas, E., Gaizauskas, R.: A light way to collect comparable corpora from the web. In: LREC (2012)

9. Cettolo, M., Girardi, C., Federico, M.: Wit3: web inventory of transcribed and translated talks. In: Proc. of EAMT, Trento, Italy, pp. 261–268 (2012)

10. Tiedemann, J.: News from opus – a collection of multilingual parallel corpora with tools and interfaces. In: Nicolov, N., Bontcheva, K., Angelova, G., Mitkov, R. (eds.) Recent Advances in Natural Language Processing (vol V), pp. 237–248. John Benjamins, Amsterdam/Philadelphia (2009)

11. Tiedemann, J.: Parallel data, tools and interfaces in opus. In: Proceedings of the 8th International Conference on Language Resources and Evaluation (LREC 2012), pp. 2214–2218

12. Marasek, K.: Ted polish-to-english translation system for the iwslt 2012. In: Proceedings of the 9th International Workshop on Spoken Language Translation IWSLT 2012, Hong Kong, pp. 126–129 (2012)

13. Koehn, P.: Statistical machine translation. Cambridge University Press (2009)

14. Zeng, W., Church, R.L.: Finding shortest paths on real road networks: the case for a*. International Journal of Geographical Information Science **23**(4), 531–543 (2009)

15. Dieny, R., Thevenon, J.J.M., Nebel, J.C.: Bioinformatics inspired algorithm for stereo correspondence. In: International Conference on Computer Vision Theory and Applications, Vilamoura - Algarve, Portugal, pp. 5–7, March 2011

16. Roessler, R.: A GPU implementation of needleman-wunsch, specifically for use in the program pyronoise 2. In: Computer Science & Engineering (2010)

Investigating Genre and Method Variation
in Translation Using Text Classification

Marcos Zampieri[1,2]([✉]) and Ekaterina Lapshinova-Koltunski[1]

[1] Saarland University, Saarbrücken, Germany
[2] German Research Center for Artificial Intelligence (DFKI), Saarbrücken, Germany
marcos.zampieri@uni-saarland.de

Abstract. In this paper, we propose the use of automatic text classification methods to analyse variation in English-German translations
from both a quantitative and a qualitative perspective. The experiments
described in this paper are carried out in two steps. We trained classifiers to 1) discriminate between different genres (fiction, political essays,
etc.); and 2) identify the translation method (machine vs. human). Using
semi-delexicalized models (excluding all nouns), we report results of up to
60.5% F-measure in distinguishing human and machine translations and
45.4% in discriminating between seven different genres. More than the
classification performance itself, we argue that text classification methods can level out discriminative features of different variables (genres and
translation methods) thus enabling researchers to investigate in more
detail the properties of each of them.

Keywords: Human and machine translation · Text classification ·
Genres

1 Introduction

Text classification is an important area of research in Natural Language Processing (NLP) and it has been applied in a wide range of tasks such as spam
detection [1] and temporal text classification [2]. From a purely engineering perspective, researchers are interested in how well classification methods can distinguish between two or more classes and what kind of features and algorithms
deliver the best performance in each task. In recent work [3,4], however, state-of-
the-art text classification methods were proposed to investigate language variation across corpora. These methods were successfully applied in the identification
of languages, varieties and dialects, as well as genres.

The present study is an attempt to use the same techniques for the identification of translation varieties – translations which differ in genres, e.g. essays,
fiction,or methods, i.e. human and machine. We train classifiers to distinguish
translated texts according to either their genre or method of translation, using
the VARTRA corpus [5], a collection of English to German translations. More
than the classification results *per se*, we use (semi-)delexicalized representations
aiming to reduce topical bias and, therefore, levelling out interesting linguistic
features that can be further used in linguistic analysis and NLP applications.

© Springer International Publishing Switzerland 2015
P. Král and V. Matoušek (Eds.): TSD 2015, LNAI 9302, pp. 41–50, 2015.
DOI: 10.1007/978-3-319-24033-6_5

2 Related Work and Theoretical Background

Genre-specific variation of translation is related to studies within register and genre theory, e.g. [6], [7], which analyse contextual variation of languages. In lexico-grammatical terms, this variation is reflected in the distribution of linguistic patterns, i.e. subject/objects, evaluative patterns, negation, modal verbs, discourse phenomena (e.g. coreference or discourse markers).

Multilingual genre analysis is concerned with the distribution of such lexico-grammatical patterns not only across genres but also across languages, comparing the settings specific for the languages under analysis, e.g. [7] on English, Nukulaelae Tuvaluan, Korean and Somali, [8] and [9] on English and German. Moreover, the latter two also consider genres in translations. Applying a quantitative approach, Neumann (2013) [9] analyses an extensive set of features and shows to what degree translations are adapted to the requirements of different genres. Other scholars [10–13], also integrate register analysis in translation studies. However, they either do not account for distributions of these features, or analyse individual texts only. De Sutter et al. (2012) [14] and Delaere & De Sutter (2013) [15] in their analysis of translated Dutch also pay attention to genre variation, but concentrate on lexical features only.

Whereas attention is paid to genre settings in human translation analysis, they have not yet been considered much in machine translation. There exist some studies in the area of SMT evaluation, e.g. errors in translation of new domains [16]. However, the error types concern the lexical level only, as the authors operate solely with the notion of domain and not genre . Domains represent only one of the genre parameters and reflect what a text is about, i.e. its topic, and further settings are thus ignored. Although some NLP studies, e.g. those employing web resources, do argue for the importance of genre conventions, see e.g. Santini et al. (2010) [17], genre remains out of the focus of machine translation. In the studies on adding in-domain bilingual data to the training material of SMT systems [18] or on application of in-domain comparable corpora [19], again, only the notion of domain is taken into consideration.

Studies involving translation methods mostly focus on translation error analysis, and human translation serves usually as a reference in MT evaluation tasks. Some of them do consider linguistic properties, or linguistically-motivated errors [20,21]. The latter one includes style errors, which is partly related to genre.

To our knowledge, the only study investigating differences between human and machine translation is Volansky et al. (2011) [22]. The authors analyse human and machine translations, as well as comparable non-translated texts. They use a range of features based on the theory of *translationese* (see [23] or [24]) expecting that the features specific for human translations can also be used to identify machine translation. Some of the translationese features were investigated using NLP techniques [25–28] similar to the ones we propose in this paper. What is most important for our study, however, is the claim by Volansky et al. (2011) [22] that some features of human translations coincide with those of machine-translated texts, whereas other features are diversifying between these two translation methods.

3 Methods

3.1 Data

For the purpose of our study we use VARTRA [5], a corpus of multiple translations from English into German. These translations were produced by: (1) human professionals (PT1), (2) human student translators (PT2), (3) a rule-based MT system (RBMT), (4) a statistical MT system trained with a large quantity of unknown data (SMT1) and (5) a statistical MT system trained with a small amount of data (SMT2). The genres available in VARTRA are: political essays (ESS), fictional texts (FIC), instruction manuals (INS), popular-scientific articles (POP), letters of share-holders (SHA), prepared political speeches (SPE), and touristic leaflets (TOU). Each subcorpus represents a translation variety, a translation setting which differs from all others in both method and genre (e.g. PT1-ESS or PT2-FIC, etc.).

Before classification was carried out, we split the corpus into sentences (of size between 12 and 24 tokens). This created 6,200 instances. The data was then split into a training (80%) and a test (20%) set.

The features used in different experiments include bag-of-words (bow), word bigrams, word trigrams and word 4-grams. The novelty of our approach is that we substitute all nouns with placeholders in some of the experiments. This results in what we call a semi-delexicalized text representation, which lies between fully delexicalized representations [3] and the classical bag-of-words or n-gram language models. Previous studies [4,29] show that named entities significantly improve the result of text classification systems, so we decided to use this semi-delexicalized representation to minimize topic variation. The decision was motivated by both our goal of investigating translation variation influenced by both genre and method, and our aim to obtain a robust classification method that could perform well on different corpora.

3.2 Algorithms

We used two algorithms in our experiments. The first is a Naive Bayes (NB) classifier using bag-of-words as features. Naive Bayes classifiers work based on an independence assumption (the presence of a particular feature of a class is not related to the presence of any other feature), which is particularly useful for supervised learning and makes them extremely fast.

The second algorithm is based on a likelihood function calculated over n-gram language models as described by Zampieri and Gebre (2014) [30]. The language models can contain characters and words (e.g. bigrams and trigrams), linguistically motivated features such as parts-of-speech (POS) or morphological categories [4], or (semi-) delexicalized models such as the one we explore here.

4 Results

In this section, we present the results we obtained in different classification experiments. For the evaluation step we used standard NLP metrics such as Precision, Recall and F-Measure. The linguistic analysis and discussion of the most important differences between both method and genre variation will be presented later in section 4.5.

4.1 Genres and Methods

The first experiment shows why it is important to use semi-delexicalized features in a dataset that represents both dimensions of variation in translation (genre and method). The question posed at this stage is simple: how different are the samples with respect to methods and genres? We use the aforementioned Naive Bayes classifier trained on (non-delexicalized) bag-of-words. In Table 1 we present the results as well as a baseline computed based on the random assignment of all documents to a particular class.

Table 1. Naive Bayes: Genres and Methods

Type	Classes	Precision	Recall	F-Measure	Baseline
Genres	7	57.4%	57.8%	57.3%	14.2%
Methods	5	35.9%	36.2%	35.3%	20.0%

The results of this preliminary experiment show that the classifier was able to distinguish between the seven translation genres with up to 57.3% F-Measure and between the five translation methods with up to 35.3% f-measure. The method is aided by named entities and content words that are domain specific and, therefore, influence the performance of the classifier. Therefore, we use placeholders to substitute nouns (both named entities and common nouns) to minimize topical bias in the following experiments. At the same time, the results of the present experment will allow us to compare classification performance of non-delexicalized vs. semi-delexicalized representations.

4.2 Translation Methods

In the next experiment, we take a closer look at the differences between five methods of translation (PT1, PT2, RBMT, SMT1 and SMT2) while minimizing topic influence, i.e. trying to distinguish between them excluding all nouns.

The data contains outputs of two different SMT systems and in this step, we decide to merge them into a unique class of SMT. This was mainly done to answer the question of whether this kind of distinction is meaningful in practical terms, and whether the outputs of SMT1 and SMT2 are significantly different.

The results (presented in Table 2) improved substantially after the grouping. In the five-class setting the f-measure obtained for class SMT1 was the lowest of all 26.4%, whereas the SMT class could obtain the best result in the four-class setting (58.5%). This indicates that the outputs of both systems contain similar features.

Table 2. Naive Bayes: Translation Method

Classes	Precision	Recall	F-Measure	Baseline
5	35.1%	35.9%	34.9%	20.0%
4	43.2%	44.9%	43.1%	25.0%

4.3   Different Genres: Different Language?

In the next step, we try to automatically distinguish between different genres represented in the dataset. For this experiment, we also use the semi-delexicalized features.

Table 3. Naive Bayes: Genres in Translation

Classes	Precision	Recall	F-Measure	Baseline
7	45.5%	46.1%	45.4%	14.2%

As seen from Table 3, the automatic distinction between the seven genres achieved ca. 45% of both Precision and F-measure. This performance is substantially above the 14.2% baseline which indicates that the genres in VARTRA are essentially different. However, we are also interested in whether the classifier's performance for genre discrimination was consistent across all translation methods. For this step, we perform genre classification within each translation method (with both SMT outputs in one class) and the result can be seen in figure 1.

The performance across all seven genres is constant regardless of the translation method applied. For example, instruction manuals (INS) followed by fiction (FIC) are the easiest genre to identify in all four translation methods, whereas

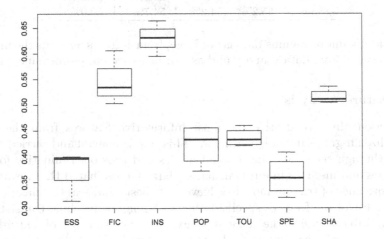

Fig. 1. Genre Distinction Across Method

speech (SPE) and essays (ESS) are consistently regarded as the most problematic ones. All the results are significantly higher than the expected baseline accuracy. The nominal values used to generate figure 1 are presented in table 4 on a scale from 0 to 1 with three decimal digits. The baseline we consider is once again the majority class, 14.2% f-measure.

Table 4. Genre Distinction Across Method

Method	ESS	FIC	INS	POP	TOU	SPE	SHA
PT2	0.399	0.533	0.595	0.372	0.421	0.346	0.536
PT1	0.314	0.606	0.664	0.456	0.425	0.371	0.507
RBMT	0.397	0.536	0.632	0.411	0.440	0.320	0.515
SMT	0.394	0.503	0.630	0.455	0.460	0.408	0.505

4.4 Human vs. Machine

In the last experiment, we investigate whether the differences between the four translation methods are weaker than between a less fine-grained classification into human and machine translation. For this step, we unify PT1 and PT2 into one class, and RBMT and SMT1 and SMT2 into the other. We also tested different sets of semi-delexicalized features, i.e. bigrams, trigrams and 4-grams to find out which allow the best classification results, see Table 5. In all three scenarios, the model performs above the expected baseline of 50.0% F-measure. The best performance, however, is obtained for the trigram model (60.5% f-measure and 61.1% precision).

Table 5. N-grams: Human x Machine

Features	Precision	Recall	F-Measure	Baseline
bigrams	53.3%	53.3%	53.3%	50.0%
trigrams	61.1%	60.0%	60.5%	50.0%
4-grams	55.2%	54.2%	54.7%	50.0%

As the amount of training data is not large, from 4-grams onwards the method seems to suffer from data sparsity and as can be expected, performance drops.

4.5 Feature Analysis

This section aims to identify the most informative features from the semi-delexicalized n-grams in our experiments. This step is manual and carried out by looking through the most informative features and thus discriminative for certain genres and methods in our translation data. We evaluated the trigrams, as the performance of trigram models achieved the best results in the classification task. The list of the features specific either to human or machine translation is shown in Table 6. Using the same strategy, we generate a list of features discriminating genre pairs (for the sake of space, we display political essays and fictional texts) in Table 7.

Table 6. Features discriminating between human and machine translations

human	machine
full nominal phrase (with def./indef. modif.)	full nominal phrase (with def./indef./poss. modif.)
personal reference (1st pers. plural)	personal reference (1st pers. sg)
extended reference (demonst.)	extended reference (pers.)
prepositional phrase with local meaning	prepositional phrase with different meanings
discourse markers with additive meaning	discourse markers with adversative meaning

Table 7. Features discriminating between political essays and fictional texts

ESS	FIC
passive constructions	active verbs
modal verbs with the meaning of volition and obligation	
to-infinitives	
prepositional phrase	predicative adjectives
demonstrative reference	personal reference
discourse markers with additive meaning	discourse markers with adversative meaning

Semi-delexicalized trigrams consist of a sequence of words and placeholders, e.g. *können PLH PLH, zu erfüllen hat, das PLH, aber*, etc. Intuitively, we try to recognize more abstract categories, i.e. modal verbs with the meaning of possibility, infinitive clauses, discourse markers with adversative function for the given trigrams. As seen from the lists, both translation methods have similar discriminating features, i.e. full nominal phrases, coreferring expressions, prepositional phrases and discourse markers. However, the differences between them can be identified on a more fine-grained level: if we take into account morphological preferences and the scope of referring expressions, the meaning of prepositional phrases and discourse markers. All these phenomena seem to be related to participants and structures of textual discourse.

The features that turn out to be specific for genres include verbs and verbal constructions, further types of phrases, and also different types of coreferring expressions and discourse markers. Genre-discriminating features are also, as in case of methods, on a more fine-grained level. However, the level of description is not on morphological, but rather on syntactic level (active vs. passive, prepositional vs. adjectival phrases). Moreover, they describe rather processes than participants of discourse. The last features coincide in both tables (additive vs. adversative construction), which means that they are informative in both genre and method classification.

Our preliminary observations on features coincide with the results of empirical analyses on genres, e.g. those obtained by Neumann (2013) [9]. For instance, the author point to personal pronouns, predicative adjectives, mental and verbal processes as indicators of narration and casual style which are specific for fictional texts. Polticial essays, which are characterized as expository texts with rather neutral style, contain relational processes, verbs of declarative mood, frequent nominalsations and almost no personal pronouns.

We believe that we need a more detailed analysis of the resulting features to have firm basis to build upon in our final conclusions on the features. For instance, the definition of mood and tense of verbal phrases, as well as their membership in a certain semantic verb class would contribute to a better specification of genres. Moreover, this step can be automatized with the help of existimg morphological tools, taggers and wordnets, which is however, beyond the scope of the present paper.

The resulting lists of features can be beneficial for not only genre classification task but also for machine translation task, as they can help to automatically differentiate between human and machine translation.

5 Conclusion and Outlook

This paper is, to our knowledge, the first attempt to use text classification techniques to discriminate methods and genres in translations and to identify their specific features and relevant systemic differences in a single study. We report results of up to 60.5% F-measure in distinguishing human and machine translations and 45.4% in discriminating between seven different genres.

We used different algorithms and sets of features to study variation in English-German translation data. For that we used not only the classical bag-of-words and n-gram language models but also the use of (semi-)delexicalized representations along with classical bag-of-words and n-gram language models, which helps us to decrease the thematic bias in classification. The aim was both the discrimination of methods and genres *per se*, and also the identification of relevant systemic differences across genres and methods of translation.

The results of our analysis can find application in both human and machine translation. In the first case, they deliver valuable knowledge on the translation product, which is influenced by the methods used in the process and the context of text production expressed by the genre. In case of machine translation, the results will provide a method to automatically identify genres in translation data thus helping to separate out-of-genre data from a training corpus.

The aforementioned practical applications of the results are part of our future work, which will also include tests with other classification algorithms such as the popular support vector machines [31] used in Petrenz and Webber (2012) for a similar task [32]. We also plan to automate the generation of more abstract categories for the informative features as well as to experiments other kinds of de-lexicalized representations such as the one used by Quiniou et al. (2012) [33]. Finally, we would like to carry out further and more detailed linguistic analysis.

References

1. Medlock, B.: Investigating classification for natural language processing tasks. Technical report, University of Cambridge - Computer Laboratory (2008)
2. Niculae, V., Zampieri, M., Dinu, L.P., Ciobanu, A.M.: Temporal text ranking and automatic dating of texts. In: 14th Conference of the European Chapter of the Association for Computational Linguistics (EACL 2014) (2014)
3. Diwersy, S., Evert, S., Neumann, S.: A semi-supervised multivariate approach to the study of language variation. Linguistic Variation in Text and Speech, within and across Languages (2014)
4. Zampieri, M., Gebre, B.G., Diwersy, S.: N-gram language models and POS distribution for the identification of Spanish varieties. In: Proceedings of TALN2013, Sable d'Olonne, France, pp. 580–587 (2013)
5. Lapshinova-Koltunski, E.: VARTRA: a comparable corpus for analysis of translation variation. In: Proceedings of the Sixth Workshop on Building and Using Comparable Corpora, Sofia, Bulgaria, pp. 77–86. ACL (2013)
6. Halliday, M., Hasan, R.: Language, context and text: Aspects of language in a social-semiotic perspective. Oxford University Press, Oxford (1989)
7. Biber, D.: Dimensions of Register Variation. A Cross Linguistic Comparison. Cambridge University Press, Cambridge (1995)
8. Hansen-Schirra, S., Neumann, S., Steiner, E.: Cross-linguistic Corpora for the Study of Translations. Insights from the Language Pair English-German. de Gruyter, Berlin, New York (2012)
9. Neumann, S.: Contrastive Register Variation. A Quantitative Approach to the Comparison of English and German. De Gruyter Mouton, Berlin, Boston (2013)
10. House, J.: Translation Quality Assessment. A Model Revisited. Günther Narr, Tübingen (1997)
11. Steiner, E.: An extended register analysis as a form of text analysis for translation. In: Wotjak, G., Schmidt, H. (eds.) Modelle der Translation - Models of Translation, pp. 235–256. Leipziger Schriften zur Kultur-, Literatur-, Sprach- und Übersetzungswissenschaft, Leipzig (1996)
12. Steiner, E.: A register-based translation evaluation. TARGET, International Journal of Translation Studies 10(2), 291–318 (1997)
13. Steiner, E.: Translated Texts. Properties, Variants, Evaluations. Peter Lang Verlag, Frankfurt/M (2004)
14. De Sutter, G., Delaere, I., Plevoets, K.: Lexical lectometry in corpus-based translation studies: combining profile-based correspondence analysis and logistic regression modeling. In: Quantitative Methods in Corpus-based Translation Studies: a Practical Guide to Descriptive Translation Research, vol. 51. John Benjamins Publishing Company, Amsterdam, pp. 325–345 (2012)
15. Delaere, I., De Sutter, G.: Applying a multidimensional, register-sensitive approach to visualize normalization in translated and non-translated Dutch. Belgian Journal of Linguistics 27, 43–60 (2013)
16. Irvine, A., Morgan, J., Carpuat, M., Daumé III, H., Munteanu, D.S.: Measuring machine translation errors in new domains. TACL 1, 429–440 (2013)
17. Santini, M., Mehler, A., Sharoff, S.: Riding the rough waves of genre on the web. In: Mehler, A., Sharoff, S., Santini, M. (eds.) Genres on the Web: Computational Models and Empirical Studies. Springer, pp. 3–30 (2010)
18. Wu, H., Wang, H., Zong, C.: Domain adaptation for statistical machine translation with domain dictionary and monolingual corpora. In: Proceedings of COLING-2008, Manchester, UK, pp. 993–1000 (2008)

19. Irvine, A., Callison-Burch, C.: Using comparable corpora to adapt MT models to new domains. In: Proceedings of the ACL Workshop on Statistical Machine Translation (WMT) (2014)
20. Popovic, M., Ney, H.: Towards automatic error analysis of machine translation output. Computational Linguistics **37**(4), 657–688 (2011)
21. Fishel, M., Sennrich, R., Popovic, M., Bojar, O.: Terrorcat: a translation error categorization-based mt quality metric. In: 7th Workshop on Statistical Machine Translation (2012)
22. Volansky, V., Ordan, N., Wintner, S.: More human or more translated? Original texts vs. human and machine translations. In: Proceedings of the 11th Bar-Ilan Symposium on the Foundations of AI With ISCOL (2011)
23. Gellerstam, M.: Translationese in Swedish novels translated from English. In: Translation Studies in Scandinavia, pp. 88–95 (1986)
24. Baker, M., et al.: Corpus linguistics and translation studies: Implications and applications. Text and technology: In honour of John Sinclair **233**, 250 (1993)
25. Baroni, M., Bernardini, S.: A new approach to the study of translationese: Machine-learning the difference between original and translated text. Literary and Linguistic Computing **21**(3), 259–274 (2006)
26. Ilisei, I., Inkpen, D., Corpas Pastor, G., Mitkov, R.: Identification of translationese: a machine learning approach. In: Gelbukh, A. (ed.) CICLing 2010. LNCS, vol. 6008, pp. 503–511. Springer, Heidelberg (2010)
27. Volansky, V., Ordan, N., Wintner, S.: On the features of translationese. Literary and Linguistic Computing (2013)
28. Ciobanu, A.M., Dinu, L.P.: A quantitative insight into the impact of translation on readability. In: Proceedings of the 3rd PITR workshop, pp. 104–113 (2014)
29. Gebre, B.G., Zampieri, M., Wittenburg, P., Heskens, T.: Improving native language identification with tf-idf weighting. In: Proceedings of the BEA, Atlanta, USA (2013)
30. Zampieri, M., Gebre, B.G.: Varclass: An open source language identification tool for language varieties. In: Language Resources and Evaluation (LREC) (2014)
31. Joachims, T.: Text categorization with support vector machines: learning with many relevant features. In: Nédellec, C., Rouveirol, C. (eds.) ECML 1998. LNCS, vol. 1398, pp. 137–142. Springer, Heidelberg (1998)
32. Petrenz, P., Webber, B.: Robust cross-lingual genre classification through comparable corpora. In: The 5th Workshop on Building and Using Comparable Corpora (2012)
33. Quiniou, S., Cellier, P., Charnois, T., Legallois, D.: What about sequential data mining techniques to identify linguistic patterns for stylistics? In: Gelbukh, A. (ed.) CICLing 2012, Part I. LNCS, vol. 7181, pp. 166–177. Springer, Heidelberg (2012)

Extracting Characteristics of Fashion Models from Magazines for Item Recommendation

Taishi Murakami, Yoshiaki Kurosawa$^{(\boxtimes)}$, Yuri Kurashita, Kazuya Mera, and Toshiaki Takezawa

Graduate School of Information Sciences, Hiroshima City University, Hiroshima, Japan
{murakami,kurosawa,kurashita,mera,takezawa}@ls.info.hiroshima-cu.ac.jp

Abstract. Many fashion magazines are available, and exclusive models in each magazine have a particular image. Readers of fashion magazines purchase fashion items by referencing the outfits worn by exclusive models. However, fashion recommendation systems for items based on the characteristics of an exclusive model do not exist on online shopping sites. Therefore, we propose an image extraction-based fashion recommendation system that considers information about the items worn by a model in a magazine. This study has performed an image extraction with a model based on this concept.

Keywords: Fashion item recommendation · Extracting characteristic · Text mining

1 Introduction

Currently, various fashion magazines are published all over the world. In Japan, more than 100 fashion magazines target different groups based on various categories such as age and taste. Presumably, people purchase and read fashion magazines to learn about latest fashion trends so that they can make purchases that reflect latest style trends.

When making this process, the reader may focus on fashion items or coordinates worn by an exclusive model. A model under contract to a particular fashion magazine is referred to as an exclusive model. Previous research [7] has suggested that information about readers' preferred models may be the most important factor influencing purchasing decisions. For example, if readers prefer a feminine *kawaii* style, they will like models with a similar style and will pay particular attention to the outfits worn by such models. Thus, readers are more likely to purchase such outfits. Therefore, an effective recommendation system would present items related to a reader's preferred model.

We focus on exclusive models and the items worn by them in fashion magazines. The proposed method uses data accumulated from magazine articles that link to Internet-based information about specific brands. Here, the objective is to construct vector models. We adopt two approachesitem name-based approach

© Springer International Publishing Switzerland 2015
P. Král and V. Matoušek (Eds.): TSD 2015, LNAI 9302, pp. 51–60, 2015.
DOI: 10.1007/978-3-319-24033-6_6

and description-based approach. The item name-based approach focuses on an item's caption or name, e.g., "leather mini skirt black." Typically, item captions include information as material (leather) and silhouette (mini). Name-based information helps distinguish particular characteristics of models. For example, although the word sexy may co-occur with a luxurious style, it would not co-occur in the description of *kawaii* items, which present a more girlish style. Therefore, by focusing on item captions, we can detect a model's particular fashion sense. The details-based approach analyzes item descriptions, e.g., "in a sensational twist, we have topped off this head-turner dress with lovely lace for an airy finish." Such descriptions include more subjective information related to impression (i.e., head-turner, lovely, and airy) than item captions. By considering such information and generating the vectors of exclusive models, it will be possible to acquire appropriate knowledge for fashion item recommendations.

2   Previous Research

Various studies on fashion recommendation systems have been published [1][4–6]. However, we focus on two studies that investigated the componential design element of fashion items such as color, material, detail and silhouette, and so on. These elements are essential for characterizing fashion items. Liu et al. proposed "the magic closet," a fashion recommendation system for particular occasions [3]. When a particular occasion such as a wedding, shopping expedition, or date is designated, the system recommends the most suitable clothes from a user's photo album or an online shop. This recommendation system focuses on particular fashion item characteristics, i.e., componential elements. For example, bright colors are appropriate for weddings and dark colors are appropriate for funerals. To select and recommend the most suitable fashion items, detailed information about a specific occasion is required.

Kamma et al. also proposed a recommendation system that focuses on componential elements[2]. Their system targets people who are not interested in fashion and have difficulty selecting appropriate outfits for specific occasions. The proposed system relies on "awareness" to recommend coordinated fashion items. The "awareness" is a fashion coordination not to wearing. In their system, a detailed description of the componential elements is important, e.g., puffed sleeves. Image keywords are selected on the basis of descriptions. The system uses six target images: classic, elegant, romantic, ethnic, sporty, and masculine. The objective of their system is to select an image that matches the componential elements of items in a user's wardrobe. For example, if the user has a top with puffed sleeves, the appropriate image would be elegant or romantic; therefore, the system would select a flared skirt because it is more elegant than most pants.

However, these studies have limitations with regard to determine a necessary componential element and its particular instance. For example, Kamma et al. dealt with puffed sleeves as an instance of detail and annotated this characteristic for items with this particular type of sleeve. Then, this characteristic was used to associate the item with such sleeves with a particular image, e.g., romantic.

The selection of a specific instance becomes an important issue. Puffed sleeves may be appropriate for detecting a romantic image; however, they do not contribute to the detection of images that are not among the six target images, e.g., a gothic image. For a gothic image, "bell sleeves" may be appropriate. However, this characteristic was not included in their study. In other words, their method cannot resolve all situations because they employed a limited number of characteristics. Thus, controlling the types of elements automatically is important in a recommendation system. We propose to extract the necessary elements from the text information linked from fashion items automatically.

Furthermore, we focus on the features associated with exclusive models. To recommend items, the system requires a combination of all items in a database. However, many recommended combinations are not particularly helpful, e.g., a white T-shirt and denim pants is an obvious combination. To avoid obvious combinations, we propose weighing the combinations worn by a particular exclusive model. Users tend to see their favorite models repeatedly wearing particular outfits and thus buy them. Focusing on this tendency and weighing combinations is important to ensure that recommendations match a user's preference. Therefore, we consider componential elements and extract some rules from numerous item descriptions accumulated from the web.

3   Proposed Method

The proposed method identifies the items worn by an exclusive model in a fashion magazine and searches for these items on online shopping sites. By using results from item names, we extract item descriptions to perform analysis using two methods.

The definite methods is shown below.

- Method 1:find a feature of the model from an item name
- Method 2:find a feature of the model from an item description

The term 'feature' of a model means features of the fashion items which she wears in the magazine in the broadest sense of the word. In our definition, various features exist because of the broadest sense. Styles such as girly are a kind of feature. The details of the item, e.g., ribbon and sleeve, are also regarded as the same. We can find a broad range of features because the diversity depends on the words included in item names and item descriptions.

3.1   Acquiring Item Name and Description

In this section, the acquisition of item names and descriptions is described. Typically, in fashion magazines, basic information, i.e., item name, brand, and price, are provided in a description next to the picture of an item. However, detailed information such as the silhouette and the material's pattern are not included. Therefore, we searched for items on online shopping sites using this basic information of the item worn by the model.

Table 1. Attribute dictionary and number of words

AD	Num. of words	AD	Num. of words	AD	Num. of words
Item	788	Pattern	217	Trend	2
Detail	657	Color	420	Size	13
Silhouette	107	Season	6	Impression	30
Technology	72	Brand	106	Other	30
Material	1,517	Country	2	Unnecessary-Unknown	138

We acquired information from the February to November 2014 issues of a women's fashion magazine for items worn by 21 exclusive models. Based on this information, we found 10,496 items by searching online shopping sites. First, we searched for the brand name and then narrowed the search to target specific information such as item category, sleeve length, color, and pattern. After determining that an item existed on an online shopping site, the item name and description were extracted.

3.2   Finding the Features of the Model from an Item Name

In method 1, for the extracted item name (Section 3.1), tagging is performed automatically using the attribute dictionary(AD) as shown in Table 1. The attribute dictionary was created using apparel-related books and a fashion terminology dictionary. The attribute dictionary contains 15 attributes, i.e., item name detail, silhouette, technology, material, pattern, color, season, brand, country, trends, size, impression, other, and unnecessary-unknown. From these attributes, 776 tags comprising attribute words were generated.

3.3   Finding the Feature of the Model from the Item Description

In method 2, we used the MeCab morphological analyzer to analyze item descriptions (Section 3.1). The nouns and adjectives obtained from the analysis were used as feature words because these parts of speech represent important features, impressions, and item atmosphere. Note that 2,732 types of feature words were extracted. However, there were significant numbers of unknown words and orthographic variants. Consequently, after creating a user dictionary, we corrected the orthographic variants and reduced the number of feature words to 2,568.

3.4   Creating a Fashion Style Vector

A fashion style is indicated by some terms such as casual and girly. Consequently, we counted the co-occurrence of content words. This can be represented by a weighted vector using each characteristic word.

When it is assumed that model A's image is boyish, the model's vector and a boyish vector should consist of a similar element. In this study, the goal was to obtain a rule such as "In speaking of boyish, it's Ms. model A." However, it is difficult to evaluate such obtained rules individually. We attempted to evaluate

Table 2. Attribute word count for each exclusive model

EM	Count	EM	Count	EM	Count	EM	Count	EM	Count	EM	Count	EM	Count
A	256	D	447	G	583	J	390	M	56	P	37	S	20
B	263	E	472	H	192	K	522	N	39	Q	35	T	10
C	286	F	217	I	391	L	370	O	44	R	30	U	9

Table 3. Exclusive models and normalization of the count each attribute words(thousand-fold)

		flare(S)	tight(S)	skirt(I)	pants(I)	skinny(I)	sweatshirt(I)	pullover(I)
	A	0.03	0.01	0.09	0.05	0.01	0.02	0.11
	B	0.04	0.03	0.06	0.05	0.03	0.00	0.10
	C	0.03	0.01	0.05	0.03	0.01	0.00	0.09
	D	0.03	0.02	0.05	0.01	0.01	0.00	0.14
	E	0.04	0.04	0.06	0.02	0.00	0.12	0.13
EM	F	0.00	0.02	0.05	0.07	0.02	0.02	0.23
	G	0.01	0.02	0.08	0.02	0.02	0.05	0.18
	H	0.09	0.03	0.13	0.06	0.01	0.03	0.13
	I	0.04	0.01	0.05	0.04	0.01	0.04	0.30
	L	0.05	0.02	0.06	0.01	0.01	0.04	0.21

fashion style vectors by considering whether there is a degree of commonality among elements.

To create a fashion style vector, many images used in the feature of the title and subtitle of a women's fashion magazine were used as reference. The resulting images can be classified as *Otona feminine*[adult-like feminine], feminine, *Otona Kawaii*[adult-like cute], cute, mature, *Otona girly*[adult-like girly], girly, *Otona casual*[adult-like casual], casual, *Otona sexy*[adult-like sexy], sexy, masculine, and boyish.

4   Experiment

Here, we discuss the results of the image evaluation obtained using the methods described in Section 3.

4.1   Method 1 Data

The attribute words obtained for each exclusive model(EM) are shown in Table 2. The attribute words in Table 2 represent numbers of attribute occurrences.

The numbers of attribute occurrences for Model M, N, ..., U are not enough to perform experiment. Therefore, in this study, experiment was performed using 1,269 item names for the top 12 exclusive models.

A normalized number of occurrences of attribute words for each exclusive model is shown in Table 3. S is Silhouette, and I is Item.

Using data in Table 3, Equations (1) and (2) were used to extract the characteristics of each exclusive model. The value of the normalized in Equations

Table 4. Total feature word count for each exclusive model

EM	Count	EM	Count	EM	Count	EM	Count	EM	Count	EM	Count	EM	Count
E	2291	D	2297	L	1733	A	1172	M	293	P	146	S	59
G	2628	I	2041	B	1540	F	1173	N	215	Q	278	T	58
K	2592	J	1857	C	1403	H	777	O	138	R	133	U	28

(1) and (2) means each of the values in Table 3. Moreover, σ means standard deviation. Supposing that our data fit to a normal distribution, approximately 95% words lie within two σ and 68% words include within one σ. Thus, these values could become criteria for distinguishing models because the rest of the distribution is rare or relatively rare case.

However, in the case of $n = 2$ in Equation (1), the number of attribute words was extremely small; thus, it was difficult to perform experiment. In addition, fewer attribute words are applicable to $n = 1$ and $n = 2$ in Equation (2). This also impedes experiment. Therefore, only data for attribute words that apply to $n = 1$ in Equation (1) were used for experiment. The extracted attribute words were used to construct a vector for each exclusive model.

$$The\ value\ of\ the\ normalized \geq Average + \sigma * n(n = 1, 2) \tag{1}$$

$$The\ value\ of\ the\ normalized \leq Average + \sigma * n(n = 1, 2) \tag{2}$$

4.2 Method 2 Data

The total count of exclusive model feature words is shown in Table 4.

From Model M to Model U, the number of occurrences of feature words is less than or equal to 500. In addition, there were fewer occurrences of these words; thus, it was difficult to perform experiment. Therefore, experiment was performed using item description data from 1,269 items.

The normalized number of occurrences of feature words for each exclusive model is shown in Table 5.

To extract the characteristics of each exclusive model, data shown in Table 5 were used with Equations (1) and (2). However, for $n = 2$ in Equation (1), the number of true attribute words was extremely small; thus, it was difficult to perform experiment. In addition, fewer attribute words apply to $n = 1$ and $n = 2$ in Equation (2), which also impedes experiment. Therefore, only attribute word data that apply to $n = 2$ were used with Equation (1). The extracted feature words were used to construct a vector for each exclusive model.

4.3 Image Score

The exclusive model vector and a fashion style vector for each image were compared. The comparison of the exclusive model vector and fashion style vector for Model A is shown in Table 6.

Using data in Table 6, the image score(IS) for each exclusive model is calculated. In the scoring method, the attribute words for each image are used to

Table 5. Exclusive models and normalization of feature word count (thousand-fold)

		flare	tight	skirt	pants	cute	ladylike	feminine
	E	0.0	0.0	0.0	0.0	12.2	0.0	4.8
	G	0.0	9.9	24.4	0.0	7.2	0.0	0.0
	K	9.3	0.0	0.0	0.0	0.0	7.7	4.2
	D	0.0	0.0	0.0	0.0	6.5	0.0	4.8
	I	0.0	0.0	0.0	15.2	8.3	0.0	0.0
EM	J	0.0	8.1	23.7	0.0	7.6	7.5	0.0
	L	8.1	6.3	21.9	0.0	0.0	8.7	0.0
	B	6.5	0.0	25.3	12.3	0.0	0.0	5.2
	C	0.0	0.0	0.0	0.0	9.3	0.0	4.3
	H	0.0	0.0	33	0.0	0.0	0.0	0.0

examine the rank at which the buyer appears. This is calculated using the weight in the following equation.

$$IS = Property\ value\ of\ 1st * 1 + Property\ value\ of\ 2nd * \frac{1}{2}$$
$$+ ... + Property\ value\ of\ nth * \frac{1}{n} \quad (3)$$

The first property value in Equations (3) corresponds the value(0.061) in the first rank in Table 6. For example, *Feminine* appears seven times (except 7th rank). By using these values, we can calculate the *IS* value about *Feminine* as follows.

$$IS = 0.061*1 + 0.041*\frac{1}{2} + 0.023*\frac{1}{3} + 0.020*\frac{1}{4} + 0.017*\frac{1}{5} + 0.016*\frac{1}{6} + 0.009*\frac{1}{8} \quad (4)$$

We calculated the image score of fashion style vector in a similar manner for the model vector described in Section 4.2. The *IS* values for each image of Model A obtained using methods 1 and 2 are shown in Table 7. CD and CR stand for Correct Data and Comparison Result, respectively.

4.4   Evaluation

An evaluation was performed using the attribute and characteristic words for items worn by 12 models, the *IS* values for each exclusive model image, and required data, which were obtained from women's fashion magazines from 2011-2014. This data were obtained by examining the image title and subtitle features listed for each exclusive model.

Table 6. Comparison of exclusive model vector and fashion style vector

Rank	Rule	Attribute value	Image		
1	Basic-Item(Trend)	0.061	Girly	Cute	Feminine
2	Tops(Item)	0.041	Girly	Feminine	Sexy
3	Blouse(Item)	0.023	Adult-like Girly	Adult-like Casual	Feminine
4	Inner(Item)	0.020	Adult-like	Casual	Feminine
5	Cocoon(Silhouette)	0.017	Feminine	Casual	
6	Flower(Pattern)	0.016	Feminine	Casual	
7	Different material(Material)	0.014	Adult-like		
7	Pointed(Silhouette)	0.014	Mannish	Casual	
8	Shirring(Detail)	0.009	Feminine		

Table 7. Image Score and Comparison result

method 1				method 2			
Image	IS	CD	CR	Image	IS	CD	CR
Feminine	0.101	1	1	Cute	0.021	0	0
Cute	0.089	0	0	Casual	0.015	0	0
Girly	0.081	1	1	Feminine	0.013	1	1
Casual	0.074	0	0	Boyish	0.012	0	0
Mannish	0.063	0	0	Adult-like	0.012	0	0
Adult-like Casual	0.028	1	1	Girly	0.011	1	1
Sexy	0.020	0	0	Adult-like Casual	0.001	1	1
Adult-like Girly	0.008	0	0	Adult-like Feminine	0.000	1	0
Adult-like	0.007	1	1	Adult-like Cute	0.000	0	1
Adult-like Feminine	0.000	1	0	Adult-like Girly	0.000	0	1
Adult-like Cute	0.000	0	1	Adult-like Sexy	0.000	0	1
Adult-like Sexy	0.000	0	1	Sexy	0.000	0	1
Boyish	0.000	0	1	Mannish	0.000	0	1

The comparison was performed using the obtained data and the IS for each image. In the correct data, if an image exists as 1, otherwise, as 0. If the correct data is 1 and the IS is not 0, then it is correct and the result of the comparison is 1. Moreover, if the correct data is 0 and the IS is 0, then it is correct and the result of the comparison is 1. Otherwise, it is incorrect, and the result of the comparison is 0. The results of comparison for Model A obtained using methods 1 and 2 are shown in Table 7.

The concordance rate and precision were calculated. The concordance rate was calculated using the total results of comparison for 13 types of image. The concordance rate is expressed by Equation (5). Precision was calculated using the total results of comparison for an image with an IS value that is not 0. Precision was calculated using Equation (6). n is a number of image IS is not 0.

$$Concordance\ rate = \frac{13\ kinds\ of\ image\ comparison\ result\ of\ total}{13} \quad (5)$$

$$Precision = \frac{n\ kinds\ of\ image\ comparison\ result\ of\ total}{n} \quad (6)$$

Table 8. Concordance rate and precision

method 1			method 2		
EM	Concordance rate	Precision	EM	Concordance rate	Precision
E	0.692	0.750	D	0.692	0.500
D	0.692	0.500	E	0.615	0.500
H	0.538	0.400	A	0.615	0.429
A	0.462	0.333	B	0.692	0.429
G	0.462	0.300	H	0.539	0.400
I	0.538	0.286	F	0.769	0.333
B	0.538	0.250	I	0.539	0.333
K	0.538	0.167	C	0.539	0.286
C	0.385	0.143	G	0.462	0.250
F	0.385	0.111	K	0.462	0.143
J	0.538	0.000	J	0.615	0.000
L	0.538	0.000	L	0.462	0.000
Ave.	0.526	0.270	Ave.	0.583	0.300

An evaluation was performed using the obtained concordance rate and precision.

4.5   Evaluation Results

Using Equations (5) and (6), we determined the concordance rate and precision. The concordance rates and precisions for each exclusive model obtained using methods 1 and 2 are shown in Table 6 in the descending order of precision.

4.6   Discussion

The proposed method did not result in good precision. We performed two experiments, and the results showed nearly the same tendency. However, the results obtained with method 2 were slightly superior to those obtained with method 1. Thus, there is room for improvement. Here, we consider two points.

The first point relates to the appearance of words for specific styles, i.e., casual and cute. These two styles (words in this case) occur frequently in our description text, and many words (characteristics) co-occur with these two styles. Note that in our evaluation, we attempted to determine IS values that are greater than 0. Therefore, a few observations, even only one time, may lead to wrong detection. If we consider the casual and cute styles, we must adopt an appropriate threshold to improve precision. Note that ISs for casual and cute are labeled incorrectly in Table 7.

The second point is what we refer to as a leakage problem. When constructing data, we attempted to link items in a magazine to real items on the web. However, we cannot link all items because they may be sold out or may not be treated in the first place. Therefore, we could not determine the clear differences between exclusive models. In other words, without increasing the amount of data to mitigate leakage, the results obtained using the accumulated data tended to be worse.

However, some good values were obtained, e.g., CR reached 0.769 for model F using method 2 (Table 8).

To address these problems, we must consider similar items will be linked to instead of the linkage items. However, this procedure is maybe subjective and annotations may differ among annotators. By carefully implementing this procedure, we hope to report better precision in future.

5   Conclusion

We attempted to use magazine data and model characteristic data to recommend fashion items. We have proposed two methodsitem-name-based method and description-based method. However, with the exception of a few cases, we did not obtain good results. In addition, we did not examine the differences between the two methods. Although our results were not satisfactory, we have identified a possible improvement that can be implemented in future studies, i.e., setting a threshold. After implementing this improvement, we hope to report a useful fashion item recommendation system.

Acknowledgments. This research is partially suported by the Center of Innovation Program from Japan Science and Technology Agency, JST and Hiroshima City University Grant for Special Academic Research (General Studies 2012-2014).

References

1. Iwata, T., Watanabe, S., Sawada, H.: Fashion coordinates recommender system using photographs from fashion magazines. In: Proceedings of the Twenty-Second International Joint Conference on Artificial Intelligence - Volume Three. IJCAI 2011, pp. 2262–2267. AAAI Press (2011)
2. Kamma, Y., Marutani, T., Kajita, S., Mase, K.: Proposal of a fashion coordinate recommendation system based on fashion image keywords. SIG Human-Computer Interaction. HCI, IPSJ SIG Technical Report **2011**(26), 1–7 (2011). (in Japanese)
3. Liu, S., Feng, J., Song, Z., Zhang, T., Lu, H., Xu, C., Yan, S.: Hi, magic closet, tell me what to wear! In: Proceedings of the 20th ACM International Conference on Multimedia. MM 2012, pp. 619–628. ACM, New York (2012)
4. Nagao, S., Takahashi, S., Tanaka, J.: Mirror appliance: recommendation of clothes coordination in daily life. In: Proceedings of the Human Factors in Telecommunication. HFT 2008, pp. 367–374 (2008)
5. Sato, A., Watanabe, K., Yasumura, M., Rekimoto, J.: suGATALOG: fashion coordination system that supports users to choose everyday fashion with clothed pictures. In: Kurosu, M. (ed.) HCII/HCI 2013, Part V. LNCS, vol. 8008, pp. 112–121. Springer, Heidelberg (2013)
6. Shen, E., Lieberman, H., Lam, F.: What am i gonna wear?: scenario-oriented recommendation. In: Proceedings of the 12th International Conference on Intelligent User Interfaces. IUI 2007, pp. 365–368. ACM, New York (2007)
7. Tomikawa, A.: The effect of model and styling in consumer information processing about magazine's image. Journal of Atomi University Faculty Of Literature **47**, A69–A89 (2012). (in Japanese)

Segment Representations in Named Entity Recognition

Michal Konkol[(✉)] and Miloslav Konopík

Department of Computer Science and Engineering, Faculty of Applied Sciences,
University of West Bohemia, Univerzitní 8, 306 14 Plzeň, Czech Republic
{konkol,konopik}@kiv.zcu.cz
http://nlp.kiv.zcu.cz

Abstract. In this paper we study the effects of various segment representations in the named entity recognition (NER) task. The segment representation is responsible for mapping multi-word entities into classes used in the chosen machine learning approach. Usually, the choice of a segment representation in the NER system is arbitrary without proper tests. Some authors presented comparisons of different segment representations such as BIO, BIEO, BILOU and usually compared only two segment representations. Our goal is to show, that the segment representation problem is more complex and that the proper selection of the best approach is not straightforward. We provide experiments with a wide set of segment representations. All the representations are tested using two popular machine learning algorithms: Conditional Random Fields and Maximum Entropy. Furthermore, the tests are done on four languages, namely English, Spanish, Dutch and Czech.

1 Introduction

Named entity recognition (NER) is a standard task of natural language processing. NER system searches for expressions of special meaning such as locations, persons, or organizations. These expressions often hold the key information for understanding the meaning of the document.

In this paper, we focus on one of many design aspects of a NER system: segment representation of multi-word entities. Many entities consist of multiple words (e.g. *Golan Heights*). If we use machine learning approach for NER, it is necessary to assign exactly one class to each token (word) in the corpus. The simplest way is to have one class for each type of named entity (and one extra type for normal words). This solution has a major limitation – it is not possible to correctly encode subsequent entities of the same type, e.g *"... the Golan Heights Israel captured from ..."* from CoNLL-2003 dataset where *Golan Heights* and *Israel* are both the *location* type. The result would look like this *"... word Location Location Location word word ..."*. Another motivation for more complex segment representations is that they can increase recognition performance (please note that we will use word *performance* in the meaning of an ability to recognize the named entities correctly not in the meaning of computational speed). For example, the recognition rules may differ for the first word

© Springer International Publishing Switzerland 2015
P. Král and V. Matoušek (Eds.): TSD 2015, LNAI 9302, pp. 61–70, 2015.
DOI: 10.1007/978-3-319-24033-6_7

and subsequent words of an entity. A segment representation that distinguishes the beginning of an entity then may help with the recognition. The idea can be further extended by more complex segment representations.

In this paper we study effects of several segment representations on multiple languages using various machine learning (ML) approaches. The experiments with multiple languages are motivated by the fact that entities are very often proper names and different languages have very different rules for writing proper names. For example in English, all words except prepositions and conjunctions usually start with uppercase letter while in Czech only the first word starts with the uppercase letter, e.g. *Česká národní banka* in Czech and *Czech National Bank* in English. We use English, Spanish, Dutch and Czech corpora for our experiments.

Experiments with multiple ML approaches are important, because the optimal representation may vary for different methods. For our experiments we have chosen the Maximum Entropy (ME) [1] as a representative of classification methods and Conditional Random Fields (CRF) [2] as a representative of sequential methods.

The rest of the paper is organized as follows. Section 2 is devoted to the description of the segment representations. Section 3 is an overview of the related work. The NER system is described in Section 4. Section 5 gives a brief overview of the corpora used in our experiments. Section 6 describes our experiments and presents and discusses the results. The last section summarizes our findings.

2   Segment Representations

As we have pointed out, there are multiple models for representing multi-word named entities (or more generally multi-word expressions). All the models (except the simplest one) use more than one tag for each type of named entity, e.g B-PERSON, I-PERSON for PERSON named entity. To our best knowledge, the most complex model uses 4 tags for each entity (plus one for not-an-entity tag). As already shown in the example, the tags are usually distinguished by a single letter prefix. The prefixes have a meaning of relative position in the named entity. The following list summarizes commonly used prefixes.

B (Beginning) Represents the first word of the entity.
I (Inside) Represents a part of the entity, which is not represented by other prefix.
L (Last, sometimes also **E**nd) Represents last word of the entity.
O (Outside or other) Represents word that is not a part of the entity.
U (Unit, sometimes also **W**ord or **S**ingle token) Represents a single word entities.

As we have said earlier, these models have two major purposes. The first one is to distinguish two subsequent entities. The model is able to do that, if it uses at least the **O**utside, **I**nside and one of the **B**egin and **E**nd tags. The second one, is to improve performance. Each tag is used as a single class in the ML methods.

It means, that each tag represents a different set of statistics that can be used in the decision process. The intuition tells us, that the statistics accumulated over the corpus may be different for the first word of the entity (**B**), the inside word (**I**) and the other cases. For example the first word of the entity has much higher probability of having the first letter uppercase in Czech. In the following sections, we will describe the commonly used models.

IO model is the name we use for the simplest representation, even though this model has no well-known or widely accepted name. Each entity is represented only by one tag, which obviously does not need any prefix. This model is unable to decode subsequent entities of the same type, but it is not as important as it may seem at first sight, because subsequent entities of the same type are rare.

BIO model (or IOB) representation decodes each entity with two tags. There are two versions of the representation. The BIO-2 uses the **B**egin tag for each first word of an entity. The BIO-1 uses the **B**egin tag for the first word, only if it follows entity of the same type. In other words, the BIO-1 uses the **B**egin tag only if it has to distinguish subsequent entities.

IEO model is similar to the BIO representation, but it replaces the **B**egin tag with the **E**nd tag. There are also two versions – IEO-1 and IEO-2. These models have the same semantics as the BIO-1 and BIO-2 models.

BIEO model (BIOE, OBIE) representation uses both **B**egin and **E**nd tags.

BILOU model (C+O) representation is the most complex model used in NER. It adds the **U**nit tag for single word entities.

Furthermore, we experiment with newly created representations IOU, BIOU and OIEU models. They extend some of the previously mentioned models with the **U** tag.

3   Related Work

The simplest segment representation (IO) was used by some of the first ML systems (e.g. [3–5]).

The CoNLL-2002 and CoNLL-2003 shared tasks used the BIO representation for annotations in their corpora (IOB-1 in 2002, IOB-2 in 2003) and many authors have adopted this model in their NER systems. The BIO model is the most commonly used model since these conferences.

The BIEO model was used in few papers [6–8], but it is very rare compared to the BIO model.

Some of the recent papers [9–11] adopted the BILOU representation probably based on the comparison in [10], where a comparison of the BIO and BILOU representation on English using CoNLL-2003 [12] and MUC-7 corpora using CRFs is provided . The BILOU representation performed better on the MUC-7 corpus on both (validation and test) data sets. On the CoNLL corpus, the BIO representation performed better on the validation set while the BILOU model performed better on the test set. The authors stated that segment representations can significantly impact the system and concluded that the BILOU model significantly outperforms the BIO model. The conclusion is only weakly supported by the results, in our opinion.

An interesting study is provided in [13]. The authors present method for using multiple segment representations together in one system. They also provide a comparison of multiple segment representations on the biomedical domain. The biomedical domain has different properties than the standard (news) corpora used in NER and cannot be compared with our results.

A similar research has been done for a different task – text chunking [14]. To our best knowledge, there are no other articles comparing segment representations in NER. We have not found any usage of representations not mentioned in this section.

4   NER System

We use two standard machine learning systems. The first one is based on Maximum Entropy (ME) and follows the description in [1]. The second one is based on Conditional Random Fields (CRF), similar to the baseline system in [15]. We use the Brainy ML library [16] for this purpose.

Both methods use the same feature set which consists of common NER features. The features are the following: words, bag of words, n-grams, orthographic features, orthographic patterns, and affixes.

5   Corpora

Our experiments are done on four languages – English, Spanish, Dutch and Czech. We use one corpus for each language.

For English, Spanish and Dutch we use the corpora from CoNLL-2002 and CoNLL-2003 shared tasks [12,17]. These corpora have approximately 300,000 tokens and use four entity types – person (PER), organization (ORG), location (LOC) and miscellaneous (MISC).

For Czech we use the CoNLL format version of Czech Named Entity Corpus 1.1 [18,19]. This corpus is smaller than the CoNLL corpora and has approximately 150,000 tokens. It uses 7 classes – time (T), geography (G), person (P), address (A), media (M), institution (I) and other (O).

All corpora use the BIO segment representation for the data. The English corpus (CoNLL-2003) uses the BIO-1 representation of segments. The rest the BIO-2. The segment representation of the corpora does not play any role in the training or evaluation as we firstly load the corpora to inner, corpus-independent representation and then transform it into training (or validation or test) data with proper segment representation for the given experiment.

6 Experiments

In all the experiments, we use the standard CoNLL evaluation with precision, recall and F-measure. We present only the F-measure because of space requirements. In the following sections we show two sets of experiments. The discussion of our results is in a separate section.

6.1 Standard Partitioning

The first set of experiments is evaluated on the original partitioning of the corpora – training, validation and test set. For our experiments, we do not need to set any parameters based on the results on validation set. The results on the validation set thus provide the same information as on the test set. This follows the same procedure as in [10].

For each combination (segment representation, ml approach) we train a model on the training data and evaluate it on the validation and test data. The results of these experiments are shown in table 1.

6.2 Significance Tests

The results of the first experiments are in many respects indecisive. For many representation pairs it is impossible to choose the better one (one is better on the test set, the other one on the validation set). Thus, we decided to perform a 10 fold cross-validation to obtain more consistent results computed on much larger data. The advantage is that our tests do not depend on a short portion of data created by manual corpus division.

The data are prepared by the following procedure. Firstly, we concatenate all the data sets for each language (ordered: training, validation, test) into the data set D_{All} and number all the sentences (s denotes the index of a sentence). For fold i, $i = 0, \ldots, 9$, the test set is $D_{Test} = \{s : s \mod 10 = i\}$ and training set $D_{Train} = D_{All} - D_{Test}$. This procedure assures uniform distribution of sentences.

Each combination (segment representation, ML approach) is then tested on each fold. We compare the different combinations using the paired Student's t-test. The results are shown in Table 2 for ME and in Table 3 for CRF. We use two confidence levels $\alpha = 0.1, 0.05$. The null hypothesis H_0 is that there is no difference between segment representations. The alternative hypothesis H_1 is that one segment representation is significantly better than the other segment representation. Each cell contains four symbols, one for each language in the order English, Spanish, Dutch, and Czech.

- The symbol $<$ (resp. $>$) means, that the row segment representation is significantly worse (resp. better) than the column representation. The H_0 hypothesis is rejected at both levels $\alpha = 0.05, 0.1$.
- The symbol $\leq$ (resp. $\geq$) means, that the row representation is significantly worse (resp. better) than the column representation. The H_0 hypothesis is rejected at the level $\alpha = 0.1$, but we fail to reject it at the level $\alpha = 0.05$.

– The symbol = is used for representations which are not significantly better or worse. We fail to reject hypothesis H_0.

6.3 Discussion

We start our discussion with the comparison to results of [10]. They compared BIO-1 and BILOU representations on the English CoNLL corpus using CRF. Our experiments have similar results. The BIO-1 representation was better on the test set, while the BILOU representation was better on the validation set. The differences slightly favor the BILOU representation, but it is unclear, if it is just a coincidence or the BILOU representation is better. This conclusion is also supported by the fact that in [10], the BILOU representation was better on the test set and worse on the validation set (in our case, better on the validation set, worse on the

Table 1. The results of our experiments on the standard partitioning of corpora.

(a) English

	ME		CRF	
	val	test	val	test
IO	86.89	78.66	88.98	83.64
IOU	**87.16**	**79.94**	88.75	83.60
BIO-1	86.89	78.27	88.98	83.61
BIO-2	86.86	79.15	88.08	83.74
BIOU	87.01	79.82	88.92	83.96
IEO-1	86.79	78.54	89.02	83.79
IEO-2	86.99	79.53	**89.25**	**84.16**
IEOU	86.91	79.85	89.10	83.62
BIEO	86.55	78.88	88.90	83.82
BILOU	86.42	79.50	88.84	83.47

(b) Spanish

	ME		CRF	
	val	test	val	test
IO	64.59	70.45	74.02	79.66
IOU	63.98	70.93	73.95	79.33
BIO-1	64.46	70.35	74.38	79.80
BIO-2	63.80	**71.37**	**74.56**	79.54
BIOU	64.04	71.27	74.15	79.18
IEO-1	**65.13**	71.03	74.27	**79.86**
IEO-2	63.12	70.33	74.45	79.50
IEOU	63.39	70.76	74.44	78.96
BIEO	63.30	70.54	74.46	79.55
BILOU	63.42	70.71	74.37	79.37

(c) Dutch

	ME		CRF	
	val	test	val	test
IO	67.25	70.09	74.31	76.34
IOU	69.28	71.85	74.39	76.62
BIO-1	67.49	70.06	**74.81**	76.53
BIO-2	68.31	70.45	74.37	76.23
BIOU	**69.56**	**72.53**	74.59	76.56
IEO-1	68.84	71.07	74.54	76.13
IEO-2	68.49	70.91	74.07	**77.17**
IEOU	69.23	72.34	73.63	76.79
BIEO	68.43	70.68	74.68	76.51
BILOU	68.78	72.08	73.82	76.82

(d) Czech

	ME		CRF	
	val	test	val	test
IO	56.93	53.48	**68.64**	68.41
IOU	57.26	54.33	68.12	68.05
BIO-1	56.16	53.45	68.50	68.90
BIO-2	56.96	54.99	68.44	69.11
BIOU	58.11	55.86	68.54	**70.26**
IEO-1	56.75	55.42	68.30	69.34
IEO-2	56.98	55.61	68.22	70.08
IEOU	58.21	56.64	67.92	69.55
BIEO	58.40	**57.21**	67.58	69.61
BILOU	**58.60**	56.73	67.41	69.21

Table 2. The significance tests for various segment representations using ME. Detailed description is provided in Section 6.2.

	IO	IOU	BIO-1	BIO-2	BIOU	IEO-1	IEO-2	IEOU	BIEO	BILOU
IO	====	====	==≥=	>===	==<<	<<=<	==<=	==≤<	>===	=≥<<
IOU	====	====	==>≥	>===	==≤=	==>=	==<=	====	>===	>===
BIO-1	==≤=	==<≤	====	>=<=	==<<	=<=<	==<=	==<<	>===	=≥<<
BIO-2	<===	<===	<=>=	====	<=≤<	<=><	<≥<=	<==≤	>>==	=>=≤
BIOU	==>>	==≥=	==>>	>=≥>	====	==>=	===>	≤===	>=>≥	≥===
IEO-1	>>=>	==<=	=>=>	>=<>	==<=	====	=><>	==<=	>>=≥	≥><=
IEO-2	==>=	==>=	==>=	>≤>=	===<	=<><	====	=≤≥<	>=>=	>===
IEOU	==≥>	====	==>>	>==≥	≥===	==>=	=≥≤>	====	>≥=≥	>>==
BIEO	<===	<===	<===	<<==	<=<≤	<<=≤	<=<=	<≤=≤	====	≤=<<
BILOU	=≤≥>	<===	=≤>>	=<=≥	≤===	≤<>=	<===	<<==	≥=>>	====

Table 3. The significance tests for various segment representations using CRF. Detailed description is provided in Section 6.2.

	IO	IOU	BIO-1	BIO-2	BIOU	IEO-1	IEO-2	IEOU	BIEO	BILOU
IO	====	>>=>	≤<<<	=<=>	===>	<==≤	=<<<	==<=	=<=≥	>==≥
IOU	<<=<	====	<<<≤	<<=<	≤<<≤	<==<	<<<<	=<<<	=<=<	==<<
BIO-1	≥>>≥	>>≥>	====	>==>	>>=>	==>=	===≤	>>==	><>>	>==≥
BIO-2	=>=<	>>=>	<==<	====	>>==	<==<	==<<	≥>≤<	>===	>===
BIOU	===<	≥>>≥	<<=<	<<==	====	<==<	≤<<<	==<<	=<==	>===
IEO-1	>==≥	>==≥	==<=	>==>	>==>	====	==<<	>=<=	>==>	>==>
IEO-2	=>>>	>>>>	===≥	==>>	≥>>>	==>≥	====	≥>≥>	=<>>	>=>>
IEOU	==>=	=>>>	<<==	≤<≥>	==>>	<=>=	≤<<<	====	=<>>	==>>
BIEO	=>=≤	=>=>	<>><	<===	=>==	<==<	=><<	=><<	====	=≥=≥
BILOU	<==≤	==>>	<==≤	<===	<===	<==<	<=<<	==<<	=≤==	====

test set). The rest of the results has similar problems. For many representation pairs, it is impossible to pick the better one.

We were not satisfied with the results of the first set of tests, because it does not compare the representations rigorously. Thus, we proposed another approach for segment representations comparison described in Section 6.2. It is based on paired Student's test and gives well-defined comparisons.

The results of the significance tests are much more convincing. On one hand, the results provide evidence, that some segment representations are better than others. On the other hand, we are still unable to decide for many representations pairs, i.e. we must treat them as equal. Given these limitations, we can create a group of representations for each language, which are at the same or better level than all the other representations. These groups are (the bold representation has the highest average F-measure):

English, ME: IOU, BIO-1, IEO-1, **IEO-2**, IEOU
English, CRF: BIO-1, **IEO-1**, IEO-2
Spanish, ME: IOU, BIO-2, BIOU, **IOE-1**, IEOU
Spanish, CRF: BIO-2, IEO-1, **BIEO**

Dutch, ME: BIOU, **IEO-2**, BILOU
Dutch, CRF: BIO-1, **IEO-2**
Czech, ME: BIOU, **IEO-1**, IEOU, BILOU
Czech, CRF: IEO-2

Surprisingly, the IOE representations perform quite good, for Czech CRF is the IOE-2 even significantly better than the rest. The BILOU representation, generally considered as the best choice, performed rather poorly. We can say, that the optimal segment representation depends on both language and algorithm. We also expect it to be dependent on the feature set.

7   Conclusion

In this paper, we provide a rigorous study of segment representations for named entities. We experiment with ten different segment representations on the English, Spanish, Dutch and Czech corpora using two machine learning approaches – maximum entropy and conditional random fields.

We performed two sets of experiments. The first one was based on the standard partitioning of CoNLL corpora. The second one exploited 10 fold cross-validation and evaluation using the paired Student's t-test. The second test provides more accurate results.

Our experiments provide an interesting evidence. The BILOU representation ended up as the worst for English using CRF, even though it was considered better than the commonly used BIO-1 by [10] and it is generally considered as one of the best representations. The results presented in [10] were similar to the results of our first set of experiments, but the second set of experiments disproved this hypothesis. The IOE-1 and IOE-2 representations seem to be the best or at least reasonable choice for almost all languages and methods. Surprisingly, these representations have not been used in NER yet.

We show that choosing the optimal segment representation for named entities is a complex problem. The optimal representation depends on the language (corpus), on the approach, and very likely on the feature set. We propose a well-defined procedure for finding the optimal representation.

Thus, the impact of the article is two fold. First, we propose a new procedure for segment representation evaluation. Second, we recommend the use of IOE-1 and IOE-2 as they provide the most promising results in our tests.

In the future, we would like to experiment with multiple feature sets and their relation to optimal segment representation. The relation of the data size and the optimal representation could be also interesting.

References

1. Borthwick, A.E.: A Maximum Entropy Approach to Named Entity Recognition. Ph.D. thesis, New York, NY, USA. AAI9945252 (1999)

2. Lafferty, J.D., McCallum, A., Pereira, F.C.N.: Conditional random fields: probabilistic models for segmenting and labeling sequence data. In: Proceedings of the Eighteenth International Conference on Machine Learning. ICML 2001, San Francisco, CA, USA, pp. 282–289. Morgan Kaufmann Publishers Inc. (2001)

3. Bikel, D.M., Miller, S., Schwartz, R., Weischedel, R.: Nymble: a high-performance learning name-finder. In: Proceedings of the Fifth Conference on Applied Natural Language Processing. ANLC 1997, Stroudsburg, PA, USA, pp. 194–201. Association for Computational Linguistics (1997)

4. Collins, M., Singer, Y.: Unsupervised models for named entity classification. In: Proceedings of the Joint SIGDAT Conference on Empirical Methods in Natural Language Processing and Very Large Corpora, pp. 100–110 (1999)

5. Béchet, F., Nasr, A., Genet, F.: Tagging unknown proper names using decision trees. In: Proceedings of the 38th Annual Meeting on Association for Computational Linguistics. ACL 2000, Stroudsburg, PA, USA, pp. 77–84. Association for Computational Linguistics (2000)

6. Cucerzan, S., Yarowsky, D.: Language independent ner using a unified model of internal and contextual evidence. In: Proceedings of, Taipei, Taiwan, pp. 171–174 (2002)

7. Mao, X., Xu, W., Dong, Y., He, S., Wang, H.: Using Non-Local Features to Improve Named Entity Recognition Recall, vol. 21. The Korean Society for Language and Information (KSLI) (2007)

8. Sun, J., Wang, T., Li, L., Wu, X.: Person name disambiguation based on topic model. In: CIPS-SIGHAN Joint Conference on Chinese Language Processing (2010)

9. Liu, X., Zhang, S., Wei, F., Zhou, M.: Recognizing named entities in tweets. In: Proceedings of the 49th Annual Meeting of the Association for Computational Linguistics: Human Language Technologies. HLT 2011, Stroudsburg, PA, USA, vol. 1, pp. 359–367. Association for Computational Linguistics (2011)

10. Ratinov, L., Roth, D.: Design challenges and misconceptions in named entity recognition. In: Proceedings of the Thirteenth Conference on Computational Natural Language Learning. CoNLL 2009, Stroudsburg, PA, USA, pp. 147–155. Association for Computational Linguistics (2009)

11. Straková, J., Straka, M., Hajič, J.: A new state-of-the-art Czech named entity recognizer. In: Habernal, I. (ed.) TSD 2013. LNCS, vol. 8082, pp. 68–75. Springer, Heidelberg (2013)

12. Tjong Kim Sang, E.F., De Meulder, F.: Introduction to the conll-2003 shared task: language independent named entity recognition. In: Proceedings of the Seventh Conference on Natural Language Learning at HLT-NAACL 2003. CONLL 2003, Stroudsburg, PA, USA, vol. 4, pp. 142–147. Association for Computational Linguistics (2003)

13. Cho, H.C., Okazaki, N., Miwa, M., Tsujii, J.: Named entity recognition with multiple segment representations. Information Processing & Management **49**(4), 954–965 (2013)

14. Shen, H., Sarkar, A.: Voting between multiple data representations for text chunking. In: Kégl, B., Lee, H.-H. (eds.) Canadian AI 2005. LNCS (LNAI), vol. 3501, pp. 389–400. Springer, Heidelberg (2005)

15. Lin, D., Wu, X.: Phrase clustering for discriminative learning. In: Proceedings of the Joint Conference of the 47th Annual Meeting of the ACL and the 4th International Joint Conference on Natural Language Processing of the AFNLP: Volume 2 - Volume 2. ACL 2009, Stroudsburg, PA, USA, pp. 1030–1038. Association for Computational Linguistics (2009)

16. Konkol, M.: Brainy: a machine learning library. In: Rutkowski, L., Korytkowski, M., Scherer, R., Tadeusiewicz, R., Zadeh, L.A., Zurada, J.M. (eds.) ICAISC 2014, Part II. LNCS, vol. 8468, pp. 490–499. Springer, Heidelberg (2014)
17. Tjong Kim Sang, E.F.: Introduction to the conll-2002 shared task: language-independent named entity recognition. In: Proceedings of the 6th Conference on Natural Language Learning. COLING 2002, Stroudsburg, PA, USA, vol. 20, pp. 1–4. Association for Computational Linguistics (2002)
18. Konkol, M., Konopík, M.: CRF-based Czech named entity recognizer and consolidation of Czech NER research. In: Habernal, I. (ed.) TSD 2013. LNCS, vol. 8082, pp. 153–160. Springer, Heidelberg (2013)
19. Ševčíková, M., Žabokrtský, Z., Krůza, O.: Named entities in Czech: annotating data and developing NE tagger. In: Matoušek, V., Mautner, P. (eds.) TSD 2007. LNCS (LNAI), vol. 4629, pp. 188–195. Springer, Heidelberg (2007)

Analyzing Text Coherence via Multiple Annotation in the Prague Dependency Treebank

Kateřina Rysová[(✉)] and Magdaléna Rysová

Faculty of Mathematics and Physics, Institute of Formal and Applied Linguistics,
Charles University in Prague, Malostranské náměstí 25, 118 00 Praha,
Czech Republic
{rysova,magdalena.rysova}@ufal.mff.cuni.cz
https://ufal.mff.cuni.cz

Abstract. Corpus-based research demonstrates an existence of a mutual interaction of bridging anaphoric relations in the text and sentence information structure. The research is carried out on large corpus data of the Prague Dependency Treebank 3.0 that contains almost 50 thousand sentences with manual annotation of both sentence information structure and bridging anaphora. We investigate in which way the bridging anaphora relations interconnect contextually bound and non-bound sentence items and how such types of connections contribute to the text coherence.

Keywords: Text coherence · Bridging anaphora · Sentence information structure · Topic-focus articulation · Prague dependency treebank

1 Introduction

The paper investigates the relation between two language phenomena: sentence information structure and bridging anaphoric text relations. Both of them have been studied as individuals in many research papers but in mutual interaction, they have been investigated only in the last recent years: so far, the theme of their interplay (on large corpus data) is elaborated especially by Hajičová [1], [2] or [3]. Hajičová principally deals with the relation between sentence information structure and coreference (and anaphora) and discourse relations. She analyzes e.g. under which circumstances, the anaphoric links lead from the sentence items that are contextually non-bound in terms of sentence information structure. In doing so, Hajičová emphasizes the need of complex text study, i.e. the need of exploration of the mentioned language phenomena in cooperation and mutual interaction because the text coherence results from the interplay of the individual intra- and inter-sentential phenomena.

For studying text coherence that covers several language areas or phenomena, it is necessary to use language data annotated on multiple language levels and planes. Nowadays, there are some corpora and computer programs enabling to see a mutual interaction of more individual language phenomena in a text at once, see [4], [5] or [6].

© Springer International Publishing Switzerland 2015
P. Král and V. Matoušek (Eds.): TSD 2015, LNAI 9302, pp. 71–79, 2015.
DOI: 10.1007/978-3-319-24033-6_8

One of the richest corpora (i.e. corpora with various types of annotation) is the Prague Dependency Treebank (PDT) [7]. It contains language annotation on morphological, analytical (surface syntactic) and tectogrammatical (deep syntactic) levels and includes, among others, manual annotation of sentence information structure, coreference and anaphoric relations, text genres and discourse relations. Therefore, the Prague Dependency Treebank is an ideal data source for our investigation of the interplay between bridging anaphoric relations and sentence information structure. In the paper, we examine the texts in Czech, but our methods may be used also for other languages in similarly annotated corpora.

2   Aim of Work

The main aim of the paper is 1) to find out whether and how the two language phenomena cooperate in a text; 2) to demonstrate linguistic and computational methods that may be used for further research of the language phenomena interplay; 3) to contribute to the general discussion on text coherence, i.e. how these two phenomena participate in text coherence.

When starting our analysis, we have concentrated on several crucial issues or questions. One of the most important is whether the bridging anaphoric links connect rather contextually bound sentence members (mutually) or rather contextually non-bound sentence members (mutually) or whether they rather interconnect both, i.e. contextually bound sentence members with the contextually non-bound sentence members.

Another aspect we focused on is whether the bridging anaphoric links operate within a Topic and Focus of the sentence in the same way.

Our assumption is that there will be more bridging anaphoric relations leading from the contextually bound sentence members (contrastive and non-contrastive, see the Section Annotation of Sentence Information Structure) looking for the connection with the previous (con)text than leading from the contextually non-bound sentence members. The reason is, in very simple terms, that the contextually bound sentence items often bring old and known information (deducible from the previous (con)text) while the contextually non-bound items often bring new and unknown (non-deducible) information.

3   Language Material – the Prague Dependency Treebank

To answer the above questions, we have analyzed the data from the Prague Dependency Treebank 3.0 (containing almost 50,000 sentences: 833,195 word tokens in 3,165 documents), a multilayer annotated corpus of Czech newspaper texts. As mentioned above, PDT contains various types of language annotations: among others, also manual annotation of sentence information structure and manual annotation of bridging anaphora.

3.1   Annotation of Sentence Information Structure

The annotation of sentence information structure in PDT is based on the theory of Functional Generative Description (FGD) [8]. During the annotation, the sentence items (nodes in a dependency tree) have been labeled as one of these three options:

a) non-contrastive contextually bound nodes (marked as t)[1];
b) contrastive contextually bound nodes (marked as c)[2];
c) contextually non-bound nodes (marked as f)[3].

For the examples of t , c and f nodes, see Figure 1. Contextually bound sentence items (contrastive or non-contrastive) are typical members of the sentence Topic. Contextually non-bound sentence items are typical members of the sentence Focus. For more details, see the annotation manual [9].

3.2   Annotation of Bridging Anaphora

Annotation of bridging anaphora (some authors use also other terms like indirect anaphora or associative anaphora) in PDT was carried out according to Nedoluzhko [10]. Bridging anaphoric relations in PDT annotation are considered the semantic or pragmatic relations between non-coreferential entities (nodes in a dependency tree) that participate in the text coherence. PDT contains the following types of such relations [11]: PART – WHOLE (e.g. *room – ceiling*), SUBSET – SET (*students – some students*), FUNCTION (*state – president*), CONTRAST (for coherence relevant discourse opposites; e.g. *this year – last year*), ANAF (for explicit anaphoric relations without coreference or one of the semantic relations mentioned above; e.g. *rainbow – that word*), REST (further underspecified group). The bridging anaphora contains both inter- and intra-sentential relations.

---

[1] Non-contrastive contextually bound expressions are expressions (both expressed and absent in the surface structure of the sentence) that introduce in the text some given information . Such expressions are repeated from the preceding text (not necessarily verbatim), they are deducible from it (e.g. using coreferential or inferential relations), or somehow related to a broader context. [9]

[2] A contrastive contextually bound expression is usually a choice from a set of alternatives. This set need not be explicitly specified in the text. A contrastive contextually bound expression can refer to a larger text segment and does not have to be deducible from the immediately preceding textual context. [...] The occurrence of a contrastive contextually bound expression is primarily determined by the thematic structure (progression) of the text. Contrastive contextually bound expressions usually occur in enumerations, at the beginning of paragraphs etc. In the spoken form of an utterance the contrastive contextually bound expression carries an optional contrastive stress. [9]

[3] Contextually non-bound expressions are expressions (both expressed and absent in the surface structure of the sentence) that represent in the text some unknown, new facts, or introduce known facts in new relations, i.e. they express information not deducible from context. [9]

Table 1. Numbers of occurrences of contextually bound (contrastive and non-contrastive: c , t) and non-bound (f) sentence items interlinked with bridging anaphoric relation (in the Prague Dependency Treebank).

	f (from)	t (from)	c (from)	To (in total)
f (to)	12,485	4,428	2,095	19,008
t (to)	7,091	4,348	1,720	13,159
c (to)	2,248	809	639	3,696
From (in total)	21,824	9,585	4,454	35,863

4   Methods

The aim of our work is to find out how the bridging relations in text correspond to the sentence information structure.

We may imagine the bridging relation as an arrow with two important aspects concerning sentence information structure: 1) where the bridging arrow leads FROM (i.e. whether rather from contextually bound or non-bound sentence items) and 2) where it leads TO (i.e. whether rather to contextually bound or non-bound sentence items).

To answer both questions, we have compiled a table (see Table 1) expressing all mutual possibilities of how many bridging anaphoric relations occur among contextually bound (non-contrastive; contrastive) and contextually non-bound sentence items.

5   Results

The main results of our analysis, i.e. the occurrences of bridging anaphoric relations connecting contextually bound and non-bound sentence items (nodes) in PDT in all possible combinations, are captured in Table 1.[4]

Table 1 demonstrates, for example, that a bridging anaphoric arrow leads from the sentence item (node) that is non-contrastive contextually bound (in PDT marked as t) to the sentence item (node) that is contextually non-bound (in PDT marked as f) in 4,428 cases.

Figure 1 shows an authentic PDT example of this combination – see Example (1) in the plain text:

(1) *Zpívají o nich v **písních**.*(f) *, např. i jedna z nejznámějších [**písní**.*(t)] –
Chajori romani – má sloku o utrpení Romů v koncentračních táborech.
(In English: They sing about them[5] in **songs**.(f), e.g. one of the most famous [**songs**.(t)] – Chajori romani – has a stanza about Gypsy suffering in concentration camps.)[6]

---

[4] For overall distributions of f , t and c nodes in PDT, see below.

[5] About the events of World War II.

[6] The preceding context is: [...] The Gypsies were affected by disaster during the World War II. [...] Events of that time are still a trauma for few gypsy survivors and their descendants.

A bridging anaphoric arrow leads from the node representing the lemma *song* (that is omitted in the surface structure but present in the deep (underlying) sentence structure) to the node representing the lemma *song* (present both in the surface and deep sentence structure).[7]

A bridging arrow starts in the node that is obviously contextually bound (known for the reader even to such extent that it is omitted from the surface word order) and points at the node that is contextually non-bound (in this case, the first occurrence of the word *song* is a part of the sentence Focus).

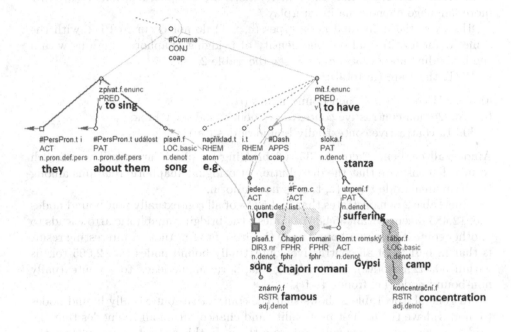

Fig. 1. PDT dependency tree representing the sentence from Example (1).

Such type of bridging relations in text is well expectable: contextually bound sentence items are connected to the preceding contextually non-bound items, i.e. the new information from one clause is repeated as the old information in the following clause where it is further elaborated. The text coherence often benefits right from this changing of the new information into the old . However, this type of text relations is not the main one, see the Table 1.

The most often type of bridging relation (in absolute numbers) is the relation between two contextually non-bound sentence items. The fact that this kind of bridging text connection is so common is quite surprising. On the other hand, although the sentence item brings new information, it can be also interlinked with other places of the text (with other sentence and text items). The general

---

[7] Another light blue bridging anaphoric arrow connects the nodes *stanza* and *song* in the picture.

Table 2. How many % of all f or t or c nodes are interlinked with a bridging anaphoric relation (in PDT).

	f (from)	t (from)	c (from)
f (to)	3.52	2.51	6.91
t (to)	2.00	2.47	5.67
c (to)	0.63	0.46	2.11

text coherence results exactly from this interlinking of various text items. In this connection, we can see that the text relations are complex relations created by more language phenomena in interplay.

However, the individual node types (c, t, f) do not occur in PDT with the same frequency. To find out the density of bridging anaphora relations within the individual node types (c, t, f), see the Table 2.
In PDT, there are (in total):

a) 354,841 contextually non-bound nodes (f);
b) 176,225 non-contrastive contextually bound nodes (t) and
c) 30,312 contrastive contextually bound nodes (c).

Among all of them, there are 35,863 bridging anaphoric arrows. The research results demonstrate that the distribution of bridging anaphoric relations among the individual node types (c, t, f) is not uniform.

The Table 2 demonstrates that in 3.52 % of all contextually non-bound nodes (i.e. 12,485 tokens within 354,841 f nodes), the bridging anaphoric arrow leads to another contextually non-bound node (i.e. from f to f). Another interesting result is that in 6.91 % of all contrastive contextually bound nodes (i.e. 2,095 tokens within 30,312 c nodes), the bridging anaphoric arrow leads to a contextually non-bound node (i.e. from c to f).

In general, the Table 2 shows that the contrastive contextually bound nodes (c nodes) have the highest probability and chance within all the nodes that the bridging anaphoric arrow will lead from them. In this respect, we may state the following main points gained from the results of our analysis:

1. The most typical bridging anaphoric connection leads from a contrastive contextually bound node to a contextually non-bound node (i.e. from c to f).
2. The second most typical bridging anaphoric connection leads from a contrastive contextually bound node to a non-contrastive contextually bound node (i.e. from c to t).
3. The third most typical bridging anaphoric connection leads from a contextually non-bound node to a contextually non-bound node (from f to f).
4. In general, the most favorite starting position for a bridging anaphoric arrow is a contrastive contextually bound sentence item (c).

Table 3 demonstrates which kinds of sentence items (from the perspective of sentence information structure) have the highest tendency to be the recipient and the sender of a bridging anaphoric relation. For example, 6.15 % within all contextually non-bound sentence items (i.e. 21,824 within 354,841) serve as a

Table 3. How many % of all f , t , c or t+c nodes are providing (to) or looking for (from) a bridging anaphoric relation (in PDT).

	f	t	c	t+c
from	6.15	5.44	14.69	6.80
to	5.36	7.47	12.19	8.16

bridging sender and 5.36 % of them (i.e. 19,008 within 354,841) as a bridging recipient . Thus, on the basis of our analysis, we came to the following points:

5. The bridging anaphoric arrow leads from every 7th and to every 8th contrastive contextually bound sentence item (c node);

6. the bridging anaphoric arrow leads from every 18th and to every 13th non-contrastive contextually bound sentence item (t node) and

7. the bridging anaphoric arrow leads from every 16th and to every 18th contextually non-bound sentence item (f node).

8. It is quite surprising that the contextually non-bound sentence items (f nodes; bringing typically the information that is non-deducible from the previous context) look for a bridging relation in the previous text even more often than non-contrastive contextually bound items (t nodes; bringing typically the information that is deducible from the previous context).

If we divide the items only into contextually non-bound and bound (without the contrastive and non-contrastive distinction), the proportion between contextually bound and non-bound nodes searching for the relation in the previous text is nearly balanced (6.15 % within all f nodes and 6.8 % within all t+c nodes are the starting position for the bridging anaphoric relation).

6 Conclusion

In PDT, we have found 35,863 bridging anaphoric relations interconnecting contextually bound or non-bound sentence items. The results of the research demonstrate that the bridging anaphoric relations are not uniformly distributed within them. Some types of sentence items (from the perspective of the sentence information structure) have a greater ability to attract bridging anaphoric relation than the other. This proves that the language phenomena of sentence information structure and bridging anaphora are closely interdependent – if the sentence item has a role of a contrastive contextually bound node in sentence information structure, there is a relatively high probability that it will be interconnected with other sentence items in the text (in sense of bridging anaphora relations).

The greatest ability to be a part of bridging anaphoric chains is proved at contrastive contextually bound sentence items. These items are the most favorite starting as well as landing positions for bridging anaphoric arrows. Among them (i.e. among the c nodes), there is the greatest density of bridging anaphora relations.

The contrastive contextually bound nodes serve as the most favorite sources of items to which the other (i.e. following) sentence items anaphorically refer (in PDT, every 8th within all c nodes serves as a recipient of bridging relations, i.e. as a landing destination of bridging anaphora arrow). At the same time, they have also the highest tendency to look for a bridging relation in the previous (con)text (in PDT, every 7th within all c nodes serves as a sender of bridging relation, i.e. as a starting destination of a bridging anaphora arrow).

Since the contrastive contextually bound items appear in the initial sentence position or near to it very often, we may assume that the sentence beginnings are very important places of text coherence realized by bridging anaphora. Therefore, the contrastive contextually bound items may be seen as a significant pillar and backbone of the text coherence expressed by bridging anaphora.

In the paper, we tried to present how we may use the multilayer corpus data to demonstrate the crucial aspects of interplay of different language phenomena like sentence information structure and bridging anaphora, which could improve or deepen our general knowledge of text coherence.

Acknowledgments. The authors acknowledge support from the Czech Science Foundation (GACR; project n. P406/12/0658) and from the Ministry of Education, Youth and Sports of the Czech Republic (project n. LM2010013 LINDAT/CLARIN and n. LH14011).

References

1. Hajičová, E., Hladká, B., Kučová, L.: An annotated corpus as a test bed for discourse structure analysis. In: Proceedings of the Workshop on Constraints in Discourse, Maynooth, Ireland, pp. 82–89. National University of Ireland, National University of Ireland (2006)
2. Hajičová, E.: On interplay of information structure, anaphoric links and discourse relations. In: Societas Linguistica Europaea, SLE 2011, 44th Annual Meeting, Book of Abstracts, Logrono, Spain, pp. 139–140. Universidad de la Rioja, Center for Research in the Applications of Language, Universidad de la Rioja, Center for Research in the Applications of Language (2011)
3. Kučová, L., Veselá, K., Hajičová, E., Havelka, J.: Topic-focus articulation and anaphoric relations: a corpus based probe. In: Heusinger, K., Umbach, C. (eds.) Proceedings of Discourse Domains and Information Structure Workshop, pp. 37–46. Edinburgh, Scotland (2005)
4. Komen, E.R.: Coreferenced corpora for information structure research. Outposts of Historical Corpus Linguistics: From the Helsinki Corpus to a Proliferation of Resources (Studies in Variation, Contacts and Change in English 10) (2012)
5. Stede, M., Neumann, A.: Potsdam commentary corpus 2.0: annotation for discourse research. In: Proceedings of the Ninth International Conference on Language Resources and Evaluation (LREC 14), pp. 925–929 (2014)
6. Chiarcos, C.: Towards interoperable discourse annotation. Discourse features in the ontologies of linguistic annotation. In: Proceedings of the Ninth International Conference on Language Resources and Evaluation (LREC 14), pp. 4569–4577 (2014)

7. Bejček, E., Hajičová, E., Hajič, J., Jínová, P., Kettnerová, V., Kolářová, V., Mikulová, M., Mírovský, J., Nedoluzhko, A., Panevová, J., Poláková, L., Ševčíková, M., Štěpánek, J., Zikánová, Š.: Prague dependency treebank **3** (2013)
8. Hajičová, E., Sgall, P., Partee, B.: Topic-focus articulation, tripartite structures, and semantic content. Kluwer, Dordrecht (1998). ISBN 0-7923-5289-0
9. Mikulová, M., Bémová, A., Hajič, J., Hajičová, E., Havelka, J., Kolářová, V., Lopatková, M., Pajas, P., Panevová, J., Razímová, M., Sgall, P., Štěpánek, J., Urešová, Z., Veselá, K., Žabokrtský, Z., Kučová, L.: Anotace na tektogramatické rovině pražského závislostního korpusu. anotátorská příručka. Technical Report TR-2005-28 (2005)
10. Nedoluzhko, A.: Rozšířená textová koreference a asociační anafora (Koncepce anotace českých dat v Pražském závislostním korpusu). Studies in Computational and Theoretical Linguistics. Ústav formální a aplikované lingvistiky, Praha (2011)
11. Nedoluzhko, A.: Generic noun phrases and annotation of coreference and bridging relations in the prague dependency treebank. In: 51st Annual Meeting of the Association for Computational Linguistics Proceedings of the 7th Linguistic Annotation Workshop & Interoperability with Discourse, Sofija, Bulgaria, pp. 103–111. Bǎlgarska akademija na naukite, Omnipress, Inc (2013)

Automatic Detection of Parkinson's Disease in Reverberant Environments

Juan Rafael Orozco-Arroyave[1,2]([✉]), Tino Haderlein[2], and Elmar Nöth[2]

[1] Faculty of Engineering, Universidad de Antioquia UdeA, Medellín, Colombia
rafael.orozco@i5.informatik.uni-erlangen.de
[2] Pattern Recognition Lab, Friedrich-Alexander-Universität Erlangen-Nürnberg
(FAU), Erlangen, Germany

Abstract. Automatic classification of speakers with Parkinson's disease
(PD) and healthy controls (HC) is performed considering a method for
the characterization of the speech signals which is based on the estimation of the energy content of the unvoiced frames. The method is tested
with recordings of three languages: Spanish, German, and Czech. Additionally, the signals are affected by two different reverberant scenarios in
order to validate the robustness of the proposed method. The obtained
results range from 85% to 99% of accuracy depending on the speech task,
the spoken language, and the recording scenario. The method shows to
be accurate and robust even when the signals are reverberated. This work
is a step forward to the development of methods to assess the speech of
PD patients without requiring special acoustic conditions.

Keywords: Parkinson's disease · Reverberant evironments · Unvoiced
frames · Multi-language

1 Introduction

Parkinson's disease (PD) is a neurological disorder that results from the death
of dopaminergic cells in the substantia nigra of the midbrain [1]. It is the second
most prevalent neurological disorder and affects about 2% of people older than 65
[2]. According to the Royal College of Physicians in London, PD patients should
have access to a set of services and therapies including specialized nursing care,
physiotherapy, and speech and language therapy, among others [3]. It is estimated
that about 90% of people with PD develop speech impairments; however, only 3%
to 4% of them receive speech therapy [4]. The symptoms observed in the speech
of PD patients include reduced loudness, a monopitch and monoloudness kind of
speech, breathy voice, and imprecise articulation, among others [4]. In addition
to the aforementioned problems in the speech of PD patients, they develop also
motor impairments that reduce their motion capabilities. The research community has shown interest in developing systems for the telemonitoring of people with
PD from speech [5–7]. However, the performance of such systems in real-life conditions, i.e. in non-controlled noise and in reverberant environments, is still an unan-
swered question. The motion problems developed by PD patients make difficult to

© Springer International Publishing Switzerland 2015
P. Král and V. Matoušek (Eds.): TSD 2015, LNAI 9302, pp. 80–87, 2015.
DOI: 10.1007/978-3-319-24033-6_9

perform their recording in places different to their house or their room in a hospital. Thus, it is necessary to develop computational tools able to perform the analysis of the speech recordings even if such records are captured in reverberant environments or in non-controlled acoustic conditions. This paper presents a method to perform the automatic classification of speakers with PD and HC from speech recordings that are altered by two different reverberant scenarios. The method is tested with recordings of three databases including people speaking three different languages (Spanish, German, and Czech). The speech tasks evaluated include isolated sentences and the rapid repetition of the syllables /pataka/, which is also called diadochokinetic (DDK) evaluation.

The rest of the paper is organized as follows. Section 2 presents the details of the experimental setup. In Section 3 the obtained results are presented, and in Section 4 the conclusions derived from this study are provided.

2 Experimental Setup

The speech recordings are affected with two different reverberant scenarios. The first one considers a reverberant room that is characterized using a microphone situated at 60 cm in front of the speaker. The second one considers the impulse response obtained from several different angles and distances with respect to the speaker's position, and two different reverberation times. The original recordings, i.e. without any reverberation procedure, are also considered. The unvoiced frames of the speech recordings are segmented automatically using the software Praat [8]. Voiced frames are not considered in this study because in previous experiments we have shown that unvoiced frames are more discriminant than voiced ones [9]. The energy content of each unvoiced frame is measured considering 12 mel-frequency cepstral coefficients (MFCCs) and 25 energy bands distributed according to the Bark scale. The automatic classification of speakers with PD and HC is performed using a support vector machine with soft margin. Figure 1 summarizes the process introduced in this paper. The stages of the process are detailed in the following subsections.

2.1 Databases

Spanish: Recordings of the PC-GITA database [10] are considered. Seven speech tasks including six isolated sentences, and rapid repetitions of /pataka/ are evaluated. The corpus contains recordings of 100 speakers (50 with PD and 50 HC). The speakers are balanced by gender and age. The age of the 25 male patients ranges from 33 to 77 (mean 62.2 ± 11.2), and the age of the 25 female patients ranges from 44 to 75 years (mean 60.1 ± 7.8). For the case of HC, the age of the 25 male ranges from 31 to 86 (mean 61.2 ± 11.3), and the age of the 25 female ranges from 43 to 76 years (mean 60.7 ± 7.7). The recording sessions were performed in a sound-proof booth at Clínica Noel in Medellín, Colombia, using a dynamic omni-directional microphone and a professional audio card. The recordings were sampled at 44.1 kHz with a resolution of 16 bits. All of the

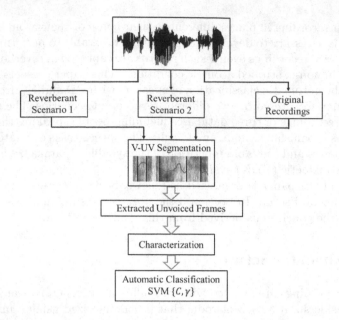

Fig. 1. Stages of the experimental setup

patients were evaluated and diagnosed by a neurologist expert. The mean values of their neurological evaluation according to the unified Parkinson's disease rating scale (UPDRS-III) and Hoehn & Yahr scale [11] are 36.7 ± 18.7 and 2.3 ± 0.8, respectively. Further details of this database can be found in [10].

German: This corpus consists of 176 native German speakers (88 PD patients and 88 HC). The set of patients includes 88 people (47 male and 41 female). The age of the male patients ranges from 44 to 82 (mean 66.7 ± 8.4), while the age of the female patients ranges from 42 to 84 years (mean 66.2 ± 9.7). The HC group contains 88 speakers (44 male and 44 female). The age of the male ranges from 26 to 83 (mean 63.8 ± 12.7), and the age of the female is from 54 to 79 years (mean 62.6 ± 15.2). The participants were recorded at the Knappschaftskrankenhaus of Bochum in Germany. The sampling frequency of the recordings is 16 kHz with a resolution of 16 bits. The speakers read five isolated sentences and performed the DDK evaluation. The mean values of their neurological evaluation according to the UPDRS-III and Hoehn & Yahr scales are 22.7 ± 10.9 and 2.4 ± 0.6, respectively. Further details of this database can be found in [12].

Czech: A total of 33 native Czech speakers were recorded (19 PD patients and 14 HC). All of the participants of this database are male. The age of the patients ranges from 41 to 60 years (mean 61 ± 12). The age of the healthy group ranges from 36 to 80 years (mean 61.8 ± 13.3). The patients were newly diagnosed with PD, and none of them had been medicated before or during the recording session. The participants were recorded in the General University Hospital in Prague, Czech Republic. The speech tasks considered in this paper include the DDK evaluation and three isolated sentences. The signals were sampled at 48 kHz with a resolution of 16 bits. The mean values of the neurological evaluations according to the UPDRS-III and Hoehn & Yahr scales are 17.9 ± 7.4 and 2.2 ± 0.5, respectively. Further details of this database can be found in [13].

2.2 Speech Tasks

The speech tasks uttered in Spanish are (1) Mi casa tiene tres cuartos, (2) Omar, que vive cerca, trajo miel, (3) Laura sube al tren que pasa, (4) Los libros nuevos no caben en la mesa de la oficina, (5) Rosita Niño, que pinta bien, donó sus cuadros ayer, and (6) Luisa Rey compra el colchón duro que tanto le gusta, (7) and the rapid repetition of /pataka/.

The speech tasks uttered in German are (1) Peter und Paul essen gerne Pudding, (2) Das Fest war sehr gut vorbereitet, (3) Seit seiner Hochzeit hat er sich sehr verändert, (4) Im Inhaltsverzeichnis stand nichts über Lindenblätentee, (5) Der Kerzenständer fiel gemeinsam mit der Blumenvase auf den Plattenspieler, and (6) the rapid repetition of /pataka/.

The speech tasks uttered in Czech are questions that differ in a couple of words among them. The set of questions is (1) Kolik máte teď u sebe asi peněz?, (2) Kolikpak máte teďka u sebe asi peněz?, (3) Kolikpak máte teďka u sebe asi tak peněz?, (4) and the rapid repetition of /pataka/. Unfortunately, we did not had access to sentences with more varied content.

2.3 Reverberation

Testing the robustness of a system for acoustic analysis with respect to different recording scenarios usually means collecting speech data in many rooms with different impulse responses. Additionally, the microphone(s) should be in different angles and distances from the speaking people who also have to be available in every location. Reverberating close-talking speech artificially with the help of pre-defined room impulse responses can reduce this effort. For this reason, the method introduced in [14] is applied here. The original audio samples from all three languages were converted to 16 kHz and 16 bit as a preprocessing step using SoX v14.3.1. In order to avoid too much clipping due to over-amplification, the volume of the Czech data is reduced to 0.98 of its original value. For the Spanish data, the factor 0.99 is used. For the German data is not necessary to apply such factor because in that case there were not clippings. Room impulse responses for reverberation were measured in a seminar room with the size $5.8\,\mathrm{m} \times 5.9\,\mathrm{m} \times 3.1\,\mathrm{m}$. The microphone was at position $(2.0\,\mathrm{m}, 5.2\,\mathrm{m}, 1.4\,\mathrm{m})$. The reverberation time could be changed from $T_{60} = 250\,\mathrm{ms}$ to $T_{60} = 400\,\mathrm{ms}$ by removing sound absorbing carpets and sound absorbing curtains from the room. 12 impulse responses were measured for loudspeaker positions on three semi-circles in front of the microphone at distances 60 cm, 120 cm, and 240 cm following the method described in [15] (Fig. 2).

For *reverberant scenario 1*, the original close-talking speech data were convolved with the impulse response measured when the loudspeaker was at 60 cm

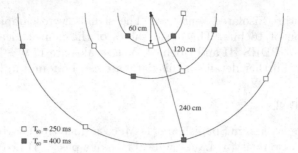

Fig. 2. Assumed speaker positions in the virtual recording rooms for artificially rever-berated data; 12 room impulse responses from different positions and with two rever-beration times (250 and 400 ms) were used.

distance right in front of the microphone. For *reverberant scenario 2*, the original data were divided into 12 parts; i.e., each part consisted of one twelfth of all recordings, as far as possible. Each of the parts was convolved with one of the 12 available impulse responses then.

2.4 Preprocessing and Characterization of Unvoiced Frames

In order to avoid possible bias introduced by the channel, i.e. microphone and sound card, mean cepstral subtraction is performed before the extraction of features from the recordings i.e., characterization. The energy content of the unvoiced frames is measured considering 25 Bark band energies (BBEs). 12 MFCCs are also calculated, as in [9]. Four low level descriptors are calculated over all feature vector, i.e. mean value, standard deviation, skewness, and kurtosis, forming a 148-dimensional feature vector per recording.

2.5 Classification

A soft margin support vector machine (SVM) with Gaussian kernel is considered to discriminate between PD speakers and HC. The complexity of the SVM (C) and the bandwidth of its kernel (γ) are optimized in a grid search with $10^{-1} < C < 10^4$ and $10^{-1} < \gamma < 10^3$. The optimization criterion is based on the accuracy on test, which could lead to slightly optimistic estimates, but considering that only two parameters are optimized, the bias should be minimal. The classifier is trained following a 10-fold cross validation strategy for the Spanish and German recordings. Each fold was chosen randomly but assuring the balance in age, gender, and the speaker independence. Due to the smaller number of recordings, leave-one-speaker-out (LOSO) cross-validation was used for the Czech data.

3 Results and Discussion

Table 1 shows the results obtained with the two reverberant scenarios and with the original recordings of the three databases. The results are presented in terms of accuracy, specificity, and sensitivity. Note that the highest accuracies on each database are obtained with the DDK evaluations. This result is in accordance with previous studies that highlighted such a task to be appropriate to assess PD speech [16]. Note also that there is an improvement in the accuracies when the signals are affected by the reverberant scenarios. This behavior can probably be explained by the reverberation process which is introducing information from voiced frames into the unvoiced regions, and thus the method is taking advantage of such "additional" information. The results reported in this paper indicate that the method works properly under reverberant conditions, so it could be used in environments where the acoustic conditions cannot be controlled.

Table 1. Results with recordings affected by two reverberant scenarios. Sent: Sentence. Rev: Reverberant. Acc: Accuracy (%), Spec: Specificity (%), Sens: Sensitivity (%)

	Spanish: Rev. Scenario 1			Rev. Scenario 2			Original Signals		
	Acc	Spec	Sens	Acc	Spec	Sens	Acc	Spec	Sens
Sent. 1	98 ± 4	98 ± 6	98 ± 6	94 ± 8	94 ± 14	94 ± 14	90 ± 9	90 ± 14	90 ± 14
Sent. 2	94 ± 7	92 ± 10	96 ± 8	91 ± 11	88 ± 17	94 ± 14	81 ± 7	84 ± 18	78 ± 15
Sent. 3	93 ± 10	100 ± 0	86 ± 19	96 ± 5	94 ± 10	98 ± 6	92 ± 13	94 ± 10	90 ± 19
Sent. 4	97 ± 7	100 ± 0	94 ± 14	99 ± 3	100 ± 0	98 ± 6	97 ± 5	98 ± 6	96 ± 8
Sent. 5	96 ± 8	100 ± 0	92 ± 17	96 ± 8	100 ± 0	92 ± 17	90 ± 9	92 ± 10	88 ± 14
Sent. 6	95 ± 9	94 ± 14	96 ± 8	97 ± 5	98 ± 6	96 ± 8	94 ± 7	94 ± 10	94 ± 14
DDK	96 ± 7	100 ± 0	92 ± 14	99 ± 3	98 ± 6	100 ± 0	99 ± 3	100 ± 0	98 ± 6
	German: Rev. Scenario 1			Rev. Scenario 2			Original Signals		
Sent. 1	95 ± 6	97 ± 5	93 ± 9	96 ± 5	98 ± 5	94 ± 6	93 ± 5	92 ± 8	94 ± 9
Sent. 2	94 ± 4	94 ± 8	94 ± 9	93 ± 6	93 ± 8	93 ± 9	86 ± 6	84 ± 14	87 ± 14
Sent. 3	91 ± 9	93 ± 8	89 ± 14	96 ± 4	98 ± 5	94 ± 9	96 ± 5	95 ± 6	97 ± 8
Sent. 4	92 ± 4	94 ± 8	90 ± 10	93 ± 7	90 ± 13	97 ± 6	97 ± 6	96 ± 8	98 ± 7
Sent. 5	93 ± 2	90 ± 6	97 ± 6	90 ± 5	92 ± 8	89 ± 9	94 ± 5	98 ± 5	91 ± 11
DDK	97 ± 4	98 ± 7	97 ± 6	97 ± 5	97 ± 6	97 ± 6	98 ± 3	99 ± 4	97 ± 6
	Czech: Rev. Scenario 1			Rev. Scenario 2			Original Signals		
Sent. 1	93 ± 17	94 ± 25	93 ± 21	90 ± 21	81 ± 41	99 ± 3	93 ± 18	89 ± 32	98 ± 10
Sent. 2	86 ± 23	99 ± 3	72 ± 46	85 ± 23	83 ± 38	88 ± 31	86 ± 23	79 ± 41	93 ± 19
Sent. 3	97 ± 10	100 ± 0	94 ± 20	93 ± 17	100 ± 0	87 ± 33	86 ± 23	88 ± 34	84 ± 37
DDK	93 ± 17	96 ± 13	90 ± 29	98 ± 9	100 ± 0	95 ± 19	94 ± 16	99 ± 3	88 ± 31

In order to show the results more compactly, Figure 3 contains the values of the Area Under the receiver operating characteristic Curves (AUC) obtained with the speech tasks of the three databases in the three scenarios (two reverberant and the original recordings). The proposed method shows to be accurate and robust in reverberant environments. The results indicate that it is possible to discriminate between speakers with PD and HC with accuracies ranging from 85% to 99% considering recordings captured in non-controlled acoustic conditions. The results with several speech tasks were higher in the reverberated scenarios. Our hypothesis is that this behavior is explained by the introduction of suprasegmental information from the voiced frames into the unvoiced regions owing to the reverberation process. Further experiments modeling the voiced/unvoiced transitions could lead to validate this hypothesis.

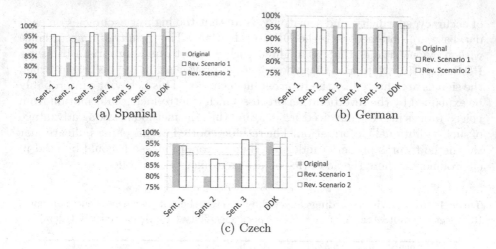

(a) Spanish   (b) German

(c) Czech

Fig. 3. AUC values obtained with each speech task of the three languages. Sent: Sentence. Rev. Scenario: Reverberant Scenario.

4 Conclusions

A method to discriminate between speakers with PD and HC is presented in this paper. The experiments consider speech recordings that are affected by two reverberant scenarios. The method consists on the automatic segmentation and characterization of the unvoiced segments. Since speech recordings of three different languages are considered, the method seems to be language-independent. Additionally, the it shows to be robust against particular plosive sounds of the considered languages, i.e., in the repetition of /pataka/, the plosive sounds /p/ and /t/ are aspirated in German but not in Czech and Spanish. This work is a step forward to develop computational tools for the assessment of speech of PD patients with non-controlled acoustic conditions.

Acknowledgments. Juan Rafael Orozco-Arroyave is under grants of COLCIENCIAS through the program "Convocatoria N⁰ 528, generación del bicentenario 2011". This work was also financed by COLCIENCIAS through the project N⁰ 111556933858. The authors thank CODI, "estrategia de sostenibilidad 2014-2015 from Universidad de Antioquia" for the support to develop this work.

References

1. Hornykiewicz, O.: Biochemical aspects of Parkinson's disease. Neurology **51**(2), S2–S9 (1998)
2. de Rijk, M.C., Launer, L.J., Berger, K., Breteler, M.M., Dartigues, J.F., Baldereschi, M., Fratiglioni, L., Lobo, A., Martinez-Lage, J., Trenkwalder, C., Hofman, A.: Prevalence of Parkinson's Disease in Europe: A collaborative study of population-based cohorts. Neurologic Diseases in the Elderly Research Group. Neurology **54**(11 Suppl 5), S21–S23 (2000)

3. Worth, P.: How to treat Parkinson's disease in 2013. Clinical Medicine **13**(1), 93–96 (2013)
4. Ramig, L.O., Fox, C., Sapir, S.: Speech treatment for Parkinson's disease. Exp. Rev. Neurother. **8**(2), 297–309 (2008)
5. Zicker, J.E., Tompkins, W.J., Rubow, R.T., Abbs, J.H.: A portable microprocessor-based biofeedback training device. IEEE Transactions on Biomededical Engineering **27**(9), 509–515 (1980)
6. Wirebrand, M.: Real-time monitoring of voice characteristics using accelerometer and microphone measurements. Master's thesis, Linköpings universitet, Linköping, Sweden (2011)
7. Vásquez-Correa, J.C., Orozco-Arroyave, J.R., Arias-Londoño, J.D., Vargas-Bonilla, J.F., Nöth, E.: New computer aided device for real time analysis of speech of people with Parkinson's disease. Rev. Fac. Ing. Universidad de Antioquia **1**(72), 87–103 (2014)
8. Boersma, P., Weenink, D.: Praat, a system for doing phonetics by computer. Glot International **5**(9/10), 341–345 (2001)
9. Orozco-Arroyave, J.R., Hönig, F., Arias-Londoño, J.D., Vargas-Bonilla, J.F., Skodda, S., Rusz, J., Nöth, E.: Automatic detection of parkinson's disease from words uttered in three different languages. In: Proceedings of the 15th INTERSPEECH, pp. 1573–1577 (2014)
10. Orozco-Arroyave, J.R., Arias-Londoño, J.D., Vargas-Bonilla, J.F., González-Rátiva, M.C., Nöth, E.: New spanish speech corpus database for the analysis of people suffering from parkinson's disease. In: Proceedings of the 9th Language Resources and Evaluation Conference (LREC), pp. 42–347 (2014)
11. Goetz, C.G., Poewe, W., Rascol, O., Sampaio, C., Stebbins, G.T., Counsell, C., Giladi, N., Holloway, R.G., Moore, C.G., Wenning, G.K., Yahr, M.D., Seidl, L.: Movement Disorder Society Task Force report on the Hoehn and Yahr staging scale: status and recommendations. Movement Disorders **19**(9), 1020–1028 (2004)
12. Skodda, S., Visser, W., Schlegel, U.: Vowel articulation in Parkinson's diease. J. of Voice **25**(4), 467–472 (2011). Erratum in J. of Voice. 2012 Mar; 25(2):267–8
13. Rusz, J., Cmejla, R., Tykalova, T., Ruzickova, H., Klempir, J., Majerova, V., Picmausova, J., Roth, J., Ruzicka, E.: Imprecise vowel articulation as a potential early marker of Parkinson's disease: effect of speaking task. Journal of the Acoustical Society of America **134**(3), 2171–2181 (2013)
14. Haderlein, T., Nöth, E., Herbordt, W., Kellermann, W., Niemann, H.: Using artificially reverberated training data in distant-talking ASR. In: Matoušek, V., Mautner, P., Pavelka, T. (eds.) TSD 2005. LNCS (LNAI), vol. 3658, pp. 226–233. Springer, Heidelberg (2005)
15. Herbordt, W.: Combination of robust adaptive beamforming with acoustic echo cancellation for acoustic human/machine interfaces. Ph.D. thesis, University Erlangen-Nuremberg, Germany, January 2004
16. Harel, B.T., Cannizzaro, M.S., Cohen, H., Reilly, N., Snyder, P.J.: Acoustic characteristics of Parkinsonian speech: A potential biomarker of early disease progression and treatment. Journal of Neurolinguistics **17**, 439–453 (2004)

Automatic Detection of Parkinson's Disease from Compressed Speech Recordings

Juan Rafael Orozco-Arroyave[1,2]([✉]), Nicanor García[2],
Jesús Francisco Vargas-Bonilla[2], and Elmar Nöth[1]

[1] Pattern Recognition Lab, Friedrich-Alexander-Universität, Erlangen, Germany
rafael.orozco@i5.informatik.uni-erlangen.de
[2] Faculty of Engineering, Universidad de Antioquia, Medellín, Colombia

Abstract. The impact of speech compression in the automatic classification of speakers with Parkinson's disease (PD) and healthy controls (HC) is tested. The set of codecs considered to compress the speech recordings includes G.722, G.226, GSM-EFR, AMR-WB, SILK, and Opus. A total of 100 speakers (50 with PD and 50 HC) are asked to read a text with 36 words. The recordings are compressed from bit-rates of 705.6 kbps down to 6.6 kbps. The method addressed to discriminate between speakers with PD and HC consists on the systematic segmentation of voiced and unvoiced speech frames. Each kind of frame is characterized independently. For voiced segments noise, perturbation, and cepstral features are considered. The unvoiced segments are characterized with Bark band energies and cepstral features. According to the results the codecs evaluated in this paper do not affect significantly the accuracy of the system, indicating that the addressed methodology could be used for the telemonitoring of PD patients through Internet or through the mobile communications network.

Keywords: Parkinson's disease · Speech compression · Speech codec · Voiced/unvoiced frames · Internet · Telemonitoring · Mobile communications network

1 Introduction

PD is the second most prevalent neurological disorder in the world, affecting about 2% of people older than 65 years [1]. It results as the dead of dopaminergic neurons in the substantia nigra of the mid brain [2]. People with PD develop several motor problems including bradykinesia, rigidity, postural instability, and resting tremor, among others. Non-motor deficits are also present in PD patients, including negative effects in sleep, cognition, and emotion [3]. Typically, the patients with PD develop dysarthric speech and the set of symptoms observed includes reduced loudness, monopitch, reduced stress, breathy and hoarse voice, and imprecise articulation, among others. Although about 90% of PD patients develop speech impairments, it is estimated that only from 3% to 4% of them receive speech therapy. The motor problems developed by PD patients make

© Springer International Publishing Switzerland 2015
P. Král and V. Matoušek (Eds.): TSD 2015, LNAI 9302, pp. 88–95, 2015.
DOI: 10.1007/978-3-319-24033-6_10

difficult to attend several clinical appointments, limiting their treatment to neurological revisions mainly focused on update the dose of medicine. The research community has been interested in solving such difficulties by developing computer aided tools to assess the speech of PD patients at home. In [4] the authors present a portable device to analyze the speech of PD patients. The device provides bio-feedback of the speech volume. A signal tone is sent to the patient if the vocal intensity is below an adjustable threshold. In [5] the authors present a computer based at-home testing device (AHTD). This device is developed to assess several symptoms of PD patients such as tremor, small and large bradykinesia, speech, reaction/movement times, among others. According to their findings, the incorporation of the AHTD in larger clinical studies is feasible and it could be used to follow the progress of the disease. In [6] the author analyzes the voice of PD patients considering measures such as the fundamental frequency of the voice, energy, and the sound pressure level. The skin vibrations are also recorded using an accelerometer. Recently, in [7] the authors present a portable device to evaluate the phonatory and articulatory capability of patients with PD. The device records sustained phonations and perform several analysis including noise content, stability, periodicity, and different articulation measures such as the triangular vowel space area (tVSA), the formant centralization ratio (FCR), and the vowel articulation index (VAI). Additionally, there are several studies that present different methodologies to assess the speech signals in order to discriminate PD speakers and healthy controls [8–11].

The use of portable devices for the assessment of PD patients at home is feasible from the technical point of view; however, it could be relatively expensive either for the patients or the health system. The Information and Communication Technologies (ICT) allow to think on doing telemonitoring of PD patients using different communication tools already existing in Internet. Although, there are several aspects in such new technologies and tools that have to be studied to analyze the feasibility of using them in real scenarios. For instance, there exist different communication systems that can be used for the remote evaluation of speech e.g., the mobile communications network, the Internet, and the landline, among others. All of these technologies compress the audio signals in order to transmit them through the communication channel. The compression rates depends on the technology and on the bandwidth available in the network.

This paper explores the impact of several codecs in the performance of the methodology presented in [10] to discriminate between speakers with PD and HC. Texts read by a total of 100 speakers were recorded, 50 with PD and 50 HC. The codecs considered in this study were used to compress the speech signals at different rates depending on the application. The set of codecs includes G.722 [12] which is used in voice over IP (VoIP) land-lines, G.726 [13] which is used to compress speech signals that are multiplexed into international trunks, GSM-EFR (Global System for Mobile Communications - Enhanced Full Rate) [14] which is used in mobile networks, AMR-WB (Adaptive Multi-Rate - Wide Band) [15] which is a relatively new standard for mobile networks, SILK [16] which is the codec used for the Skype® calls, and Opus [17] which is used in VoIP calls made trough Internet and in several applications with audio streaming.

The rest of the paper is organized as follows. Section 2 provides the details of the methodology and the experiments addressed in this study. Section 3 includes the results obtained in the experiments, and finally Section 4 provides the conclusions derived from this study.

2   Experimental Setup

The methodology addressed here comprises four steps. (1) The speech recordings are compressed with six different codecs. For the sake of comparisons, the original recordings i.e., without any compression, are also considered. (2) The signals are preprocessed and the voiced/unvoiced (v/uv) speech frames are segmented. (3) voiced and unvoiced frames are characterized separately, and (4) the discrimination between speech of PD patients and HC is performed using a support vector machine (SVM). Further details of each step are provided in the next subsections.

2.1   Speech Recordings

A total of 100 native Spanish speakers are considered, 50 with PD and 50 HC. All of the patients were diagnosed a neurologist expert. The age of the patients ranged from 33 to 77 (mean 61 ± 9) and the age of the healthy speakers ranged from 31 to 86 (mean 60 ± 9). All of the participants were asked to read a text with 36 words. The sampling frequency was 44.1 kHz with 16 bits of resolution. The people was recorded in a sound-proof booth, with a dynamic omnidirectional microphone and a professional audio card. Note that if these recordings would be transmitted over a network, the bit-rate would be $44.1 \times 16 = 705.5$ kbps. Further details of the database can be found in [18].

2.2   Encoding - Compression

The codecs used in this study compress the speech signals in order to reduce the bit-rate and thus to make a more efficient use of the network resources e.g., bandwidth. A total of six codecs are considered. G.722 and G.726 are based on the adaptive differential pulse code modulation (ADPCM) [19] method. While GSM-FR, AMR-WB, SILK, and Opus are based on the analysis-by-synthesis concept. A brief description of each codec is provided below.

ADPCM: In this method the difference between the original signal $x(n)$ and the predicted signal $\tilde{x}(n)$ is quantized. The prediction process is based on a linear prediction (LP) filter, thus the parameters of the LP filter correspond to the model of the vocal tract. The difference between the predicted signal and the original $(d(n))$ corresponds to the excitation. The parameters of the LP filter and the excitation signal are encoded. This procedure is summarized in Figure 1. A brief description of the G.722 and G.726 codecs is provided below.

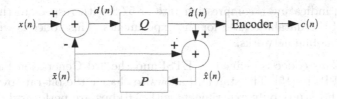

Fig. 1. General process of ADPCM. Q: quantizer, P: LP filter

G.722: This codec is defined by the International Telecommunications Union (ITU) in [12]. The spectrum of the signal is divided into two parts which are quantized independently. This codec is used in VoIP calls where high bandwidth is available e.g., land-lines and local area networks. In this case the recordings are re-sampled at 8kHz and the quantization is performed with 16 bits, thus the bit-rate is 64 kbps.

G.726: This codec is defined by the ITU [13]. It is mainly used in international trunks. It encodes the speech signal at different bit-rates. This paper only includes experiments with 16 kbps with a sampling rate of 8 kHz, which means that only 2 bits are used for the quantization of the difference $d(n)$.

Analysis-by-Synthesis: This method consists on an iterative process where the error $e(n)$ between the original signal $x(n)$ and the resulting from a synthesis model ($\hat{x}(n)$) is minimized. The parameters of the synthesis filter and the excitation signal are coded. This procedure is summarized in Figure 2. A brief description of the GSM-FR, AMR-WB, SILK, and Opus codecs is provided below.

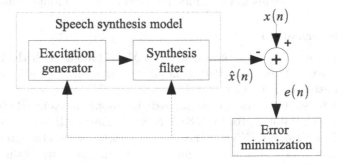

Fig. 2. General process of analysis-by-synthesis.

GSM-EFR: This codec is defined by the European Telecommunications Statandards Institute (ETSI) in [14]. It is based on the algebraic code excited liner prediction (ACELP) encoding scheme. The bit-rate of the signals is decreased to 12.2 kbps, indicating a compression rate of 57.8 with respect to the original recordings of the database used for the experiments. This codec is widely used in the GSM mobile networks.

AMR-WB: This codec is defined by ETSI and the 3rd Generation Partnership Project (3GPP) in [15]. The standard allows to change the bit-rate over frames, however in this study only experiments with 6.6 kbps are performed. The compression rate in this case is 106.9 with respect to the bit-rate of the original recordings. This codec is being used in new implementations of GSM/UMTS mobile networks to improve the voice quality.

SILK: This codec was developed by Skype® Limited. It can encode the speech signal at variable bit-rates ranging from 6 kbps to 40 kbps [16]. For the experiments addressed in this study a sampling frequency of 24 kHz with an average bit-rate of 25 kbps is used.

Opus: This codec is defined by the Internet Engineering Task Force (IETF) in its request for comments (RFC) 6716 [17]. It is based on the SILK codec and also supports variable bit-rates which in this case range from 8 kbps to 40 kbps. This codec supports variable sampling rates. In this paper only experiments with bit-rates of 64 kbps are reported. This bit-rate is chosen in the assumption of applications with high-speed Internet connection.

2.3 Pre-processing and Voiced/Unvoiced Segmentation

The recordings are normalized in amplitude and mean cepstral subtraction is applied to avoid possible bias introduced by the channel i.e., microphone and sound card. The segmentation of voiced and unvoiced frames is performed in Praat [20]. Voiced and unvoiced segments are grouped separately. Each frame is windowed using Hamming windows with 40 ms length and 20 ms time shift. Frames shorter than 40 ms were excluded as well as pauses longer than 270 ms.

2.4 Characterization

The *Voiced frames* are characterized with 12 MFCC along with their first and second derivatives (Δ and $\Delta\Delta$). Perturbation measures such as absolute and relative values of jitter and shimmer, and the variability of F_0 are also included. Additionally, four noise measures are considered: Harmonic-to-Noise Ratio (HNR), Glottal-to-Noise Excitation Ratio (GNE), Noise-to-Harmonic Ratio (NHR), and Normalized Noise Energy (NNE). The *Unvoiced features* are characterized with 12 MFCC, Δ, and $\Delta\Delta$. The energy content of the unvoiced frames is measured over 25 band scaled according to the Bark scale [21]. The mean value, standard deviation, kurtosis, and skewness are calculated from each feature vector.

2.5 Classification

A support vector machine (SVM) with soft margin is used to discriminate between PD and healthy speakers. The margin parameter C and the bandwidth of the Gaussian kernel γ are optimized through a grid-search with

$10^{-3} < C < 10^4$ and $10^{-1} < \gamma < 10^3$. The selection criterion was based on the accuracy obtained in the test set. The SVM is trained following a 10–fold cross-validation strategy. All of the folds were formed randomly but assuring the balance in age, gender, and the speaker independence.

3 Results

The results are presented in terms of the accuracy obtained in the classification process. The standard deviation measured among the 10 folds in the validation process is also indicated. Table 1 includes the results with the unvoiced frames. Note that most codecs do not affect significantly the accuracy of the classifier. Only the results on GSM-EFR and Opus are slightly reduced.

Table 1. Classification results obtained with unvoiced frames (values in %). SR: sampling rate [kHz], BR: bit-rate [kbps], BBE: Bark band scales, All: merging all features.

Codec	SR	BR	MFCC+Δ+$\Delta\Delta$	BBE	All
Original	44.1	705.6	97 ± 7	95 ± 7	97 ± 5
G.722	16	64	97 ± 5	95 ± 7	99 ± 3
G.726	8	16	97 ± 10	94 ± 7	95 ± 10
GSM-EFR	8	12.2	94 ± 7	93 ± 10	96 ± 5
AMR-WB	16	6.6	96 ± 8	98 ± 6	95 ± 9
SILK	24	25	98 ± 4	95 ± 5	96 ± 7
Opus	variable	64	93 ± 10	95 ± 10	94 ± 11

The results in Table 2 show that most of the codecs do not affect significantly the accuracy of the classifier using measures extracted from the voiced segments. However, when the G.726 or SILK codecs are used, the accuracies of the perturbation measures increase with respect to those obtained with the original recordings. It seems like the modifications of the speech spectrum performed by these two codecs are affecting the frequencies above 500 Hz, but not modifying the frequencies around the fundamental frequency, which is the basis to estimate the perturbation features.

Table 2. Classification results obtained with Voiced frames (values in %). SR: sampling rate [kHz], BR: bit-rate [kbps], All: merging all features.

Codec	SR	BR	MFCC+Δ+$\Delta\Delta$	Noise	Perturbation	All
Original	44.1	705.6	86 ± 8	77 ± 12	76 ± 8	84 ± 11
G.722	16	64	86 ± 11	79 ± 11	77 ± 8	87 ± 8
G.726	8	16	83 ± 12	74 ± 8	80 ± 15	81 ± 11
GSM-EFR	8	12.2	86 ± 8	80 ± 9	82 ± 9	88 ± 6
AMR-WB	16	6.6	81 ± 7	76 ± 10	75 ± 13	79 ± 13
SILK	24	25	84 ± 8	70 ± 13	83 ± 13	82 ± 11
Opus	variable	64	88 ± 6	76 ± 7	77 ± 7	86 ± 7

The results are summarized in Figure 3. Note that the highest accuracies are obtained with the unvoiced features in all of the cases. It seems that there is almost no negative impact of the codification methods in the performance of the system.

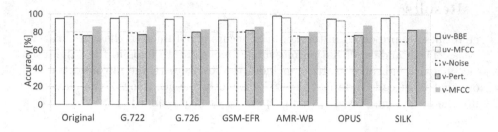

Fig. 3. Accuracy for each codec with each set of characteristics.

4   Conclusion

Speech recordings of 50 patients with PD and 50 HC are compressed considering six speech codecs widely used in different commercial applications through Internet or through the mobile network. The impact of such codecs in the automatic discrimination of speakers with PD and HC is evaluated in this paper. According to the results, the impact of the audio-compression in the accuracy of the system is minimal. Although the results indicate that the methodology addressed here could be used for telemonitoring PD patients through Internet or the mobile communications network, it is worthy to note that we did not consider the effects introduced by the communications channel, i.e., scenarios with loss of packets during the communication are not considered. Further experiments with recordings captured through Internet or through the mobile network are required to obtain more conclusive results.

References

1. de Rijk, M.C.: Prevalence of Parkinson's disease in Europe: A collaborative study of population-based cohorts. Neurology **54**, 21–23 (2000)
2. Hornykiewicz, O.: Biochemical aspects of Parkinson's disease. Neurology **51**(2), S2–S9 (1998)
3. Logemann, J.A., Fisher, H.B., Boshes, B., Blonsky, E.R.: Frequency and cooccurrence of vocal tract dysfunctions in the speech of a large sample of Parkinson patients. Journal of Speech and Hearing Disorders **43**, 47–57 (1978)
4. Rubow, R., Swift, E.: A microcomputer-based wearable biofeedback device to improve transfer of treatment in Parkinsonian dysarthria. Journal of Speech and Hearing Disorders **50**(2), 178–185 (1985)

5. Goetz, C.G., et al.: Testing objective measures of motor impairment in early Parkinson's disease: feasibility study of an at-home testing device. Movement Disorders **24**(4), 551–556 (2009)
6. Wirebrand, M.: Real-time monitoring of voice characteristics using accelerometer and microphone measurements. Master's thesis, Linkping University, Linkping, Sweden (2011)
7. Vásquez-Correa, J.C., Orozco-Arroyave, J.R., Arias-Londoño, J.D., Vargas-Bonilla, J.F., Nöth, E.: New computer aided device for real time analysis of speech of people with Parkinson's disease. Fac. Ing. Univ. Antioquia **1**(72), 87–103 (2014)
8. Rusz, J., Cmejla, R., Ruzickova, H., Klempir, J., Majerova, V., Picmausova, J., Roth, J., Ruzicka, E.: Acoustic assessment of voice and speech disorders in Parkinson's disease through quick vocal test. Movement Disorders **26**(10), 1951–1952 (2011)
9. Bocklet, T., Steidl, S., Nöth, E., Skodda, S.: Automatic evaluation of Parkinson's speech - acoustic, prosodic and voice related cues. In: Proceedings of the 14th INTERSPEECH, pp. 1149–1153 (2013)
10. Orozco-Arroyave, J.R., Hönig, F., Arias-Londoño, J.D., Vargas-Bonilla, J.F., Skodda, S., Rusz, J., Nöth, E.: Automatic detection of Parkinson's disease from words uttered in three different languages. In: Proceedings of the 15th INTERSPEECH, pp. 1473–1577 (2014)
11. Orozco-Arroyave, J.R., Hönig, F., Arias-Londoño, J.D., Vargas-Bonilla, J.F., Nöth, E.: Spectral and cepstral analyses for Parkinson's disease detection in Spanish vowels and words. Expert Systems, 1–10 (2015) (to appear)
12. International Telecommunication Union (ITU): 7 kHz audio-coding within 64 kbit/s. Recommendation ITU-T G.722, Std. (2012)
13. International Telecommunication Union (ITU): 40, 32, 24, 16 kbit/s Adaptive Differential Pulse Code Modulation (ADPCM). Recommendation ITU-T G.726, Std. (1990)
14. Digital cellular telecommunications (Phase 2+); Enhanced Full Rate (EFR) speech transcoding; (GSM 06.60 version 8.0.1 Release 1999), European Telecommunications Standards Institute (ETSI) Std., November 2000
15. Adaptive Multi-Rate Wideband (AMR-WB) speech Codec; Transcoding functions (3GPP TS 26.190 version 12.0.0 Release 12), 3rd Generation Partnership Project (3GPP) Std., October 2014
16. Internet Engineering Task Force (IETF): SILK Speech Codec, Std. (2010)
17. Internet Engineering Task Force (IETF): Definition of the Opus Audio Codec. RFC 6716, Std. (2012)
18. Orozco-Arroyave, J., Arias-Londoño, J., Vargas-Bonilla, J., González-Rátiva, M., Nöth, E.: New Spanish speech corpus database for the analysis of people suffering from Parkinson's disease. In: Proceedings of the 9th Language Resources and Evaluation Conference (LREC), pp. 342–347 (2014)
19. Rabiner, L.R., Schafer, R.W.: Introduction to digital speech processing, 4th edn. Now Publishers Inc., Hanover (2007). vol. 1, no. 1–2
20. Boersma, P., Weenink, D.: PRAAT, a system for doing phonetics by computer. Glot International **5**(9/10), 341–345 (2001)
21. Zwicker, E., Terhardt, E.: Analytical expressions for critical-band rate and critical bandwidth as a function of frequency. Journal of Acoustical Society of America **68**(5), 1523–1525 (1980)

Time Dependent ARMA for Automatic Recognition of Fear-Type Emotions in Speech

J.C. Vásquez-Correa[1]([envelope]), J.R. Orozco-Arroyave[1,2], J.D. Arias-Londoño[1],
J.F. Vargas-Bonilla[1], L.D. Avendaño[3], and Elmar Nöth[2]

[1] Faculty of Engineering, Universidad de Antioquia UdeA, Medellín, Colombia
jcamilo.vasquez@udea.edu.co
[2] Pattern Recognition Lab, Friedrich-Alexander-Universität,
Erlangen-Nürnberg, Germany
[3] Laboratory for Stochastic Mechanical Systems and Automation (SMSA),
Department of Mechanical and Aeronautical Engineering,
University of Patras, Patras, Greece

Abstract. The speech signals are non-stationary processes with changes
in time and frequency. The structure of a speech signal is also affected
by the presence of several paralinguistics phenomena such as emotions,
pathologies, cognitive impairments, among others. Non-stationarity can
be modeled using several parametric techniques. A novel approach based
on time dependent auto-regressive moving average (TARMA) is proposed
here to model the non-stationarity of speech signals. The model is tested
in the recognition of "fear-type" emotions in speech. The proposed approach is applied to model syllables and unvoiced segments extracted from
recordings of the Berlin and enterface05 databases. The results indicate
that TARMA models can be used for the automatic recognition of emotions in speech.

Keywords: Non-stationary signals · Speech emotion recognition ·
Continuous speech · Time dependent ARMA models

1 Introduction

The affective state of humans can be detected through speech [1]. The applications of this technology are mostly in learning, entertainment, and dialogues in
call centers [2]. In the last few years the interest of the research community has
been focused on the detection of "fear-type" emotions such as anger, disgust,
and fear, which appear in abnormal situations when the human integrity is at
risk [3].

One of the main challenges in speech emotion recognition is to find suitable
features to represent the affective state of the speaker. The characterization of
emotions in speech has been focused on prosodic, spectral and cepstral features,
and voice quality measures [2]. In [4] the authors apply six different large scale
acoustic feature sets including several prosodic and spectral features to characterize recordings of the Berlin [5], and enterface05 [6] emotional databases.

© Springer International Publishing Switzerland 2015
P. Král and V. Matoušek (Eds.): TSD 2015, LNAI 9302, pp. 96–104, 2015.
DOI: 10.1007/978-3-319-24033-6_11

The authors use Support Vector Machines (SVM) with a linear kernel function for classification, and use Leave One Speaker Out (LOSO) cross validation to test the system. The best results reported are around 85% in Berlin and 76% in enterface05 databases. In [7] the authors use Mel-frequency cepstral coefficients (MFCCS), and their first and second derivatives to characterize emotions in speech. They use Berlin [5], and enterface05 [6] databases, and classify the emotions using a Deep Neural Network with a Hidden Markov Model (DNN-HMM). The reported accuracies are 77.92%, and 53.89% for Berlin, and enterface05 databases. In [8] the authors characterize emotional speech using features related to non-linear dynamics (NLD) to model the non-linear effects produced in emotional speech. They use recordings of the Berlin database [5]. The classification is performed with an artificial neural network, and report a global accuracy of 75.40%.

Besides the modeling approaches reported in the literature, the components of the phonetic inventory of human languages are characterized by different non-stationary processes such as word accents, diphthongs, and syllables, which have particular characteristics in time, and frequency [9]. The speech production process involves also several physiological aspects such as turbulent noise, caused by an air escape through the glottis, and the laryngeal tensions involved in breathy, and whisper phonation, which may carry important paralinguistic information related to the emotion of the speaker [10]. These processes produce a non-stationary behavior in speech signal, which cannot be characterized properly using the conventional acoustic features derived from cepstral, spectral, and perturbation measures due to the assumption of local stationarity [11]. In order to model these phenomena, non-stationary modeling should be applied.

The non-stationary analysis allows both to evaluate the time-dependence, and to represent the spectral evolution of the signal [12]. Non-stationary models can be classified as parametric and non-parametric [12]. Non-parametric methods are based on the representation of energy as a simultaneous function of time and frequency. These methods include the short time Fourier Transform (STFT), the Wigner-Ville distribution, and the wavelet based methods, among others. Parametric methods are based on parameterized representations of the time dependent auto-regressive moving average (TARMA) models which are able to represent abrupt changes in the spectral evolution of the signals [12]. Such methods can be classified into three approaches according to the "structure" of their parameters, (1) the unstructured parameter evolution methods which are characterized by low parsimony and slow tracking on the dynamics, (2) the stochastic parameter evolution methods, which are characterized by slow and medium tracking of dynamics, and (3) the deterministic parameter evolution which are characterized by high parsimony and fast or slow tracking depending on the estimated parameters [12]. Figure 1 summarizes the classification of parametric and non-parametric methods for non-stationary signal modeling.

TARMA models have been applied on the modeling and simulation of earthquake ground motion [13], modeling and detection of damage in mechanical structures with time-dependent dynamics [12] [14], and modeling of speech and

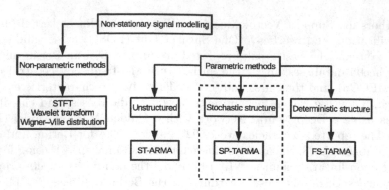

Fig. 1. Classification of methods for non-stationary signal modelling. ST-TARMA: short time TARMA; SP-TARMA: smoothness prior TARMA; FS-TARMA: functional series TARMA.

other bio-signals [11] [12] [15]. These previous attempts have demonstrated the usefulness of TARMA models as representations of non-stationary processes, and makes them very appealing for the automatic classification of emotions in speech. However, TARMA models have not been used for this particular purpose.

In previous works, the recognition of "fear-type" emotions in speech was performed using features extracted from voiced and unvoiced segments using the wavelet packet transform [16]. The present paper is focused on the use of SP-TARMA to model the transitions between vocal sounds and consonants in syllables, and to analyze the non-stationary evolution of unvoiced segments in speech. Voiced segments are not included in this study since they can be considered quasi-stationary signals. The estimated features include low level descriptors (LLD) of the coefficients obtained from the non-stationary model, and conventional features of speech such as MFCCs estimated from the model predictions. The experiments are performed using recordings of the Berlin [5], and enterface05 [6] databases. The classification is performed using a Gaussian Mixture Model (GMM) adapted from a Universal Background Model (GMM-UBM), and a SVM as a second stage of classification. The rest of paper is as follows: Section 2 contains the description about the non-stationary modeling using SP-TARMA, the feature estimation, and classification. Section 3 describes the experimental framework including the databases and the results. Finally section 4 contains the conclusions derived from this study.

2   Materials and Methods

Figure 2 contains the general scheme of the proposed methodology. It consists of four stages: (1) Unvoiced segments and syllables are segmented from the speech recordings. (2) The segmented speech frames are modeled using SP-TARMA. (3) Two different feature sets are estimated to characterize the emotional content of speech, the first one includes LLD estimated from the coefficients of the SP-TARMA model, and the second one contains conventional features as MFCCs calculated from the model prediction. (4) A GMM is used to model the emotional content of the speech signals, and the final decision is taken using a SVM trained with the posterior probabilities of the GMM.

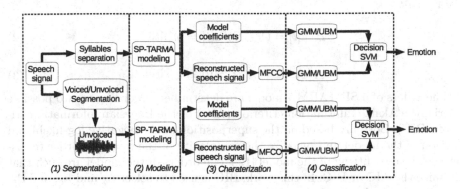

Fig. 2. General scheme of the methodology

2.1 Segmentation

Two different segmentation processes are considered. (1) The voiced and unvoiced frames are segmented using Praat [17], and (2) the syllables contained in the recordings are segmented. For the Berlin database the labels provided in the database are used, while the syllables of the recordings in the enterface05 database are segmented using Praat.

2.2 SP-TARMA Modeling

A TARMA(n_a, n_c) model is defined by Equation 1, which includes the autoregressive (AR) and the moving average (MA) components. n_a and n_c are the orders of the AR and MA models. t is the discrete time, $x[t]$ is the non-stationary signal to be modeled, $e[t]$ is an unobservable "innovations" sequence with zero mean, and time-dependence variance $\sigma_e^2[t]$, and $a_i[t]$, $c_i[t]$ are the parameters of AR, and MA models [12].

$$x[t] + \underbrace{\sum_{i=1}^{n_a} a_i[t] \cdot x[t-i]}_{AR\ part} = e[t] + \underbrace{\sum_{i=1}^{n_c} c_i[t] \cdot e[t-i]}_{MA\ part} \tag{1}$$

Stochastic parameter evolution TARMA imposes an stochastic structure in the time-dependence of the parameters. In this case the evolution of the parameters a_i, c_i is subjected to stochastic smoothness constraints. The constraints are referred to smoothness priors TARMA (SP-TARMA). In this case the model is referred to SP-TARMA(n_a, n_c, k). Where k denotes the order of the difference equations that describe the evolution of the parameters as is shown in Equations 2 and 3. Where $w_{ai}[t]$ and $w_{ci}[t]$ are Gaussian sequences with possibly time-dependent variances. B is the back-shift operator, which operates at $B^k a[t] = a[t - k]$.

$$\Delta^k a[t] = (1 - B)^k a[t] = w_{ai}[t] \tag{2}$$

$$\Delta^k a[t] = (1 - B)^k a[t] = w_{ci}[t] \tag{3}$$

The orders of a SP-TARMA model are determined according to two possible criteria, the Akaike information criterion (AIC) or the Bayesian information criterion (BIC). Both are based on the superposition of the negative log-likelihood function of the model and penalize the complexity of the model in order to discourage the over-fitting of the model [12]. The orders of the model are such that minimize the AIC or BIC. In this work, BIC is used to select n_a and n_c, and the minimum residual sum squares (RSS) is used to select the parameter k.

2.3   Feature Estimation

Two sets of features are estimated on the speech frames. One set comprises seven LLD calculated from the AR and MA coefficients of the SP-TARMA model, i.e. $a_i[t]$ and $c_i[t]$, respectively. The set of LLD includes mean value, standard deviation, skewness, kurtosis, maximum, minimum, and the log-energy. Along with the LLD the set of features contains also the order of the model, i.e. n_a, n_c, and k. The second set of features includes 12 MFCCs which are calculated from the signals that are reconstructed using the SP-TARMA model. MFCC estimated from the model predictions can have reduced noise content, compared to the estimates obtained from the raw signal. The MFCCs are calculated upon windows with 40ms length and 20ms time shift.

2.4   Classification

The classification is performed using a GMM-UBM approach. The background model is adapted using maximum a posteriori (MAP) adaptation process [18]. A GMM can be defined as a probabilistic model represented by a linear combination of several multivariate Gaussian components. The UBM is trained using the Expectation Maximization (EM) algorithm [18] using recordings from all classes, i.e. emotions. Then the specific GMM for each class is adapted using the MAP method. Finally, given a sample $X = \mathbf{x}_1, \mathbf{x}_2, \cdots, \mathbf{x}_T$, where $\mathbf{x}_i$ is the feature vector extracted from the segment i, the decision about to which class belongs each speech recording is taken by evaluating the maximum Log-Likelihood according to Equation 4, where $p(\mathbf{x}_i|\Theta)$ is the posterior probability of sample $\mathbf{x}_i$ for the model Θ.

$$LL(X, \Theta) = \sum_{i=1}^{T} log(p(\mathbf{x}_i|\Theta)) \tag{4}$$

The posterior probabilities obtained from the GMM based on the features obtained from the coefficients of the model is combined with posterior probabilities obtained from features calculated on the model prediction. The probabilities are used as new features to train a SVM with a Gaussian kernel. The SVM is tested following a one vs. all strategy. The validation process is performed using leave-one-group-speaker-out cross-validation (LOGSO-CV). For the GMM-UBM. The number of Gaussians is optimized in a grid-search from 2 to 8. The parameters C and γ of the SVM are optimized also through a grid-search up to powers of ten with $10^{-1} < C < 10^4$ and $10^{-2} < \gamma < 10^2$. The selection criterion was based on the accuracy obtained on the test set.

3   Experimental Framework and Results

3.1   Datasets

Berlin emotional database [5]: it contains 534 voice recordings of 10 speakers who acted 7 different emotions. The recordings were sampled at 16KHz. In this paper three of the seven emotions of the database are considered: anger, disgust, and fear.

Enterface05 database [6]: this database contains 1317 audio-visual recordings of 42 speakers. In this paper three of the six emotions included in the database are considered: anger, disgust, and fear. Each subject was instructed to listen six short stories. After each story the subject had to react to the situation by speaking predefined phrases that fit into the short story.

3.2   Experimental Setup

Two different experiments were carried out: (1) several "fear-type" emotions including anger, disgust, and fear are recognized from speech recordings of the two databases. (2) Recordings of the Berlin database with neutral emotion are discriminated from recordings with anger, disgust, or fear. Additionally, positive and negative emotions are discriminated from recordings of the enterface05 database.

3.3   Results and Discussion

Table 1 shows the accuracy of the classification of the "fear-type" emotions using different sets of features calculated from the SP-TARMA models. Note that the best results are obtained using syllables instead of unvoiced segments. There is an improvement of 17.9% with the fusion of probabilities in the Berlin database. In enterface05 database the improvement is 5%. Note also that the second classification stage provides an improvement in the general accuracy rate relative to the first classification step, specially in the Berlin database.

Table 1. Global accuracies (%) for features estimated from the SP-TARMA models on unvoiced (UV) segments and on syllables.

	Berlin		enterface05	
	UV	Syllables	UV	Syllables
LLD from TARMA	62 ± 7	70 ± 9	40±4	41 ± 6
MFCC on reconstructed segments	59±11	78 ± 8	55 ± 5	59 ± 5
Fusion of prob.	65 ± 9	82 ± 14	55 ± 5	60 ± 8

Table 2. Confusion Matrixes. GWA: Global Weighted Accuracy.

	Berlin			enterface05		
	Fear	Disgust	Anger	Fear	Disgust	Anger
Fear	80.3	10.1	8.9	58.3	16.2	16.8
Disgust	15.2	76.7	8.9	21.4	59.8	21.6
Anger	4.5	13.3	82.2	20.3	24.0	61.6
GWA		82.0			59.7	

Table 2 contains the confusion matrixes with the best results obtained on the Berlin and enterface05 databases. In both cases, the best result is obtained with the feature set calculated from syllables with the fusion of probabilities.

Table 3 contains the accuracy of the classification between neutral and the three "fear-type" emotions of the Berlin database. The results classifying positive and negative emotions of the enterface05 database are also provided. The results are presented in terms of accuracy (Acc), sensitivity (Sens), and specificity (Spec). The highest accuracies are obtained also with the SVM. With the Berlin database the highest accuracy obtained using the unvoiced segments is 85.6%, and using syllables is 86.3%. In the enterface05 database the highest accuracy obtained using the unvoiced segments is 68.2%, and using syllables is 67.9%.

Table 3. Results classifying neutral vs. "fear-type" in the Berlin database and positive vs. negative in the enterface05 database. Syll: syllables. UV: unvoiced.

		Berlin			enterface05		
	Frames	Acc	Sens	Spec	Acc	Sens	Spec
LLD from TARMA	UV	80.1	82.9	75.3	55.3	55.5	54.8
MFCC on reconstructed segments	UV	77.8	80.2	70.6	54.0	62.1	37.5
Fusion of probabilities	UV	85.6	86.4	78.2	68.2	68.2	67.9
LLD from TARMA	Syll	84.2	85.0	78.2	56.0	60.9	46.2
MFCC on reconstructed segments	Syll	76.3	79.2	67.4	53.9	55.0	31.0
Fusion of probabilities	Syll	86.3	86.9	81.5	67.9	68.24	63.6

4   Conclusion

A new method for the characterization of speech signals is presented in this paper. The method allows to model the non-stationarity structure of the speech signals. Unvoiced frames extracted from continuous speech along with the vocal sounds and consonants in syllables are modeled through the proposed approach. The method is tested in three classification experiments, one multi-class and two bi-class. The multi-class consisted on the recognition of three different "fear-type" emotions (fear, disgust, and anger) in speech recordings of two databases, and the bi-class tasks consisted on the discrimination between neutral and the "fear-type" emotions and the classification of positive and negative emotions in speech signals.

The results indicate that the method is more appropriate to model syllables instead of unvoiced frames. For future work, the combination of features related to parametric non-stationary analysis, and the classical acoustic features must be addressed in order to improve the accuracies in emotion recognition and other speech processing tasks. The extraction of more features from the TARMA methods seems to be a promising approach to model the non-stationary structure of the speech.

Acknowledgments. Juan Rafael Orozco Arroyave is under grants of Convocatoria 528 para estudios de doctorado en Colombia 2011 financed by COLCIENCIAS. The authors thank to CODI, estrategia de sostenibilidad 2014-2015 from Universidad de Antioquia for the support for the development of this work. This work is partially funded also by COLCIENCIAS through the project number 111556933858.

References

1. Schuller, B., Batliner, A.: Computational Paralinguistics: Emotion, Affect and Personality in Speech and Language Processing. Wiley (2014)
2. Schuller, B., Batliner, A., Steidl, S., Seppi, D.: Recognising Realistic Emotions and Affect in Speech: State of the Art and Lessons Learnt from the First Challenge. Speech Communication 53(9–10), 1062–1087 (2011)
3. Clavel, C., Vasilescu, I., Devillers, L., Richard, G., Ehrette, T.: Fear-type emotion recognition for future audio-based surveillance systems. Speech Communication 50(6), 487–503 (2008)
4. Eyben, F., Batliner, A., Schuller, B.: Towards a standard set of acoustic features for the processing of emotion in speech. Proceedings of Meetings on Acoustics 9(1), 1–12 (2012)
5. Burkhardt, F., Paeschke, A., Rolfes, M., Sendlmeier, W., Weiss, B.: A database of german emotional speech. In: Proc of the INTERSPEECH 2005, pp. 1517–1520 (2005)
6. Martin, O., Kotsia, I., Macq, B., Pitas, I.: The enterface 2005 audio-visual emotion database. In: Proceedings of the 22nd International Conference on Data Engineering Workshops. ICDEW 2006, pp. 8–15 (2006)

7. Li, L., Zhao, Y., Jiang, D., Zhang, Y., Wang, F., Gonzalez, I., Valentin, E., Sahli, H.: Hybrid deep neural network-hidden markov model (DNN-HMM) based speech emotion recognition. In: Proceedings of the 2013 Humaine Association Conference on Affective Computing and Intelligent Interaction (2013)312–317
8. Henríquez, P., Alonso, J.B., Ferrer, M.A., Travieso, C.M., Orozco-Arroyave, J.R.: Nonlinear dynamics characterization of emotional speech. Neurocomputing **132**, 126–135 (2014)
9. Tüske, Z., Drepper, F.R., Schlüter, R.: Non-stationary signal processing and its application in speech recognition. In: Workshop on Statistical and Perceptual Audition, Portland, OR, USA, September 2012
10. Ishi, C.T., Ishiguro, H., Hagita, N.: Analysis of the roles and the dynamics of breathy and whispery voice qualities in dialogue speech. EURASIP J. Audio, Speech and Music Processing **2010** (2010)
11. Funaki, K.: A time-varying complex AR speech analysis based on GLS and ELS method. In: Eurospeech, pp. 1–4 (2001)
12. Poulimenos, A., Fassois, S.: Parametric time-domain methods for non-stationary random vibration modelling and analysis a critical survey and comparison. Mechanical Systems and Signal Processing **20**(4), 763–816 (2006)
13. Fouskitakis, G.N., Fassois, S.D.: Functional series TARMA modelling and simulation of earthquake ground motion. Earthquake Engineering & Structural Dynamics **31**(2), 399–420 (2002)
14. Avendaño Valencia, L.D., Fassois, S.D.: Generalized stochastic Constraint TARMA models for in-operation identification of wind turbine non-stationary dynamics. Key Engineering Materials **569**, 587–594 (2013)
15. Rudoy, D., Quatieri, T.F., Wolfe, P.J.: Time-varying autoregressive tests for multiscale speech analysis. In: INTERSPEECH, pp. 2839–2842 (2009)
16. Vásquez-Correa, J.C., Garcia, N., Vargas-Bonilla, J.F., Orozco-Arroyave, J.R., Arias-Londoño, J.D., Quintero, O.L.: Evaluation of wavelet measures on automatic detection of emotion in noisy and telephony speech signals. In: 2014 International Carnahan Conference on Security Technology (ICCST), pp. 1–6, October 2014
17. Boersma, P., Weenink, D.: Praat, a system for doing phonetics by computer. Glot International **5**(9/10), 341–345 (2001)
18. Reynolds, D.A., Quatieri, T.F., Dunn, R.B.: Speaker verification using adapted Gaussian mixture models. Digital Signal Processing **10**(1–3), 19–41 (2000)

Using Lexical Stress in Authorship Attribution of Historical Texts

Lubomir Ivanov(✉) and Smiljana Petrovic

Computer Science Department, Iona College, 715 North Avenue,
New Rochelle, NY 10801, USA
{livanov,spetrovic}@iona.edu

Abstract. This paper presents some early results from a comprehensive project, whose goal is to investigate the use of intonation and lexical stress in authorship attribution. We demonstrate how lexical stress patterns extracted from written text can be used to train a variety of machine learning algorithms to perform attribution of texts of unknown or disputed authorship. Specifically, we apply our methodology to a collection of 18th century American and British political writings, and demonstrate how combining lexical stress with other lexical features can significantly improve the attribution results.

Keywords: Lexical stress · Authorship attribution · Machine learning

1 Introduction

Authorship attribution is the problem of determining, with a high degree of confidence, the identity of the creator(s) of a particular text of unknown or disputed authorship. Attribution is a challenging, multi-faceted problem with a long history dating as far back as 15th century [1]. Modern computers and the development of machine learning, data mining, and natural language processing have led to a broad interest in authorship attribution in areas as diverse as literature, digital rights, and forensic linguistics [2].

Historically, establishing the authorship of texts has been done by human experts with extensive knowledge of the life, literary work, and the socio-economic, political, and philosophical views of potential authors. Relying on knowledge, experience, and, sometimes, intuition, these experts examine statements in the texts, and cross-reference historical, political, or ideological facts to advance or oppose an authorship hypothesis.

With the rapid development of machine learning methods, the availability of digitalized texts, and the computational power to perform extensive content analysis, automated, computer-based authorship attribution has become a powerful alternative / complement to the traditional human expert attribution. The power of automated authorship attribution comes from its ability to analyze

© Springer International Publishing Switzerland 2015
P. Král and V. Matoušek (Eds.): TSD 2015, LNAI 9302, pp. 105–113, 2015.
DOI: 10.1007/978-3-319-24033-6_12

writers' styles, uncovering features used by authors consistently and without a conscious thought. Automated authorship attribution is objective and not influenced by the subjective beliefs of the human expert. Automated authorship attribution techniques have been successfully applied to many texts of uncertain or disputed authorship, including works by William Shakespeare [3,4], Jane Austen [5], and Greek prose [6]. Perhaps the most well-known authorship attribution work is that of Mosteller and Wallace on the Federalist papers [7]. Other successful authorship attribution studies of American political writings of the late 18th century include attribution/de-attribution of works of Thomas Paine, Anthony Benezet, and other political writers of that time period [8,9].

Despite the power of automated authorship attribution, the most reliable and convincing results continue to be those obtained through a combination of computational stylistic analysis and human expert interpretation of the documents' contents. This collaborative approach is at the core of our team's strategy: We have developed a powerful computational methodology for performing automated authorship attribution, the results of which are cross-validated by humanities researchers with expertise in 18th century history, literature, and political science.

The focus of this paper is a new aspect of our methodology - the use of lexical stress in automated authorship attribution. Lexical stress has been extensively studied by linguists and psychologists [10–12]. In [13], an attribution methodology is considered based partly on lexical stress, but the results are inconclusive for authorship attribution.

In the remainder of this paper we discuss our approach to extracting lexical stress patterns from written text and using them for training machine learning algorithms to recognize author styles. We also explore combining the lexical stress approach with more traditional attribution methods such as function words and part-of-speech attribution. The results of our research are presented and analyzed, pointing out the strengths and weaknesses of lexical stress attribution. Finally, we present directions for future work, including combining lexical stress and intonation analysis.

2   Lexical Stress

2.1   Motivation

When analyzing the writings of Thomas Paine, a colleague commented that "there is a distinct rhythm - almost a melody - to his writing that nobody else has". This informal, intuitive observation led to our effort to quantify the notion of "melody in text" and to use it for authorship attribution. Lexical stress is a prosodic feature, which describes emphasis placed on syllables in words. Unlike stress in other languages, lexical stress in English can be placed on different parts of a word. This variable stress can be used to phonemically distinguish between words and compound nouns. In speech, stress involves a louder/longer pronunciation of the stressed syllables, usually accompanied by a change in voice pitch. By using these vocal aspects of stress, a speaker can express emotions or attitudes towards the topic of speech. Lexical stress is, of course, a much weaker indicator of emotion compared to sentence intonation. However, by selecting appropriately stressed words, a skillful writer can evoke much emotion in his or her audience. Therefore, investigating lexical stress as an indicator of authorship is a reasonable first step in a larger study of the use of prosodic features for automated authorship attribution.

2.2 Extracting Lexical Stress from Text

To evaluate the effectiveness of lexical stress for authorship attribution, we need to extract the stress patterns from the selected written texts of known authorship, and use them for training machine learning algorithms. Once trained, these machine learning algorithms can be used to assign an unattributed text to one of the authors used during training, based on lexical stress patterns extracted from an unattributed text.

To extract lexical stress patterns from English words, we used the widely popular Carnegie Mellon University (CMU) pronunciation dictionary. At present, the dictionary contains 133778 distinct words, each of which is transcribed for pronunciation into syllables with lexical stress indicated by three numeric values: 0 - no stress, 1 - primary stress, 2 - secondary stress. From the CMU Dictionary we constructed a map, which associates with each word its lexical stress pattern. For example, the CMU dictionary word-spelling pair "AARDVARK AA1 R D V AA2 R K" is converted to a map entry "AARDVARK 12", whereas the word-spelling pair "MULTIMEDIA M AH2 L T IY0 M IY1 D IY0 AH0" is converted to "MULTIMEDIA 20100". Numerical word patterns such as dates and years are ignored. The word-stress-pattern map is stored in a file, which is easily and quickly loaded in memory when processing text.

The text collection we use in our experiments includes 126[1] attributed and 50 unattributed documents. The attributed documents have been authored by the writers in Table 1. In order to train the machine learning algorithms to perform attribution, we used the attributed documents to prepare a set of training and testing files, each consisting of a lexicographically ordered set of pairs of stress patterns and their relative document frequency, i.e. the number of times the pattern occurs in the file divided by the total number of words in the file (see Table 2). The pre-processing program replaces each word in the original text file with its corresponding lexical stress pattern. If a word is not found in the CMU pronunciation dictionary, it is replaced by a "-1". Numerical patterns are eliminated. However, all punctuation is preserved and interpreted as lexical patterns. The actual pattern-frequency pairs are created either by our pre-processing program or, in our later experiments, by the JGAAP [15] software, which we use in the machine learning phase of the experiments.

---

[1] Multiple Franklin and Jefferson texts are grouped into 9 files.

Table 1. Authors of Attributed Documents

Author	Num. of Documents
John Adams	10
Anthony Benezet	5
James Boswell	3
James Burgh	7
Edmund Burke	6
Nicolas de Condorcet	2
John Dickinson	4
Phillip Francis	4
Benjamin Franklin	9^1
Francis Hopkinson	9
Thomas Jefferson	8^1
Thomas Macaulay	6
William Moore	5
William Ogilvie	4
Thomas Paine	8
Richard Price	7
Joseph Priestley	3
Benjamin Rush	6
John Wilkes	5
John Witherspoon	9
Mary Wollstonecraft	6

Table 2. Sample File with Patterns/Frequencies

Pattern	Frequency
"0"	0.1876484561
"01"	0.0485239226
"01020"	0.0003393281
"012"	0.0003393281
"02010"	0.0013573125
"1"	0.4340006787
"10"	0.1598235494
"10020"	0.0003393281
"102"	0.0061079063
"1020"	0.0091618595
"12"	0.0033932813
"1200"	0.0006786563
"20010"	0.0006786563
"201"	0.0006786563
"2010"	0.0057685782
"201020"	0.0003393281
"21"	0.0013573125
"2100"	0.0027146250
"21200"	0.0003393281
"220100"	0.0006786563
....	....

3   Machine Learning Algorithms

3.1   Learning Methods

Selecting the proper set of learning algorithms is a critical step in the automated authorship attribution of text. We use two learning methods: support vector machines with sequential minimal optimization (SMO) and multi-layer perceptron neural networks with error backpropagation (MLP).

Support Vector Machines (SVM) seek a hyperplane in the n-dimensional input space, which best separates points corresponding to different candidate authors. The best separator is the hyperplane that maximizes the distance to the closest training data points of different authors. To attribute a disputed document, we evaluate on which side of the hyperplane the point corresponding to that document lies. We used the Platt's sequential minimal optimization method (SMO) implemented by Weka [14], as part of the JGAAP [15] software. SVMs work particularly well on high dimensional vector spaces.

Artificial Neural Networks (ANN) are computational structures, which imitate the way organic neural networks operate. A multi-layer perceptron with error backpropagation (MLP) is an ANN, which consist of computational elements called perceptrons, organized into layers: an input layer, which preprocesses the input, an output layer, which produces the output vector, and one or more "hidden" layers used to carry out non-linear separation. The output of each perceptron is thresholded using some activation function - a signmoid function in our experiments. Our inputs are elements of a vector of lexical stress pattern frequencies. At each training step, the output vector is compared to a predetermined desired output vector, which has a non-zero value only for the vector element corresponding to the true author of a training document. The difference between the actual and the desired output - the error vector - is propagated back through the network, and used to adjust the interconnection weights. The process of feeding a training input forward and the error backwards is repeated until the overall computational error is reduced below an acceptable threshold or a specified number of iterations is completed. At that point training is terminated and the network is ready to be used for attribution. A sample MLP is shown in Figure 1.

3.2   Features Used

Our initial experiments considered the relative frequencies of patterns in documents. In further experiments we also studied N-grams of lexical stress patterns. Pattern N-grams are sequences of N consecutive patterns from a given text. We experimented with pattern-2-grams, pattern-3-grams, and pattern-4-grams.

In some experiments, we combined lexical stress with several traditional features: Function words (fw) express grammatical relationships with other words in a sentence, including articles, prepositions, pronouns and conjunctions. We used the 70 function words from the Mosteller-Wallace Federalist Papers study [7]. The part-of-speech feature (pos) identifies words as nouns, verbs, adjectives, adverbs etc. To tag words in our texts as parts-of-speech, we used the maxent tagger developed by the Stanford NLP group [16]. We also used word N-grams and part-of-speech N-grams (N=2, 3, 4).

3.3   Evaluating Performance

In order to check the accuracy of a learning method, the available documents are divided into training and testing sets. The training set is used to build the model, and then the model is tested on the remaining documents. In our work we adopted a "leave-one-out" validation: n-1 of the available n documents are used for training, and validation is carried out using the remaining one document. The procedure is repeated n times, so every document is used at some point for validation. The percentage of correctly classified documents constitutes the "leave-one-out" accuracy of the method. Two other measures of method performance are precision and recall. The precision of a method is the fraction of documents attributed to an author, which are indeed his/her work. The recall of a method is the fraction of an author's documents attributed to him/her.

3.4   Weighted Voting

When a human expert attempts to attribute a text of unknown or disputed authorship, he/she usually examines a broad set of lexical, stylistic, syntactic, and content features. It is logical to apply the same strategy in automated attribution: We use a methodology, which combines different learning methods and features through weighted voting [9].

Each method independently selects and supports one author. The intensity of support of a method is proportional to its leave-one-out accuracy, so that methods performing more accurately on the training documents contribute more in voting for their choice. The overall support for each author is calculated by adding the supports the author received from methods that choose him/her. Thus, the accuracy-weighted method selects the author with the highest overall support. Only methods that have an accuracy higher than a preset threshold in the leave-one-out validation are considered. In our experiments, the accuracy-weighted method often outperformed any individual method.

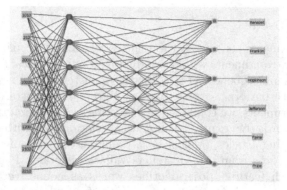

Fig. 1. Multiplayer Perceptron Generated from WEKA [14]

4   Authorship Attribution with Lexical Stress Pattern Vectors

4.1   Experiments with Lexical Stress Patterns Only

We have carried out a large number of experiments both with subsets as well as with the entire corpus of 18th century documents available to us. The initial experiments used only six randomly-selected authors: Adams, Benezet, Dickinson, Hopkinson, Jefferson, and Paine. The results were very encouraging: 76% accuracy of classification and a high average precision and recall (0.791 and 0.767 respectively). However, experiments with the full set of 21 authors and all 126 documents revealed that the method works well only for some authors. This surprising observation seems to confirm the original intuitive hypothesis of our colleague, who claimed that some authors have a more "melodic" writing style. It appears that, whether consciously or subconsciously, some authors tend to choose words with particular lexical stress patterns much more frequently. While the true "melody" of text/speech is conveyed through intonation, it appears that lexical stress does play a role in the emotive expressiveness of text.

We used a number of different pattern selection strategies - stress patterns of words, stress pattern 2-grams, 3-grams, and 4-grams. The classification accuracy was the highest (67%) for pattern-2-grams learned using SMO. Weighted combination had 66% accuracy and its recall and precision increased for most authors. Some authors, like Condorset and Ogilvie scored 100% in both recall and precision, and several other authors had almost as strong results. However, a few authors remained problematic: Benezet had a recall of 44%, Witherspoon - 33%, and Boswell scored 0% on both recall and precision. The reason for this disparity is unclear. Our initial assumption was that the small number of available documents is affecting the results for some authors (e.g. Boswell has only 3 documents). However, our corpus contains nine Witherspoon texts and he still scored pretty low. On the other hand, Priestly scored 100% recall and 75% precision with only 3 document in the corpus. While the length of the documents may affect the results of the attribution, it is pretty clear that author predilection towards words with specific types of lexical patterns undoubtedly affects the attribution results.

4.2 Combining Stress with Other Lexical Features

Our results demonstrate that the weighted accuracy approach produces much stronger results compared to using traditional lexical features only or lexical stress only (Tables 3, 5, 4). The best single traditional feature accuracy achieved was 75%. Combining traditional methods, we were able to achieve an accuracy of 82%. By adding lexical-stress, the overall accuracy improved further to 88%.

Table 3. Lexical Stress Accuracy

Lexical Stress Features	Accuracy
SMO-pattern-2-gram	67%
SMO-pattern-3-gram	58%
SMO-pattern-4-gram	43%
SMO-single-pattern	50%
MLP-pattern-2-gram	60%
MLP-pattern-3-gram	58%
MLP-pattern-4-gram	46%
MLP-single-pattern	56%

Table 4. Accuracy of Weighted Combinations

Features	Accuracy
Lexical Stress Only	66%
Traditional Features Only	82%
Combined	88%

Table 5. Traditional Features Accuracy

Traditional Features	Accuracy
SMO-function-words	70%
SMO-single-pos	57%
SMO-pos-2-gram	68%
SMO-pos-3-gram	56%
SMO-pos-4-gram	52%
SMO-word-2-gram	65%
SMO-word-3-gram	50%
SMO-word-4-gram	30%
SMO-single-word	75%
MLP-function-wods	71%
MLP-single-pos	60%
MLP-pos-2-gram	70%
MLP-pos-3-gram	63%
MLP-pos-4-gram	52%
MLP-word-2-gram	65%
MLP-word-3-gram	49%
MLP-word-4-gram	25%
MLP-single-word	75%

The ability of lexical-stress features to recognize some authors provided a valuable contribution: the weighted combination of traditional features improved accuracy by 7%; the accuracy including lexical stress was 13% higher than the highest accuracy among individual methods.

5   Conclusion and Future Work

We have clearly demonstrated that lexical stress, in combination with other lexical features, can produce strong attribution results. There are, however, a number of limitations to this methodology. The primary limitation is the availability of an accurate lexical stress dictionary. To the best of our knowledge, the CMU pronunciation dictionary is the only non-commercial resource available for extracting lexical stress. While fairly comprehensive, the CMU dictionary lacks many words, which were in common use in the 18th century, but have since fallen out of use. Moreover, since the first voice recordings of spoken English did not appear until the mid-19th century, there is really no way of knowing the historically accurate lexical stress patterns used during 18th century. Thus, our work is based on a present-day stress patterns of written words from a much earlier historical period.

Another issue arises due to fact that some words may have multiple pronunciation patterns with stress placed on different syllables. Our current research is aimed at finding an approach, which account for multiple word stress patterns when used for attribution. Two other major research directions include the development of algorithms for extracting intonation and alliteration from text, and using them in automated authorship attribution. Finally, in collaboration with colleagues from the Humanities, we are applying our methodology to our ever-growing collection of attributed and unattributed texts in an effort to determine the authorship of important political documents from the time of the American and French Revolutions.

Acknowledgments. We would like to gratefully acknowledge the contributions of our colleague Gary Berton from the Institute for Thomas Paine Studies at Iona, whose expertise and insights have been invaluable in our work.

References

1. Love, H.: Attributing Authorship: An Introduction. Cambridge University Press, Cambridge (2002)
2. Stamatatos, Efstathios: A survey of modern authorship attribution methods. Journal of the American Society for Information Science and Technology **60**(3), 538–556 (2009)
3. Lowe, D., Matthews, R.: Shakespeare vs. Fletcher: A Stylometric Analysis by Radial Basis Functions. Computers and the Humanities **29**, 449–461 (1995)
4. Matthews, R., Merriam, T.: Neural computation in stylometry: An application to the works of Shakespeare and Fletcher. Literary and Linguistic Computing **8**(4), 203–209 (1993)

5. Burrows, J.: Computation into Criticism: A Study of Jane Austen's Novels and an Experiment in Method. Clarendon Press, Oxford (1987)
6. Morton, A.Q.: The Authorship of Greek Prose. Journal of the Royal Statistical Society (A) **128**, 169–233 (1965)
7. Mosteller, F., Wallace, D.: Inference and Disputed Authorship: The Federalist. Addison-Wesley, Reading (1964)
8. Petrovic, S., Berton, G., Schiaffino, R., Ivanov, L.: Authorship attribution of Thomas Paine works. In: Interdisciplinary Social Sciences Conference, Prague, Czech Republic (2013)
9. Petrovic, S., Berton, G., Schiaffino, R., Ivanov, L.: Authorship attribution of Thomas Paine works. International Conference on Data Mining DMIN 2014, pp. 182–188. Springer, Heidelberg (2014)
10. Chomsky, N., Morris, H.: The sound pattern of English. New York (1968)
11. Fudge, E.: English word-stress. London (1984)
12. Creel, S., Tanenhaus, M., Aslin, R.: Consequences of Lexical Stress on Learning an Artificial Lexicon. Journal of Experimental Psychology: Learning, Memory, and Cognition **32**(1), 15–32 (2006)
13. Argamon, S., Whitelaw, C., Chase, P., Raj Hota, S., Garg, N., Levitan, S.: Stylistic Text Classification Using Functional Lexical Features. Journal of the American Society for Information Science and Technology **58**(6), 802–822 (2007)
14. Hall, M., Frank, E., Holmes, G., Pfahringer, B., Reutemann, P., Witten, I.: The WEKA Data Mining Software: An Update. SIGKDD Explorations **11**(1) (2009)
15. Juola, P.: Authorship Attribution. Foundations and Trends in Information Retrieval **1**(3), 233–334 (2006)
16. Toutanova, K., Klein, D., Manning, C.: Feature-rich part-of-speech tagging with a cyclic dependency network. In: HLT-NAACL, pp. 252–259 (2003)

Eye Gaze Analyses in L1 and L2 Conversations: Difference in Interaction Structures

Koki Ijuin[1], Yasuhiro Horiuchi[1], Ichiro Umata[2], and Seiichi Yamamoto[1]([✉])

[1] Department of Information Systems Design, Doshisha University,
1-3 Miyakodani, Tatara, Kyotanabe-shi, Kyoto, Japan
seyamamo@mail.doshisha.ac.jp
[2] National Institute of Information and Communication Technology,
Institute2, 2-2-2 Hikaridai, Seika-cho, Soraku-gun, Kyoto, Japan

Abstract. The importance of conversation in a second language (L2) during international collaboration continues to increase, but the features of non-verbal communications such as eye gaze and gestures in L2 conversation have not been clarified to the extent they have in the native language (L1). This study provides quantitative analyses of both speakers' and listeners' eye gaze activities to examine their differences in interaction structures between L1 and L2. Our analyses clarify that the listeners gaze more at speakers in conversations in L2 than in L1 and that the speaker gazes more at the next speaker in conversations in L2 than in L1 when they join small-party conversations. These analyses demonstrate that interaction structures in L2 are different from those in L1.

Keywords: Eye gaze · Second language conversation · Multimodal communication

1 Introduction

The rapid development of transportation systems and information technologies has given people more opportunities for worldwide communications. The importance of conversation in a second language (L2) has been increasing more than ever, so analysis of how mutual understanding works in an L2 conversation has taken on the same importance as its operation in a mother tongue (L1) conversation.

Language use is a form of joint action carried out by groups of people acting in coordination. Their joint actions involve not only verbal activities but also non-verbal ones such as eye gaze, gestures, body posture, and nodding to achieve a common grounding process, i.e., to form the basis of mutual understanding [1,2]. Eye gaze has particularly important functions such as signaling interpersonal attitudes, augmenting speech contents, controlling the synchronization of speech, and distraction through avoiding excess input of information, as well as acquiring feedback information from conversation partners [3–8]. The studies cited here, however, have mainly been conducted for conversations in L1.

© Springer International Publishing Switzerland 2015
P. Král and V. Matoušek (Eds.): TSD 2015, LNAI 9302, pp. 114–121, 2015.
DOI: 10.1007/978-3-319-24033-6_13

Eye gaze plays an important role in second language conversations as well. Previous research has reported that eye gaze activities and facial expression play an important role in monitoring both partners' understanding in the conversation repair process [9].

Veinott et al. [10] found that non-native speaker pairs benefited from video in route-guiding tasks in the field of computer supported collaborative work (CSCW) while native speaker pairs, in contrast, did not. They argued that this was because video helped the non-native pairs to negotiate a common ground, whereas it did not serve this purpose for the native pairs. Their study revealed that video images of the conversation partners helped them to establish mutual understanding in their second language conversations, although it was still not clear which element in the video information contributed to establishing the common ground. This research suggested that gaze activities in conversations in L2 had different functions from those in L1.

Our previous research also revealed that the participants gazed more at speakers in conversations in L2 than in L1 when they joined small-party conversations [11–14]. Several possible reasons have arisen to explain the differences between eye gaze activities in conversations in L2 and L1, among them: (1) participants monitored their understanding of what was being said to make repairs if necessary, (2) participants used visual information to help in perceiving the auditory information, and (3) participants gave a polite acknowledgement of the speakers' effort in producing speech with difficulty. The actual reason, however, is still not clear, and it's reasonable to assume that not a single factor but rather a combination of several factors causes this phenomenon.

Content analyses of utterances are needed to validate these possibilities. However, our previous study showed that listeners gazed at the next speaker before speaking started for more utterances in L2 than in L1 conversations [15]. These results suggest that the participants could better predict the next speaker in conversations in L2 than in L1. This might be because the conversational flow in L2 is simpler than conversations in L1 due the participants' relatively lower level of language expertise. Speakers tried to make conversations easier by monitoring the other participants' understanding and by using visual information to more fully perceive the auditory information.

This paper compares eye gaze activities of both speakers and listeners in detail to examine what factors cause the phenomenon of listeners gazing more at the speaker in L2 conversations than in L1 conversations.

The paper is structured as follows. In Section 2, we briefly describe the multimodal corpus used in this research. We report our analytical results on eye gaze in Section 3 and present a discussion in Section 4. We conclude in Section 5 with a brief summary.

2   Mutimodal Corpus

We used a multimodal corpus [14] of conversations by three participants in these analyses. The main features of the corpus are briefly explained as follows.

2.1 Participants

A total of 60 university students between the ages of 18 and 24 were recruited to develop the multimodal corpus. They were divided into twenty groups of three participants each. All participants were native Japanese speakers who had learned English as a second language. They were not acquainted with each other before the meeting held for data collection. The Test of English for International Communication (TOEIC) score was used to determine participants' English communicative skill. Participants' scores ranged from 450 to 985 (990 being the highest score that could be attained). Each member of a given group was ranked into one of three degrees of English expertise based on their TOEIC score, namely, Rank1 (high proficiency), Rank2 (middle) and Rank3 (low).

2.2 Experimental Setup

Three participants sat 1.5 m apart in a triangular formation around a table. Each participant sat in the same position for all four trials. Three sets of NAC EMR-9 eye trackers and headsets with microphones recorded their eye gazes and voices. The participants talked about two predetermined themes in English (as their L2) and in Japanese (as their L1). Each group had two conversations in each language.

2.3 Procedure

Two conversational topics were assigned before each trial. The first was a free-flowing one where participants chatted about their favorite foods or animals. The second was a goal-oriented task where they collaborated on deciding what to take to a deserted island or the mountains. The orders of the conversation topics and the languages were set randomly to counterbalance any order effect. Each group had six-minute conversations on the free-flowing and goal-oriented topics in both languages.

2.4 Annotations

The time information of the utterances and the eye gazes were annotated. Instances of the annotation feature GazeObject were manually annotated based on the gaze path given by the eye tracker; whose values included Gaze at the person to the right, Gaze at the person to the left, Gaze to other (gaze to objects besides the person to the right or left), and NoGaze (gaze was not detected). Gaze events are defined as gazing at some object, that is, participants focus their visual attention on a particular object for a certain period of time [16]. The annotation was done using the EUDICO Linguistic Annotator (ELAN) [17] developed at the Max Planck Institute as an annotation tool. Cohen's kappa coefficient of segmenting gaze events was 0.83.

2.5   Transcription

The transcription of utterances in L2 is a difficult task even for native speakers. After all conversations had finished, the recorded voices in the conversations in L2 were transcribed by the participants themselves and checked with a bilingual assistant. The transcription procedures were specified by the authors. For example, when the speaker was laughing or hesitating, the span had to be surrounded by an exclamation point, as in !laugh!. Words also had to be bound by hash marks, as in #Tokyo# or #sushi# when speakers uttered a proper noun or a word in Japanese. We developed a tool for linking annotated tags of utterances and their transcribed data.

3   Analyses

We quantitatively analyzed the activities of eye gaze in L1 and L2 conversations from the following perspectives.

3.1   Analysis 1: Listener's Eye Gaze Activities

Previous research [12–14] has suggested that speakers are gazed at more by listeners in L2 conversations than in L1 conversations. Based on the definition described in the previous research [15], we used the listener's gazing ratio to analyze the eye gaze activities of listeners while the other participants were speaking. Listener's Gazing Ratio is defined as

$$Listener's\,Gazing\,Ratio = DLG_j(i)/D(i), \tag{1}$$

where $D(i)$ is the duration of the i-th utterance and $DLG_j(i)$ is the duration when the j-th participant is gazing at the speaker in the i-th utterance. The average and SD values of the Listener's Gazing Ratio for each language and topic are listed in Table 1.

We conducted an ANOVA test with the language difference and the topic difference being within-subject factors and the difference between Listener's Gazing Ratios of participants of higher expertise to participants of lower expertise and those of participants of lower expertise to participants of higher expertise being the between-subject factors. A significant main effect of language difference was found ($F_{(1,117)} = 107.75$, $p < .001$). These results show that the speakers were gazed at more in L2 conversations than in L1 conversations. This phenomenon was found in both free-flowing and goal-oriented conversations in L2.

3.2   Analysis 2: Speaker's Eye Gaze Activities

Previous research [12–14] has also suggested that speakers gaze at listeners in conversations in L2 almost as much as in L1 conversations. We used the speaker's

Table 1. Basic statistics of eye gaze activities

Features in conversation	Average ± standard deviation			
	Free(JPN)	Free(ENG)	Goal(JPN)	Goal(ENG)
$Listener'sGazingRatio$	0.47 ± 0.14	0.58 ± 0.15	0.44 ± 0.16	0.57 ± 0.17
$Speaker'sGazingRatio$	0.28 ± 0.13	0.28 ± 0.14	0.28 ± 0.16	0.28 ± 0.16

gazing ratio to analyze the eye gaze activities of speakers during speaking. Speaker's Gazing Ratio is defined as

$$Speaker'sGazingRatio = DSG_j(i)/D(i), \qquad (2)$$

where $D(i)$ is the duration of the i-th utterance and $DSG_j(i)$ is the duration when the speaker is gazing at the j-th participant in the i-th utterance. The average and SD values of the Speaker's Gazing Ratio for each language and topic are also listed in Table 1.

We conducted an ANOVA test with the language difference and the topic difference being within-subject factors and the difference between Speaker's Gazing Ratios of participants of higher expertise to participants of lower expertise and those of participants of lower expertise to participants of higher expertise being the between-subject factors. The results do not show any significant main effect of language difference ($F_{(1,117)} = 0.1$).

3.3   Analysis 3: Eye Gaze Activities Among the Participants

In order to investigate the difference of speakers' and listeners' gazing activities in conversations in L1 and L2, we analyzed eye gaze activities among the participants in line with conversational flow. That is to say, we analyzed gazing activities between current speakers and next speakers, between current speakers and listeners (not next speakers), and between next speakers and listeners. Eye Gazing Ratio among participants is defined as

$$GazingRatioAmongParticipants = DG_{jk}(i)/D(i), \qquad (3)$$

where $D(i)$ is the duration of the i-th utterance and $DG_{jk}(i)$ is the duration when participant j is gazing at participant k in the i-th utterance. Participants j and k can be the current speaker, next speaker, or listener.

Figure 1 shows the average of gazing ratios among participants categorized by the relation between the gazing and gazed persons. Table 2 lists their averages and standard deviations.

We conducted an ANOVA test with the language difference, conversation topic difference and gazing-at-object difference being within-subject factors. The results show a significant main effect of gazing-at-object difference ($F_{(1,19)} = 151.0$, $p < .01$) and a significant first-order interaction between language difference and gazing-at-object difference ($F_{(1,19)} = 22.4$, $p < .01$). We can easily observe the following phenomena from these analyses:

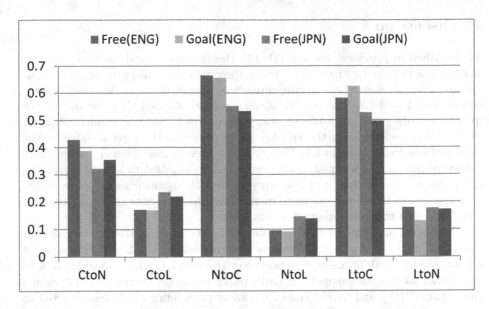

Fig. 1. Average of gazing ratios among participants for free-flow and goal-oriented conversation in L1 and L2. The symbol CtoN represents the condition where the gazing person is a current speaker and the gazed person is a next speaker. Other categories represent similar conditions based on notations where the symbols C, N, and L represent current speakers, next speakers, and listeners, respectively.

Table 2. Basic statistics of mutual eye gaze activities

Gazing person - gazed person	Average ± standard deviation			
	Free(JPN)	Free(ENG)	Goal(JPN)	Goal(ENG)
Current speaker - next speaker	0.32 ± 0.08	0.43 ± 0.07	0.35 ± 0.09	0.39 ± 0.12
Current speaker - listener	0.24 ± 0.11	0.17 ± 0.09	0.22 ± 0.06	0.17 ± 0.09
Next speaker - current speaker	0.55 ± 0.10	0.66 ± 0.08	0.53 ± 0.11	0.66 ± 0.13
Next speaker - listener	0.15 ± 0.10	0.09 ± 0.07	0.14 ± 0.05	0.09 ± 0.06
Listener - current speaker	0.53 ± 0.11	0.58 ± 0.10	0.50 ± 0.11	0.62 ± 0.13
Listener - next speaker	0.18 ± 0.06	0.18 ± 0.05	0.17 ± 0.08	0.13 ± 0.09

1a. Both the next speaker and the listener gazed at the current speaker more than other participants in both conversations in L1 and L2.

1b. However, the averages of gazing ratios by both the next speakers and the listener were larger in conversations in L2 than those in L1.

2a. The ratio of current speakers gazing at next speakers was more than that of gazing at the listener in both conversations in L1 and L2.

2b. However, the average of gazing ratios of current speakers gazing at next speakers in conversations in L2 was larger than that in L1.

4   Discussion

As described in previous research [11–14], the listeners gazed more at speakers in conversations in L2 than in L1 when they joined small-party conversations and the average of speakers' gazing ratios was almost the same between conversations in L1 and L2. The results of our analyses showed that the object the speaker mainly gazed at, however, was different in L2 conversations and in L1 conversations: specifically, the speaker gazed more at the next speaker in the conversations in L2 than in L1. These results suggest that listeners gazed more at the speaker in L2 conversations to use visual information to help in perceiving the auditory information, and also suggest that the speaker gazed more at the next speaker in L2 conversations to monitor his/her understanding of what was being said and to make repairs if necessary.

The results also suggest that the interaction structure in L2 may be different from that in L1. That is to say, speakers in conversations in L2 may have more of a tendency to speak to a single person than those in L1, where speakers talk to both other participants equally. A multi-party conversation consists of "ratified participants" [18], and participants with lower proficiency might be relegated to "side participant" status. This observation suggests that the participants' relatively lower language expertise let speakers to choose simpler interaction structure in conversations in L2 than in L1, although content analyses of utterances are needed to validate these conjectures.

5   Conclusion

We compared the gazing activities of speakers in small-party conversations in L1 and L2 and found that the speakers gazed more at the next speaker in conversations in L2 than in L1. These results suggest a difference between the interaction structure in L1 and L2, maybe because of the participants' relatively lower language proficiency in the latter. We are currently conducting detailed tagging including discourse tagging to the multimodal corpus to clarify difference of gazing activities in conversations in L1 and L2.

Acknowledgments. We would like to express our deepest gratitude to Emeritus Professor Masuzo Yanagida of Doshisha University for his invaluable comments and helpful discussion. This research was supported in part by a grant from the Japan Society for the Promotion of Science (JSPS) (No.15K02738).

References

1. Clark, H., Brennan, S.: Grounding in communication. Perspectives on Socially Shared Cognition, 222–233 (1991)
2. Clark, H.: Using language. Cambridge University Press (1996)
3. Argyle, M., Dean, J.: Eye-contact, distance and affiliation. Sociometry **28**, 289–304 (1965)

4. Kendon, A.: Some functions of gaze-direction in social interaction. Acta Psychologica **26**, 22–63 (1967)

5. Argyle, M., Lallijee, M., Cook, M.: The effects of visibility on interaction in a dyad. Human Relations **21**, 3–17 (1968)

6. Argyle, M., Ingham, R., Alkema, F., McCalin, M.: The different functions of gaze. Semiotica **7**, 19–327 (1973)

7. Argyle, M., Lefebvre, L.M., Cook, M.: The meaning of five patterns of gaze. European Journal of Social Psychology **4**, 125–136 (1974)

8. Kleinke, C.L.: Gaze and eye contact: a research review. Psychological Bulletin **100**, 78–100 (1974)

9. Hosoda, Y.: Repair and relevance of differential language expertise in second language conversations, pp. 25–50 . Oxford University Press (2006)

10. Veinott, E., Olson, J., Olson, G., Fu, X.: Video helps remote work: speakers who need to negotiate common ground benefit from seeing each other. In: Proceedings of the Conference on Computer Human Interaction, CHI 1999, pp. 302–309. ACM Press, PA (1974)

11. Kabashima, K., Jokinen, K., Nishida, M., Yamamoto, S.: Multimodal corpus of conversations in mother tongue and second language by same interlocutors. In: Proceedings of the 4th Workshop on Eye Gaze in Intelligent Human Machine Interaction, Volume Article No. 9, pp. 302–309 (2012)

12. Yamamoto, S., Taguchi, K., Umata, I., Kabashima, K., Nishida, M.: Differences in interactional attitudes in native and second language conversations: Quantitative analyses of multimodal three-party corpus. In: Proceedings of the 35th Annual Meeting of the Cognitive Science Society, pp. 3823–3828 (2013)

13. Umata, I., Yamamoto, S., Ijuin, K., Nishida, M.: Effects of language proficiency on eye-gaze in second language conversations: toward supporting second language collaboration. In: ICMI 2013, pp. 413–420 (2013)

14. Yamamoto, S., Taguchi, K., Ijuin, K., Umata, I., Nishida, M.: Multimodal corpus of multiparty conversations in l1 and l2 languages and findings obtained from it. Language Resources and Evaluation (2015). doi:10.1007/s10579-015-9299-2

15. Ijuin, K., Taguhci, K., Umata, I., Yamamoto, S.: Eye gaze analyses in l1 and l2 conversations: From the perspective of listeners' eye gaze activity. In: Understanding and Modeling Multiparty, Multimodal Interactions - Workshop at ICMI 2014 (2014)

16. Jokinen, K., Furukawa, H., Nishida, M., Yamamoto, S.: Gaze and turn-taking behavior in casual conversational interaction. ACM Transactions on Interactive Intelligent Systems **3**(2), 12:1–12:30 (2013)

17. ELAN. http://tla.mpi.nl/tools/tla-tools/

18. Goffman, E.: Replies and responses. Language in Society **5**(3), 257–313 (1976)

Word Categorization of Corporate Annual Reports for Bankruptcy Prediction by Machine Learning Methods

Petr Hájek[(✉)] and Vladimír Olej

Institute of System Engineering and Informatics, Faculty of Economics and Administration, University of Pardubice, Pardubice, Czech Republic
{Petr.Hajek,Vladimir.Olej}@upce.cz

Abstract. The language of company related documents is recognized as being an important indicator of future financial performance. This study aims to extract various word categories from corporate annual reports and examine their effect on bankruptcy prediction. We show that the language used by bankrupt companies is characterized by stronger tenacity, accomplishment, familiarity, present concern, exclusion and denial. Bankrupt companies also use more modal, positive, uncertain and negative language. We used neural networks, support vector machines, decision trees and ensembles of decision trees to predict corporate bankruptcy. The prediction models utilized both financial indicators and word categorizations as input variables. We show that both general dictionary and financial dictionary categories can significantly improve the accuracy of the prediction models.

Keywords: Bankruptcy prediction · Word categorization · Sentiment analysis · Machine learning · Meta-learning

1 Introduction

Filing for bankruptcy is considered the most serious form of corporate financial distress. Much research is aimed at elucidating the mechanisms of bankruptcy prediction using financial determinants such as profitability, liquidity, and debt ratios. Bankruptcy prediction is realized as a two-class problem, classifying bankrupt and non-bankrupt companies. These approaches have evolved from the use of multivariate statistical models to recent use of artificial intelligence (AI) and machine learning methods, see e.g. [1] for a review. Neural networks (NNs) [2] and support vector machines (SVMs) [3] represent the most frequent AI methods applied for corporate bankruptcy prediction. Recently, there have been attempts to use ensembles of AI methods to improve prediction accuracy [4–6].

Even though AI methods significantly outperformed traditional statistical methods in previous research [5], they were not capable to fully explain the complex dynamical relations between historical financial determinants and future

© Springer International Publishing Switzerland 2015
P. Král and V. Matoušek (Eds.): TSD 2015, LNAI 9302, pp. 122–130, 2015.
DOI: 10.1007/978-3-319-24033-6_14

bankruptcy. Recently, the text analysis of company related documents has gathered increasing interest due to its ability to help better understand various financial events, such as fraud, stock return and volatility [7]. In related literature, [8] developed a financial ontology from the extracted concept scores to predict corporate bankruptcy. Shirata et al. [9] report that extracting phrases from annual reports may be an effective predictor of corporate bankruptcy. Sentiment analysis of news articles was employed by [10] to demonstrate that topic-specific negative sentiment is more important for future credit rating changes in comparison with positive sentiment. Similarly, recent studies [11–13] report that financial performance prediction can be improved with sentiment information extracted from company related documents. Despite this interest, no one as far as we know has studied the effect of word categories on bankruptcy prediction. The contribution of our work is to explore the consequences of using both quantitative (i.e. financial ratios) and qualitative (word categories) information in predicting corporate bankruptcy. As a source of information on word categories, we use the narrative texts from annual reports since they are considered one of the most important external documents that reflect corporate financial performance and strategy.

The remainder of this paper has been organized in the following way. The next section deals with data collection and their preprocessing. Section 2 also examines the differences between bankrupt and non-bankrupt companies regarding the chosen word categorizations. Section 3 presents the results of predicting bankruptcy using NNs, SVMs, classification trees and the ensembles of classification trees. Our conclusions are drawn in the final section.

2 Data Preprocessing

First, we used the BankruptcyData.com database to detect bankrupt U.S. companies for the period 2004-2013. Second, we collected financial data for the bankrupt companies from the Value Line database three years before bankruptcy. The set of financial indicators was based on extensive literature on corporate bankruptcy prediction. As a result, we obtained a total set of 386 bankrupt companies. Third, we collected data for the corresponding set of 386 of non-bankrupt companies. In the selection process, we attempted to preserve similar distributions of companies across years, size and industries. Note that the original dataset is strongly imbalanced and that this issue can be also addressed by other sampling methods [14].

In the next step, we collected annual reports (10-Ks filings) for both the bankrupt and non-bankrupt companies from U.S. Securities and Exchange Commission EDGAR System. More specifically, we extracted only the most important textual section from the documents, i.e. Item 7. Management's Discussion and Analysis of Financial Condition and Results of Operations. Again, only the annual reports three years before bankruptcy were involved in the data collection. Tokenization and lemmatization of the documents were performed to obtain a set of term candidates which were subsequently compared with two word categorizations: (1) financial dictionary developed by [7] and (2) general dictionary Diction 5.0 developed by [15]. The financial dictionary covered the following word categories: negative, positive, uncertainty, litigious, modal strong and modal weak. In case of the positive category, negation was detected to give the category accuracy. Additional thirty categories were obtained using the Diction 5.0 in order to cover the general tone of language, including ambivalence, tenacity, satisfaction, inspiration or concreteness. Next, the *tf.idf* term weighting scheme was applied to each term to obtain its relative frequency. Finally, the average importance of each word category was calculated.

The list of the word categories together with their average values for each class (non-bankrupt/bankrupt) is presented in Table 1. Student's paired *t*-test was performed to show significant differences in the frequencies of each word category. Regarding the Diction 5.0 word categories, the results show that the language used by bankrupt companies was characterized on one hand by stronger tenacity (connoting confidence and totality), accomplishment, familiarity (most common words), present concern, exclusion (sources and effects of social isolation) and denial (negative functions words and terms designating null sets) and, on the other hand, by weaker blame (social inappropriateness, downright evil and unfortunate circumstances), hardship (natural disasters and hostile actions), spatial awareness, concreteness (tangibility and materiality), past concern (past-tense forms of the verbs), rapport (affinity, assent, deference and identity), diversity and liberation. Bankrupt companies also used more modal (weak modal such as "almost" or "could" and strong modal such as "always" or "definitely"), positive, uncertain and negative language. These findings suggest that bankrupt

Table 1. Mean *tf.idf* values of sentiment attributes.

Attribute	Non-Bankr./Bankr.	Attribute	Non-Bankr./Bankr.
Numerical	0.273/0.281	Present concern	0.230/0.244*
Ambivalence	0.232/0.240	Human interest	0.279/0.275
Tenacity	0.669/0.769***	Concreteness	0.356/0.334**
Levelling	0.251/0.261	Past concern	0.281/0.262**
Collectives	0.332/0.319	Centrality	0.277/0.290
Praise	0.273/0.261	Rapport	0.275/0.256*
Satisfaction	0.236/0.248	Cooperation	0.287/0.298
Inspiration	0.310/0.302	Diversity	0.313/0.288**
Blame	0.252/0.214***	Exclusion	0.260/0.286**
Hardship	0.327/0.301**	Liberation	0.291/0.244***
Aggression	0.265/0.267	Denial	0.250/0.294**
Accomplishment	0.290/0.306**	Motion	0.318/0.299
Communication	0.303/0.304	Litigious	0.296/0.285
Cognition	0.302/0.318	Strong modal	0.189/0.225***
Passivity	0.266/0.262	Weak modal	0.243/0.275***
Spatial awareness	0.379/0.339***	Positive	0.250/0.290***
Familiarity	0.336/0.361*	Uncertainty	0.215/0.232**
Temporal	0.279/0.280	Negative	0.229/0.262***

*** statistically significant differences at $p=0.01$, ** at $p=0.05$, * at $p=0.1$.

companies attempted to emphasize their persistence and achievements, using more common and uncertain language.

In addition to word categorizations, we used financial indicators as bankruptcy predictors. These included profitability ratios (return on equity, return on capital, effective tax rate), liquidity (cash / total assets), leverage (market debt / total capital, book debt / total capital), assets structure (fixed assets / total assets, intangible assets / total assets), business situation (effective tax rate, revenue growth last year) and market value ratios (earnings per share, price to book value, price to earnings per share (both current and forward), value line beta, high to low stock price, dividend yield, payout ratio, standard deviation of stock price, institutional and insiders holdings).

Feature selection was the last step of data preprocessing. We performed this procedure using correlation-based filter [16] to eliminate the redundant variables. Thus, the dimensionality of the feature space was reduced and the accuracy of the algorithms could be improved [17]. Table 2 shows the list of attributes after feature selection. Again, Student's paired t-test was performed to show the differences between non-bankrupt and bankrupt classes.

Table 2. Descriptive statistics (mean±stdev) of selected input attributes.

	Attribute	Non-bankrupt	Bankrupt	t-value
x_1	Tenacity	0.67±0.50	0.77±0.70	-2.71***
x_2	Denial	0.25±0.25	0.29±0.35	-2.46**
x_3	Strong modal	0.19±0.15	0.22±0.22	-3.22***
x_4	Forward Price / Earnings	12.26±15.32	1.75±6.24	5.31***
x_5	Value line beta	1.17±0.34	1.09±0.52	2.80***
x_6	Standard deviation of stock price	0.41±0.32	0.93±0.64	-15.31***
x_7	Effective tax rate	0.26±0.16	0.10±1.04	3.74***
x_8	Cash / Total assets	0.09±0.13	0.08±0.18	1.31
x_9	Intangible assets / Total assets	0.19±0.21	0.10±0.19	6.80***
x_{10}	Dividend yield	0.03±0.07	0.01±0.09	4.47***
x_{11}	Insider holdings	0.07±0.12	0.03±0.08	3.08***
	N	386	386	

*** statistically significant differences at $p=0.01$, ** at $p=0.05$, * at $p=0.1$.

3   Experimental Results

In our experiments, we examined various structures of the following methods for bankruptcy prediction: multilayer perceptron (MLP), SVM, decision trees (J48), alternating decision trees (ADTrees) [18], Naïve Bayes decision trees (NBTrees) [19], random forest (RanFor) [20], and meta (ensemble) algorithms, i.e. rotation forest (RF_{J48}) [21] and random subspace method (RSS_{J48}) [22]. To avoid overfitting, all experiments were performed using 10-fold cross-validation. The structures and parameters of the learning algorithms were found using grid search procedure.

MLP was trained using the backpropagation algorithm with momentum. The following parameters of the MLP were examined: the number of neurons in the hidden layer = (attributes + classes) / 2, learning rate = {0.05, 0.1, 0.2, 0,3}, momentum = {0.05, 0.1, 0.2, 0.3}, and the number of epochs = {5, 10, 20, 100, ... , 800}.

SVMs use kernel functions to separate the hyperplane between two classes by maximizing the margin between the closest data points. This is done in a higher-dimensional space where the data become linearly separable. We used the SVMs trained by the sequential minimal optimization algorithm. The classification performance of the SVMs was tested for the following user-defined parameters: kernel functions = polynomial, $\gamma = \{0.001, 0.01, 0.1, 0.2\}$, the level of polynomial function = {1, 2, 3}, complexity parameter $C = \{1, 2, 4, 8, 16, 32, 64\}$, round-off error $\varepsilon = \{1.0E\text{-}10, 1.0E\text{-}12, 1.0E\text{-}14\}$, and tolerance parameter = {0.001, 0.01, 0.1}.

The group of J48, ADTree, NBTree and RanForest algorithms uses a tree representation assigning a class to an object based on its attributes. J48 classifier uses an error based pruning algorithm. The user can choose a confidence value to be used when pruning the tree. An attribute with the best value of the splitting criterion is assigned to each root and intermediate node. The following parameters of J48 were examined to obtain the best classification performance: confidence factor = {0.1, 0.15, ... , 0.55}, minimum number of instances per leaf = {1, 2, ... , 5} and number folds = 3.

ADTrees represents a generalization of decision trees, voted decision trees and voted decision stumps. The algorithm uses a boosting procedure to produce predictions based on a majority vote over a number of decision trees. The classification performance of ADTrees depends on the number of boosting iterations = {5, 10, 20, ... , 60}. In NBTrees algorithm, root nodes use decision tree classifier and leaf nodes use Naïve Bayes classifier. In RanFor, every tree is prepared by randomly selected objects from dataset. Thus, the accuracy and prediction power can be improved because it is less sensitive to outlier objects. For RanFor, the number of variables to select from at each node is set to $log_2(n) + 1$. The classification performance of the RanFor depends on the number of trees = {5, 10, 20, 30, 40, 50}.

We also employed a wide range of meta (ensemble) algorithms for bankruptcy prediction, namely multiboosting, adaboosting, bagging, dagging, RF and RSS, see e.g. [23] for their comparative review. From this set of algorithms, the highest accuracy for our bankruptcy prediction problem was achieved by RF and RSS. RF fixes the number of variables in each subset. The following parameters of RF were examined: the number of iterations = {5, 10, 20, ... , 60}, max group = 3, min group = 3. RSS classifier includes a large number of trees generated systematically using pseudorandom selections of the subsets of variables from subspace. The following parameters of RSS were examined to achieve the best classification performance: the number of iterations = {5, 10, 20, ... , 100}, and subspace size = {0.1, 0.2, ... , 0.9}. Table 3 presents the best settings of the above mentioned algorithms.

Table 3. Best setting of learning parameters.

Method	Parameters for the best classification performance
MLP	The number of neurons in the hidden layer = (attributes + classes) / 2, learning rate = 0,3, momentum = 0.2, number of epochs = 700.
SVM	Kernel functions = polynomial, $\gamma = 0.001$, level of polynomial function = 1, complexity $C = 64$, round-off error $\varepsilon = 1.0E\text{-}12$, tolerance parameter = 0.001.
J48	Confidence factor = 0.25, min. number of instances per leaf = 2, number folds = 3.
ADTree	The number of boosting iterations = 20.
NBTree	-
RanFor	The number of trees = 30.
RF$_{J48}$	The number of iterations = 50, max group = 3, min group = 3.
RSS$_{J48}$	The number of iterations = 90, subspace size = 0.5.

The quality of prediction was measured by the standard classification performance criteria [24]: accuracy Acc [%], true positives (TP rate), false positives (FP rate), F-measure (F-m), and the area under the receiver operating characteristic (ROC) curve. Table 4 shows the detailed classification performance on the bankruptcy prediction dataset (the average values are reported from the 10-fold cross-validation). NBTree performed best with 95.1 % total accuracy. This method achieved high accuracies for individual classes as well (94.1 % for non-bankrupt and 96.9 % for bankrupt class). This finding corroborates

Table 4. Comparison of classification performance over the chosen methods for non-bankrupt (NB) and bankrupt (B) classes with word categories.

	MLP		SVM		J48		ADTree	
Acc. [%]	86.0771		86.6416		91.3452		93.9793	
Class	NB	B	NB	B	NB	B	NB	B
TP rate	0.914	0.767	0.904	0.801	0.932	0.881	0.944	0.933
FP rate	0.233	0.086	0.199	0.096	0.119	0.068	0.067	0.056
F-m	0.893	0.800	0.896	0.813	0.932	0.881	0.952	0.918
ROC	0.885	0.885	0.852	0.852	0.923	0.923	0.981	0.981
	NBTree		RanFor		RF$_{J48}$		RSS$_{J48}$	
Acc. [%]	**95.1082**		94.2615		92.5682		94.0734	
Class	NB	B	NB	B	NB	B	NB	B
TP rate	0.941	0.969	0.957	0.917	0.944	0.894	0.954	0.917
FP rate	0.031	0.059	0.083	0.043	0.106	0.056	0.083	0.046
F-m	0.961	0.935	0.955	0.921	0.942	0.897	0.954	0.918
ROC	0.989	0.989	0.982	0.982	0.979	0.979	0.985	0.985

Table 5. Classification performance without word categories.

	MLP NN		SVM		J48		ADTree	
Acc. [%]	85.0423		86.5475		91.1571		92.0978*	
Class	NB	B	NB	B	NB	B	NB	B
TP rate	0.905	0.754	0.942	0.731	0.913	0.909	0.938	0.891
FP rate	0.246	0.095	0.269	0.058	0.091	0.087	0.109	0.062
F-m	0.885	0.785	0.899	0.798	0.929	0.882	0.938	0.891
ROC	0.901	0.901	0.836	0.836	0.934	0.934	0.972	0.972
	NBTree		RanFor		RF$_{J48}$		RSS$_{J48}$	
Acc. [%]	92.3801*		92.3801*		91.6275		**93.0386**	
Class	NB	B	NB	B	NB	B	NB	B
TP rate	0.939	0.896	0.948	0.881	0.935	0.883	0.938	0.917
FP rate	0.104	0.061	0.119	0.052	0.117	0.065	0.083	0.062
F-m	0.940	0.895	0.941	0.894	0.934	0.885	0.945	0.905
ROC	0.975	0.975	0.969	0.969	0.973	0.973	0.979	0.979

* statistically significant drop in accuracy at $p=0.05$.

those obtained in other studies where ensembles of AI methods outperformed single methods. To assess the effect of word categorizations on prediction performance, we performed additional experiments using only financial indicators $x_4, x_5, ..., x_{11}$ as input variables (here we do not present the best settings of the classifiers due to limited space). Table 5 shows that the performance of all classifiers worsened, for ADTree, NBTree and RanFor even significantly. This can be explained by increase in the variability of ensemble base learners. We employed Student's paired t-test at $p = 0.05$ to test the differences in accuracy.

4 Conclusion

To sum up, our study has argued that the indicators based on word categories extracted from corporate annual reports can be effectively employed in bankruptcy prediction models. In this paper we have investigated the differences in language used by bankrupt and non-bankrupt companies. The evidence from this study suggests that both general and domain-specific dictionaries provide important information on imminent bankruptcy. Although the feature selection procedure led to only three word categories used in the prediction models, it has to be noted that most of the word categorizations were significantly correlated and, thus, important variables could be discarded in feature selection. It was also shown that both positive and negative sentiment categories are used by bankrupt companies more frequently in order to mitigate concerns of stakeholders. At the same time, our results suggest that bankrupt companies attempt to look confidently and goal-oriented. They also use more common and less concrete language. Compared to previous studies based solely on financial indicators, the results of this research offer a much richer understanding of the role of annual reports in corporate bankruptcy prediction.

A number of caveats need to be noted regarding the present study. The present study was focused on the annual reports of U.S. companies and, thus, the findings might not be transferable to other countries. Therefore, we propose that further research should be undertaken in other countries. It should also be directed toward rather specific financial industry [25].

Acknowledgments. This work was supported by the scientific research project of the Czech Sciences Foundation Grant No: 13-10331S.

References

1. Kirkos, E.: Assessing Methodologies for Intelligent Bankruptcy Prediction. Artificial Intelligence Review **43**(1), 83–123 (2015)
2. Huang, S.M., Tsai, C.F., Yen, D.C., Cheng, Y.L.: A Hybrid Financial Analysis Model for Business Failure Prediction. Expert Systems with Applications **35**(3), 1034–1040 (2008)
3. Chaudhuri, A., De, K.: Fuzzy Support Vector Machine for Bankruptcy Prediction. Applied Soft Computing **11**(2), 2472–2486 (2011)
4. Alfaro, E., García, N., Gámez, M., Elizondo, D.: Bankruptcy Forecasting: An Empirical Comparison of AdaBoost and Neural Networks. Decision Support Systems **45**(1), 110–122 (2008)
5. Verikas, A., Kalsyte, Z., Bacauskiene, M., Gelzinis, A.: Hybrid and Ensemble-based Soft Computing Techniques in Bankruptcy Prediction: A Survey. Soft Computing **14**(9), 995–1010 (2010)
6. Heo, J., Yang, J.Y.: AdaBoost Based Bankruptcy Forecasting of Korean Construction Companies. Applied Soft Computing **24**, 494–499 (2014)
7. Loughran, T., McDonald, B.: When is a Liability not a Liability? Textual Analysis, Dictionaries, and 10-Ks. The Journal of Finance **66**(1), 35–65 (2011)
8. Cecchini, M., Aytug, H., Koehler, G.J., Pathak, P.: Making Words Work: Using Financial Text as a Predictor of Financial Events. Decision Support Systems **50**(1), 164–175 (2010)
9. Shirata, C.Y., Takeuchi, H., Ogino, S., Watanabe, H.: Extracting Key Phrases as Predictors of Corporate Bankruptcy: Empirical Analysis of Annual Reports by Text Mining. Journal of Emerging Technologies in Accounting **8**(1), 31–44 (2011)
10. Lu, H.M., Tsai, F.T., Chen, H., Hung, M.W., Li, S.H.: Credit Rating Change Modeling Using News and Financial Ratios. ACM Transactions on Management Information Systems **3**(3), 14 (2012)
11. Lu, Y.C., Shen, C.H., Wei, Y.C.: Revisiting early warning signals of corporate credit default using linguistic analysis. Pacific-Basin Finance Journal **24**, 1–21 (2013)
12. Hájek, P., Olej, V.: Evaluating sentiment in annual reports for financial distress prediction using neural networks and support vector machines. In: Iliadis, L., Papadopoulos, H., Jayne, C. (eds.) EANN 2013, Part II. CCIS, vol. 384, pp. 1–10. Springer, Heidelberg (2013)
13. Hajek, P., Olej, V., Myskova, R.: Forecasting Corporate Financial Performance using Sentiment in Annual Reports for Stakeholders' Decision-Making. Technological and Economic Development of Economy **20**(4), 721–738 (2014)

14. Zhou, L.: Performance of Corporate Bankruptcy Prediction Models on Imbalanced Dataset: The Effect of Sampling Methods. Knowledge-Based Systems **41**, 16–25 (2013)
15. Hart, R.P.: Redeveloping DICTION: theoretical considerations. In: West, M.D. (ed.) Theory, Method, and Practice in Computer Content Analysis, pp. 43–60 (2001)
16. Hall, M.A.: Correlation-based Feature Selection for Machine Learning. Doctoral dissertation, The University of Waikato (1999)
17. Hajek, P., Michalak, K.: Feature Selection in Corporate Credit Rating Prediction. Knowledge-Based Systems **51**, 72–84 (2013)
18. Freund, Y., Mason, L.: The alternating decision tree learning algorithm. In: 16th Int. Conf. on Machine Learning, pp. 124–133, Bled, Slovenia (1999)
19. Kohavi, R.: Scaling up the accuracy of naive-bayes classifiers: a decision-tree hybrid. In: Second International Conference on Knowledge Discovery and Data Mining, pp. 202–207 (1996)
20. Breiman, L.: Random Forests. Machine Learning **45**(1), 5–32 (2001)
21. Rodriguez, J.J., Kuncheva, L.I., Alonso, C.J.: Rotation Forest: A New Classifier Ensemble Method. IEEE Transactions on Pattern Analysis and Machine Intelligence **28**(10), 1619–1630 (2006)
22. Ho, T.K.: The Random Subspace Method for Constructing Decision Forests. IEEE Transactions on Pattern Analysis and Machine Intelligence **20**(8), 832–844 (1998)
23. Banfield, R.E., Hall, L.O., Bowyer, K.W., Kegelmeyer, W.P.: A Comparison of Decision Tree Ensemble Creation Techniques. IEEE Transactions on Pattern Analysis and Machine Intelligence **29**(1), 173–180 (2007)
24. Powers, D.M.W.: Evaluation: from Precision, Recall and F-measure to ROC, Informedness, Markedness and Correlation. Journal of Machine Learning Technologies **1**(2), 37–63 (2011)
25. Hájek, P., Olej, V., Myšková, R.: Predicting financial distress of banks using random subspace ensembles of support vector machines. In: Silhavy, R., Senkerik, R., Oplatkova, Z.K., Prokopova, Z., Silhavy, P. (eds.) Artificial Intelligence Perspectives and Applications. AISC, vol. 347, pp. 131–140. Springer, Heidelberg (2015)

Novel Multi-word Lists for Investors' Decision Making

Renáta Myšková[1] and Petr Hájek[2(✉)]

[1] Faculty of Economics and Administration, Institute of Business Economics
and Management, University of Pardubice, Pardubice, Czech Republic
Renata.Myskova@upce.cz
[2] Faculty of Economics and Administration, Institute of System Engineering
and Informatics, University of Pardubice, Pardubice, Czech Republic
Petr.Hajek@upce.cz

Abstract. The language of firm-related documents is recognized as
being an important indicator of transparent firm culture and management access to stakeholders. This study aims to analyze annual reports
of selected U.S. firms during 2008-2010 from the investor's perspective.
We examine whether investment indicators correspond to the tone (sentiment) of management comments in annual reports. To overcome the
limitations of domain-specific single-word dictionaries, we develop positive and negative multi-word dictionaries. We present the results separately for two sectors, manufacturing and services. We show that the
multi-word dictionaries correlate better with the indicators of investment
activity, in particular with those related to long-term investment.

Keywords: Word list · Sentiment analysis · Annual report · Investor

1 Introduction

Long-term firm performance depends on investments necessary for renewal,
upgrading, and expansion of firm assets. The investment should be continu-
ally undertaken by the top management and shareholders. Investment decisions
are reflected not only in economic indicators but also in the perceptions of other
stakeholders, especially potential investors. It is assumed that a growing value of
assets, together with good financial performance, indicates future firm prosper-
ity. From the stakeholders' point of view, it is therefore essential to be informed
about the investment intentions of firm management. Annual reports represent
an important tool to communicate management's investment strategy to the
stakeholders. Although several scholars note that managers may be tempted to
manipulate investors' judgements by excessive positive statements [1], official
releases from insiders are generally considered as a valuable source of internal
knowledge [2].

A social psychology perspective is also important in annual managerial corporate reporting. Three complementary behaviour scenarios were detected by [3]:

© Springer International Publishing Switzerland 2015
P. Král and V. Matoušek (Eds.): TSD 2015, LNAI 9302, pp. 131–139, 2015.
DOI: 10.1007/978-3-319-24033-6_15

self-presentational dissimulation, impression management by means of enhancement, and retrospective sense-making. Firm investment activity is thus reported not only by numerical financial indicators, but above all by verbal comments in annual reports. This information enables stakeholders to evaluate both current and future investment activity, as well as the management's attitude to risk. Consequently, the perceptions of stakeholders may affect firm market value. That is why increasing attention has recently been given to word categories used not only in annual reports [4,5], but also in press releases [6,7], news stories [8–10], or analysts' reports [11]. Dictionary-based (bag-of-words) approaches have been predominantly used in the related literature on financial decision making, for example in financial distress prediction [12–14]. Earlier literature has shown that general dictionaries such as General Inquirer / Harvard or DICTION are not appropriate for the analysis of financial texts because terms important in the context of financial disclosure are often omitted [15]. Therefore, several domain-specific financial dictionaries have been proposed to address this issue [4,6]. For example, Loughran and McDonald [4] report that most of the negative terms listed in the General Inquirer / Harvard dictionary are not negative in a financial context. The dominance of finance-specific dictionaries have been demonstrated on various financial decision-making problems [4,16].

One of the major limitations of finance-specific dictionaries is the use of single terms (unigrams), although it is often the context that determines the correct word categorization. To bridge this gap, we propose two word lists (positive and negative) specific for investment decision making. We demonstrate that, when compared with commonly used domain-specific dictionaries, our dictionary may better correlate with chosen investment indicators. We show that this finding is true for firms in both manufacturing and services industries.

The remainder of this paper has been organized as follows. The next section introduces our research methodology. Section 3 describes data collection and their preprocessing, and section 4 examines the correlations between the proposed word lists and investment indicators. Our conclusions are drawn in the final section.

2   Research Methodology

This paper aims to analyze the annual reports of selected U.S. firms listed on the New York stock exchange or NASDAQ stock exchange, respectively. We focused on the relationship between firm investment and verbal comments in annual reports during the period 2008-2010. We assumed that the tone of the textual information given to stakeholders corresponds to firm investment activity. Investment indicators were calculated from uniformly structured financial statements. Investment activity is an important subject of managerial comments in Item 7 (MD&A) of the annual reports.

To ensure comparability of firms, the criteria for selection were established as follows: (a) firm size (market capitalization greater than $10 million), (b) classification of economic activity (manufacturing and services), and (c) financial performance (Z-score ≥ 1.8). The criteria are important for several reasons. The firm size determines the range of barriers to business activities, both internal and external [17]. Moreover, investment activity is specific to each industry and it also depends on current financial conditions. To evaluate the overall financial condition, we employed Altman Z-score [18]. This bankruptcy model uses a combination of financial ratios and provides an assessment of current and future (medium-term) financial conditions. We discarded those firms with $Z < 1.8$, indicating a high risk of bankruptcy. This should ensure that all included firms have some investment potential and their management is motivated to present its future investment intentions. We further assume that larger firms have a higher share of investment assets and that the need for firm investment is associated with the economic activity, being larger for manufacturing firms investing in production facilities. Therefore, we hypothesize that manufacturing firms will use the investment sentiment in annual reports more frequently than those from the services industry. Taken together, we assume that higher investment activity is positively correlated with the tone of managerial comments.

To verify our hypotheses, we first performed comparative financial analysis focused on investment assets. The intensity of using investment assets was calculated as a ratio of long-term (fixed) assets (tangible and intangible) and total assets. This ratio indicates both percentage weight of investment assets and the investment policy of the management. To better evaluate the development of investment activity, we also measured the change in the investment ratios. Decreasing values of these indicators indicate either disinvestment policy or a more intensive use of fixed assets and cost-effective renewal investment. The growing values of these indicators are, in contrast, taken positively, although related to long-term tied funds.

To evaluate the source of funds, we calculated the ratio of equity to total assets. From the investment point of view, this leverage ratio is considered positive in the range $0.4 - 0.7$ (a higher share of equity promotes firm access to funds), but it is industry sensitive. The firms in this study showed average values between 0.5 and 0.6 in all monitored years. Another indicator of growth potential is the fixed assets coverage ratio, which is the ratio of operating cash flow to fixed assets. This indicator is important for manufacturing firms in particular. The value of the indicator should not be lower than 1, but this may be excusable given a high growth of investment.

3   Data Collection and Description

Annual reports (Item 7) for the years 2008-2010 were collected for 680 firms from the U.S. Securities and Exchange Commission EDGAR System. The firms were selected according to the above mentioned criteria. Given the strong dependency of investment indicators on industry classification, the data set was divided into manufacturing (430 firms) and services (250 firms). These industries indeed differed in all measured investment indicators (Table 1). Financial stability (Z-score) was higher for manufacturing firms, but both industries showed a positive trend (despite the financial crisis which is likely to become evident with certain time lag). As expected, the ratio of fixed assets to total assets was higher for manufacturing firms. The higher share of equity to total capital of manufacturing firms can be explained as a result of both depreciation policy and efforts to accumulate profit for further development. Worsening economic conditions, on the other hand, caused a decrease in cash flow for the manufacturing industry. The services industry also showed a higher growth of investments between 2008 and 2010, suggesting a higher investment activity in this period.

Further, we performed tokenization and lemmatization of the documents to obtain a set of term candidates which were subsequently compared with positive and negative word categorizations. In case of the positive categories, negation was detected to give the category accuracy. For comparative purposes, we used two state-of-the-art finance-specific word categorizations developed by [6] and [4]. The authors in [6] examined the relation between the frequency of word categories and the stock market reaction (abnormal market returns) to press releases about earnings. They explain this result by prospect theory, which argues that positive language in financial reports influences investors' thinking towards increases relative to reference points. The latter study shows that general word lists misclassify common words in financial text [4]. The lists with the most frequent words from the word categories together with their $tf.idf$ weights are presented in Table 2 (averages across 2008-2010).

Table 1. Mean values of investment indicators. FA are fixed assets, TA are total assets, E is equity, CF is cash flow, and growth denotes change between the years 2008 and 2010.

	manufacturing			services		
	2008	2009	2010	2008	2009	2010
Z-score	3.29	3.68	3.83	2.57	3.01	3.28
FA/TA	.257	.273	.273	.217	.226	.228
E/TA	.596	.544	.567	.580	.517	.542
CF/FA	.703	.846	.517	1.14	0.75	1.00
FA/TA growth	-	-	.081	-	-	.177
FA growth	-	-	.077	-	-	.169

In contrast, our word lists cover both single words (unigrams) and multi-words (N-grams, where $N = 1,2,3$) (Table 3). The multi-words lists were created in two steps. First, we used a statistical relevance scheme (χ^2 statistic) to test whether the variation of the $tf.idf$ frequency of the given terms is statistically significant (at $p=0.05$) between firms with Z-score$\geq$1.8 (included in further experiments, see Section 2) and firms with Z-score$<$1.8. Second, we used expert knowledge to categorize the (multi)-words into positive and negative financial connotation (we discarded those with unclear tone). In total, our word lists included 185 positive and 76 negative (multi)-words from the financial investment domain. Table 3 shows that the crucial investment words (investment and

Table 2. Top 10 most frequent words (*tf.idf* frequencies) from [6] and [4].

	Henry [6]		Loughran and McDonald [4]	
	positive	negative	positive	negative
1	deliver (0.74)	weaken (0.61)	allianc (0.72)	downgrad (0.77)
2	grew (0.63)	threat (0.53)	innov (0.69)	deficit (0.74)
3	solid (0.62)	drop (0.53)	leadership (0.62)	nonperform (0.73)
4	reward (0.61)	worsen (0.52)	stabl (0.62)	abandon (0.73)
5	leader (0.61)	depress (0.50)	reward (0.62)	hazard (0.72)
6	certanti (0.58)	difficulti (0.50)	resolv (0.61)	unfund (0.69)
7	stronger (0.55)	lowest (0.46)	collabor (0.61)	defect (0.69)
8	accomplish (0.52)	unfavor (0.45)	superior (0.59)	divest (0.69)
9	strengthen (0.52)	fall (0.44)	satisfact (0.59)	crisi (0.68)
10	succeed (0.50)	fell (0.41)	attain (0.59)	refinanc (0.68)

Table 3. Top 15 most frequent (multi)-words (*tf.idf* frequencies) in our word lists.

	positive	negative
1	gross profit (1.14)	wast (0.77)
2	investment portfolio (0.86)	violat (0.56)
3	innov (0.67)	lost (0.55)
4	rate of return (0.66)	breach (0.53)
5	investment policy (0.52)	fall (0.44)
6	reinvest (0.51)	losses on securities (0.44)
7	invested capital (0.49)	abus (0.43)
8	dividends per share (0.44)	investment risk (0.34)
9	paid dividends (0.41)	forfeit (0.31)
10	advantag (0.41)	accrued dividends (0.29)
11	highly liquid investments (0.39)	margin (0.17)
12	dividend yield (0.39)	uneven (0.17)
13	long-term investments (0.38)	trespass (0.17)
14	expans (0.38)	cause losses (0.12)
15	new investments (0.28)	cumulative dividend (0.10)

dividend) appear in both lists at the same time. It is only the collocation in the context that enables a positive or negative tone to be assigned.

4   Relationship Between Word Lists and Investment Indicators

Despite the differences presented in the previous section, Table 4 shows that our word lists are strongly correlated with the compared dictionaries. Interestingly, positive correlations were also obtained between positive and negative word lists. This may be due to the effort of management to balance the negative (positive)

Table 4. Pearson's correlation coefficients between word lists in 2010. All correlation coefficients are statistically significant at $p=0.05$, LM is the word list from [4], H is the word list from [6], and INV is the investor's multi-word list developed here.

	LM-pos	LM-neg	H-pos	H-neg	INV-pos
LM-neg	.512				
H-pos	.663	.377			
H-neg	.462	.493	.429		
INV-pos	.482	.457	.441	.395	
INV-neg	.281	.588	.287	.386	.354

Table 5. Pearson's correl. coefficients between word lists and investment indicators from 2010. Statistically significant correlation coefficients at $p=0.05$ are in bold.

	Z-score		FA/TA		E/TA	
	manuf.	serv.	manuf.	serv.	manuf.	serv.
LM-pos-2010	**.146**	**.157**	.007	.125	-.008	**.153**
LM-pos-2009	.070	**.207**	.022	-.007	-.053	**.217**
LM-pos-2008	.108	**.191**	-.058	-.048	-.031	**.234**
LM-neg-2010	-.072	**.162**	**.491**	**.187**	-.090	.127
LM-neg-2009	-.088	**.178**	**.429**	**.171**	-.116	.105
LM-neg-2008	-.101	**.210**	**.379**	.111	**-.129**	**.148**
H-pos-2010	**.219**	**.179**	-.039	-.068	.105	**.186**
H-pos-2009	**.196**	**.195**	-.067	**-.167**	.064	**.169**
H-pos-2008	.046	**.214**	-.040	**-.257**	-.076	**.205**
H-neg-2010	.014	.020	.117	.103	-.073	.092
H-neg-2009	.002	-.042	.116	.058	-.048	-.021
H-neg-2008	-.016	-.090	**.164**	-.034	-.085	-.021
INV-pos-2010	**.166**	**.245**	**-.135**	**-.128**	.024	**.206**
INV-pos-2009	**.154**	**.201**	-.100	**-.178**	.040	**.190**
INV-pos-2008	**.158**	**.196**	**-.136**	-.088	.018	**.242**
INV-neg-2010	**-.130**	**-.166**	**.330**	**.168**	**-.148**	-.042
INV-neg-2009	-.085	**-.212**	**.191**	**.232**	**-.152**	-.056
INV-neg-2008	-.061	**-.161**	**.190**	Ò78	-.093	-.044

tone by the opposite one in order to achieve the objectivity impression, which may lead to more positive perceptions with stakeholders. However, our word lists were less correlated between themselves when compared with the other dictionaries. This suggests that multi-words better captured the tone in the context of collocated investment terms.

Tables 5 and 6 show the correlation coefficients between word lists and investment indicators. Our dictionary provides promising results for the Z-score in particular. This is probably due to the fact that the chosen investment ratios have lower explanatory power in the year-on-year evaluation. In contrast, the

Z-score model is assumed to provide a high predicative accuracy for up to three years. As expected, the correlations were stronger for years near to investment indicators' values (2010). What is also striking are the significantly negative correlations between INV-neg and Z-score, suggesting that the negative tone was also important in predicting financial condition. The higher share of investment (FA/TA) was associated with a more negative tone, in particular for the manufacturing industry. This result suggests that firms from the services industry better support their investment achievements with verbal comments. This was true for all compared dictionaries. Therefore, the assumption that manufacturing firms accompany their investment results and intentions with corresponding sentiment was not confirmed. However, the comparison of the dictionaries shows that the investment indicators correlate with the tone of annual reports obtained by all dictionaries. As expected, both higher shares of equity and fixed assets coverage were accompanied by more positive tones of annual reports. The finding that investment growth was not taken positively in managerial comments was also supported by positive correlations between FA/TA and FA growth and negative tone. Another explanation for this may be the concerns about the investment risks in worsening economic conditions and decreasing rates of return of investment in the whole U.S. economy.

Table 6. Pearson's correl. coefficients between word lists and investment indicators from 2010. Statistically significant correlation coefficients at $p=0.05$ are in bold.

	CF/TA		FA/TA growth		FA growth	
	manuf.	serv.	manuf.	serv.	manuf.	serv.
LM-pos-2010	.016	.115	-.072	-.023	.111	-.015
LM-pos-2009	-.033	**.166**	-.031	.027	.058	.058
LM-pos-2008	.014	.073	-.044	-.023	.043	.022
LM-neg-2010	**-.162**	.107	.116	**.165**	**.329**	.117
LM-neg-2009	-.125	**.160**	.107	**.180**	**.229**	**.139**
LM-neg-2008	-.118	.068	.078	**.158**	**.171**	.089
H-pos-2010	.018	.088	-.108	.041	.058	.040
H-pos-2009	-.004	.130	-.100	-.018	.032	.051
H-pos-2008	-.036	**.186**	-.011	-.066	.019	-.058
H-neg-2010	-.102	-.014	-.045	-.073	.088	.028
H-neg-2009	**-.148**	.081	-.026	-.044	-.030	.008
H-neg-2008	**-.160**	.108	-.024	.026	-.047	.027
INV-pos-2010	.043	**.195**	**-.168**	.016	-.086	**.161**
INV-pos-2009	.043	**.188**	-.125	-.003	-.043	**.143**
INV-pos-2008	.032	.077	-.072	.031	-.071	**.216**
INV-neg-2010	**-.240**	**.136**	.129	-.043	.046	-.024
INV-neg-2009	**-.189**	**.210**	**.156**	-.050	.028	-.016
INV-neg-2008	**-.178**	**.181**	.123	.116	.013	.021

5 Conclusion

To sum up, our study has argued that positive and negative word categories extracted from corporate annual reports provide significant information normally presented in investment indicators. In this paper we have investigated the degree of positively and negatively colored words when evaluating corporate investments in manufacturing firms and firms providing services. The evidence from this study suggests that manufacturing firms insufficiently present their investment activities. It has to be noted that the word categorizations employed in this study were significantly correlated. It was also shown that the dictionary developed for investment in decision-making support helps to evaluate better the financial results contained in annual reports. Compared with previous studies based solely on single words, the investor's multi-word categorization may be more accurate in predicting future investment activity, in particular for those indicators related to long-term investment.

However, the present study was focused on the annual reports of U.S. firms and a rather specific time period, which casts doubt upon the transferability of its findings to other countries and time periods. Moreover, the applied *tf.idf* weighting scheme strongly affected the obtained results. As a result, several words with the highest weights determined the overall sentiment. Therefore, we propose that further research should be undertaken considering alternative weighting schemes.

Acknowledgments. This work was supported by the scientific research project of the Czech Sciences Foundation Grant No: 13-10331S.

References

1. Bowen, R.M., Davis, A.K., Matsumoto, D.A.: Emphasis on Pro Forma versus GAAP Earnings in Quarterly Press Releases: Determinants, SEC Intervention, and Market Reactions. The Accounting Review **80**(4), 1011–1038 (2005)
2. Bushman, R.M., Smith, A.J.: Financial Accounting Information and Corporate Governance. Journal of Accounting and Economics **32**(1), 237–333 (2001)
3. Merkl-Davies, D.M., Brennan, N.M., McLeay, S.J.: Impression Management and Retrospective Sense-making in Corporate Narratives: A Social Psychology Perspective. Accounting, Auditing & Accountability Journal **24**(3), 315–344 (2011)
4. Loughran, T., McDonald, B.: When is a Liability not a Liability? Textual Analysis, Dictionaries, and 10-Ks. The Journal of Finance **66**(1), 35–65 (2011)
5. Li, F.: The Information Content of Forward-Looking Statements in Corporate Filings - A Naïve Bayesian Machine Learning Approach. Journal of Accounting Research **48**(5), 1049–1102 (2010)
6. Henry, E.: Are Investors Influenced by How Earnings Press Releases are Written? Journal of Business Communication **45**(4), 363–407 (2008)
7. Price, S.M., Doran, J.S., Peterson, D.R., Bliss, B.A.: Earnings Conference Calls and Stock Returns: The Incremental Informativeness of Textual Tone. Journal of Banking & Finance **36**(4), 992–1011 (2012)

8. Huang, A.H., Zang, A.Y., Zheng, R.: Evidence on the Information Content of Text in Analyst Reports. The Accounting Review **89**(6), 2151–2180 (2014)
9. Garcia, D.: Sentiment During Recessions. The Journal of Finance **68**(3), 1267–1300 (2013)
10. Twedt, B., Rees, L.: Reading between the Lines: An Empirical Examination of Qualitative Attributes of Financial Analysts' Reports. Journal of Accounting and Public Policy **31**(1), 1–21 (2012)
11. Liu, B., McConnell, J.J.: The Role of the Media in Corporate Governance: Do the Mmedia Influence Managers' Capital Allocation Decisions? Journal of Financial Economics **110**(1), 1–17 (2013)
12. Hájek, P., Olej, V.: Evaluating sentiment in annual reports for financial distress prediction using neural networks and support vector machines. In: Iliadis, L., Papadopoulos, H., Jayne, C. (eds.) EANN 2013, Part II. CCIS, vol. 384, pp. 1–10. Springer, Heidelberg (2013)
13. Hájek, P., Olej, V.: Predicting firms' credit ratings using ensembles of artificial immune systems and machine learning – an over-sampling approach. In: Iliadis, L., Maglogiannis, I., Papadopoulos, H. (eds.) AIAI 2014. IFIP AICT, vol. 436, pp. 29–38. Springer, Heidelberg (2014)
14. Hajek, P., Olej, V., Myskova, R.: Forecasting Corporate Financial Performance using Sentiment in Annual Reports for Stakeholders' Decision-Making. Technological and Economic Development of Economy **20**(4), 721–738 (2014)
15. Kearney, C., Liu, S.: Textual Sentiment in Finance: A Survey of Methods and Models. International Review of Financial Analysis **33**, 171–185 (2014)
16. Henry, E., Leone, A.J.: Measuring Qualitative Information in Capital Markets Research (2009). SSRN: http://ssrn.com/abstract=1470807
17. Šúbertová, E., Kinčáková, M.: Podpora Podnikania pre Male a Stredné Podniky. Bratislava, Ekonóm (in Slovak) (2014)
18. Altman, E.I.: Predicting Financial Distress of Companies: Revisiting the Z-Score and ZETA® Models. Stern School of Business, New York University (2000)

Topic Classifier for Customer Service Dialog Systems

Manex Serras[1](✉), Naiara Perez[1], M. Inés Torres[2](✉), Arantza Del Pozo[1],
and Raquel Justo[2]

[1] HSLT Department, Vicomtech-IK4 Research Center, Mikeletegi 57,
San Sebastian, Spain
{mserras,nperez,adelpozo}@vicomtech.org
http://www.vicomtech.org

[2] Speech Interactive Research Group, Universidad del País Vasco UPV/EHU, Depto.
Electricidad y Electrónica, Fac. Ciencia y Tecnología, Sarriena s/n, Bilbao, Spain
{manes.torres,raquel.justo}@ehu.es
http://www.ehu.eus/en/web/speech-interactive/

Abstract. Using dialog systems to automatize customer services is
becoming a common practice in many business fields. These dialog systems are often required to relate the users' issues with a department of
the company, which is especially hard when each department covers a
wide range of topics. This paper proposes an entropy-based classifier to
support the dialog system's dialog manager in the decision-making. As
the classifier is implemented in a feedback-available scenario, an extra
class has been inserted with which uncertain cases are labeled, in order
to retrieve more information from the user. This way, robust decisions
are always ensured. The classifier's input is the sequence of semantic
units decoded from the user turn, extracted from technical records in
this paper. This allows the designers to introduce domain specific knowledge and reduce the classifier's workload. Experiments show that the
classifier achieves a high precision, slightly improving some SVM and
Bayes classifiers.

Keywords: Topic classification · Dialogue systems · Semantic
grammars · Entropy-based classification

1 Introduction

Automation of customer support is an increasingly convenient option for busi-
nesses that have to invest a lot of resources in this service. Dialog Systems (DS)
are often used for this purpose, as they can provide a natural user-machine interaction [1,2]. This paper stems from a problem encountered while designing a DS
for a local information technology company's technical support service, namely,
that the topic of the customer's problem must be identified with a specific department of the company, each covering a wide-range of topics. The solution given
is a DS-embedded topic classifier. Introducing classifiers in DSs has proved successful not only in call-routing [3], but in many other tasks as well, such as

© Springer International Publishing Switzerland 2015
P. Král and V. Matoušek (Eds.): TSD 2015, LNAI 9302, pp. 140–148, 2015.
DOI: 10.1007/978-3-319-24033-6_16

user intention detection [4–6]. Although a lot of research has been done recently on statistical classification and DS management [7–10], the nature of the corpus available in our project does not allow implementing these methods. When facing a topic classification issue such as that presented in this paper, some commonly used approaches are tf-idf and/or latent Dirichlet allocation [11], but the need of using methods that go beyond words and deal with concepts has long been manifested [12].

The topic classifier proposed here is based on Semantic Units (SUs) decoded from the user turn with a semantic parser, and works on entropy measurements of certainty, yielding a classification only when enough information has been retrieved from the user. Such behavior is possible because it is embedded in a DS. Classification is carried out as follows:

1. The received user input is processed using a semantic parser. Our current implementation employs the CMU-Phoenix parser [13]. The SUs extracted by the parser are referred to as Observed SUs (OSUs), and conform the classifier's input, which is forced to base its decision on key-concepts.
2. The classifier finds the combination of the OSUs that minimizes the entropy of the class set, composed by the different departments of the company. In order to allow robust decisions, an extra class named No-Understanding (NU) has been introduced for those cases where the entropy is too high, so the DS knows it must continue retrieving information from the user.

Section 2 describes the specific task for which the classifier has been designed and the corpus used for its development. Section 3 is concerned with the generation and description of the semantic grammars. Section 4 explains how the entropy of the OSUs has been modeled, and section 5 shows the algorithm of classification. In section 6 the proposed classifier is compared with SVMs and Bayes classifiers. Finally, section 7 presents the conclusions reached and suggests some future guidelines.

2 The Task and Corpus

The task consists in classifying the problem of the customer into one of the departments of the company. Preliminarily, two departments have been chosen: Finance and Human Resources. These are class 0 and class I of the classifier, respectively. 13393 technical records of issues consulted compose the knowledge source. They include e-mails sent by customers explaining their problems, and notes taken by the technicians that attended to them. The domain specific language contains many shortened forms and numeric expressions that vary greatly in their phrasing.

Table 1 shows quantitative data of the corpus. The vocabulary consists of 14011 words. 8130 of them (58.03%) can be found in class 0, and 9867 (70.42%) in class I. 71.55% of the vocabulary is class specific, the remaining 28.45% appears in both. Records tend to be shorter in class 0, although there is a great variability in length overall.

Table 1. A quantitative description of the corpus

	Total	Class 0	Class I
Records	13393	6430	6963
Running Words	348262	135441	212821
Vocabulary Size	14011	8130	9867
Class Specific Words	-	4144	5881
Record Length	27.03 ±849.52	22.07 ±708.79	31.60 ±935.26

3 Semantic Parsing as a Previous Step to Classification

This work is based on the assumption that different words and expressions
can provide equal information within a given domain, provided they have the
same interpretation, i.e. those that do correspond to surface representations or
phrasings of the same SU in a semantic grammar. Thus, different phrasings of
shortened forms and numeric expressions are clustered together in one SU. For
instance: 'one nine two', 'one ninety two' and 'a hundred ninety two' belong to
the SU [192]. Inflected forms of verbs, nouns and adjectives are treated the
same way. This procedure brings two main advantages: a) it reduces the workload of the classifier considerably; and, b) it prevents the classifier from modeling
useless data, facilitating the generalization of the domain.

In order to measure the impact of the approach above, we have compared the
performance of the classifier both with a manually crafted Semantic Grammar
(SG) and an automatically generated Word-level Grammar (WG) for the task
domain. The following subsections explain the procedures followed to create each
type of grammar and provide brief quantitative descriptions of each one.

3.1 The Semantic Grammar

Two lists have been used as guidance for the generation of the SG: word frequency lists and lists of class specific words. Based on the observation of the
corpus and these lists, a set of 557 key-concepts of the domain has been derived
manually by a linguist. This is the SG's set of SUs. Then, different phrasings have
been grouped in each SU: those found in the corpus, and others that potential
users might employ. Overall, 554 of the SUs of the grammar have been observed,
of which 41.70% are class specific OSUs (see Table 2). 584 records are null, i.e.,
they have zero OSUs.

3.2 Word-Level Grammars

WGs have been generated automatically by a program developed specifically
for this task. The program does not cluster words according to their semantic
proximity; hence, WGS do not have SUs but words, as many as words are in
the vocabulary of the given corpus after discarding digits, single characters, and

stop-words, each word containing itself as unique surface representation. Two types of WGs have been developed: a) a Closed WG (CWG) made out of the whole corpus, in order to examine what is modeled, and what is not; and b) a set of 10 Open WGs (OWG) developed with 85% of the corpus in order to be able to measure the generalizability of the classifier trained with a WG in a 10-fold cross validation scenario.

The CWG has 11288 different words, 20 times the SUs in the SG. 2270 words are typographic errors or useless combinations of digits and characters, at least 639 are proper names and place names, and 245 are words in languages other than Spanish. This amounts for the 28.38% of the whole grammar. Consequently, a lot of noise is modeled when training the classifier with WGs, which could lead to overfitting and unreliable classification. Table 2 shows the results of parsing the corpus the CWG. 11110 words have been observed, of which 71.65% are class specific words.

Table 2. Results of parsing the corpus with the SG and the CWG

	Parsing with the SG			Parsing with the CWG		
	Total	Class 0	Class I	Total	Class 0	Class I
Null Records	584	301	283	17	12	5
OSUs / Words	554	441	436	11110	6410	7850
Class Specific OSUs / Words	-	118	113	-	3260	4700
OSUs / Words per Record	5.01	4.58	5.40	10.48	9.41	11.47

The OWGs have an average of 10223 words. Table 3 shows the average results of parsing both the training and the test sets with each of the OWGs. Notice that the amount of null records is much lower with WGs than with the SG. This is hardly surprising, since, as has been explained, WGs contain almost every word of the corpus from which they are created, and the SGs only contain those relevant for classification.

Table 3. Average results of parsing the training dataset and the testing dataset with the OWGs

	Parse of the Training Set			Parse of the Testing Set		
	Total	Class 0	Class I	Total	Class 0	Class I
Records	11385	5466	5919	2008	964	1044
Null Records	14	10	4	4	3	1
Words Observed	10223	5876	7207	3869	1762	2106
Class Specific Words	-	3016	4347	-	880	1223
Words per Record	10.50	9.41	11.50	9.88	8.92	10.77

4   Entropy Modeling

The process explained in this section applies indifferently to WGs and the SG, so both words and OSUs will be referred to as concepts.

Given a dataset S with a total of n observed concepts and k classes, $C = \{c_1, c_2, \cdots, c_n\}$ is defined as the set of all our concepts and $\Omega = \{\omega_1, \cdots, \omega_k\}$ as our class set.

Being S our training set, each record of our task s_i $i = 1, \cdots, |S|$ is represented as a vector of $c \in C$ $s_i = (c_{i_1}, c_{i_2}, ..., c_{i_{|s_i|}})$ where c_{i_j} $j = 1, \cdots, |s_i|$ are the observed concepts in the record s_i, and $S = (s_1, s_2, \cdots, s_{|S|})$. This representation is done under the assumption that the information obtained by each concept is not modified by its frequency in the sample, so repetitions are not taken into account. Each $s_i \in S$, has its corresponding class $\omega \in \Omega$.

The entropy of each $c_i \in C$ is calculated according to our class set Ω. Previously, we need the conditional probability of the concept given the class. Then, for each $\omega_j \in \Omega$ and $c_i \in C$ we estimate:

$$\hat{P}(c_i|\omega_j) = \frac{N(c_i, \omega_j)}{\sum_{s=1}^{k} N(c_i, \omega_s)} j = 1, \ldots, k$$

Where $\hat{P}(c_i|\omega_j)$ is the estimation of the conditional probability of c_i given the class ω_j. For each c_i we have $\hat{P}_i = [\hat{P}(c_i|\omega_1), \ldots, \hat{P}(c_i|\omega_k)]$ a set of estimated conditional probabilities from which the entropy of the concept c_i regarding to our class set is obtained using the logarithm of base k.

$$H_i = H(c_i) = -\sum_{j=1}^{k} \hat{P}(c_i|\omega_j) log_k(\hat{P}(c_i|\omega_j))$$

In this stage of the algorithm the first thresholding is done. The threshold θ_1 is set manually and the entropy of each concept is compared to it. Those whose entropy is below θ_1 are considered discriminant, while the other concepts are labeled as non-discriminants. We denote C_{disc} the subset of discriminant concepts and C_{nodisc} the subset of non-discriminant ones. It is important to notice that the combination of two non-discriminant concepts can result in a discriminant one. For each $c_i' \in C_{nodisc}$ we estimate:

$$\hat{P}(c_i', c_m'|\omega_j) = \frac{N((c_i', c_m') \in \omega_j)}{\sum_{s=1}^{k} N((c_i', c_m') \in \omega_s)} \quad j = 1, \ldots, k \; ; \; m = i+1, \ldots, |C_{nodisc}|$$

These are the estimated joint probabilities of the concept c_i' with the rest of concepts one by one according to the given class ω_j. For each (c_i', c_m') we have a set of estimated conditional probabilities $\hat{P}'_{i,m} = [\hat{P}(c_i', c_m'|\omega_1), \ldots, \hat{P}(c_i', c_m'|\omega_k)]$. For each probability vector $\hat{P}'_{i,m}$ we can obtain its entropy $H'_{i,m}$, which gives us the entropy of both joint variables according to the class set.

With this, the symmetric Information Gain Matrix (IGM) is defined. This matrix has value one in the position (i, j) if $H'_{i,j}$ is lower than $H(c_i)$, zero otherwise.

So, with the probabilities of the variables, the partition of the set C of discriminants and non-discriminants and the IGM, our model is fixed.

5   Classification

When a new parsed input $\boldsymbol{y} = (c_{\boldsymbol{y}_1}, c_{\boldsymbol{y}_2}, \cdots c_{\boldsymbol{y}_{|\boldsymbol{y}|}})$ is observed, where $c_{\boldsymbol{y}_j}$ $j = 1, \cdots, |\boldsymbol{y}|$ are the concepts observed in $\boldsymbol{y}$. The classification is done in three steps:

- **Step 1:** If any $c_i \in \boldsymbol{y}$ is in the discriminant set C_{disc} and being the value of minimum entropy:

$$H(c_k) = min\{H(c_i) \; : \; c_i \in \boldsymbol{y} \wedge c_i \in C_{disc} \}$$

  The class assigned to the observation $\boldsymbol{y}$ is the $\omega_j \in \Omega$ which satisfies:

$$\hat{\omega} = arg \; max_{\omega_j \in \Omega} \; P(c_k | \omega_j)$$

- **Step 2:** If all $c_i \in \boldsymbol{y}$ belongs to C_{nodisc} a vector of combinations of these concepts is generated with length two to five. The number of possible combinations is:

$$\binom{|\boldsymbol{y}|}{2} + \binom{|\boldsymbol{y}|}{3} + \binom{|\boldsymbol{y}|}{4} + \binom{|\boldsymbol{y}|}{5}$$

  All the combinations which grant no information are ignored using the IGM.

Algorithm 1. Use of the Information Gain Matrix

1: **procedure Pick a new combination of length 2 $(c_{\boldsymbol{y}_i}, c_{\boldsymbol{y}_j})$**
2:    Set $p_1 = c_{\boldsymbol{y}_i}$ and $p_2 = c_{\boldsymbol{y}_j}$.
3:    Check in the IGM intersection of the concepts p_1 and p_2.
4:     If zero then Delete all combinations which contains these concepts and goto 1.
5:     Else:
6:        Calculate the entropy of the combination and store it.
7:        Append the next concept to the combination: $c_{\boldsymbol{y}_k}$.
8:        Set $p_1 = c_{\boldsymbol{y}_j}$ and $p_2 = c_{\boldsymbol{y}_k}$
9:        If the combination length is 5 then store its entropy and goto 1.
10:        goto 3.
11: Repeat until all combinations are evaluated or deleted.

If the combination with lowest entropy is below a manually defined non-understanding threshold θ_2, the initial observation $\boldsymbol{y}$ is classified in the class ω_k which maximizes the conditional probability of the sample given the class.

- **Step 3:** If there is no class assigned in the previous steps, a NU value is returned.

6   Experiments and Results

In order to evaluate the performance of the classification algorithm, Cross Validation tests have been performed both on the SG and OWGs, using 10 folds and leaving the 15% of the records out for test. The behavior of the classifier is determined by in which step each record was classified. As the grammars used have a direct impact in the classification, information about the observed concepts, model generation time and classifier behavior is shown in Table 4 for the SG and OWGs.

Table 4. Comparison between grammars and classifier behavior

| | Concepts | | Generation | % of records classified | | |
	Total	% discriminant	time	Step 1	Step 2	Step 3
SG	554	75.27	t	88.5	3.6	7.9
OWG	10223	76.80	$18.36t$	91.5	5.2	3.3

The observed behavior and the percentage of discriminant concepts are nearly the same, but the OWGs are slower than the SG, as they contain more concepts.

The performance of the proposed classifier has been compared to that of other standard classifiers in Cross Validation test. For this purpose, SVM classifiers with Linear and RBF kernel functions (SVM-L and SVM-RBF) and Naive Bayes classifiers with Gaussian and Multinomial distributions (G-NB and M-NB) have been chosen. The metrics used are precision, accuracy and recall. Both, classifiers and metrics are fully implemented in the Scikit-learn Python library [14]. To perform these comparisons, all the null records have been excluded.

Taking into account that an extra class (the NU class) is only generated in the proposed classifier for those records which grants insufficient information, the initial results (CrossV) are not fully comparable to the rest of the classifiers. To compensate this, the proposed classifier has been forced to classify all the records in a class different than the NU, even if the decision taken is not robust enough (Forced CrossV). The results obtained both with the SG and OWGs are shown in Table 5. The proposed classifier slightly improves the accuracy and recall compared to the other algorithms.

Using OWG the best results are obtained using the M-NB classifier but the overall results are worse than using SG. In addition, when OWGs are used, a vast amount of unnecessary information is taken into account, reducing the quality of our model.

Table 5. Metrics of the proposed classifier vs standard classifiers

	Using the SG					
	CrossV	Forced CrossV	SVM-L	SVM-RBF	G-NB	M-NB
Precision	**0.97** $\pm$ 4·10^{-4}	0.93 $\pm$ 0.001	0.95 $\pm$ 0.001	**0.97** $\pm$ 0.001	0.95 $\pm$ 0.004	0.95 $\pm$ 0.001
Accuracy	0.88 $\pm$ 0.001	**0.94** $\pm$ 0.001	0.93 $\pm$ 0.001	0.86 $\pm$ 0.002	0.93 $\pm$ 0.001	0.94 $\pm$ 0.001
Recall	0.88 $\pm$ 0.001	**0.96** $\pm$ 0.001	0.93 $\pm$ 0.001	0.75 $\pm$ 0.004	0.9 $\pm$ 0.003	0.93 $\pm$ 0.001
F1	0.92 $\pm$ 6·10^{-4}	**0.94** $\pm$ 0.001	**0.94** $\pm$ 0.001	0.84 $\pm$ 0.002	0.93 $\pm$ 0.001	0.93 $\pm$ 0.001
NU	0.09 $\pm$ 0.01	–	–	–	–	–
	Using the CWGs					
	CrossV	Forced CrossV	SVM-L	SVM-RBF	G-NB	M-NB
Precision	0.93 $\pm$ 0.001	0.91 $\pm$ 8·10^{-4}	0.93 $\pm$ 0.001	0.52 $\pm$ 8·10^{-4}	0.91 $\pm$ 0.002	**0.94** $\pm$ 0.001
Accuracy	0.89 $\pm$ 0.001	0.91 $\pm$ 0.001	0.92 $\pm$ 0.001	0.52 $\pm$ 8·10^{-4}	0.72 $\pm$ 0.001	**0.93** $\pm$ 0.001
Recall	0.89 $\pm$ 0.001	0.91 $\pm$ 0.001	0.92 $\pm$ 0.001	1 $\pm$ 0.0	0.51 $\pm$ 0.003	**0.94** $\pm$ 0.001
F1	0.91 $\pm$ 0.001	0.91 $\pm$ 0.001	0.92 $\pm$ 0.001	0.68 $\pm$ 7·10^{-4}	0.66 $\pm$ 0.002	**0.94** $\pm$ 0.001
NU	0.04	–	–	–	–	–

7   Conclusions and Future Work

The proposed algorithm slightly improves the performance of the actual standard classifiers, is better-suited for qualitative variable domains, and has real time decision making capabilities. Moreover, it provides a good measure unit to know whether we need to retrieve more information from the user in feedback-available scenarios. When seeking the highest precision possible, the classifier is not forced to make a decision, at the expense of accuracy and recall; when seeking the highest accuracy and recall, the classifier is forced to make a decision, at the expense of precision.

As for the SG, using key-concepts instead of words as input of the classifier yields improved performance and avoids overfitting. Moreover, the time taken to generate the model of the classifier is far shorter than that needed with WGs. However, writing SGs does require a considerable amount of work. The worthiness of expending this effort should be weighed according to the needs of each situation.

As future work, we aim to test this classification algorithm in higher entropy domains, with different decision making criteria, still based on the entropy regarding to the class set. It would be interesting to test other probabilistic distributions in our modeling, such as the Multinomial Naive Bayes, which is the one that got the best results among the tested standard classifiers.

References

1. Gorin, A.L., Riccardi, G., Wright, J.H.: How may i help you? Speech Commun. **23**(1–2), 113–127 (1997)
2. Chu-Carroll, J., Carpenter, B.: Vector-based natural language call routing. Comput. Linguist. **25**(3), 361–388 (1999)

3. Dinarelli, M., Stepanov, E., Varges, S., Riccardi, G.: The luna spoken dialogue system: Beyond utterance classification. In: 2010 IEEE International Conference on Acoustics Speech and Signal Processing (ICASSP), pp. 5366–5369 (March 2010)
4. Li, Q., Tur, G., Hakkani-Tur, D., Li, X., Paek, T., Gunawardana, A., Quirk, C.: Distributed open-domain conversational understanding framework with domain independent extractors. IEEE Institute of Electrical and Electronics Engineers (December 2014)
5. Inoue, R., Kurosawa, Y., Mera, K., Takezawa, T.: A question-and-answer classification technique for constructing and managing spoken dialog system. In: 2011 International Conference on Speech Database and Assessments (Oriental COCOSDA), pp. 97–101 (October 2011)
6. Mu, Y., Yin, Y.: Task-oriented spoken dialogue system for humanoid robot. In: 2010 International Conference on Multimedia Technology (ICMT), pp. 1–4. IEEE (October 2010)
7. Young, S., Gasic, M., Thomson, B., Williams, J.: Pomdp-based statistical spoken dialog systems: A review. Proceedings of the IEEE **101**(5), 1160–1179 (2013)
8. Gupta, N., Tur, G., Hakkani-Tur, D., Bangalore, S., Riccardi, G., Gilbert, M.: The at&t spoken language understanding system. IEEE Transactions on Audio, Speech, and Language Processing **14**(1), 213–222 (2006)
9. Ghigi, F., Torres, M.I.: Decision making strategies for finite state bi-automaton in dialog management. In: 2015 IEEE International Workshop Series on Spoken Dialogue Systems Technology (IWSDS) (January 2015)
10. Griol, D., Hurtado, L.F., Segarra, E., Sanchis, E.: A statistical approach to spoken dialog systems design and evaluation. Speech Communication **50**(89), 666–682 (2008)
11. Morchid, M., Dufour, R., Bousquet, P.M., Bouallegue, M., Linares, G., De Mori, R.: Improving dialogue classification using a topic space representation and a gaussian classifier based on the decision rule. In: 2014 IEEE International Conference on Acoustics, Speech and Signal Processing (ICASSP), pp. 126–130 (May 2014)
12. Dyer, M.: Connectionist natural language processing: A status report. In: Sun, R., Bookman, L. (eds.) Computational Architectures Integrating Neural and Symbolic Processes. The Springer International Series In Engineering and Computer Science, vol. 292, pp. 389–429. Springer, US (1995)
13. Ward, W.: Extracting information in spontaneous speech. In: The 3rd International Conference on Spoken Language Processing (ICSLP 1994) (September 1994)
14. Pedregosa, F., Varoquaux, G., Gramfort, A., Michel, V., Thirion, B., Grisel, O., Blondel, M., Prettenhofer, P., Weiss, R., Dubourg, V., Vanderplas, J., Passos, A., Cournapeau, D., Brucher, M., Perrot, M., Duchesnay, E.: Scikit-learn: Machine learning in Python. Journal of Machine Learning Research **12**, 2825–2830 (2011)

Defining a Global Adaptive Duration Target Cost for Unit Selection Speech Synthesis

David Guennec$^{(\boxtimes)}$, Jonathan Chevelu, and Damien Lolive

IRISA - University of Rennes 1, Lannion, France
{david.guennec,jonathan.chevelu,damien.lolive}@irisa.fr
https://www-expression.irisa.fr/

Abstract. Unit selection speech synthesis systems generally rely on target and concatenation costs for selecting a best unit sequence. These costs, though often considering contextual features, mainly include local distances that are accumulated afterwards. In this paper, we describe a new duration target cost that takes a whole sequence into account. It aims at selecting a sequence globally good, instead of a very good sequence almost everywhere but having a few local duration cost leaps that are counter-balanced by other units. The problem of weighting this new duration cost with other sub-costs is also investigated. Experiments showed this new measure performed well on sentences featuring duration artefacts, while not deteriorating others.

Keywords: Target cost · Cost function · Neural networks · Corpus-based TTS · Unit selection

1 Introduction

While new Statistical Parametric Speech Synthesis based TTS techniques are currently emerging, like DNN-based TTS, unit selection and HSMM-based synthesis remain the two most influential methods investigated so far, along with hybrid techniques that try to get the best from both worlds. HMM-based parametric approaches, for which HTS [1] is the main system, are quite recent and have been the framework for many academic work in recent years. These methods offer advanced control on the signal and produces extremely intelligible speech but generated voice lacks naturalness. The historical approach, unit selection, is a refinement of concatenative synthesis [2–7].

In the formulation of the unit selection problem, a unit is a list of contiguous segments (from a speech corpus) fitting a portion of the target sequence of phonemes to synthesize. To discriminate units that fit requirements expressed via the target sequence, the usual method [3] is to search a unit graph with a search algorithm, evaluating the context matching degree (target cost) and the risk of creating an artefact if concatenating the unit (concatenation cost) via balanced cost functions. Alternative ways exist though. For instance, one can also achieve unit selection with genetic algorithms and selection and crossover operators are used along with fitness measures [8].

© Springer International Publishing Switzerland 2015
P. Král and V. Matoušek (Eds.): TSD 2015, LNAI 9302, pp. 149–157, 2015.
DOI: 10.1007/978-3-319-24033-6_17

Speech created using unit selection features naturalness and prosodic quality unmatched by other methods, as it basically concatenate speech actually produced by a human being. For this reason, most industrial TTS systems mainly use either pure unit selection approaches or hybrid ones. However, unit selection offers less control than statistical parametric methods, especially over prosody. Moreover, artefacts may appear in the synthesized signal and penalize intelligibility. While obtaining good speech output with neutral voice is (almost) a solved problem with unit selection, getting prosody right for natural and expressiveness is entirely another matter. Prosody modification methods after selection - like TD-PSOLA for adapting duration - are an option, but for now none has been convincing. The possibility of influencing selection to choose units that are the closest to the required prosody remains. A good state of the art for expressive speech synthesis is made in [9]. As phonetic durations are subject to a lot of changes when considering voices with different levels of expressiveness, controlling duration gets particularly important. Lastly, decision trees have been the most widely used method to predict duration, for instance, in systems like HTS, with only a few mentions to using a target duration cost (e.g. in [10]) within unit selection cost function. Recent approaches where DNNs replace HTS decision tree can also be mentioned [11].

In this article, we propose a new way of computing duration target cost, not only based on the assumption that we want to get units as close as possible to a predicted duration. Thus, we try to find the units that stay the closest to requested duration by optimizing the mean duration error with respect to the previous units. Hence, it prevents inadequate units in terms of duration from being selected if other units are available while not forcing a path with homogeneous durations. The main idea is that it is better to have units globally longer or shorter than to have only one or two units with a big duration error in the synthesized speech. The paper is organized as follows. The TTS system used as a basis for experimentation is presented in section 2. Proposed target cost, along with the underlying duration model are presented in section 3. Experimental evaluation on french corpora including both objective assessments of the model and the target cost (4.2) and subjective evaluation by listeners (4.3) are presented in section 4. Conclusions and future work are presented in section 5.

2   The TTS System

In this work, we use the IRISA corpus-based TTS system[12] as a basis for our experiments. The cost function is built following the traditional equation [3]:

$$U^* = \underset{U}{argmin} \, (W_{tc} \sum_{n=1}^{card(U)} w_n C_t(u_n)$$

$$+ \, W_{cc} \sum_{n=2}^{card(U)} v_n C_c(u_{n-1}, u_n)) \tag{1}$$

where U^* is the best unit sequence according to the cost function and u_n the candidate unit trying to match the nth target unit in the candidate sequence U. In this work, the considered unit size is the diphoneme. $C_t(u_n)$ is the target cost and $C_c(u_{n-1}, u_n)$ is the concatenation cost. W_{tc}, W_{cc}, w_n and v_n are weights for adjusting magnitude for all the parameters. Sub-costs are weighted using corresponding mean cost values in the TTS corpus to compensate magnitudes of all sub-costs. Our concatenation cost is composed of amplitude, MFCC and F0 distances. Current target cost is composed of the new duration cost alone. The algorithm accesses the corpus via a set of preselection filters, preventing units that do not match them to be added to the graph. Their purpose is twofold. First, it considerably prunes the graph explored by the unit selection algorithm, making selection process faster. Second, it serves as a set of binary target cost functions relying on the assumption that if a unit doesn't respect the required set of features, it cannot be used for selection. The preselection filters should therefore be seen as part of the cost for a node. In our system, when no corpus unit respects a given set of preselection filters, the set is relaxed (removing features that seem the least helpfull one by one) until units are found. This mechanism ensures finding a path in all cases provided that the corpus has a full covering of diphonemes. The set of preselection filters we use in this work is the following:

1. Unit label (cannot be relaxed).
2. Is the unit a Non Speech Sound (cannot be relaxed)?
3. Is the phone in the last syllable of its sentence?
4. Is the phone in the last syllable of its breath group?
5. Is the current syllable in word end?

3 An Adaptive Duration Target Cost

3.1 Neural Network

Prediction of phoneme duration has a long history in the TTS field. It was first performed by creating expert hand-made rules that were integrated in rules-based (formant synthesis) and concatenation synthesizers. Over last years, decision trees have been the most widely used method to predict duration, for instance, in systems like HTS. In particular, the use of neural networks for phoneme duration prediction starts in the early '90s. A TTS system using a set of ANNs (one for each phoneme) trained on cepstral coefficients can be cited [13]. A TDNN (Time Delay Neural Network) has also proven to be very efficient for predicting duration, though the learning set was small [14]. In following years, major improvements in the technique were obtained mainly by increasing the number of input features and the size of the learning corpus. The advantage of neural networks is that, contrary to decision trees, they do not cluster predicted values (at least when properly trained). When the network faces an unknown set of features, the predicted value is not the assimilated result for the closest feature set, which can result in much better results [15]. Recent work in speech synthesis is now focusing on deep approaches (DNNs, DBNs, DRNs). For duration prediction, we did not think such deep approaches were necessary. Thus, we use a MLP (Multi-Layer Perceptron) with batch gradient descent. Input data is composed of a set of 50 features by phoneme, mainly phonetic and linguistic parameters. We also take into account the contextual information for the two preceding and following phonemes. Thus, the network has a topology of 250 input neurons, 1 rectified linear hidden layer of 512 neurons and one output linear Gaussian neuron (directly predicting durations in ms as other measures like $log\ ms$ were not performing better). These parameters were the best among the different configurations tested.

3.2   Duration Target Cost

The proposed duration target cost aims at influencing selection so that selected units are, on average, at the same distance of the predicted unit durations. Defining the cost that way means we prefer a sequence moderately close to predicted values, but homogeneous in the repartition of the duration distance among units, to a sequence of perfect elements featuring one unit with dramatic cost. The cost for target unit n in the sequence U (see eq. 1) is as follows:

$$D_e = |D_t(u_n) - D(u_n)| \tag{2}$$
$$C_d(u_n) = |\Delta(u_{n-1}) - D_e| \tag{3}$$
$$\Delta(u_n) = \frac{\Delta(u_{n-1}) * (n-1) + D_e}{n} \tag{4}$$

with Δ_{u_n} being the mean distance to predicted duration for previous target units in the sequence (from u_1 to u_n), $D_t(u_n)$ the target duration for unit u_n, $D(u_n)$ the duration of u_n and $C_d(u_n)$ the target duration cost for unit u_n.

Equation (2) computes the local cost between the target duration and the current unit. This cost is then used to compute the duration target cost in equation (3), which takes into account the mean distance to predicted duration for all the previous units. Finally, the mean duration error is updated using equation (4). Thus, the quality of the current unit depends on the quality of previous units. In other words, it means that if u_n is longer (resp. shorter) that desired, the target cost will be low if the previous units are also longer (resp. shorter). This way, we want to keep the consistency between the different units which might be better than inconsistency and perhaps produce a credible speaking rate slow-down or speed-up.

4   Experiments

We have conducted experiments aiming at (i) testing the accuracy of our ANN, (ii) measuring the impact of the new target cost on the unit selection algorithm and (iii) subjectively assessing the improvement in produced speech.

4.1   Corpus Description

Two corpora were used, both as learning sets for ANNs and TTS voices. All are in french language. The first one, *Audiobook*, is extracted from an highly expressive audiobook. The speaker is a male and the mean F0 value for voiced segments is low, at only 87Hz in the corpus. Data was automatically annotated using the process described in [16]. *Audiobook learning* has 353, 691 phonemes and 22, 727 Non Speech Sounds and is 10 hours long. Its diphoneme covering is not full (78%) but all the most commonly used diphonemes are present. The second corpus, named *IVS*, was recorded for TTS purposes within an Interactive Vocal System with a hand-made recording script which aim was to cover all diphonemes present in French and comprises most used words in the telecommunications field. It features a neutral Female voice sampled at 16kHz (lossless encoding, 1 channel) with a mean F0 at 163Hz for voiced segments. The corpus is composed of 7, 662 utterances, 239, 260 phonemes and 20, 424 Non Speech Sounds for 7h05' speech and is manually annotated. Both corpora are managed using the ROOTS toolkit [17].

200 utterances were removed from each corpus to create four 100 utterances corpora: *Audiobook test, Audiobook validation, IVS test* and *IVS validation*. Test corpora are used during the train process on ANNs to control training quality at each epoch. Validation sets were used to verify the efficiency of the model after training.

Finally, we used 100 french sentences (i.e. sometimes featuring more than 1 utterance) corpus extracted from a wide variety of audiobooks, featuring very different styles, many being far from *IVS* and *Audiobook*'s styles. It served as our TTS test corpus. All TTS generations in our experiments used these as target sentences. In the following, it will be referred to as *TTS test* corpus.

4.2   Objective Analysis

Neural Network. The mean RMS error for *IVS* voice is slightly better (RMS=24.24, std=9.07) than for *Audiobook* (RMS=26.58, std=6.61). Pearson scores show that predictions are strongly correlated to real values, and the probability of error on the Pearson score is extremely weak. A detailed analysis on a per phoneme basis shows that the worst phonemes are those having very few representations in the learning corpus, for each voice. For instance, /ɲ/ has only 2 realizations in the *Audiobook* corpus, and only one in *Audiobook validation*. Finally, when looking at real and predicted centroids for each phoneme, most of them are very close, if not identical. Given these results, which we consider as fair, and knowing we do not need extremely accurate predictions as they are solely used to influence selection, these models have been kept as is.

Behavior of the Cost Function. To evaluate the impact of duration cost and its interactions with concatenation costs, we considered all $\{W_{tc}, W_{cc}\}$ couples in the $[0, 100]$ interval with a pace of 10. For each weight configuration, we generated the 100 sentences in our *TTS test* corpus. Sentence (not utterance) based measures were extracted for each configuration. In this section, we will only discuss these measures on *IVS* voice, but exactly the same patterns are observed on *Audiobook* voice. Only small variation in magnitudes are observed between the two voices. It is important to point out that costs presented here are obtained *without* applying W_{tc} and W_{cc} weights. Magnitudes due to these weights have been removed to get raw costs.

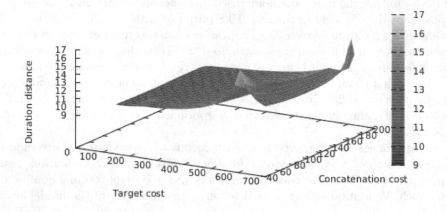

Fig. 1. Duration delta between model predictions and synthesized durations evolution when target and concatenation costs vary. Distance, per phoneme, is given in ms. Data computed using synthesis from *TTS test* corpus.

Figure (1) shows the evolution of the mean delta per phoneme in *ms* between predictions by the network and final produced durations in relation to target and concatenation costs magnitudes. As it can be seen, the general trend is that distance increases when the target cost increases, which shows a good functioning of our target cost. Moreover, when getting the worst target cost, the delta largely increases. An unexpected result is the relation between the delta and concatenation cost when target cost is high which seems to suggest that concatenation cost excludes units with worst duration, improving the delta. When concatenation cost increases again, the delta dramatically increases again too. We can further note that duration delta at high target costs and low concatenation costs, while being good, remains much higher than the delta we get at lower target costs (this time independently of concatenation cost).

This result led us to think it would be worth investigating the behavior of a system where the duration target cost would be activated only on certain conditions, like for high concatenation cost or when confronted to a drastic relaxation of preselection filters.

4.3   Subjective Evaluation

Based on precedent measures, we selected the configuration $\{W_{tc} = 30, W_{cc} = 70\}$ for listening tests. This choice was motivated by the low variability in terms of duration costs when getting over $W_{tc} = 30$ and the fact that concatenation cost alteration at this level is low. The same reasoning led us to $W_{cc} = 70$. In consequence, listening tests were performed using two system configurations: the baseline system, with weight configuration $\{W_{tc} = 0, W_{cc} = 100\}$ which we call *Uncontrolled*; and the configuration incorporating our duration distance, $\{W_{tc} = 30, W_{cc} = 70\}$, called *Controlled*.

We performed two AB tests involving 13 testers for the first and 11 for the second (half of which were experts) on the *Uncontrolled* and *Controlled* systems. Tests follow recommendations in [18]. Three answers were proposed to the listeners: A, B and "Indifferent". Both *Audiobook* and *IVS* voices were mixed in each test.

The first test presented 20 stimuli for each voice, taken randomly in the TTS test set. The testers were asked to assess the rhythm of speech and select the best system. On raw results, systems were getting almost as much votes (43% for *Uncontrolled* and 38% for *Controlled* with overlapping intervals). We spotted extremely different scales of notation among testers, with none seeming to have the same way of performing the test. Thus, no hard conclusion can be derived from this test. Nonetheless, it suggests the two systems are on par. It is important to underline that post-analysis of the stimuli presented for this test showed that very few samples had strong duration incoherences.

An important point is that *IVS* corpus featuring only neutral voice, duration artefacts are less serious and less frequent. On the contrary, *Audiobook*, being very expressive, features much more minor duration issues. Major duration problems are also much more frequent.

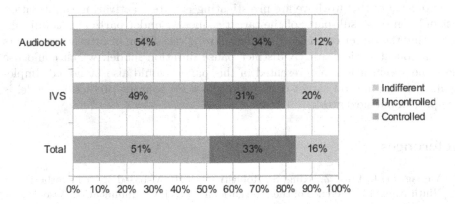

Fig. 2. AB test results. *Uncontrolled* featuring duration artefacts is opposed to *Controlled* system. First and second row are a decomposition of the third one. *Controlled* is clearly preferred.

The second test was focused on sentences having audible duration artefacts. 22 different sentences featuring duration artefacts (of various amplitudes but all being audible) were extracted from *Uncontrolled* synthesis (11 for each voice). They were confronted to their equivalent with *Controlled* system. The testers were asked to say which system has the most natural voice. The testers were also asked to pay particular attention to rhythm (but not exclusively).

Results for this second test are presented on figure 2. First row shows results for *Audiobook* voice only, second for *IVS* only while the third one is the global result. In this test, *Controlled* is strongly preferred by testers, especially for *Audiobook* voice which is normal as it is the voice the most likely to generate artefacts. It was also interesting to see that testers all followed the same trend, placing *Controlled* ahead with different levels of preference. Experts especially had a strong preference for *Controlled* when using expressive voice *Audiobook*, and less for *IVS*.

Given these results, it can be derived that our target costs behaves well in enhancing durations when needed and only when needed, while not deteriorating synthesis on other aspects.

5   Conclusion

In this paper, we presented a new duration target cost for unit selection. This cost aims at selecting the whole unit sequence that best minimizes duration distance with predicted values rather than choosing the sequence containing units that individually minimize a duration distance. This is intended to avoid cases like excellent synthesis penalized by few very bad units. Experiments showed that this new measure performs well on speech samples that feature durations issues, especially on expressive voices. Furthermore, the new measure does not seem to affect synthesized samples that have good durations from the beginning. An extension of this work we are investigating is to test activating the duration cost only on some sub-parts of the target sequence, under particular conditions suggesting the target cost is needed like a strong relaxation of preselection filters or high concatenation cost. A distinct pause duration model, which could use the same specifications as presented in this paper should also be added. Implementing an intonation target cost relying on a F0 contour prediction model is also part of our next work.

References

1. Yamagishi, J., Ling, Z., King, S.: Robustness of HMM-based speech synthesis. In: Ninth Annual Conference of the International Speech Communication Association, pp. 2–5 (2008)
2. Sagisaka, Y.: Speech synthesis by rule using an optimal selection of non-uniform synthesis units. In: Proc. of ICASSP, pp. 679–682. IEEE (1988)
3. Black, A., Taylor, P.: Chatr: a generic speech synthesis system. In: Proc. of Coling, Association for Computational Linguistics (1994)

4. Hunt, A., Black, A.: Unit selection in a concatenative speech synthesis system using a large speech database. In: Proc. of ICASSP, pp. 373–376. IEEE (1996)
5. Taylor, P., Black, A., Caley, R.: The architecture of the festival speech synthesis system. In: Proc. of the ESCA Workshop in Speech Synthesis, pp. 147–151 (1998)
6. Breen, A., Jackson, P.: Non-uniform unit selection and the similarity metric within bts laureate tts system. In: Proc. of the ESCA Workshop on Speech Synthesis, pp. 373–376. Citeseer (1998)
7. Clark, R., Richmond, K., King, S.: Multisyn: Open-domain unit selection for the festival speech synthesis system. Speech Communication, 317–330 (2007)
8. Kumar, R.: A genetic algorithm for unit selection based speech synthesis. In: Eighth International Conference on Spoken Language Processing (2004)
9. Schröder, M.: Expressive Speech Synthesis: Past, Present, and Possible Futures. In: Affective Information Processing, pp. 111–126. Springer, London (2009)
10. Alías, F., Formiga, L., Llorá, X.: Efficient and reliable perceptual weight tuning for unit-selection text-to-speech synthesis based on active interactive genetic algorithms: A proof-of-concept. Speech Communication, 786–800 (May 2011)
11. Hashimoto, K., Oura, K., Nankaku, Y., Tokuda, K.: The effect of neural networks in statistical parametric speech synthesis. In: IEEE International Conference on Acoustics, Speech and Signal Processing, pp. 4455–4459 (2015)
12. Guennec, D., Lolive, D.: Unit selection cost function exploration using an A* based text-to-speech system. In: Sojka, P., Horák, A., Kopeček, I., Pala, K. (eds.) TSD 2014. LNCS, vol. 8655, pp. 432–440. Springer, Heidelberg (2014)
13. Tuerk, C., Robinson, T.: Speech synthesis using artificial neural networks trained on cepstral coefficients. In: Proc. of EUROSPEECH, pp. 4–7 (1993)
14. Karaali, O., Corrigan, G., Gerson, I.: Speech synthesis with neural networks. In: Proc. of World Congress on Neural Networks, pp. 45–50 (1996)
15. Taylor, P.: The target cost formulation in unit selection speech synthesis. In: Proc. of Stress, pp. 2038–2041 (2006)
16. Boeffard, O., Charonnat, L., Le Maguer, S., Lolive, D., Vidal, G.: Towards fully automatic annotation of audio books for tts. In: Proc. of LREC, pp. 975–980 (2012)
17. Chevelu, J., Lecorvé, G., Lolive, D.: Roots: a toolkit for easy, fast and consistent processing of large sequential annotated data collections. In: Proc. of LREC, pp. 619–626 (2014)
18. ITU-T: Itu-t recommendation p. 800: Methods for subjective determination of transmission quality (1996)

Adaptive Speech Synthesis of Albanian Dialects

Michael Pucher[1]([⊠]), Valon Xhafa[2], Agni Dika[2], and Markus Toman[1]

[1] Telecommunications Research Center Vienna, Vienna, Austria
{pucher,toman}@ftw.at
http://www.ftw.at
[2] Department of Computer Engineering, University of Prishtina, Prishtina, Kosova
{valon.xhafa,agni.dika}@uni-pr.edu
http://www.uni-pr.edu/

Abstract. In this paper, we show how adaptive modeling within the statistical parametric speech synthesis framework can be applied to Albanian dialects. We develop speaker dependent voices for the Tosk and Gheg dialect and adapt models for the Gheg dialect from the Tosk models. We show that the adapted Gheg models outperform the speaker dependent Gheg model on an intelligibility and dialect classification task. Furthermore we show that the speaker dependent Tosk model outperforms a formant based synthesizer on an intelligibility, dialect classification and pair-wise comparison task. This formant based synthesizer is the only publicly available synthesizer for Albanian at the moment. We also show that our Gheg and Tosk synthesizers are as intelligible as natural speech. The method where one dialect is modeled through adaptation of a closely related other dialect can be applied to language varieties in general, where the background variety and adapted variety can be chosen based on pragmatic considerations like speaker or data resource availability.

Keywords: Speech synthesis · Albanian · Adaptation · Dialect

1 Introduction

Adaptive parametric HMM-based speech synthesis [1,2] allows for the usage of a background model or average model to improve synthetic voice quality with small amounts of adaptation data. The authors of [3] applied the adaptive framework to the synthesis of Austrian German dialects. In this paper, we use an Albanian Tosk dialect background model to improve an Albanian Gheg dialect adapted voice.

The adaptive approach that works with small amounts of data is especially interesting for languages such as Albanian where not so many language resources are available. Today the only open-source available synthesizer for Albanian is a formant-based synthesizer that is still in a "provisional" development stage [4]. The synthesizers developed for this study are based on open-source components [4,5] and we plan to release open-source synthesizers for Albanian.

© Springer International Publishing Switzerland 2015
P. Král and V. Matoušek (Eds.): TSD 2015, LNAI 9302, pp. 158–164, 2015.
DOI: 10.1007/978-3-319-24033-6_18

The method we use here can be used for any variety pairs of phonetically closely related varieties (dialects, sociolects, and accents). We can use the variety were it is easier to collect a larger amount of data to train a background model. Then we can adapt the variety where it is more difficult to obtain large amounts of data. It can be difficult to obtain data when speakers are difficult to find, or language resources like lexica, grapheme-to-phoneme rules, or texts for recording scripts are not available. In our case we choose the Tosk dialect, which is also the basis for Standard Albanian, as background model since it was easier to obtain the large amount of text data that is needed to select a recording script by solving the respective set-cover problem. Furthermore, we could use a slightly modified grapheme-to-phoneme conversion from an existing synthesizer for Tosk [4].

2   The Albanian Language

The Albanian language belongs to the family of Indo-European languages. In the tree of languages, the Albanian language does not share any descent connection with other member languages of this family and is presented as a separate branch that grows from the root of the tree. This classification is based on phonological, morphological and other features [6].

There are two basic dialect forms of the Albanian language: Gheg and Tosk [7,8]. Furthermore there are also mixtures between Gheg and Tosk dialects, like Arbëresh and Arvanitika [7]. In countries where Albanian is spoken the official language is Standard Albanian, based on the Tosk dialect [9]. It is used in institutions, newspapers, and books. Gheg is spoken in more informal settings. For these reasons we adapted Gheg from Tosk, using the Tosk language in our background model. The Tosk dialect has 7 vowels and 29 consonants. The Gheg dialect uses long and nasal vowels, which are absent in Tosk [7].

Albanian almost has a one-to-one correspondence between letters and phones, which make grapheme-to-phoneme conversion easier, compared to some other languages.

3   Grapheme-to-Phoneme Conversion

Grapheme-to-phoneme (G2P) rules for the Tosk dialect were taken from an existing speech synthesizer [4]. Some errors of the G2P rules had to be corrected for Tosk and additional rules had to be introduced for Gheg as shown in Table 1. In general the G2P problem is relatively simple to solve for Albanian since there is an almost one-to-one mapping between letters and phones.

The additional rules for Tosk and Gheg shown in Table 1 are context independent, i.e. the characters "dh" are replaced everywhere by "D". Table 1 also shows the International Phonetic Alphabet (IPA) symbols in brackets. Due to phonological differences between Tosk and Gheg we need to introduce additional rules for Gheg. The Gheg dialect has the nasal vowels "ê" and "â". These nasal vowels are mapped to "nA" and "nE" in our phone set.

Table 1. Additional G2P rules for Tosk and Gheg (left) and Gheg (right).

c → ts [ts]	ô→ nO
ll → L [l]	û→ nU
dh → D [d]	â → nA [ā]
sh → S [ʃ]	ê → nE [ɛ̃]

4   Recording Script

First we collected a large amount of sentences from books, newspapers and other sources to have a database from which we can select a set of sentences that fulfills specific phonetic criteria. Sentences with proper nouns were removed from the corpus to avoid any problems with irregular pronunciation in the training data. Furthermore, we excluded sentences that are longer than 20 words, which would make the reading during the recordings too difficult. This corpus without sentences containing proper nouns and long sentences consisted of 16041 sentences.

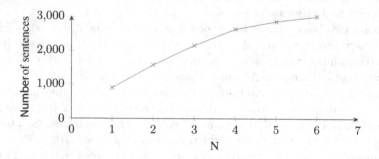

Fig. 1. Selected sentences for different minimum number of diphones for Tosk.

For selecting the recording script from the collected corpus we used the modified G2P rules from Section 3 to phonetically transcribe the selected 16041 sentences. Then we used a greedy algorithm to select the smallest set that contains all diphones N times, where N was in the range of $1 - 6$. Figure 1 shows the number of selected sentences for different number of N for the Tosk dialect. This algorithm always adds the sentences to the set that contains the largest number of new diphones. We used a greedy algorithm for this, since the set-cover problem is known to be NP-complete [10]. Our greedy algorithm will not necessarily find the smallest set of sentences containing all diphones N times.

5   Recording

The recording was done in a semi-professional soundproof recording room. We recorded 3010 sentences from one female Tosk speaker (containing all Tosk diphones 6 times) and 412 sentences from one female Gheg speaker (containing all Gheg diphones 1 time). We recorded more material from the Tosk speaker since we wanted to use Tosk as background language. To have good quality recordings, we recorded a small amount of the corpus every day for three weeks. The microphone used was a vocal Sennheiser microphone with a pop filter. After recording, low frequency noise was removed. The recording of a whole recording session was automatically split into utterance size audio files.

6   Voice Building

We built a speaker dependent model from the data of a Tosk female speaker and a Gheg female speaker. Adapted models for the Gheg speaker were trained through adaptation from the Tosk model. For the speaker dependent and adaptive training we used the training scripts from HTS 2.3-alpha [5]. Clustering questions included phone identity, phonetic, and articulatory features for current, previous two, and following two phones as well as word features. A flat-start model was used for forced alignment of the training data. The overall quality of the adapted voice could be improved by using more speakers in the background/average voice model [1, 2, 11].

7   Evaluation

We evaluated the different synthetic and natural voices through a subjective listening test that consisted of an intelligibility test and a pair-wise comparison to evaluate the quality of the voices.

7.1   Design

For the evaluation we compared the methods for Tosk synthesis and the methods for Gheg synthesis. All used voices are shown in Table 2. For each of the dialects we had one speaker.

10 listeners participated in the evaluation with age between 19 and 23. We had 5 male and 5 female listeners.

Table 2. Voices used in the evaluation.

Dialect / Method	Recorded	Speaker-dependent	Adapted	Formant-based
Tosk	✓	✓		✓
Gheg	✓	✓	✓	

In the first part of the evaluation each listener had to listen to each of the 20 prompts for Tosk and 20 prompts for Gheg (40 samples in total). Listeners were asked to write down what they have heard (intelligibility test) and if they think it was the Tosk or Gheg dialect. Figure 2 shows a screenshot of the application that was used for this part of the evaluation.

In the second part of the evaluation listeners had to listen to pairs of prompts and had to judge which one of the prompts is better in terms of overall quality.

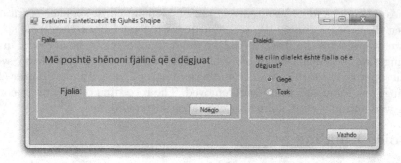

Fig. 2. Application used for evaluation of Word-Error-Rate (WER) and dialect rating.

7.2 Results

Word-Error-Rate (WER): Table 3 shows the Word-Error-Rate results for the different methods that were computed by finding the minimum number of edit operations that are necessary to convert the correct transcript into the transcribed version divided by the total number of words in the correct transcript.

Table 3. Word-Error-Rate (WER) for the different methods in %.

Error / Method	Tosk rec.	Tosk spk. dep.	Tosk formant	Gheg rec.	Gheg spk. dep.	Gheg adapt.
WER	4.7	7.8	23.5	8.2	9.9	11.9

For the Tosk synthesizers we can see that the formant based synthesizer is worse than the HMM-based synthesizer and that there is only a small difference between HMM-based synthesizer and recorded speech. This shows that the HMM-based synthesizer outperforms the formant synthesizer on the intelligibility task.

For the Gheg synthesizers there are only small differences between the speaker dependent and adapted version.

Dialect Rating: Table 4 shows the Dialect Classification Error (DCE) that was computed as the ratio of wrongly classified samples and total samples.

For the Tosk dialect we can see that the formant based synthesizer had the worst performance in terms of modeling the dialect accurately. 16.1% of the samples were wrongly classified as Gheg dialect samples. The speaker dependent HMM-based voice has similar performance to the recorded samples.

For Gheg the Dialect Classification Error (DCE) was in general higher than for Tosk, which might be related to our listeners' competence. Here we can see that the adapted HMM-based voice has the lowest error followed by the recordings and the speaker dependent voice.

Table 4. Dialect classification error (DCE) for the different methods in %.

Error / Method	Tosk rec.	Tosk spk. dep.	Tosk formant	Gheg rec.	Gheg spk. dep.	Gheg adapt.
DCE	4.6	4.4	16.1	14.7	19.4	10.7

Interestingly the speaker-adapted voice is more often found to produce Gheg samples than the original Gheg samples. This result shows that there is a difficulty with classifying the Gheg dialect, which either comes from the language competence of our Gheg speakers or from the Gheg classification competence of our listeners.

Pair-Wise Comparison: Figure 3 shows the results for the pair-wise comparison task. For Tosk the HMM-based speaker dependent voice is significantly better than the formant voice ($p < 0.05$ according to a paired T-test). The recordings are not surprisingly significantly better than both synthetic voices. Thereby we can show that the HMM-based method outperforms the formant-based method also for this task.

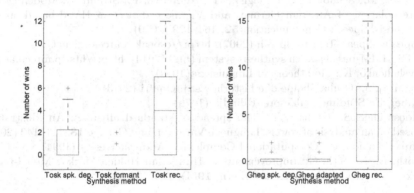

Fig. 3. Results of pair-wise comparison between different methods for Tosk (left) and Gheg (right).

For the Gheg voices we can see no significant differences between the synthetic voices. Only the recorded voice is significantly better in terms of overall quality.

8   Conclusion

We have developed a state-of-the-art HMM-based synthesizer for the Albanian Tosk and Gheg dialects. We showed that our speaker dependent Tosk model outperforms an existing formant-based synthesizer on an intelligibility, dialect classification and pair-wise comparison task. We also saw that the speaker-adapted Gheg model outperformed the Tosk model on a dialect classification and pair-wise comparison task, while there was no performance difference for the pair-wise comparison task. We also showed that the synthesizers are as intelligible as natural speech, which makes them usable in many application scenarios such as spoken dialog systems, web readers, and speech output for blind users.

Acknowledgments. This work was supported by the Austrian Science Fund (FWF): P23821-N23. The Competence Center FTW Forschungszentrum Telekommunikation-Wien GmbH is funded within the program COMET - Competence Centers for Excellent Technologies by BMVIT, BMWA, and the City of Vienna. The COMET program is managed by the FFG.

References

1. Yamagishi, J., Tamura, M., Masuko, T., Tokuda, K., Kobayashi, T.: A training method of average voice model for HMM-based speech synthesis. IEICE Trans. Fundamentals **E86-A**(8), 1956–1963 (2003)
2. Yamagishi, J., Kobayashi, T.: Average-voice-based speech synthesis using HSMM-based speaker adaptation and adaptive training. IEICE Trans. Inf. & Syst. **E90-D**(2), 533–543 (2007)
3. Pucher, M., Schabus, D., Yamagishi, Y., Neubarth, F., Strom, V.: Modeling and interpolation of Austrian German and Viennese dialect in HMM-based speech synthesis. Speech Communication **52**, 164–179 (2010)
4. eSpeak. eSpeak Text-to-Speech (2007). http://espeak.sourceforge.net/
5. HTS. HMM-based speech synthesis system (hts) (2014). http://hts.sp.nitech.ac.jp/
6. Tyshchenko, K.: Metatheory of Linguistics (1999)
7. Demiraj, S.: Gjuha Shqipe dhe historia e saj. Onufri (2013)
8. Çabej, E.: Studime gjuh'esore. Rilindja (1976)
9. Moosmüller, S., Granser, T.: The spread of standard albanian: An illustration based on an analysis of vowels. Language Variation and Change **18**, 121–140 (2006)
10. Papadimitriou, C.: Computational Complexity. Addison Wesley (1994)
11. Tóth, B., Németh, G.: Improvements of Hungarian Hidden Markov Model-based text-to-speech synthesis. Acta Cybern. **19**(4), 715–731 (2010)

Language-Independent Age Estimation from Speech Using Phonological and Phonemic Features

Tino Haderlein[1]([✉]), Catherine Middag[2], Florian Hönig[1], Jean-Pierre Martens[2], Michael Döllinger[3], Anne Schützenberger[3], and Elmar Nöth[1]

[1] Lehrstuhl für Mustererkennung (Informatik 5), Friedrich-Alexander-Universität Erlangen-Nürnberg (FAU), Martensstraße 3, 91058 Erlangen, Germany
Tino.Haderlein@cs.fau.de
http://www5.cs.fau.de

[2] Vakgroep voor Elektronica en Informatiesystemen (ELIS), Universiteit Gent, Sint-Pietersnieuwstraat 41, 9000 Gent, Belgium

[3] Phoniatrische und pädaudiologische Abteilung in der HNO-Klinik, Klinikum der Universität Erlangen-Nürnberg, Bohlenplatz 21, 91054 Erlangen, Germany

Abstract. Language-independent and alignment-free phonological and phonemic features were applied for automatic age estimation based on voice and speech properties. 110 persons (average: 75.7 years) read the German version of the text "The North Wind and the Sun". For comparison with the automatic approach, five listeners estimated the speakers' age perceptually. Support Vector Regression and feature selection were used to compute the best model of aging. This model was found to use the following features: (a) the percentage of voiced frames, (b) eight phonological features, representing vowel height, nasality in consonants, turbulence, and position of the lips, and finally, (c) seven phonemic features. The latter features might be relevant due to altered articulation because of dentures. The mean absolute error between computed and chronological age was 5.2 years (RMSE: 7.0). It was 7.7 years (RMSE: 9.6) for an optimistic trivial estimator and 10.5 years (RMSE: 11.9) for the average listener.

Keywords: Age estimation · Phonological features · Phonemic features · SVR

1 Introduction

In speech science, increasing attention is given to age-dependent characteristics of speech, as life expectancy and the percentage of elderly population are growing fast, especially in Europe and North America. Better understanding of aging effects on speech performance will provide better insight into models of the anatomical, physiological, and linguistic consequences of aging. The accuracy of

© Springer International Publishing Switzerland 2015
P. Král and V. Matoušek (Eds.): TSD 2015, LNAI 9302, pp. 165–173, 2015.
DOI: 10.1007/978-3-319-24033-6_19

a model for vocal aging can be tested by classifying a speaker's age automatically. A person with a large discrepancy between chronological and perceived or computed age should be examined more in detail in order to reveal possible symptoms of beginning diseases. The analysis of healthy speech may provide key contributions to the early diagnosis of neurodegenerative disorders, such as shown for Parkinson's disease [1]. In other scenarios, age-specific recognition systems can be applied, when the user's age has been estimated automatically, for instance by choosing specific speed, volume, or music for system prompts.

The focus of many papers in that field is on the disambiguation of only a few age classes of practical relevance. Our work concentrates on the development of a large feature vector that allows to estimate an adult speaker's age as precisely as possible.

Phonological and phonemic features capture many voice and also speech properties. They were successfully used for language-independent detection of voice quality and speech intelligibility and can even be used to visualize these aspects [2–4]. Hence, we regarded them also suitable for automatic estimation of a person's age from speech.

This paper is organized as follows: Section 2 introduces the speech data used for the experiments, Sect. 3 describes the features computed from the data and the Support Vector Regression for creating the aging model. The results will be discussed in Sect. 4.

2   Test Data and Subjective Evaluation

110 German persons (31 men, 79 women) without voice or speech problems and between 50 and 94 years of age participated in this study. The average age was 75.7 years with a standard deviation of 9.6 years (Fig. 1, Table 2). They were recruited from senior community centers, senior meetings, and assisted living facilities. Persons receiving voice-related medical treatment, in need of skilled nursing care and/or with relevant cognitive limitations (e.g. dementia) were excluded from the study [5]. Each person read the phonetically rich text "Der Nordwind und die Sonne" ("The North Wind and the Sun", [6]), which is frequently used in medical speech evaluation in German-speaking countries. It contains 108 words (71 distinct) with 172 syllables. The data were recorded with a sampling frequency of 16 kHz and 16 bit amplitude resolution. The study respected the World Medical Association (WMA) Declaration of Helsinki on ethical principles for medical research involving human subjects and has been approved by the ethics committee of the University Erlangen-Nürnberg (FAU).

One female and four male raters evaluated the audio data perceptually by assigning the age to each speaker after listening to the respective audio sample. One male rater was speaking German as a second language, the others were native German speakers. They did not know about the distribution and the range of age in the data in advance.

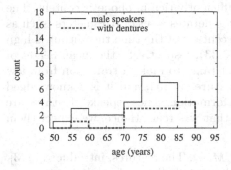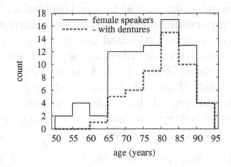

Fig. 1. Distribution of age and dentures among the 31 male and 79 female speakers; for one woman, no information about the dental status was available.

3   Features Computed from the Speech Data

Since it is expected that speech of elderly persons is not only affected by voice aging but also by changes in articulation, we use phonological and phonemic features to capture these effects. They were designed for Flemish, but in recent studies [2,4] they have been also successfully used for German. The pre-processing stage returns 12 Mel-frequency cepstral coefficients (MFCCs) and an energy value for each 25 ms speech frame (frame shift: 10 ms). From this spectro-temporal representation of the acoustic signal, speaker features are extracted which constitute a compact characterization of the speech of the tested person. Based on the stream of MFCCs, two text-independent feature extraction methods, focusing on phonological and phonemic aspects, have been explored.

Alignment-free phonological features (ALF-PLFs): First described in [7], these features follow from a tracking of the temporal evolutions of the individual outputs of an artificial neural network that was trained (see [3,7] for more details) to generate 14 phonological properties per frame. These properties describe:

- vocal source: voicing
- manner of articulation: silence, consonant-nasality, vowel-nasality, turbulence (referring to fricative and plosive sounds)
- place of consonant articulation: labial, labio-dental, alveolar, velar, glottal, palatal
- vowel features: vowel height, vowel place, vowel rounding

Every phonological property is analyzed by two sub-networks. One of them determines whether the property is relevant at a given time (e.g. it is not relevant to investigate vowel place during utterance of a consonant); the other one determines whether the characteristic (e.g. "labial") is actually present or not. The hypothesis is that temporal fluctuations in the network outputs can reveal articulatory deficiencies, regardless of the exact phonetic content of the text that was read, at least as long as this text is sufficiently rich in phonetic content. The temporal analysis of each network output generates a set of parameters, such as the mean and standard deviation, the percentage of the time the output is high (above 0.66), intermediate or low (below 0.33), respectively, the mean height of the peaks (maxima), and the mean time it takes to make a transition from low to high. The overall number of output features is 504, and it is acknowledged that several of them may carry similar information. These speaker features are computed without knowledge of the text that was read. Hence, we expect them to be text-independent.

Alignment-free phonemic features (ALF-PMFs): The features, introduced in [3], were originally based on the hypothesis that intelligibility degradation is corre-lated with problems in realizing a certain *combination* of phonological classes that is needed for the production of a certain phone. Therefore, the ALF-PMFs follow from a plain analysis of posterior phone probabilities. Considering all frames for which the maximal probability is assigned to a particular phone, one computes the mean and standard deviation of that probability, and the mean of the peaks (maxima) and the valleys (minima) found in its temporal evolu-tion. In addition, the percentage of the time a frame is assigned to the phone, and the mean probability of this phone over all frames are computed. Clearly, these features are computed without any knowledge of the text that was read and can therefore be expected to be text-independent. There are 495 different ALF-PMFs.

All the neural networks for the computation of ALF-PLFs and ALF-PMFs had been trained with Flemish speech data and were now used with German test data. Their general independence of the language had been shown before [2,4].

Prosodic features (AMPEX): They originate from a holistic analysis of the frame-level volume, fundamental frequency, and voicing evidences. This analysis can be conducted on arbitrary speech, irrespective of the language that is spoken. The frame-level prosodic features are converted into 8 AMPEX features [8]. The voicing evidence and the signal loudness (see [9,10]) are used to label the frames as voiced/unvoiced and as speech/silence, and to locate pauses, defined as intervals of more than 200 ms long. Based on these classifications, the AMPEX feature extractor computes the features listed in Table 1. They can be grouped into voicing-related parameters (e.g. the percentage of speech frames classified as voiced) and F_0-related features (e.g. average jitter of the fundamental frequency F_0 in voiced frames). They were computed for the whole length of each speech sample. In earlier studies, supplementing phonological features with these F_0- and-voicing related speaker characteristics enhanced intelligibility prediction [3]. We assumed that they may also support automatic age estimation.

Support Vector Regression (SVR): In order to determine the best subset of all phonological, phonemic, and prosodic features to model the chronological age, Support Vector Regression (SVR, [11]) was used. The underlying SVM used a linear kernel. The complexity constant C for the SVR was set to 0.01 after a short series of experiments with heuristic changes to C by powers of 10. Each

Table 1. The AMPEX features (for details, see [10])

feature	description
PVF	percentage of all frames in the recording that were labeled voiced
PVS	percentage of speech frames that were labeled voiced
AVE	average voicing evidence in voiced frames
PVFU	percentage of voiced frames with an unreliable F_0
Jit	average F_0-jitter in voiced frames
Jc	average F_0-jitter in voiced frames with a reliable F_0
VL90	90th percentile (in seconds) of the voiced fragment durations
Tmax	duration (in seconds) of the longest speech fragment (not interrupted by a pause)

training example for the regression consisted of a set of features (the inputs) and a chronological age (the target output). The sequential minimal optimization algorithm (SMO, [11]) of the Weka toolbox [12] was applied in a 10-fold cross-validation manner.

For the selection of the attributes, the Greedy Stepwise algorithm was applied. The standard settings were not changed. All input features were standardized (mean value: $\mu=0$, standard deviation: $\sigma=1$) for the analysis. For the final regression, the most relevant features were used, precisely those who had been selected between 7 and 10 times during the 10 folds of the process.

4 Results and Discussion

For the final set of the aging model (Table 3), one AMPEX feature (PVF) was selected. Eight phonological features were in the best feature set. The most relevant (selected 9 and 10 times) are:

- *highlow_presence_meanmin:* the mean minimum probability of the vowel height throughout all vowels in the text
- *highlow_presence_tneg:* the mean duration (in number of frames) of a segment in which a low vowel is present
- *consonantnasality_presence_meanneg:* the mean probability of nasality in a consonant that sounds non-nasal
- *consonantnasality_relevance_negdelta1:* the time needed to do a transition from consonant to vowel

Concerning the vowel height, or the vowel trapezium in general, there is no proof that the measured effects are caused by anatomical changes due to aging. Instead of the voice, altered articulation may be the reason. Earlier studies reported that the pronunciation of phones can change during time due to changes in the language, or in the way one speaker uses a language [13,14]. Hence, also the vowel space, i.e. the area enclosed by the vowel trapezium, can change. Older speakers of English were reported to undergo a shift in the speaker space roughly along a diagonal in the phonetic height × backness plane [15]. It is not sure so far

Table 2. Chronological age, perceived age for single raters and their average, and automatically determined age (SVR); $\mu(|e|)$ denotes mean absolute error, min/max e denotes minimal/maximal error; perceptual ratings were given in integer numbers (without decimal places)

Age	all speakers				women				men			
	μ(age)	$\mu(|e|)$	min e	max e	μ(age)	$\mu(|e|)$	min e	max e	μ(age)	$\mu(|e|)$	min e	max e
	75.7	—	—	—	76.3	—	—	—	74.0	—	—	—
rater 1	72.9	7.2	−34	27	73.0	7.1	−34	18	72.8	7.6	−14	27
rater 2	74.5	6.6	−22	22	75.6	6.6	−22	20	71.9	6.7	−15	22
rater 3	57.3	19.1	−42	11	57.9	18.9	−42	11	55.9	19.7	−39	11
rater 4	63.5	13.3	−34	25	63.0	13.9	−34	9	64.7	12.0	−24	25
rater 5	63.4	13.9	−31	24	63.7	14.1	−31	15	62.6	13.5	−25	24
rater avg.	66.3	10.5	−24.0	19.2	66.6	10.4	−24.0	7.4	65.6	10.7	−19.2	18.4
SVR	76.0	5.2	−24.8	14.8	76.2	5.4	−19.0	14.0	76.5	5.8	−24.4	7.2

that these findings can be generalized to other languages, however. Nevertheless, we regard our results important for age estimation from speech – as opposed to age estimation from voice which uses less information.

Transitions, as represented by *consonantnasality_relevance_negdelta1*, might in older people be slower. The negative sign in the regression weight (Table 3) supports this assumption. The weights are very low, however. Different weights for men and women in the formulae, especially for *highlow_presence_meanmin*, might be related to the different F_0 or to a more rapidly falling F_0 in women as result of aging processes.

One feature in the set is related to the position of the lips in vowels (mean of negative relevance of *roundedspread*). Three features are related to turbulence in the voice.

The most relevant phonemic features (Table 3) refer to minimal, maximal, and mean probabilities of some phones. The mean and maximum probabilities of /s/ over all frames where it was recognized, and the percentage of positive presence values for /Z/ (as in French 'journal'; SAMPA notation) may indicate altered articulation due to dentures. This is an important aspect to consider when estimating age from speech. The same reason may hold for the occurrence of /n/, /v/, and /l/ (*l_meanpos* is the mean positive value for the presence of /l/). /Y/ denotes the short u-umlaut. Its role in the set is currently unclear.

The absolute error between the automatically estimated and the chronological age was 5.2 years for all speakers together (root mean square error RMSE: 7.0), for men and women separately, it was slightly higher (Table 3). A trivial estimator optimizing with respect to the mean absolute error would have had an error of 7.7 years (RMSE: 9.6). Hence, our results show an error which is lower by more than two years. The average human rater estimated with an absolute error of 10.5 years (RMSE: 11.9). The greatest mismatch occurred for a woman who was estimated 42 years younger by rater 3.

Table 3. Regression weights for the best feature set when applied to all speakers, to women and men separately, respectively. The human-machine correlation and the respective errors between computed and chronological age are given in the lower part of the table

feature	type	chosen	all	women	men
PVF	AMPEX	7	-0.097	-0.121	-0.009
Y_min	ALF-PMF	7	-0.097	-0.052	-0.019
$L\_meanpos$	ALF-PMF	7	-0.175	-0.118	-0.090
n_min	ALF-PMF	8	-0.169	-0.163	-0.054
s_mean	ALF-PMF	8	-0.066	-0.060	-0.071
s_max	ALF-PMF	9	-0.131	-0.127	-0.069
v_min	ALF-PMF	7	-0.055	-0.074	-0.050
$Z\_posperc$	ALF-PMF	9	0.138	0.147	0.088
$consonantnasality\_relevance\_negdelta1$	ALF-PLF	10	-0.048	-0.034	-0.023
$consonantnasality\_presence\_meanneg$	ALF-PLF	10	0.177	0.154	0.066
$highlow\_presence\_tneg$	ALF-PLF	9	-0.114	-0.106	-0.027
$highlow\_presence\_meanmin$	ALF-PLF	10	0.173	0.183	0.016
$roundedspread\_relevance\_meanneg$	ALF-PLF	7	0.129	0.149	0.050
$turbulence\_relevance\_meanmax$	ALF-PLF	7	-0.076	-0.033	-0.088
$turbulence\_presence\_mean$	ALF-PLF	7	-0.024	-0.044	-0.051
$turbulence\_presence\_tneg$	ALF-PLF	8	0.099	0.101	0.061
correlation r to chronological age	—	—	0.72	0.73	0.69
mean abs. error to chronological age	—	—	5.2	5.4	5.8
RMSE to chronological age	—	—	7.0	7.0	8.2

The correlation of the automatically estimated and the chronological age was r=0.72 for all speakers, r=0.73 for men, and r=0.69 for women only. The average human rating showed a correlation of r=0.65 to the 'real' age of the whole speaker group. No single rater performed as good as the machine (rater 1: r=0.41; 2: r=0.66; 3: r=0.59; 4: r=0.43; 5: r=0.30). The inter-rater agreement on all the data, i.e. the correlation of one rater against the average of the others, was r=0.69 for three of the raters, r=0.62 for rater 1, and r=0.48 for rater 5. The chronological, perceived, and computed age are also shown in Fig. 2. The smaller range of the computed values is caused by the error minimization during the SVR training. Without the somewhat optimistic feature selection, the human-machine correlation was r=0.52 (mean error: 6.5 years, RMSE: 8.6) for all speakers.

Studies on age estimation have been presented before. For instance, Schötz [16] used prosodic and spectral features, such as F_0, formants, energy, jitter, shimmer, and duration, for age estimation by Classification And Regression Trees (CARTs). The average error in the classified age was above 15 years on 24 speakers from two age classes (18 to 31 and 60 to 82 years). F_0, the formants F_1 to F_4, and prosodic features were also used together with MFCCs on the University of Florida Vocal Aging Database [17]. The mean absolute error of listeners was 6.4 years, and the error of the machine was 10.0 years for gender-independent classification. Different years of age were not represented continuously in the

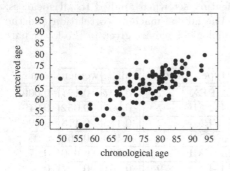 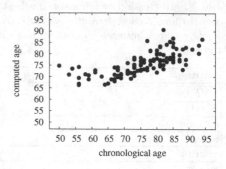

Fig. 2. Average chronological vs. perceived *(left graphics)* and computed age *(right graphics)*

data, however, but in three separate age groups with gaps between them. This may also be the reason why the humans were better in their evaluation than the machine. In our experiments, there was a monomodal distribution of age.

Minematsu et al. [18] reported correlations between perceived and computed age of up to r=0.88, but on audio data showing a clear trimodal age distribution. Their approach was based on Gaussian Mixture Models (GMMs). In a study of Bocklet et al. with children in preschool and primary school age (average: 8.3 years), a system based on GMMs and SVR showed a mean absolute error of 0.8 years and a maximal error of 3 years [19]. The ratio of error and average age was smaller in our system; however.

This study showed the potential of the presented features for language- and gender-independent estimation of age from speech data. The method may be helpful for clinical screening tests and for applications based on automatic speech recognition in general.

Acknowledgments. This work was partially funded by the Else Kröner-Fresenius-Stiftung (Bad Homburg v. d. H., Germany; grant 2011_A167) and by Hessen Agentur/ Hessen ModellProjekte (Wiesbaden, Germany; project 463/15-05). Further supporters were "Kom op tegen Kanker", the campaign of the Vlaamse Liga tegen Kanker VZW, and The Netherlands Cancer Institute/Antoni van Leeuwenhoek Hospital (Amsterdam).

References

1. Rusz, J., Cmejla, R., Ruzickova, H., Ruzicka, E.: Quantitative acoustic measurements for characterization of speech and voice disorders in early untreated Parkinson's disease. J. Acoust. Soc. Am. **129**, 350–367 (2011)
2. Middag, C., Bocklet, T., Martens, J.-P., Nöth, E.: Combining phonological and acoustic ASR-free features for pathological speech intelligibility assessment. In: Proc. Interspeech, ISCA, pp. 3005–3008 (2011)
3. Middag, C.: Automatic Analysis of Pathological Speech. PhD thesis, Ghent University, Ghent, Belgium (2012)

4. Haderlein, T., Middag, C., Maier, A., Martens, J.-P., Döllinger, M., Nöth, E.: Visualization of intelligibility measured by language-independent features. In: Sojka, P., Horák, A., Kopeček, I., Pala, K. (eds.) TSD 2014. LNCS, vol. 8655, pp. 547–554. Springer, Heidelberg (2014)

5. Schneider, S., Plank, C., Eysholdt, U., Schützenberger, A., Rosanowski, F.: Voice Function and Voice-Related Quality of Life in the Elderly. Gerontology **57**, 109–114 (2011)

6. International Phonetic Association (IPA): Handbook of the International Phonetic Association. Cambridge University Press, Cambridge (1999)

7. Middag, C., Saeys, Y., Martens, J.-P.: Towards an ASR-free objective analysis of pathological speech. In: Proc. Interspeech, ISCA, pp. 294–297 (2010)

8. Moerman, M., Pieters, G., Martens, J.-P., van der Borgt, M.-J., Dejonckere, P.: Objective evaluation of the quality of substitution voices. Eur. Arch. Otorhinolaryngol. **261**, 541–547 (2004)

9. van Immerseel, L., Martens, J.-P.: AMPEX Disordered Voice Analyzer [computer program]. Digital Speech and Signal Processing research group, Ghent University, Ghent, Belgium. http://dssp.elis.ugent.be/downloads-software (last visited May 28, 2015)

10. van Immerseel, L.M., Martens, J.-P.: Pitch and voiced/unvoiced determination with an auditory model. J. Acoust. Soc. Am. **91**, 3511–3526 (1992)

11. Smola, A.J., Schölkopf, B.: A Tutorial on Support Vector Regression. Statistics and Computing **14**, 199–222 (2004)

12. Witten, I.H., Frank, E.: Data Mining: Practical Machine Learning Tools and Techniques, 2nd edn. Morgan Kaufmann, San Francisco (2005)

13. Harrington, J., Palethorpe, S., Watson, C.I.: Does the Queen speak the Queen's English? Nature **408**, 927–928 (2000)

14. Watson, P.J., Munson, B.: A comparison of vowel acoustics between older and younger adults. In: Proc. ICPhS XIV, pp. 561–564. International Phonetic Association (2007)

15. Harrington, J., Palethorpe, S., Watson, C.I.: Age-related changes in fundamental frequency and formants: a longitudinal study of four speakers. In: Proc. Interspeech, ISCA, pp. 2753–2756 (2007)

16. Schötz, S.: Prosodic and non-prosodic cues in human and machine estimation of female and male speaker age. In: Bruce, G., Horne, M. (eds.) Nordic Prosody: Proceedings of the IXth Conference, pp. 215–223. Lund, Sweden (2004)

17. Spiegl, W., Stemmer, G., Lasarcyk, E., Kolhatkar, V., Cassidy, A., Potard, B., Shum, S., Song, Y.C., Xu, P., Beyerlein, P., Harnsberger, J., Nöth, E.: Analyzing features for automatic age estimation on cross-sectional data. In: Proc. Interspeech, ISCA, pp. 2923–2926 (2009)

18. Minematsu, N., Sekiguchi, M., Hirose, K.: Automatic estimation of perceptual age using speaker modeling techniques. In: Proc. Eurospeech, ISCA, pp. 3005–3008 (2003)

19. Bocklet, T., Maier, A., Nöth, E.: Age determination of children in preschool and primary school age with GMM-based supervectors and support vector machines/regression. In: Sojka, P., Horák, A., Kopeček, I., Pala, K. (eds.) TSD 2008. LNCS (LNAI), vol. 5246, pp. 253–260. Springer, Heidelberg (2008)

LecTrack: Incremental Dialog State Tracking with Long Short-Term Memory Networks

Lukáš Žilka[✉] and Filip Jurčíček

Faculty of Mathematics and Physics, Institute of Formal and Applied Linguistics,
Charles University in Prague, Malostranské náměstí 25,
11800 Prague, Czech Republic
lukas@zilka.me, jurcicek@ufal.mff.cuni.cz

Abstract. A dialog state tracker is an important component in modern spoken dialog systems. We present the first trainable incremental dialog state tracker that directly uses automatic speech recognition hypotheses to track the state. It is based on a long short-term memory recurrent neural network, and it is fully trainable from annotated data. The tracker achieves promising performance on the *Method* and *Requested* tracking sub-tasks in DSTC2.

Keywords: Dialog systems · Recurrent neural network · Dialog state tracking

1 Introduction

A dialog state tracker is an essential component of modern spoken dialog systems. It maintains the user's goals throughout the dialog by looking at the automatic speech recognition (ASR) results of her utterances. For example, in the restaurant information domain, the dialog state tracker tracks what kind of food the user wants and which price range is she looking for, and provides this information as a probability distribution over *food* and *price_range*: P(food, price_range) The dialog state tracker also needs to deal with speech recognition errors and tries to reduce their impact on the dialog [1].

The state-of-the-art dialog state trackers [2–6] achieve their performance by learning from annotated data, and they were shown to work well in the restaurant

F. Jurčíček—This research was partly funded by the Ministry of Education, Youth and Sports of the Czech Republic under the grant agreement LK11221, core research funding, grant GAUK 2076214 of Charles University in Prague. This research was (partially) supported by SVV project number 260 224. This work has been using language resources distributed by the LINDAT/CLARIN project of the Ministry of Education, Youth and Sports of the Czech Republic (project LM2010013). Cloud computational resources were provided by the MetaCentrum under the program LM2010005 and the CERIT-SC under the program Centre CERIT Scientific Cloud, part of the Operational Program Research and Development for Innovations, Reg. no. CZ.1.05/3.2.00/08.0144. We gratefully acknowledge the support of NVIDIA Corporation with the donation of the Titan Z GPU used for this research.

© Springer International Publishing Switzerland 2015
P. Král and V. Matoušek (Eds.): TSD 2015, LNAI 9302, pp. 174–182, 2015.
DOI: 10.1007/978-3-319-24033-6_20

information domain in the dialog state tracking challenge DSTC2 [7]. However, they possess several undesirable traits. First, they can only track the dialog state turn-by-turn (as opposed to a more complicated word-by-word approach), which limits the responsivity of the dialog system. Second, some of the trackers rely on the results from a spoken language understanding (SLU) component [8], which brings an additional component into the dialog system that needs to be trained and tuned. Third, elaborate and complicated tracking models of the trackers are difficult to reproduce and maintain. We aim to address these problems in the model proposed in this paper.

The contribution of this paper is our novel dialog state tracker, which we refer to as LecTrack[1]. This paper aims towards building more responsive and simpler dialog systems by proposing the first trainable dialog state tracker which naturally operates incrementally, word-by-word, and can directly learn from annotated dialogs, removing the need for an SLU unit. The word-by-word mode of tracking allows the dialog manager to be more responsive with the users. Simplicity comes from the fact that the whole dialog state tracker can be automatically optimized from data by a standard backpropagation algorithm, without requiring the user to manually tune opaque hyper-parameters or task-specific pre-processing.

Our LecTrack tracker is based on the long-short term memory recurrent neural network (LSTM RNN) [9]. We have chosen this approach due to several reasons: First, LSTMs were shown to be effective for learning sequence mappings in automatic speech recognition [10], machine translation [11], protein structure prediction [12], and many other sequence classification tasks. The length of the sequences successfully modelled by LSTMs is comparable to the length of the word sequences in the spoken dialog systems. Second, the sequential nature of the dialog naturally fits the LSTM's recurrent mode of operation. And finally, as the tracker processes the input, it incrementally builds an intermediate representation of the dialog. It has been shown that good intermediate representations help generalization [13]. The success of LSTM on multiple complicated and diverse tasks promises to be exploitable also in dialog state tracking.

The paper is organized as follows. First, we give a basic description of the task (Section 2). In Section 3, the model of the LSTM dialog state tracker is described with its training procedure. Section 4 shows how it performs in the benchmarks. Then, in Section 5, we discuss the results and qualities of the LSTM dialog state tracker. Finally we discuss some related work from the literature (Section 6) and conclude with Section 7.

2 Dialog State Tracking

The task of dialog state tracking is to monitor progress in the dialog and provide a compact representation of the dialog history in the form of a *dialog state* [7,14]. Because of uncertainty in the user input, statistical dialog trackers maintain a probability distribution over all dialog states, called the *belief state*.

---

[1] (L)STM R(ec)urrent Neural Network Dialog State (Track)er.

As the dialog progresses, the dialog state tracker updates this distribution given new observations.

In this paper, we define the dialog state at time t as a vector $s_t \in C_1 \times ... \times C_k$ of k dialog state components, sometimes called slots in the literature. Each component $c_i \in C_i = \{v_1, ..., v_{n_i}\}$ takes one of the n_i values. Our dialog state tracker maintains a probability distribution over s_t factorized by the dialog state components:

$$P(s_t|w_1, ..., w_t) = \prod_i p(c_i|w_1, ..., w_t; \theta)$$

Note that all models $p(c_i|\cdot)$ share a substantial portion of the parameters, as detailed in the next section, so despite the fact that the predictions are factorized and thus independent, they were optimized to minimize a joint objective function and therefore naturally model the dependence between the dialog state components.

3   LSTM Dialog State Tracker

Here we define our novel LSTM dialog state tracking model. Its task is to map a sequence of words in the dialog $w_1, ..., w_t$ to predictions for each of the k dialog state components $(p_t^{(1)}, ..., p_t^{(k)})$. Each $p_t^{(i)}$ is a vector corresponding to a probability distribution over the values of the i-th dialog state component. For example, $p_t^{(area)}$ is a probability distribution over values $\{north, south, east, west\}$ at the time t.

3.1   Model

Our dialog state tracking model can be seen as an encoder-classifier model, where an LSTM is used to encode the information from the input word sequence into a fixed-length vector representation. Given this representation, the classifier predicts a value for each of the dialog state components (as a probability distribution).

Formally we have an encoder that maps an input word and a previous hidden state to a new hidden state, $Enc(w, q_{t-1}) = q_t$, and a classifier that maps a hidden state to a prediction, $C(h_t) = p_t$. To encode the whole dialog, the encoder is applied sequentially on the input sequence of words. In our system, we have one encoder Enc and multiple classifiers C, one for each dialog state component (e.g. $C^{(food)}$, $C^{(area)}$, ...).

LecTrack tracker is composed of two major components: the encoder Enc and the classifier C:

$$\textbf{LecTrack} : a_1, ..., a_n \rightarrow p_1, ..., p_k$$
$$\forall i \in 1, ..., k : p_i = \text{C}_i(\text{Enc}(E \cdot a_1, ..., E \cdot a_n, h_0, c_0))$$

Here, n is the length of the input sequence, k the number of the dialog state components, $a_1, ..., a_n$ is the input word sequence encoded in a one-hot encoding, E

is a word embedding matrix, and $h_0 = c_0 = \mathbf{0}$ are zero vectors. As shown, each token a_i is mapped to its corresponding embedding vector through the embedding matrix E, $w_i = E \cdot a_i$. The model of the encoder Enc is a standard LSTM RNN [9]. In case of a recursive application of Enc, we write $Enc(w_1, ..., w_n, h, c)$ instead of $Enc(w_n, ... Enc(w_1, h, c))$ to simplify the notation. The model of the classifier C is a neural network with a single layer composed of rectified linear units and a softmax output layer

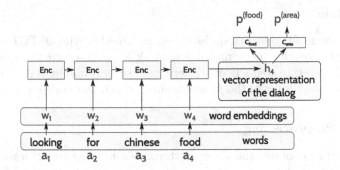

Fig. 1. A demonstration of the LSTM Dialog State Tracker applied to a user utterance "looking for chinese food". The encoding LSTM model Enc is sequentially applied to each input word and at the end, its hidden state is used to feed to the state component classifiers.

3.2 Training

The model is trained using the standard cross-entropy criterion [15] in the vanilla stochastic gradient descent scenario:

$$l(a_1, ..., a_n, y_1, ..., y_k; \theta) = -\sum_{i=0}^{k} \log \text{LecTrack}(a_1, ..., a_n)_{y_i}^i$$

Here, $\text{LecTrack}(.)_n^m$ denotes the probability of the n-th value in the m-th dialog state component.

After each optimization epoch, we monitor the performance[2] of the model on a held-out set D. When the performance stops increasing for several iterations, we terminate the training and select the best-performing model.

4 Experiments

4.1 Dataset

To train and evaluate our model, we use the DSTC2 [7] data, which is a common data set for dialog state tracking evaluation. The DSTC2 data consists of about

---

[2] See the experiments section for the description of the featured metrics.

3,000 dialogs from the restaurant information domain, each dialog is 10 turns long on average. The data is split into training, development and test sets. This data allows us to measure the performance of our tracker on turn-based dialogs. Ideally we would run the evaluation on a data set where we could also measure the incremental capabilities of the tracker, but to the best of our knowledge, no such data set is publicly available yet, and we shall address this in our future work.

4.2 Baseline

A baseline system for this domain has been provided by the DSTC2 organizers. It uses the SLU results and confidence to rank hypotheses for the values of the individual dialog state components. There were several baselines described in [7] and we report the results of the *focus* baseline, which was the best among them.

4.3 Data Preprocessing

Each dialog turn contains the system utterance and the user utterances, which we need to serialize into a stream of words as the input to our model. The system utterance undergoes a simple preprocessing detailed below, and the user utterance is directly fed to the model word-by-word without any further preprocessing. There is no difference between the system and user utterance in the eyes of our model, both are seen together as one long sequence of words.

System Input: To get the the system input, we perform a simple preprocessing. We flatten the system dialog acts of the form `act_type(slot_name=slot_value)` into a sequence of two tokens t_1, t_2, where $t_1 = $ (act_type, slot_name) and $t_2 = $ slot_value. For example `request(slot=food)` is flattened as $(request, slot), food$, which the model then sees as a word sequence of length two.

User Input: For the sake of simplicity, we use only the best live-ASR[3] hypothesis and ignore the rest of the n-best list. We plan to extend our model for processing multiple ASR hypotheses in the near future.

Out-of-Vocabulary Words are randomly mixed into the training data to give the model a chance to cope with unseen words: At training time, a word in the user input word is replaced by a special out-of-vocabulary token with probability α. At test time, this token is used for all unknown words.

---

[3] There are batch and live ASR results in the DSTC2 data. We use the live ones and refer to them as live-ASR.

4.4   Experimental Methodology

We follow the DSTC2 methodology [7] and measure the accuracy and L2 norm of the joint slot predictions. The joint predictions are grouped into the following groups, and the results of each group is reported separately: Goals, Requested, Method. For each dialog state component in each dialog the measurements are taken *at the end of each dialog turn*, provided the component has already been mentioned in some of the SLU n-best lists in the dialog[4]

4.5   Results

The results of LecTrack on the DSTC2 data are summarized in Table 1. For the groups *Method* and *Requested* LecTrack's accuracy is better than the baseline and comes closer to the state-of-the-art. Within these groups the handcrafted preprocessing present in the baseline and the state-of-the-art models is not as effective as for the *Goal* group.

We hypothesize that the accuracy on the *Goal* group does not achieve the state of the art because of two reasons. First, LecTrack needs to see examples for each value of each dialog state component. But the distribution of the individual values in the data has a heavy tail, and thus the baseline method and the state-of-the-art methods that use various kinds of hand-crafted abstraction to make the data denser beat LecTrack. Second, our model does not fully utilize the information in the n-best lists, thus loses useful information in the uncertain cases where more hypotheses than the first one are useful.

For the frequently seen values from the group *Goal* the performance of Lec-Track is much better than the baseline, as is shown in Table 2. We looked at the sub-goal *food* and compared the classification accuracy of its individual values. The top of the table contains 9 values, which occur more than 100 times in the test set, as the representatives of the classes that are well-represented in the data; the bottom of the table contains the representatives of the under-represented classes, and we selected values which occur at least 10 times in the test set to get meaningful accuracy estimates. For the well-represented classes, LecTrack's performance is stable and usually beats the baseline by a large margin, however for the under-represented classes LecTrack's performance is much worse than the baseline. This suggests that some form of abstraction should improve the results for the under-represented cases and improve the overall performance.

To keep the model simple we did not use any form abstraction, such as gazetteers to preprocess our data, and only used 1-best hypothesis as an input. Gazetteers offer a cheap solution to data sparsity for English but are difficult to gather and maintain for other languages where one word can have many forms. In our future experiments we plan to introduce some form of abstraction. Also, it is not obvious from the machine learning literature how an n-best list could be used in our model to improve the performance. This is another aspect that will be addressed in our future experiments.

---

[4] Note we do not use the SLU n-best list in our model at all, but we adapt this metric to be able to compare to the other trackers in DSTC2.

Table 1. Performance on the DSTC2 data.

	Dev						Test					
	Goal		Method		Requested		Goal		Method		Requested	
model	Acc.	L2	Acc.	L2	Acc.	L2	Acc.	L2	Acc.	L2	Acc.	L2
baseline	0.61	**0.63**	0.83	0.27	0.89	0.17	0.72	0.46	0.90	0.16	0.88	0.20
LecTrack	0.62	0.79	0.87	0.24	0.95	0.09	0.60	0.79	0.91	0.17	0.96	0.07
state-of-the-art [2]	**0.71**	0.74	**0.91**	**0.13**	**0.97**	**0.05**	**0.78**	**0.35**	**0.95**	**0.08**	**0.98**	**0.04**

Table 2. Accuracy for the most frequent values for the *food* dialog state component which have at least 100 test examples in the test set, and for some that contain between 10 to 20 examples in the test set.

	chinese	indian	korean	asian o.	don't care	european	italian	spanish	thai	...	traditional	steakhouse	romanian	german
baseline	0.53	0.49	0.67	0.54	0.98	0.61	0.41	0.69	0.14		0.17	0.14	0.35	0.28
LecTrack	0.82	0.79	0.93	0.86	0.88	0.80	0.79	0.73	0.64		0.17	0.07	0.21	0.07

5   Discussion

Our LSTM dialog state tracker is capable of learning from raw dialog text, annotated with true dialog state component values at some time-steps. No spoken language understanding unit is needed to pre-process the input for our model. In addition, the model performance does not suffer if the input word sequences are long, which is in accordance with other LSTM applications [11].

Our model naturally handles the inter-slot dependence by projecting the input sequence into a fixed-length vector from which all the dialog state component predictions are made. However, the predictions are made independently for all of the state components and the joint distribution is not explicitly modelled.

In noisy conditions, waiting for silence is very limiting for the dialog system. The tracker's ability to process the input incrementally can overcome this issue and signal to the dialog manager when the incoming speech starts to make sense. This can lead to more human-like and interactive dialogs and simpler dialog managers. Our model was designed to be able to predict at arbitrary time in the dialog the full distribution over the dialog state components, and this mode of operation costs no additional computation as opposed to other trackers.

6   Related Work

The only incremental dialog system in the literature that we are aware of is [16]. In this paper, the authors describe an incremental dialog system for number dictation as a specific instance of their incremental dialog processing framework. To track the dialog state, they use a discourse modelling system, which keeps track of the confidence scores from the semantic parses of the input. The semantic parses are produced by a grammar-based semantic interpreter with a hand-coded context-free grammar. While their system is mostly handcrafted, ours is trained using annotated dialog data, so we do not need the handcrafted grammar and an explicit semantic representation of the input.

Using RNN for dialog state tracking has been proposed before [3,17]. The dialog state tracker in [3] uses an RNN, with a very elaborate architecture, to track the dialog state turn-by-turn. Similarly to our model, their model does not need an explicit semantic representation of the input. However, unlike our model, they use tagged n-gram features, which allows them to perform better generalization on rare but well-recognized values. Our model is capable of such generalization, too, but it needs more data. We refrain from using the tagged features because they introduce a preprocessing effort, and we are interested in a model that can learn from the data directly without assuming any corre-spondence between the names and values of the dialog state components and their surface forms that occur in the dialog (e.g. that value "chinese" of the dialog state component "food" will typically be represented as "chinese food" in the dialog). In English dialog systems, it might be perceived as an unnecessary complication not to leverage these tagged features, but when we consider other languages, where a word often has a lot of forms, it pays off, because the effort spent on producing good quality tagged features is non-trivial.

7 Conclusion

We presented a first trainable incremental dialog state tracker that directly uses automatic speech recognition hypotheses to track the state. It is based on a long short-term memory recurrent neural network and fully trainable from the dialog utterances annotated at certain points in time by the dialog state information. It represents the history of the whole dialog as a low-dimensional real vector, which is on its own used for the prediction of the whole dialog state. We evaluated our dialog state tracker on the data from Dialog State Tracking Challenge 2, where we showed that it achieves a promising performance on the *Method* and *Requested* tracking sub-tasks. We believe that the simplicity, ease of use, and the incremental tracking capability of LecTrack make it a first good step on the way towards more responsive dialog systems.

References

1. Williams, J., Raux, A., Ramachandran, D., Black, A.: The dialog state tracking challenge. In: Proceedings of the SIGDIAL 2013 Conference, pp. 404–413 (2013)
2. Williams, J.D.: Web-style ranking and slu combination for dialog state tracking. In: 15th Annual Meeting of the Special Interest Group on Discourse and Dialogue, p. 282 (2014)

3. Henderson, M., Thomson, B., Young, S.J.: Word-based dialog state tracking with recurrent neural networks. In: Proceedings of SIGdial (2014)
4. Lee, B.J., Lim, W., Kim, D., Kim, K.E.: Optimizing generative dialog state tracker via cascading gradient descent. In: 15th Annual Meeting of the SIGDD, p. 273 (2014)
5. Smith, R.W.: Comparative error analysis of dialog state tracking. In: 15th Annual Meeting of the Special Interest Group on Discourse and Dialogue, p. 300 (2014)
6. Sun, K., Chen, L., Zhu, S., Yu, K.: The sjtu system for dialog state tracking challenge 2. In: 15th Annual Meeting of the Special Interest Group on Discourse and Dialogue, p. 318 (2014)
7. Henderson, M., Thomson, B., Williams, J.: The second dialog state tracking challenge. In: 15th Annual Meeting of the Special Interest Group on Discourse and Dialogue, p. 263 (2014)
8. Wang, Y.Y., Deng, L., Acero, A.: Spoken language understanding. IEEE Signal Processing Magazine 22(5), 16–31 (2005)
9. Hochreiter, S., Schmidhuber, J.: Long short-term memory. Neural Computation 9(8), 1735–1780 (1997)
10. Graves, A., Schmidhuber, J.: Framewise phoneme classification with bidirectional lstm and other neural network architectures. Neural Networks 18(5), 602–610 (2005)
11. Sutskever, I., Vinyals, O., Le, Q.V.: Sequence to sequence learning with neural networks. In: Advances in Neural Information Processing Systems, pp. 3104–3112 (2014)
12. Sønderby, S.K., Winther, O.: Protein secondary structure prediction with long short term memory networks. arXiv preprint arXiv:1412.7828 (2014)
13. Gülçehre, Ç., Bengio, Y.: Knowledge matters: Importance of prior information for optimization. arXiv preprint arXiv:1301.4083 (2013)
14. Zilka, L., Marek, D., Korvas, M., Jurcicek, F.: Comparison of bayesian discriminative and generative models for dialogue state tracking. In: Proc. of the SIGDIAL 2013 Conf. (2013)
15. Rubinstein, R.Y., Kroese, D.P.: The cross-entropy method: a unified approach to combinatorial optimization, Monte-Carlo simulation and machine learning. Springer Science & Business Media (2004)
16. Skantze, G., Schlangen, D.: Incremental dialogue processing in a micro-domain. In: Proc. of the 12th Conf. of EC-ACL, pp. 745–753. Association for Computational Linguistics (2009)
17. Henderson, M., Thomson, B., Young, S.J.: Deep neural network approach for the dialog state tracking challenge. In: Proceedings of SIGdial (2013)

Automatic Construction of Domain Specific Sentiment Lexicons for Hungarian

Viktor Hangya[✉]

Department of Computer Algorithms and Artificial Intelligence,
University of Szeged, Szeged, Hungary
hangyav@inf.u-szeged.hu

Abstract. Sentiment analysis has become an actively researched area recently, which aims to detect positive and negative opinions in texts. A good indicator for the polarity of a given text is the number of words in it that have positive or negative meanings. The so called sentiment lexicons are lists containing words together with their polarities. In this paper we present methods for creating sentiment lexicons automatically. We use these lexicons in sentiment analysis tasks on general and domain-specific Hungarian corpora. We compare the efficiency of sentiment lexicons from different domains and show the importance of using domain-specific sentiment lexicons for different sentiment analysis tasks.

Keywords: Sentiment analysis · Sentiment lexicon · Natural language processing

1 Introduction

People have the opportunity to share their thoughts with others thanks to the increased popularity of social media. A big amount of data is created daily which contains people's opinions about various topics like products, celebrities and companies. In the past few years sentiment analysis has become popular not only in scientific research but also in economy as well. The task of sentiment analysis is to determine the polarity of an opinion in a given document.

The task can be considered as a classification problem in which documents have to be classified into positive or negative classes according to the opinions they contain. When a big amount of annotated data is available, a good method is to create supervised machine learning based classifiers which rely on words from the documents. But this method can be inaccurate if the size of the annotated corpus is insufficient because word types have low frequency in it. To overcome the lexical sparsity problem, sentiment lexicons can be useful, which contain a predefined set of positive and negative words. This knowledge can be used to extract various features besides n-grams. Many general purpose sentiment lexicons are available for English [1,2] but there are much fewer for Hungarian.

A simple method for creating lexicons is to translate one from a foreign language. However, this method has some disadvantages. First, the process can be

© Springer International Publishing Switzerland 2015
P. Král and V. Matoušek (Eds.): TSD 2015, LNAI 9302, pp. 183–190, 2015.
DOI: 10.1007/978-3-319-24033-6_21

time consuming and expensive. Furthermore, polysemous words can have different polarities in different text domains, thus domain-specific lexicons are needed. Most of the available foreign lexicons are for general use. In this paper, we show that domain-specific lexicons are more useful for sentiment analysis on domain-specific corpora. Furthermore, we propose language independent techniques for creating lexicons automatically. By incorporating texts from a given domain we created lexicons from scratch which are useful for extracting features for sentiment analysis tasks. We also created semi-automatic methods which can extend a given seed lexicon by using word similarity.

We evaluated lexicons by employing them in sentiment analysis tasks. For this we used two Hungarian text databases, one is a domain specific corpus which contains reviews about IT products, the other contains texts from news [3] related to various topics, like sports, politics, etc. We show that it is important to use lexicons from the same domain as the texts which we classify into polarity classes.

2 Sentiment Lexica

The most important indicators of sentiments are sentiment words, which can express positive and negative opinions. Three main approaches exist for creating sentiment lexicons: manual, dictionary-based and corpus-based [4]. The manual approach is time-consuming thus expensive. Dictionaries contains synonym and antonym sets for words. Dictionary-based approaches use this knowledge to automatically collect sentiment words starting from a manually created seed lexicon. The goal of the corpus-based approaches is to employ knowledge which can be found in a set of documents. Seed lexicons can be extended by using rules, e.g. adjectives on both sides of a conjunction in a sentence have the same polarity. If the documents are also labeled with positive and negative labels, statistical methods can be used to create lexicons from scratch. In this work, we propose new methods from all three main approaches.

An important fact in using sentiment lexicons is the domain from which the texts come from because some words can have different polarities in different domains. Consider the following example:

- The usage of this mixer is easy and it is very **silent**.
- For this price it's too **silent** for me, I thought it will be louder.

The first example is from the domain of kitchen devices, where *silent* has a positive meaning. In contrast, the second example is from the speakers domain, where *silent* is a negative quality. From this, it can be seen that in a sentiment analysis task choosing lexicons from the appropriate domain is important. There is a need for an automatic method which can create domain-specific lexicons, because there are no lexicons for every domain and creating them manually is expensive and requires an expert in that domain.

In the following we present methods for creating and adapting sentiment lexicons and also highlight their positive and negative aspects.

2.1 Translating a Foreign Lexicon

In our experiments we used a manually translated lexicon for comparison reasons. In English there are many general purpose lexicons, i.e. SentiWordNet [1] or MPQA [2]. We had access to an English lexicon already used by a reputation monitoring system, which we translated to Hungarian. The translated lexicon contained 3322 word forms each with its polarity level from the [−5, 5] interval, where -5 is the most negative value and 5 is the most positive. The translation was carried out manually and we used all of the possible Hungarian translations of a given word.

The method has some disadvantages. First, translation can be time-consuming thus expensive especially if the original lexicon is big. Translating polysemous words can be difficult too, because it is unclear which meaning to use. For example, the word *terrific* (*awesome, horrible*) has two meanings with opposite polarities. By using the polarity value from the original lexicon the correct meaning can be guessed, but not in all cases. The word *cool* (*cold, awesome*) can have two meanings both with positive polarity but not with the same intensity. Most of the existing lexicons are for general use, so during the translation process we had to consider the domain in which the translated lexicon will be used and in the case of some words the original polarity value should be altered.

2.2 Bootstrapping Sentiment Lexicon

To overcome the above mentioned problems we implemented methods which can automatically create sentiment lexicons. The first method is a corpus-based which exploits a document set which is annotated with polarity labels. The annotation can be done in various ways. The most accurate one is by annotating it by hand. It can be done automatically as well, using an existing sentiment analysis system. For example, this system can be a simple n-gram based model trained on text from another genre. The automatic method can yield a noisy annotation but a large amount of data filters noise. The polarity of a given word can be computed using pointwise mutual information [5]. The method gives a polarity value for all words in the corpus which reflects the positiveness and negativeness of the given word in that domain. Additionally, we scale these values into the [−5, 5] interval. In the following we will refer to lexicons created with this method with the name **pmi**.

2.3 Extending Lexicons

In this section we propose dictionary-based methods for extending seed lexicons. The input seed lexicon contains only a low number of words with their polarity values. The extension is based on similarity measures between words, more precisely we add words to the extended lexicons which are similar to those which are already in the seed lexicon. By using a similarity measure which reflects the aspects of a domain we not only extend the input lexicon, but also adapt it to the given domain.

To assemble the input seed lexicon we created a semi-automated method. We trained an n-gram based sentiment analysis system with maximum entropy classifier on the training portion of the given corpus. Using the trained model it is possible to extract those words that are most likely to occur in positive and negative texts, respectively. From these we used 20 words for both polarity classes in the seed lexicon. Again we scaled the polarity values into the $[-5, 5]$ interval.

WordNet. The input of the first extension method is a seed lexicon and a wordnet in a given language. WordNets are large lexical databases which contain words grouped into sets of synonyms (synsets). Synsets are linked by means of conceptual-semantic and lexical relations. Our first similarity measure over words is based on wordnets. Our hypothesis is that the polarity of a word and all of its synonyms is equal. The extension process is as follows. Initially each synset has a polarity value of 0. We iterate over all words in the seed lexicon and assign the actual seed word's polarity value to those synsets in which it appears. Additionally, we used the relation between synsets, namely which sets have similar or opposite meanings. For this we used the *similar_to* and *hyponym* relations in the wordnet. We assign the polarity value or its inverse of a synonym set to all related synsets depending on the relation type. If a synset is related to multiple synsets with non 0 polarity value, we calculate their average. In the last step, we add all the words with the appropriate value to the extended lexicon which are in a synset with a polarity value different than 0. A word type can be in multiple synsets with different polarity values. For example the word *terrific* is included in the following positive and negative synsets {*wonderful,* terrific, fantastic} and {*terrifying,* terrific}, where the seed words are *wonderful* and *terrifying.* In such cases we calculated the average of these polarity values.

The method can be run iteratively, the output of a step can be used as the input of the next one further expanding the lexicon in each step. An important fact is that some words can be added to the extended lexicon with wrong polarity values in an iteration step. For example, in the IT domain if the *silent* word is used as a positive seed word, the *uncommunicative* will be added as positive to the extended lexicon which does not have any polarity in this domain. Because of this, after some iteration step the extended lexicon becomes too noisy. Furthermore, wordnets are general lexical resources, thus the extracted word similarities are not domain dependent. For this reason it is important to start with a seed lexicon which is already domain-specific, this way the extension is aware of the specifics of a given domain. For our experiments we used the Hungarian WordNet [6].

Word Clusters. We developed another word similarity measure which is more aware of the specifics of the given domain. For this we used the Brown clustering algorithm [7]. It is a hierarchical clustering of words based on the context in which they occur. The input of this method is a seed lexicon as before and an unlabeled corpus from a given domain. Similar to the previous method, our hypothesis is that the polarities of words in the same cluster are equal. We build clusters on the unlabeled dataset. The initial step of the algorithm is to assign 0 polarity value to all clusters. The next step is to iterate over all words in the seed lexicon and assign the polarity value of the actual seed word to the cluster which contains it. If a cluster contains multiple seed words, we calculate their average value. Lastly, we add words with the appropriate polarity value to the extended lexicon which are in a cluster with not 0 value. The method has one parameter which is the number of clusters to use. If we use a small number of clusters, words which are not similar can be in the same cluster, which causes that the extension assigns wrong polarity values to some words. Inversely, if we use too many clusters, just a small number of new words will be added to the new lexicon. The main advantage of this method in contrast with the wordnet based one is that the clustering algorithm which uses domain-specific texts can capture word similarities which are specific to the given domain. This way it is capable of domain-adapting the input lexicon.

3   Data

In this section we present the used corpora. For our experiments we used two databases: one with texts from news sites and one with IT related product reviews.

The *OpinHuBank* [3] is a corpus created directly for sentiment analysis tasks and contains texts from a general domain using various Hungarian news sites, blogs and forums about sports, politics, economics, etc. Each text instance is an at least 7 token long sentence. The sentences were annotated by 5 annotators with positive, negative or neutral labels. We only used the ones with polarity, more precisely those which were annotated at least by three positive or three negative labels. This way we got 882 positive and 1629 negative sentences in the **opinhu** corpus.

We created a domain specific corpus out of IT product reviews. For this we used the content of a Hungarian site called *árukereső*[1]. This site contains reviews about a wide range of products from which we only used the ones from the PC and electronic products (TV, digital cameras, etc.) categories. The reviewers on this site have to provide pros and cons when writing a review. We used these as positive and negative texts respectfully. Furthermore, we applied filtering on the texts in such a way that we only kept reviews that are one sentence long. The resulting **prodrev** database consists of 3573 positive and 3149 negative sentences.

4   Results

The goal of this work was to create methods to automatically assemble sentiment lexicons which are useful in sentiment analysis tasks. To comparatively evaluate

---

[1] www.arukereso.hu

lexicons, we defined a sentiment analysis task in which we classify sentences into positive and negative classes and the system was strongly built upon the lexicons. We used a maximum entropy classifier with lemmatized unigrams and lexicon based features. We define the usefulness of a lexicon given a corpus with the accuracy of the classifier system which uses that lexicon. The higher the accuracy, the more useful the lexicon is. In the following we consider a word as *sentiment word* if it is included in the given lexicon and its absolute polarity value is at least 1. A sentiment word is positive or negative depending on the sign of its polarity value. The lexicon based features are the following (an example can be seen in Table 1):

- the sentiment words in the text (in their original form)
- the overall values of positive and negative words respectively
- the overall values of sentiment words
- pairs made of the polarity of a sentiment word and its preceding or following lexical neighbor

Table 1. An example sentence and the features extracted from it. The sentiment word in the sentence is *better*, which has 5.0 polarity value.

Sentence:	The laptop's display has **better** parameters!
Lemmatized unigrams:	the, laptop, display, have, good, parameter, !
sentiment words:	better
Overall values:	POSITIVE=5.0, NEGATIVE=0.0, POLARITY=5.0
Neighbors:	has_POSITIVE, POSITIVE_parameter

Table 2. Extension of seed lexicon with wordnet (wn) and cluster based methods. The accuracies on opinhu and prodrev corpora were measured using 10-fold cross-validation.

opinhu-seed	86.2	prodrev-seed	90.7
opinhu-seed-wn-1	86.4	prodrev-seed-wn-1	90.8
opinhu-seed-wn-2	85.9	prodrev-seed-wn-2	90.5
opinhu-seed-wn-3	86.3	prodrev-seed-wn-3	90.8
opinhu-seed-wn-4	86.0	prodrev-seed-wn-4	90.9
opinhu-seed-cluster-15	86.7	prodrev-seed-cluster-18	90.8
opinhu-seed-cluster-15-t3	86.8	prodrev-seed-cluster-19-t3	90.8

The result of the systems using the lexicon extending techniques can be seen in Table 2. In the case of both the opinhu and prodrev databases, we created a seed lexicon with the semi-automatic method which was presented earlier. The tables show the accuracy of the sentiment analysis systems, which was calculated using 10-fold cross-validation. The notation *wn* indicates the usage of the wordnet based word similarities and the number after that gives the number of iterations we ran. In the case of the opinhu corpus we achieved the highest increase in accuracy with 1 iteration while in the case of prodrev 4 iterations was the best. In both cases, after the 5th iteration the lexicons became too noisy and the results begun to decrease. In the last two rows of the tables, the results of the clustering based extension can be seen. The number at the end of the lines shows the level where the cluster hierarchy was cut and $t3$ indicates that we filtered out words from the lexicon which have a frequency of at most 3 in the corpus. This technique was better in the case of the opinhu corpus, and slightly worse in the case of prodrev.

Table 3. Achieved accuracies using different lexicons on opinhu and prodrev.

	opinhu	prodrev
baseline-opinhu	86.1	70.1
baseline-prodrev	61.6	90.0
opinhu-seed-cluster-15-t3	86.8	90.1
prodrev-seed-wn-4	86.2	90.9
translated	88.4	90.2
opinhu-pmi	96.3	90.0
prodrev-pmi	84.3	91.9
prodrev2-pmi	-	91.0

In Table 3, the results of the baseline systems which used only lemmatized unigrams as features can be seen for both corpora, along with the best extended lexicons, the bootstrapped (*pmi*) lexicons and the manually *translated* lexicon (Section 2.1). Two baseline systems had been created, the first was trained on the opinhu corpus and the second on prodrev. The results show that the system not being trained on the same domain as the test corpus resulted in a significantly lower accuracy score. Furthermore, it can be seen that an increase can be achieved with the extending techniques comparing with the baselines if the lexicon is in the appropriate domain. If not, this increase is much smaller. The *translated* lexicon caused 2.3% increase in the opinhu corpus and only 0.2% in the prodrev database, which is less than the effect of the extended lexicons. The reason for this is that opinhu is not domain-specific and the lexicon which was translated was assembled for a similar text genre. The prodrev corpus is IT specific thus needs a lexicon from the same domain.

The bottom 3 rows of Table 3 shows the results for the bootstrapped lexicons. The prefix of each line indicates the annotated corpus which was used to create the lexicon. In those cases where the corpus used for the creation of the lexicon is the same as the corpus on which the sentiment analysis system was evaluated, the results show a theoretical maximum. This maximum shows the accuracy which can be achieved if we have a perfect lexicon for that corpus. It can be seen that these lexicons are not useful for the other domains as they can even decrease the results as well (prodrev-pmi lexicon on the opinhu corpus). We also tried to create a lexicon using texts from the domain of prodrev. For this we created the *prodrev2* corpus, which consists of those positive and negative reviews from the árukereső site that are not one sentence long (shorter and longer). Using this we managed to outperform the lexicons based on the extension methods.

5   Conclusions

In this work we focused on how to create sentiment lexicons automatically, which are useful in sentiment analysis tasks. We presented a technique to create lexicons from scratch by using annotated texts. We also gave methods for extending and adapting lexicons by using two types of word similarity measures. The input of these methods is a small seed lexicon (which we created semi-automatically) and/or (un)labeled domain-specific texts. Our results empirically underpin that it is important to use lexicons which are aware of the specificities of the domain on which the sentiment analysis system operates and by using a lexicon from a different domain the results can even be decreased. Although we achieved an increase in accuracy with the automatically created lexicons on the opinhu corpus, the best results were given by the manually assembled (and translated) lexicon. From this we can conclude that the manually created lexicons are better, but they are much more expensive and it is hard to create one for all domains, thus automatic methods are needed. In the IT specific domain we managed to reduce the errors by 10%. The results show that the proposed automatic methods are useful for increasing the performance of sentiment analysis systems in all domains.

References

1. Baccianella, S., Esuli, A., Sebastiani, F.: SentiWordNet 3.0: an enhanced lexical resource for sentiment analysis and opinion mining. In: Proceedings of the Seventh International Conference on Language Resources and Evaluation, LREC 2010 (2010)
2. Wilson, T., Wiebe, J., Hoffmann, P.: Recognizing contextual polarity in phrase-level sentiment analysis. In: Proceedings of the Conference on Human Language Technology and Empirical Methods in Natural Language Processing, pp. 347–354. Association for Computational Linguistics (2005)
3. Miháltz, M.: OpinHuBank: szabadon hozzáférhető annotált korpusz magyar nyelvű véleményelemzéshez. In: IX. Magyar Számítógépes Nyelvészeti Konferencia, pp. 343–345 (2013)
4. Liu, B.: Sentiment analysis and opinion mining. Synthesis Lectures on Human Language Technologies 5(1), 1–167 (2012)
5. Turney, P.D., Littman, M.L.: Measuring praise and criticism: Inference of semantic orientation from association. ACM Transactions on Information Systems (TOIS) 21(4), 315–346 (2003)
6. Miháltz, M., Hatvani, C., Kuti, J., Szarvas, G., Csirik, J., Prószéky, G., Váradi, T.: Methods and results of the Hungarian WordNet project. In: Proceedings of the Fourth Global WordNet Conference, GWC 2008. Citeseer (2008)
7. Brown, P.F., Desouza, P.V., Mercer, R.L., Pietra, V.J.D., Lai, J.C.: Class-based n-gram models of natural language. Computational Linguistics 18(4), 467–479 (1992)

Investigation of Word Senses over Time Using Linguistic Corpora

Christian Pölitz[1]([⊠]), Thomas Bartz[1], Katharina Morik[1], and Angelika Störrer[2]

[1] Artificial Intelligence Group, TU Dortmund University,
Otto Hahn Str. 12, 44227 Dortmund, Germany
christian.poelitz@tu-dortmund.de
[2] Germanistische Linguistik, Mannheim University,
Schloss, Ehrenhof West, 68131 Mannheim, Germany

Abstract. Word sense induction is an important method to identify possible meanings of words. Word co-occurrences can group word contexts into semantically related topics. Besides the pure words, temporal information provide another dimension to further investigate the development of the word meanings over time. Large digital corpora of written language, such as those that are held by the CLARIN-D centers, provide excellent possibilities for such kind of linguistic research on authentic language data. In this paper, we investigate the evolution of meanings of words with topic models over time using large digital text corpora.

Keywords: Word sense induction · Topic models · Time · Linguistic corpora

1 Introduction

Finding polysemy of words is an important linguistic analysis. For instance the word *bank* has multiple meanings in English. It can be used in the context of a credit institute or a river bank. Identifying these meanings not only helps understanding language better, it also can be used to filter out words in certain meanings from given contexts. The latter for instance is important to investigate usage of certain words in certain meanings.

Word sense induction (WSI) is a technique to find possible meanings of words based on automatic analysis methods. The most prominent methods in WSI are clustering of words, contexts of words or finding groups based on co-occurrence statistics. To automatically find possible meanings, large textual data set are usually used. In this paper, we investigate how good linguistic corpora are suited for WSI. We use Latent Dirichlet Allocation as introduced by Blei et al. [1] (LDA) for WSI on so called key words in context (KWIC) lists. A KWIC list contains snippets, usually some sentences, that contain a word for which we want to identify possible meanings. For convenience, throughout the paper we use the term document for the snippets in the KWIC lists. Linguistic infrastructure projects like Clarin-D provide excellent linguistic resources to retrieve such KWIC lists and to perform such linguistic research, see for instance McEnery et al. [8].

© Springer International Publishing Switzerland 2015
P. Král and V. Matoušek (Eds.): TSD 2015, LNAI 9302, pp. 191–198, 2015.
DOI: 10.1007/978-3-319-24033-6_22

Besides the pure identification of different meanings of words, the investigation of the development of these meanings over time is also an important linguistic task. For instance the English word *cloud* has recently got the new semantic context of cloud computing. Such emergences of new word meanings appear often over time. Interesting questions in this context are whether the meaning becomes the dominant meaning of the word or do several meanings coexist. Further in lexicography, the evolution of word meanings is important to construct descriptive examples to update existing dictionary entries as in Engelbert and Lemnitzer [4]. We use the continuous time topic model by Wang and McCallum [12] which is an extension of LDA to model also the temporal dimension of the word meanings.

In this paper, we investigate the development of meanings over time for the German language on the dictionary of the German language: "Wörterbuch der deutschen Sprache" (DWDS). The DWDS core corpus of the 20th century (DWDS-KK), constructed at the Berlin-Brandenburg Academy of Sciences and Humanities (BBAW), contains approximately 100 million running words, balanced chronologically (over the decades of the 20th century) and by text genre (belles-lettres, newspaper, scientific and functional texts). The newspaper corpus Die ZEIT (ZEIT) covers all the issues of the German weekly newspaper Die ZEIT from 1946 to 2009, approximately 460 million running words (see [5] by Geyken; [7] by Klein and Geyken).

2   Related Work

The induction of semantic meaning by usage patterns in the area of automatic analysis for linguistics is already well researched. An early statistical approach was completed by Brown et al. [3], Navigili [9] provides a comprehensive overview on the current research. Brody and Lapata [2] have shown, that they obtained the best results with the help of Latent Dirichlet Allocation [1]. In addition, they expanded their method to take into consideration various other context features besides the pure word occurrences (e.g. part of speech tags, syntax, etc.). Originally, LDA was used for thematic clustering of document collections. Navigli and Crisafulli [10] have already shown this to also be useful for the disambiguation of small text snippets, for example when clustering the search results from a web search engine. Rohrdantz et al. [11] showed the benefits of this method as a basis for the visualization of semantic change of example words from an English newspaper corpus, allowing them to observe the emergence of new meanings and reconstruct their development over time.

3   Topic Models

Topic models are statistical models that group documents and words from a document collection into so called topics. The words and documents that are associated with a topic are statistically related based on co-occurrences of words. Latent Dirichlet Allocation (LDA) as introduced by Blei et al. [1] has been successfully used for the estimation of such topics. In LDA, it is assumed that the words in a document are drawn from a Multinominal distribution that depends on latent factors, later interpreted as topics. We briefly summaries the generative process of document as the following:

1. For each topic z:
   (a) Draw $\theta_z \sim Dir(\beta)$
2. For each document d:
   (a) Draw $\phi_d \sim Dir(\alpha)$
   (b) For each word i:
       i. Draw $z_i \sim Mult(\phi_d)$
       ii. Draw $w_i \sim Mult(\theta_{z_i})$

Assuming a number of topics, we draw for each of them a Multinominal distribution of the words in this topic from a Dirichlet distribution $Dir(\beta)$ with metaparameter β. For each document we draw a Multinominal distribution of the topics in this document from a Dirichlet distribution $Dir(\alpha)$ with metaparameter α. Finally, for each word in the document we draw a topic with respect to the topic distribution in the document and a word based on the word distribution for the drawn topic. The metaparameter α and β are prior probabilities of the Multinominal distributions drawn from the Dirichlet distribution. These priors are the expected word probabilities in a topic before we have seen any data.

The generation of the LDA Topic Model is usually done by Variational Inference, as in the original work by Blei et al. [1], or via Gibbs samplers, as for proposed by Griffiths et al. [6]. We use Gibbs sampler to sample topics directly from the topic distribution. Integrating θ and ϕ out, we get for the probability of a topic z_i, given a word w in a document d and all other topic assignments:

$$p(z_i|w, d, z_1, \cdots z_{i-1}, z_{i+1}, \cdots z_T)$$
$$\propto \frac{N_{w,z_i} - 1 + \beta}{N_{z_i} - 1 + W \cdot \beta} \cdot (N_{d,z_i} + \alpha) \tag{1}$$

We denote $N_{w,z}$ the number of times topic z has been assigned to word w, $N_{d,z}$ the number of times topic z has been assigned to any word in document d, N_z the number of times topic z has been assigned to any word, W the number of words in the document collection and T the number of topics.

After a sufficient number of samples from the Gibbs sampler we get estimates of the word distributions for the topics and the topic distributions for the documents:

$$\theta_{w|t} = \frac{N_{w,t} + \beta}{N_t + W \cdot \beta} \tag{2}$$

$$\phi_{d|t} = \frac{N_{d,t} + \alpha}{N_d + T \cdot \alpha} \tag{3}$$

4   Topic Models over Time

While the standard topic models group only words and documents in semantically related topics, we are further interested in the distribution of the topics over time. In order to extract the distribution of word senses over time, we use topic models that consider temporal information about the documents. Each document has a time stamp. These time stamps are assumed to be Beta distributed. This Beta distribution is integrated in an LDA topic model. Wang and McCallum [12] introduced this model to investigate topics over time. Throughout this paper we call this method topics over time LDA. The generative process given by the Enumeration 2 is extended such that for each word w_i in each document, we also draw a time stamp $t_i \sim Beta(\psi_{z_i})$ with $\psi_{z_i} = (\alpha, \beta)$ the shape parameters of the Beta distribution. The shape parameters are estimated by the method of moments. After each Gibbs iterations the parameters are estimated in the following way: For each topic z we estimate the mean $\hat{m}$ and sample variance s^2 of all time stamps from the documents that have been assigned this topic. By the method of moments, we set $\alpha = \hat{m} \cdot (\frac{\hat{m} \cdot (1-\hat{m})}{s^2} - 1)$ and $\beta = (1 - \hat{m}) \cdot (\frac{\hat{m} \cdot (1-\hat{m})}{s^2} - 1)$ for each topic. Integrating the time stamp as Beta distributed random variable, we get for the probability of a topic z_i, given a word w in a document d with time stamp t and all other topic assignments:

$$p(z_i | w, d, t, z_1, \cdots z_{i-1}, z_{i+1}, \cdots z_T)$$

$$\propto \frac{N_{w,z_i} - 1 + \beta}{N_{z_i} - 1 + W \cdot \beta} \cdot (N_{d,z_i} + \alpha) \cdot \frac{(1 - t_d)^{\alpha - 1} \cdot t_d^{\beta - 1}}{Beta(\alpha, \beta)} \tag{4}$$

where the last term comes from the density of the Beta distribution at time stamp t_d and $Beta(\alpha, \beta)$ the Beta function.

5   Experiments

We perform experiments on the DWDS corpus for two German words with multiple meanings over time. The words "Platte" (with meanings board / disc / hard disc / plate / conductor), and "Ampel" (with meanings traffic light / a coalition of German parties (the social democrats (red), the liberals (yellow) and the green party).

   In the first experiment, we investigate how good possible different meanings can be found with respect to time. We compare the standard LDA with topics over time LDA. We use 10 topics and performed 2000 Gibbs iterations. In Figure 1 we plot the distribution of the extracted topics over time for both methods. We use the DWDS core corpus in the experiment to retrieve KWIC lists from the documents from 1900 till 2000. Standard LDA does not consider time, but we can accumulate the probability proportions of the documents for the topics grouped by time periods and plot them.

   From the two distributions we see that using topics over time LDA, we get a much clearer distinction of the topics over time. We can directly read off

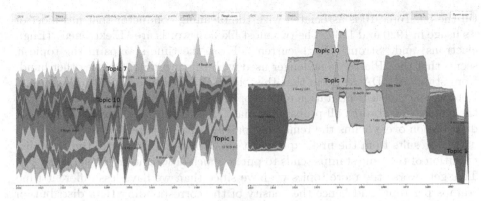

Fig. 1. The distribution of 10 topics extracted from the KWIC lists of the word "Platte" (board/disc/hard disc/plate/conductor). Left: Standard LDA; Right: LDA with topics over time.

the topics and the temporal period when this topic was prominent. From the standard LDA, we get a much more diffuse distribution of the topics over the time. The results indicate three possible main meanings that clearly separate over time. These topics are summarize in Figure 5. There we show the most likeliest words for the topic and the distributions of the time stamps as histogram. First, in topic 1 we find computer related words as most likely. The distribution of the time stamps shows a peak between 1990 and 2000. Before this period, this topic has not appeared. For topic 7, the most probable words indicate the meaning of a photographic plate for the word "Platte". The two most likeliest words are "Abb" which is short for "Abbildung" (Engl. picture) and "zeigt" (Engl. to show). The distribution of the time stamps shows a major usage of this meaning

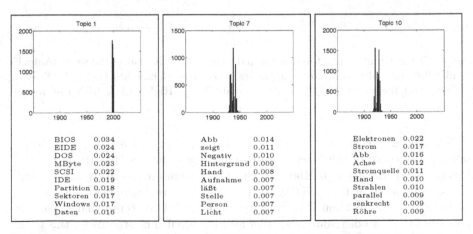

Fig. 2. Three topics extracted from the KWIC lists for the word "Platte" using topics over time LDA. Top: histograms of the time stamps in the topics. Bottom: most likely words in the topics.

till the 50. Topic 10 is associated with the meaning conductor that has most of its usage in 1920 and 1930. The two most likeliest words are "Elektronen" (Engl. electrons) and "Strom" (Engl. current). From the time stamps in the topic it seems that the "Platte" is no longer used with this meaning. On the other hand, from standard LDA we seem that this "Platte" is still used with this meaning. Here we see the clear limitation of the topics over time LDA: the separation into the time spans absorbs all probability mass. This mean, the density of the Beta distribution overwhelms the remaining parts of the topic probabilities.

The results from the first experiment show that the density of the Beta distribution of the time stamps tends to put to much of weight on the single topics. This gets worse the more topics we have since than we have less different time stamps per topic and hence the density of the corresponding Beta distribution gets very large at these time stamps.

To investigate this further, we perform another experiment with only 2 topics. On the Zeit corpus, we investigate the development of possible meanings from the word "Ampel". Figure 3 shows the distribution over time for the two topics. Compared to the previous experiment, we still get a very strict separation between the topics with respect to time. In Figure 4 we further investigate the distribution of the time stamps in the two topics. Additional to the histograms of the time stamps, we also show the fitted Beta distributions from each of the topics and the histogram of the time stamps over all topics.

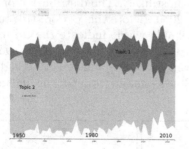 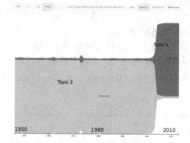

Fig. 3. The distribution of 2 topics extracted from the KWIC lists of the word "Ampel" (traffic light/a coalition of German parties (the social democrats (red), the liberals (yellow) and the green party)). Left: Standard LDA; Right: LDA with topics over time.

The histograms show also a strict separation of topic 1 and topic 2. Only a look on the curve of the fitted Beta distribution indicates that topic 2 is still present today. Investigating the number of documents respectively time stamps per year, we see that from 2000 on we have much more documents. This means, for topic 2, we have a much larger variance in the time stamp that makes the density of the Beta distribution smaller for individual time stamps for this topic. In topic 1 on the other hand, there are many times stamps from a small time period. This makes the Beta density much larger for those time stamps compared to the time stamps from topic 1.

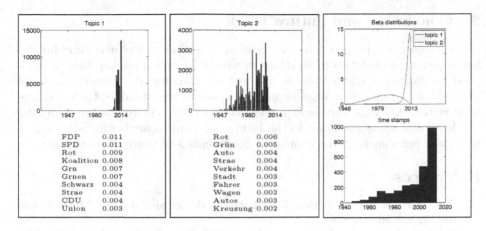

Fig. 4. Two topics extracted from the KWIC lists for the word "Ampel" using topics over time LDA. Top: histograms of the time stamps in the topics. Bottom: most likely words in the topics. Right top: fitted Beta distributions in the topics. Right bottom: histogram of time stamps over all topics.

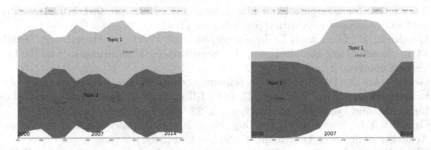

Fig. 5. The distribution of 2 topics extracted from the KWIC lists from 2000 till 2014 of the word "Ampel" (traffic light/a coalition of German parties (the social democrats (red), the liberals (yellow) and the green party)). Left: Standard LDA; Right: LDA with topics over time.

In Figure 5, we show the results of LDA and topic over time LDA on the same date as before but we filter out all KWIC list entries that have a time stamp from before 1999. Both figures show that the topic 2 still is present after year 2000. Topic 1 is very prominent from 2008 to 2010. Due to election periods and coalition talks between parties, these times make sense. Further, it also is clear that "Ampel" in the meaning of traffic light will no get lost. Topics over time LDA separates the topics still better and now also keeps topic 2 over the time.

To conclude, standard LDA can be used to identify the emergence of new topics to a certain degree. Topics over time LDA on the other hand separates topics over time quite quite, but sometimes to strict.

6   Conclusion and Future Work

In this paper, we proposed to use topic models that model also time for the investigation of evolution of meanings of words. In such a setting, time is modelled as addition random variable that is Beta distributed. We performed two extensive experiments on large linguistic corpora to test standard LDA and topics over time LDA for word sense induction over time. The results are promising but leave also some questions. In the future, we want to further investigate how to handle imbalances in the number of documents and words over the time.

References

1. Blei, D.M., Ng, A.Y., Jordan, M.I.: Latent dirichlet allocation. J. Mach. Learn. Res. **3**, 993–1022 (2003)
2. Brody, S., Lapata, M.: Bayesian word sense induction. In: Proceedings of the 12th Conference of the European Chapter of the Association for Computational Linguistics, EACL 2009, pp. 103–111. Association for Computational Linguistics, Stroudsburg (2009)
3. Brown, P.F., Pietra, S.A.D., Pietra, V.J.D., Mercer, R.L.: Word-sense disambiguation using statistical methods. In: Proceedings of the 29th Annual Meeting on Association for Computational Linguistics, ACL 1991, pp. 264–270. Association for Computational Linguistics, Stroudsburg (1991)
4. Engelberg, S., Lemnitzer, L.: Lexikographie und Wörterbuchbenutzung. Number 14 in Stauffenburg-Einführungen; 14; Stauffenburg-Einführungen. Stauffenburg-Verl., Tübingen, 2. aufl. edition (2004)
5. Geyken, A.: The DWDS corpus. A reference corpus for the german language of the twentieth century. In: Fellbaum, C. (ed.) Idioms and Collocations. Corpus-Based Linguistic and Lexicographic Studies, pp. 23–40. Continuum, London (2007)
6. Griffiths, T.L., Steyvers, M.: Finding scientific topics. Proceedings of the National Academy of Sciences **101**(Suppl. 1), 5228–5235 (2004)
7. Klein, W., Geyken, A.: Das Digitale Wörterbuch der Deutschen Sprache (DWDS) **26**, 79–96 (2010)
8. Mautner, G.: Tony mcenery, richard xiao and yukio tono, corpus-based language studies: an advanced resource book. Routledge, London (2006). pp. xix, 386. pb. Language in Society, 37:455–458, 7 2008
9. Navigli, R.: Word sense disambiguation: A survey. ACM Comput. Surv. **41**(2), 10:1–10:69 (2009)
10. Navigli, R., Crisafulli, G.: Inducing word senses to improve web search result clustering. In: Proceedings of the 2010 Conference on Empirical Methods in Natural Language Processing, EMNLP 2010, pp. 116–126. Association for Computational Linguistics, Stroudsburg (2010)
11. Rohrdantz, C., Hautli, A., Mayer, T., Butt, M., Keim, D.A., Plank, F.: Towards tracking semantic change by visual analytics. In: Proceedings of the 49th Annual Meeting of the Association for Computational Linguistics: Human Language Technologies: Short Papers, HLT 2011, vol. 2, pp. 305–310. Association for Computational Linguistics, Stroudsburg (2011)
12. Wang, X., McCallum, A.: Topics over time: a non-markov continuous-time model of topical trends. In: Proceedings of the 12th ACM SIGKDD International Conference on Knowledge Discovery and Data Mining, KDD 2006, pp. 424–433. ACM, New York (2006)

Dependency-Based Problem Phrase Extraction from User Reviews of Products

Elena Tutubalina[✉]

Kazan (Volga Region) Federal University, Kazan, Russia
tutubalinaev@gmail.com

Abstract. Capturing knowledge from customer reviews about products is an important object of interest for a company. This paper describes an approach to target extraction from user reviews of products. In contrast to other works, based on machine learning approaches, our system is defined by syntactic and semantic connections between possible targets and problem indicators. We present an approach where domain-specific targets are extracted using a problem phrase structure with dependency trees and semantic knowledge from a lexical database. The algorithm achieves an average F1-measure of 77%, evaluated on reviews from four different domains (reviews of electronic products, automobiles, home tools, and baby products). The F1-measure ranges from 76% for the reviews about baby products to 79% for automobile reviews.

Keywords: Opinion mining · Text classification · Information extraction · Mining defects with products

1 Introduction

Sentiment analysis of customer feedback has many possible benefits for a company that provides a service or a product; discovering problems from reviews covers an important case of opinion mining.

Problem extraction can be divided into two basic steps: (i) extraction of problem trigger (or problem indicator) and (ii) extraction of target for a given problem trigger. We define problem triggers as words and phrases that contain explicit links to a problem (words such as *problem, issue*, etc.) or implicit links to a problem (words such as *after, sometimes*, etc.). Targets are common entities, problems discussed by users in reviews of a particular domain.

Problem trigger detection and extraction of problem phrases have been studied in several papers ([1], [2], [3]). Recent studies on problem phrase extraction proposed different approaches to identify problem phrases: using a supervised classifier based on lexical and syntactic patterns [1], other works used unsupervised approaches based on the analysis of sentences' clauses, manually created dictionaries, and other heuristics.

In this study, we propose a method for the target extraction, discovering connections between a product description (the target of a problem phrase) and

© Springer International Publishing Switzerland 2015
P. Král and V. Matoušek (Eds.): TSD 2015, LNAI 9302, pp. 199–206, 2015.
DOI: 10.1007/978-3-319-24033-6_23

a trigger that describe a problem. The task is to identify which noun phrases (NPs) referred to a problem target in a sentence that contains at least one problem trigger. The target can be a product (*laptop*, *printer*, etc.) or part of a product (*display*, *SD card*, etc.). Nevertheless, we suppose that each problem trigger has a connection with a target. In contrast to other works, we propose a straightforward approach using syntactic and semantic connections between a problem trigger and mentions of a target.

Our contributions in this work can be summarized as follows: (i) we propose an unsupervised method for extracting problem phrases and targets that has been applied in different domains; (ii) we explore the use of a lexical resource to detect targets that are highly related with the review domain and use them to improve problem phrase classification.

2   Related Work

Extracting information from unstructured text has received much attention in sentiment analysis ([4], [5], [6], [7]), event detection, subjectivity detection ([8], [9]) and public sentiment tracking. Traditional approaches in opinion mining are based on extracting phrases containing adjectives or adverbs and on high-frequency noun phrases to detect product aspects [4]. These approaches are limited due to lower results on extracting low-frequency aspects. Another group of related work explore extracting information about subjectivity, based on dependency relations to classify deeply nested clauses [8]. State-of-the-art papers have implemented probabilistic topic models, such as Latent Dirichlet Allocation (LDA), for fine-grained multiaspect analysis tasks [10]. However, topic models achieve lower performance on multiaspect sentence classification than Support Vector Machine model in different domains [10].

Problem detection and extraction of problem phrases from texts are less studied. Ivanov and Tutubalina [2] used a clause-based approach to problem-phrase extraction from user reviews of products. They achieved a recall of 77% and a precision of 74% for user reviews about electronic products. The authors reported that clause-based approach performed well compared to the simple baseline given by supervised machine-learning algorithms. However, error analysis has shown that there are cases where authors classify sentences to a problem class when they are not related to any problem with particular products. The current task of our research is identifying the targets of problem phrases to reduce classification errors. Gupta [1] studied the extraction of problems only with AT&T products and services from English tweets using a maximum entropy classifier. Gupta reported F-measure of 75% for identification of the target phrase. Our method is based on grammatical domain-independent relationships in a sentence.

3   Target Phrase Extraction

In this section, we describe our method for extracting problem phrases, related to the targets, from customer reviews. The approach is composed of two steps. The first step is related to problem-phrase identification and consists of the following:

- collecting problem indicators (i.e., words or phrases that indicate a problem).
- extracting from a parse tree of a sentence[1] a problem phrase that contains problem indicators from manually created dictionaries.

The second step of our method relates to the classification step. It consists of the following:

- generating a set of possible targets by combining problem indicators with mentions about products in given problem phrases. In this step, we use dependencies with the problem indicator in the same problem phrase.
- detecting domain-specific targets from the set using WordNet-based semantic relatedness measure.
- classifying the sentence as a problem sentence if the approach finds at least one combination of the problem indicator with the domain-specific target.

3.1   Dependency Relations for Target Extraction

The second step uses sentence dependencies to determine connections between possible targets and problem indicators. We propose that existing phrase dependencies contain common contextual information from the indicator. The phrase describing the problem is a combination of the target and the problem indicators. The rationale behind dependency-based extraction of targets is that only syntactically and semantically rich mentions of targets are extracted, thus reducing noise in the extracted set of nouns.

To identify connections between the problem indicator and the targets of problem phrase, we use direct and indirect dependency relations between two words that are defined in [7]. A direct dependency indicates that one word directly depends on another word. Indirect dependency indicates that one word depends on another word through some additional words, which we call *successors*. A successor is an additional word that connects to a problem indicator and replaces a problem indicator in relation with a target. Sentences 1–2 show examples with indirect dependency relations of the selected types, taken from the review sentences. PW refers to a problem indicator, and T refers to a target of a problem phrase, and S refers to a successor of a problem indicator.

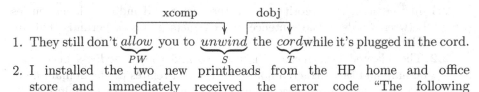

1. They still don't *allow* you to *unwind* the *cord* while it's plugged in the cord.
2. I installed the two new printheads from the HP home and office store and immediately received the error code "The following printhead has a problem".

---

[1] For parsing we use the Stanford parser (http://nlp.stanford.edu/software/lex-parser.shtml).

3.2 Calculating Semantic Relatedness of Problem Targets

In general, a product review is a set of those product parts with which a customer has problems. We use semantic information from WordNet to find the domain-specific target of a problem. Domain-specific targets are objects that have important meanings in a particular domain. We decide to use Wordnet as a standard lexical semantic resource due to synsets' fine-grained definitions.

We consider several relations such as hyponymy, hypernymy, meronymy and holonymy between the target and the WordNet term to find *semantic relatedness* between the product review domain and the target. Semantic relatedness is used to determine how much two targets are semantically connected by using the relations between them. We use an adapted Lesk algorithm according to the synset hierarchies provided in WordNet (i.e., type-of and part-of relations). The common idea of the Lesk algorithm is that related senses can be identified by finding overlapping words in their definitions ([11]). For example, user reviews about electronic products clearly relates the words *button* and *computer*, but WordNet doesn't have a direct link between their synsets. The glosses of these two synsets have similar words and synsets explicitly related to *button* are also related to *computer's* synsets. The synset of *electric switch* (which is the hypernym synset of *button*) is related to the synset of *peripheral* (which is the meronym synset of *computer*).

We briefly describe relatedness calculation. The overlap score $overlap(g_1, g_2)$ is calculated as $\sum_n m^2$ for n phrasal m-word overlaps, where g_1 and g_2 are definitions for the words w_1 and w_2, respectively.

3.3 Dependency-Based Approach

We suppose, that the domain-specific problem phrase is a combination of the target and the indicators. The method may be briefly described as follows:

Step 1. In this step, the algorithm looks for verb phrases headed by an action verb with a negation or looks for problem words from manually created dictionaries (without related negations)[2].

Step 2. Given a set of the found problem indicators, the method looks for possible targets for each problem indicator.

2.1 In this step, the algorithm uses the direct and indirect dependencies between words in the sentence and extracts nouns or gerunds that are related to the indicator.

2.2 The adapted Lesk measure is used to find the relatedness between a domain and the extracted noun. If the noun relates to the domain more than to general word *product*, we mark the noun as the domain-specific target and extract a pair (indicator, target). We describe the algorithm in algorithm 1.

Step 3. The algorithm marks the sentence as a problem sentence if at least one pair (indicator, target) is extracted in step 2.

---

[2] The ProblemWord dictionaries are previously described in [2].

Algorithm 1. Pseudo-code for extracting domain-specific targets according to the problem indicators in the sentence

---

1 **Function** `lookupForProblemsWithTargets`(*s, domain_term*)

    Input: s – the input sentence, domain_term – the common domain-specific term

    Output: PWTs – set of pairs (problem indicator, target)

2     $PWTs \leftarrow \emptyset$

    `/* detecting the problem indicators in the sentence      */`

3   PWs = lookupForPW(s);

    `/* get the typed dependencies of the sentence      */`

4   DRs = (getGrammStructure(s)).typedDependenciesCollapsed(true)

    foreach *pw in PWs* **do**

5   targets=lookupForRelatedTargets(pw, DRs)

6   **foreach** t_i *in targets* **do**

            `/* calculate semantic relatedness of the target to the`
            `common domain term and to the domain term product      */`

7   **if** $relScore(domain\_term, t_i) \geq relScore(domain\_term, "product")$ **then**

8   PWTs = PWTs $\cup$ {pair(pw, t_i)}

9   **return** *PWTs*

4　Evaluation and Experiments

For our experiments, we collected a testing corpus from [3] and annotated sentences from Amazon reviews[3] for three categories of products in a similar way. The distribution of sentence classification in the corpus is presented in Table 1. Each sentence does not have any particular label for targets and contains at least one problem indicator that the approach can find. We propose that the problem phrase with the problem indicator always has targets, but not necessarily domain-specific targets. The approach identifies problem phrases correctly if targets are extracted in connection with the problem indicators. We view this as text classification for our performance metrics.

Table 2 presents domain-specific targets for each domain from the sentences, that have the highest or lowest scores, based on the semantic relatedness measure. These results suggest that there is no a single common dictionary for detecting targets from different reviews about products. For example, words such as *sheet* and *Internet* can be associated with products from *baby* and *electronic* domains, respectively, though they are not the targets in the reviews about tools.

We used rule-based methods to compare the performance metrics, as follows:

1. We consider the targets extracted by direct dependencies (we did not use any lexical knowledge).

---

[3] The dataset is available at https://snap.stanford.edu/data/web-Amazon.html

Table 1. Summary of customer review dataset after three MTurk runs.

Product domain	No. of sentences	No. of sentences in problem class	No. of sentences in no-problem class
Electronics	720	498	222
Car	998	827	171
Home improvement tools	850	611	239
Baby	1,143	780	363

Table 2. Domain-specific targets, that have highest or lowest scores.

Domain	Domain term	Targets, extracted using our approach					
		with highest scores			with lowest scores		
Electronics	computer	computer	cables	printer	toner	corner	purchase
		windows	program	discs	ribbon	doubt	requirement
		hardware	driver	router	future	magenta	description
Car	car	car	auto	gas	tech	fun	reviewer
		motorcycle	shifter	truck	logo	laptop	diagnosis
		accelerator	wiper	bike	xenon	laser	sunlight
Home tools	tools	equipment	drill	shaft	coupon	fuses	Internet
		controller	hook	tap	site	window	sheet
		switch	grip	knife	veneer	wheel	strobe
Baby	stroller or carriage	stroller	device	wheel	guide	mobile	clue
		mechanism	fabric	cloth	coffee	drawer	sunshade
		material	seat	toy	piano	stairs	technolody

2. We consider all targets extracted by direct and indirect dependencies as problem domain-specific targets.
3. We consider domain-specific targets extracted by direct and indirect dependencies and calculate the WordNet semantic relatedness of the targets. We extract the targets if the selected target does not have a lexical meaning such as time or person and relates to the domain more than to the general word *product*.
4. In addition to domain-specific targets, extracted using WordNet and semantic relatedness with the domain, we consider compound targets, that do not have related synsets in WordNet. These targets are also extracted by direct and indirect dependencies.

For the binary classification, we computed precision, recall, and F1-measure. Performance metrics are calculated on the dataset that contains sentences from four product types (domains): *computer electronics*, *cars*, *home tools*, and *baby* products. The performance metrics are provided in Table 3.

The average recall and the average precision of problem-phrase extraction, related to domain-specific targets, are 0.78 and 0.77, respectively. The best F1-measure is 0.79 for domain-specific targets in the car review dataset, based on direct and indirect dependencies, WordNet synsets, and the Lesk measure of lexical semantic relatedness. We compare our results with the clause-based method

Table 3. Performance metrics of the dependency-based approach.

Method name	Electronics			Cars			Home tools			Baby		
	P	R	F1	P	R	F1	P	R	F1	P	R	F1
Direct Dependencies (DD)	.73	.70	.71	.83	.67	.74	.75	.68	.71	.67	.62	.65
DD+Indirect Dependencies (ID)	.73	.93	**.82**	.83	.92	**.87**	.72	.93	**.81**	.66	.92	.77
DD+ID+Semantic relatedness (SR)	.74	.77	.76	.86	.69	.77	.78	.70	.74	.70	.77	.74
DD+ID+SR+Targets without synsets	.75	.82	.78	.85	.74	.79	.78	.73	.76	.69	.84	.76
Clause-based approach	.96	.71	.81	.91	.83	.86	.87	.72	.79	.91	.68	**.78**

that used to detect whether a sentence contains a problem mention or not [2]. As shown in Table 3, the performance metrics of the rule-based method based on direct and indirect dependencies are very close to those that have been shown by the clause-based approach. However, the dependency-based approach that incorporates domain knowledge from WordNet and computes semantic relatedness between synsets, does not show any significant improvement over the baseline due to difficult notion of a problem with domain-specific targes.

4.1   Difficulties in Evaluation

After the analysis of errors, we define four types of problem phrases about products depending how users denote the target. A problem sentence about products from user reviews includes the problem indicator and:

- the product target, which is indicated by the problem phrase in the text.
  3. *These drill bits do not perform as advertised on TV.*
- the product target as a key component that makes the product functional (*keyboard, hard drive*, etc.).
  4. *Windows 8 also does not allow us to install a HP 4000 laser printer.*
- the product target, which is well-known according to the problem indicator. The following sentence is directed to a problem with a printer.
  5. *I could not print after one month of use.*
- the product target, which is not established as a particular object and is an undetermined problem.
  6. *Even though our baby was light, it was still hard to carry around the baby.*

Example 5 requires knowledge from a domain to determine, whether *I could not print* means *printer does not work*. In example 6, the sentence, describing a problem, corresponds to a car seat. Our result analysis indicates that users describe the whole situation, not only reporting about a problem with the product. While humans can understand which product causes the difficult situation in example 6, our approach marks these sentences as sentences without the targets about baby products. These types of problem phrases decrease the recall (from 92% to 84% in the *baby* domain).

Another group of errors is connected with the WordNet knowledge. WordNet is limited and does not include many proper nouns (e.g., *Macbook, Honda*) related to a particular domain (e.g., *computer electronics, automobile*) in contrast with domain-specific dictionaries. Our approach does not determine the proper targets, that do not have Wordnet synsets, and these errors decrease the recall.

5 Conclusion

In this paper, we propose an unsupervised method for extracting a problem phrase and its target from user reviews of products. Without using domain-specific knowledge about products, we focus our attention on dependency-based syntactic information between the target and the problem indicators in the text. We use the adapted Lesk algorithm to reduce targets that are not semantically related to a product domain. Our approach performs well in different domains (from an F1-measure of 76% to 79%); the average value of F1-measure is about 77%. The results indicate that the proposed approach could improve the performance of the problem classification of user reviews about products. For future works, we plan to use our approach for texts in the Russian language.

Acknowledgments. This work was supported by Russian Science Foundation (Project 15-11-10019).

References

1. Gupta, N.: Extracting phrases describing problems with products and services from twitter messages. Computacin y Sistemas. **17**, 197–206 (2013)
2. Tutubalina, E., Ivanov, V.: Clause-based approach to extracting problem phrases from user reviews of products. In: Proceedings of the AIST 2014 (2014)
3. Solovyev, V., Ivanov, V.: Dictionary-based problem phrase extraction from user reviews. In: Sojka, P., Horák, A., Kopeček, I., Pala, K. (eds.) TSD 2014. LNCS, vol. 8655, pp. 225–232. Springer, Heidelberg (2014)
4. Turney, P.D.: Thumbs up or thumbs down?: semantic orientation applied to unsupervised classification of reviews. In: Proceedings of the 40th Annual Meeting on Association for Computational Linguistics, pp. 417–424 (2002)
5. Liu, B., Hu, M.: Mining and summarizing customer reviews. In: Proceedings of the Tenth ACM SIGKDD, pp. 168–177 (2004)
6. Lu, B.: Identifying opinion holders and targets with dependency parser in chinese news texts. In: Proceedings of the NAACL HLT 2010 Student Research Workshop, pp. 46–51. Association for Computational Linguistics (2010)
7. Bu, J., Chen, C., Qiu, G., Liu, B.: Opinion word expansion and target extraction through double propagation. Computational linguistics, 9–27 (2011)
8. Hwa, R., Wilson, T., Wiebe, J.: Just how mad are you? finding strong and weak opinion clauses. In: AAAI, pp. 761–769 (2004)
9. Cardie, C., Breck, E., Choi, Y.: Identifying expressions of opinion in context. In: AAAI, pp. 2683–2688 (2007)
10. Cardie, C., Lu, B., Ott, M., Tsou, B.K.: Multi-aspect sentiment analysis with topic models. In: 2011 IEEE 11th International Conference Data Mining Workshops (ICDMW), pp. 81–88 (2011)
11. Patwardhan, S., Banerjee, S., Pedersen, T.: Using measures of semantic relatedness for word sense disambiguation. In: Gelbukh, Alexander (ed.) CICLing 2003. LNCS, vol. 2588, pp. 241–257. Springer, Heidelberg (2003)

Semantic Splitting of German Medical Compounds

Claudia Bretschneider[1,2(✉)] and Sonja Zillner[1,3]

[1] Siemens AG, Corporate Technology, Munich, Germany
{claudia.bretschneider.ext,sonja.zillner}@siemens.com
[2] Center for Information and Language Processing,
University Munich, Munich, Germany
[3] School of International Business and Entrepreneurship,
Steinbeis University, Berlin, Germany

Abstract. Compounding is widespread in highly inflectional languages with a quarter of all nouns created by composition. In our field of study, the German medical language, the amount of compounds significantly outnumbers this figure with 64 %. Thus, their correct splitting is a high-impact preprocessing step for any NLP-based application. In this work we address two challenges of medical decomposition: First, we introduce the consideration of unknown constituents in order to split compounds that were not recognized as such so far. Second, our approach builds on the corpus-based approach of Koehn and Knight and adds semantic knowledge from domain ontologies to increase the accuracy during disambiguation of the various split options. Using this first-of-a-kind semantic approach in a study on decomposition of German medical compounds, we outperform the existing approaches by far.

Keywords: Compound splitting · Medical NLP · Semantics · Ontology

1 Introduction

In languages used in technical domains such as the medical language, the splitting of compound terms (also known as *decomposition*) is a major issue in text preprocessing. Compounds are words that are built by concatenating single base forms to a new, single word [1], like compounding *krank* (Eng.: ill) and *Haus* (Eng.: house) to *Krankenhaus* (Eng.: hospital). It is necessary to split a language's compounds into its single constituents in order to build Machine Translation (MT), Information Retrieval (IR) or Information Extraction (IE) applications on top of the preprocessed text.

In languages that are rich in morphological forms, such as German, Swedish or Dutch, a huge amount of compounds is used – about 25 % of the nouns used are compounds [2]. In domains such as medicine compounding is even more widespread: Our study shows that in a medical corpus 63.69 % of all nouns are compounds, because of the long tradition of compounding in the medical language with its Greek roots [3].

© Springer International Publishing Switzerland 2015
P. Král and V. Matoušek (Eds.): TSD 2015, LNAI 9302, pp. 207–215, 2015.
DOI: 10.1007/978-3-319-24033-6_24

Current state-of-the-art approaches either rely on statistical information from corpora or make use of lexicon information or linguistic knowledge in order to find constituents and the correct splitting option for a given compound. However, for technical domains like medicine, these approaches are only partially applicable. This is because the corpus or standard lexicons used do not contain such special sublanguage vocabulary as single constituents for compound splitting. Furthermore, relevant domain ontologies for non-English vocabularies are only partially translated, e.g. the RadLex ontology relevant for our study only contains German translations for 25 % of the terms.

At the same time, Langer [1] demands that the decomposition should include a disambiguation of different splitting options that results in the semantically most reasonable one. However, the currently available approaches ignore this semantic disambiguation completely and, e.g. split *Dünndarmschlingen* (Eng.: slings in small intestine) to *Dünn–darm–schlingen* (Eng.: small – slings – in intestine) instead of *Dünndarm–schlingen* (Eng.: small intestine – slings). While the first option leads to a semantic misinterpretation, the second one is more reasonable for the medical world.

Hence, we propose a decomposition approach that integrates the semantics from domain ontologies into the existing corpus-based compound splitting algorithm to choose the most appropriate splitting option. This leads to an approach that is applicable for any technical domain that requires semantic information for improved compound splitting.

The paper's contribution is fourfold: First, we describe how we generate splitting options based on Koehn and Knight's approach (Section 3). Second, we introduce *unknown constituents* to decomposition, which enables splitting of compounds that remained unsplit so far (Section 3.4). Third, we show how domain-specific knowledge can be incorporated for disambiguation (Section 3.5). Fourth, we evaluate our approach using German medical nouns using precision and recall measures (Section 4).

2 Related Work

Langer [1] is the first one to recognize the importance of compound splitting for correct semantic analysis and considers the semantic class of the head term when selecting the semantically most reasonable splitting option. Brown [4] conducted a first, but rather simplistic approach to medical compound splitting by comparing terms of parallel corpora using the Levenshtein distance. Hence, German compounds are split when the corresponding (English) multi-word terms are similar.

By splitting compounds into its constituents, an improved performance can be observed for IR and MT tasks. Several papers have investigated the effectiveness of compound splitting in comparison to stemming for IR: Braschler et al. [5] split based on lexical base forms and observe an increase in precision to up 20%. They stress that decomposition contributes more to performance than stemming, but requires a far more sophisticated method. Monz et al. [2] observe the same effect on Dutch, Italian and German with their greedy approach of splitting a compound to the smallest units possible. In MT tasks, the importance of decomposition has been research with highly inflectional languages: The most popular approach by Koehn and Knight [6] is based on corpus terms and their frequencies to find split options and disambiguate the most probable. Stymne [7] transfers the approach for translating German and English parallel corpora. Popovic et al. [8] compare linguistic and corpus-based approaches and conclude that linguistic approaches tend to oversplit because of the missing frequencies. Fritzinger et al. [9] combine the linguistic and corpus-driven approach into a hybrid that outperforms the isolated approaches.

3   Semantic Compound Splitting Approach

Our compound splitting approach operates in five steps and relies on two resources: The *corpus* serves as basis for extracting constituent candidates and generating split options. For their disambiguation we rely on the semantic knowledge of *domain ontologies*. We adapt the compound splitting algorithm of Koehn and Knights's [6], but optimize major steps such as the generation of split options and and their disambiguation.

3.1   Extract Constituent Candidates from Corpus

In the first step, we create a list of possible constituents from a given corpus. First, the text is tokenized in order to acquire valid constituents to generate splitting options. We use spaces and hyphens for tokenization. Thus, all compounds written with hyphens such as *Abdomen-CT* are already split and its parts are regarded as individual tokens. Second, as in the original approach, statistical indicators for later disambiguation are gathered. Therefore, the frequency of each previously extracted token with length three or longer is counted. In further processing these tokens are regarded as *known constituents* and are used for creation of all probable split options.

3.2   Generate Corpus-Based Split Options

The generation of likely splitting candidates makes use of the known constituents from the previous step combined with additional filler morphs. At this point, each combination of known constituents and fillers that fully covers the term is regarded as a valid split option. One candidate represents the input term remaining unsplit. The original algorithm limits the fillers to $+s$ and $+es$, which are presumed to be the most common ones. In order to overcome the shortcoming that some compounds are not split, we expand the fillers used by those introduced by Langer [1] shown in Table 1. Furthermore, the medical language as a derivative of the Latin and Greek language uses additional fillers, such as $+a$ and $+o$, that need to be integrated (shown in Table 2).

Table 1. Fillers cited from Langer [1]

Type	Suffixes
Additions	+s +n +en +nen
	+e +es +er +ien
Truncations	-e -en
Umlaute	+"/+e +"/+er
Combinations	-us/+en -um/+en -um/+a
	-a/+en -on/+en -on/+a -e/+i

Table 2. Additional, medical language specific fillers used in the domain-specific compound splitting process

Type	Suffixes
Addition	+ial
Combinations	-um/+o -um/+al
	-a/+o -o/+a
	-al/+a -eus/+id

3.3 Dismiss Split Options Based on POS Tags and Suffixed

In order to avoid nonsensical splittings like splitting *folgenden* (Eng.: following) to *folgen* (Eng.: consequences) + *den* (Eng.: the), each constituent of the probable splittings is POS tagged. We use the TreeTagger [10] for POS tagging and use the resulting tags for validating splitting options that contain only content words such as nouns, adverbs, adjectives, and verbs; others that include prepositions or determiners are dismissed. Hence, split options containing constituents with POS tags other than ADJA, ADJD, ADV, NN, NE, PTKNEG, VVFIN, VVINF, VVIZU, VVPP, VAFIN, VAIMP, VAINF, VAPP, VMFIN, VMINF, VMPP are no longer regarded as valid. In the example *folgenden* is not split as the constituents include a determiner, which is not a valid constituent. Furthermore, as the tagger does not tag suffixes with a dedicated tag, we include the recognition of the German nominalization suffixes *-keit*, *-heit*, *-ung*, *-nis* and define them also as excluded constituents. That is why splitting *Krank–heit* (Eng.: ill–ness) to an adjective and a suffix constituent is regarded as irregular.

3.4 Generate Split Options that Include Unknown Constituents

If an input term cannot be fully decomposed by combining known constituents and fillers, the whole term remains unsplit. This can happen even though the input term is a compound. E.g. the compound *Herzthoraxrelation* cannot be subdivided into its constituents *Herz–Thorax–Relation*, even though *Herz* and *thorax* are recognized as known constituents, because the word *Relation* is not used in the given corpus. We introduce a handling for these cases: If at least one starting constituent is known from the corpus, the remaining but unrecognized tail part is assumed to be an *unknown constituent*. We introduce the notion of *unknown constituents* to address valid constituents that do not appear as distinct terms in the corpus but are part of a compound. With their consideration additional, so far unrecognized split options of a compound are created.

3.5 Disambiguation of Split Options

For the disambiguation of the most appropriate split option from the previously created set, we introduce semantic information about the constituents.The consideration of this knowledge is novel and has not been taken into account

Table 3. Three split options s_n for the compound *Beckenbodenmuskel*. The number in parenthesis indicates the constituent's frequency. Underlined constituents are RadLex terms with indicated RIDs.

s_1	Beckenbodenmuskel(2)		
	RID378		
s_2	Beckenboden(1)		muskel(2)
	RID377		RID13196
s_3	Becken(568)	boden(38)	muskel(2)
	RID2507	RID5959	RID13196

Table 4. Hierarchical relations from the RadLex ontology used for disambiguation. A denotes the superior ontology concept to sub-concept B.

A Has_Part, Has_Constitutional_Part, B Has_Subtype, Has_Regional_Part

B Part_Of, Regional_Part_Of, Consti- A tutional_Part_Of, Is_A, Anatomical_Site, subClassOf

for the decomposition task so far. In particular for domains such as medicine, whose vocabulary is tightly associated with its semantics, this approach is of special interest. We implement a 3-step process that compares existing split options pairwise. Details are explained using the split options of the compound *Beckenbodenmuskel* (pelvic floor muscle) shown in Table 3.

Step #1 – Disambiguate using Semantic Relations. We disambiguate splitting options based on the semantic relations between constituents. These relations are taken from domain ontologies and describe hierarchies between the constituents. For our study on medical compounds we use the relations from the RadLex ontology that define any kind of hierarchical relation (see Table 4). For the example, we extracted the constituent relations shown in Figure 1. If the constituents of compounds are semantically related, we use this information for their disambiguation. Constituents are related if (1) one constituent is a substring of the other and (2) they relate via the defined ontology relations (with an edge length of up to three). We define that the split option with the super-

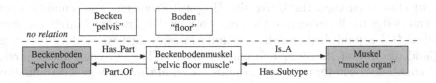

Fig. 1. RadLex relations of constituents in *Beckenbodenmuskel*. Superclass concepts in gray.

$$\text{argmax}_S \frac{|c_v|}{|c|}$$

$$\text{argmax}_S \prod_{c \in S} (count(c_i) + 1)^{\frac{1}{n}}$$

Fig. 2. Formula to calculate the preferred split s with highest amount of vocabulary constituents c_v compared to all constituents c

Fig. 3. Smoothed formula to calculate the geometric mean of a given compound based on constituent counts c_i

class constituent is the preferred one, because the subclass concept is much too specific to remain compound. When disambiguating s_1 and s_2, we recognize that all superclass concepts are constituents of s_2, so that this split option is preferred to s_1; from s_1 and s_3, the split option that contains the more general constituent (s_3) is preferred.

Step #2 – Disambiguate using Domain Vocabulary. However, if the constituents of two split options do not semantically relate (as s_2 and s_3), the second disambiguation step needs to include vocabulary information: If a constituent is part of the given vocabulary, it is assumed to be more important than constituents that are out of vocabulary. The compound with the higher amount of vocabulary constituents is preferred (see formula in Figure 2). Since both s_2 and s_3 have the same amount of RadLex vocabulary constituents (1.0), the usage of a third disambiguation step is necessary.

Step #3 – Disambiguate using Geometric Mean. Our fall back is the algorithm described by Koehn. It has proven to be valid to find the statistically most appropriate splitting option by calculating the geometric mean of the constituents' frequencies. However, with the consideration of unknown constituents, whose frequency is zero, the formula needs adjustment. Therefore, add-one smoothing is applied (see Figure 3). For the disambiguation of s_2 and s_3 we calculate their mean scores. As a result, s_3 (40.53) is preferred to s_2 (2.45). Finally, the disambiguation prefers s_3 to s_2 to s_1 and determines s_3 as most semantically reasonable option to split the compound to *Becken–boden–muskel*.

4   Evaluation

4.1   Evaluation Resources

We evaluate the approach by using all nouns from a medical corpus. This corpus contains 2,713 German radiology reports of lymphoma patients and was provided by our clinical partner, the University Hospital Erlangen. The semantic medical knowledge to disambiguate the split options delivers the RadLex ontology. RadLex [11] is published to deliver a uniform controlled vocabulary for indexing and retrieval of radiology information sources. The current English version 3.11 contains 34,446 terms, which are integrated in the ontology with numerous hierarchical relations (a selection of relevant relations is listed in Table 4).

4.2   Evaluation Technique

Each of the 7,332 nouns from the corpus is annotated by a human annotator with the information whether it is a compound or not. If the noun is a compound, the valid split option is provided in addition. The percentage of compounds 63.8 % (4,675 nouns) exceeds the generally assumed amount of compounds of 25 % in German texts by far. We evaluate the results using precision, recall and accuracy metrics as defined in Table 5. The baseline is built using Koehn's original approach. We compile different versions of (1) fillers, (2) unknown constituents and (3) semantic disambiguation applied for compound splitting (Table 6). The additional challenge of handling a German corpus and its associated German ontology version is described as (4)th case.

Table 5. Definition of evaluation metrics: Precision, recall and accuracy

correct split:	is compound and was split correctly	**precision:** $\dfrac{\text{correct split}}{\text{correct split} + \text{wrong faulty} + \text{wrong split}}$
wrong split:	is base form but was split	**recall:** $\dfrac{\text{correct split}}{\text{correct split} + \text{wrong faulty} + \text{wrong not}}$
wrong not:	is compound but was not split	**accuracy:** $\dfrac{\text{correct split}}{\text{correct split} + \text{wrong}}$
wrong faulty:	is compound but was split wrongly	

Table 6. Overview of the evaluation steps and measures clustered by (1) fillers applied, (2) unknown constituents, (3) disambiguation step and (4) translated vocabulary

	Filler	Unknown	Disambiguation indicator	Precision	Recall	Accuracy
(1) K+K	Koehn	–	geometric mean	0.7895	0.0032	0.0032
Langer	Langer	–	geometric mean	0.8626	0.3170	0.3129
Medical	Medical	–	geometric mean	0.8641	0.3251	0.3207
(2) Unknown	Medical	yes	geometric mean	**0.8910**	0.6473	0.6315
(3) Known	Medical	yes	domain vocabulary	0.8161	0.7889	0.7344
Semantic	Medical	yes	semantic relations	0.8153	**0.7976**	0.7397
(4) Translated	Medical	yes	semantic relations	0.8294	0.7924	**0.7444**

4.3   Evaluation Results

In order to see whether Koehn's hypothesis holds that most **filler usage** can be covered by the suffixes $+s$ and $+es$, we first test the effect adding Langer's and later the medical fillers. The fillers additionally integrated into the decomposition process significantly increase the recall, i.e. more constituents are successfully recognized and correctly split. Although additional (medical) fillers boost the measures only to a small extend, we recommend to use them if available, as only little effort is needed to add these.

In a next step, also **unknown constituents** are considered in the decomposition process. The splitting of compounds with unknown tailing constituents delivers a significant increase in recall. At the same time, precision slightly increases. Other approaches so far ignored these options because the respective constituents are not included in the resource they use (corpus or lexicon). In this work, we proved that the splitting of compound that also include unknown constituents can be integrated and is beneficial even though there are still open issues in this technique.

As a third step, we evaluate how the different **disambiguation steps** perform for the medical corpus. We integrate domain semantics using ontologies and increased the number of recognized and successfully split compounds by more than 14 %. However, this increase comes along with the tradeoff of decreasing precision.

Finally, dealing with **German compounds** we have to consider a special challenge: Most domain ontologies are developed for the English language and translations are only partially introduced as non-English ontology concept labels. The RadLex ontology also contains German labels for the concepts, but the number of translations reaches a mere 25 %. In another work, this lack of terminology was targeted with a corpus-based translation approach for ontologies and integrates additional 558 translations into the RadLex ontology as additional vocabulary [12]. We assumed that additional translations lead to an increase in successful splitting, because more splitting options can be successfully disambiguated. The numbers show that there is indeed an increase, but the significance of a complete translated ontology can only be assumed because the additional translations only account for additional translation coverage of 1.6 % (558/34,446).

4.4   Discussion and Future Work

The analysis of the evaluation results reveals two major issues, whose implementation brings additional increase in compound splitting, but are left as future work. First, even though the recognition of compounds with unknown (tailing) constituents is introduced, still the whole compound remains unsplit if the head constituent is not known from the corpus. Hence, we plan to integrate an additional feature that considers unknown starting and known tail constituents, when splitting compounds. Second, for the semantic disambiguation of the splitting options, the coverage of the ontology is of upmost importance. E.g., the correct splitting of *Bauch–wand* depends on the availability of both concepts *Bauch* and *Bauchwand* and their relation within the ontology. However, as the term *Bauch* is not translated in RadLex 3.11, 32 compounds of our study are split wrong or are not split at all – just for one translation missing in the ontology.

Finally, we show in our study that the proposed approach is applicable for the medical domain, however the approach can be adapted for any other given domain. The only requirement imposed is that semantic information on the domain terminology is available in form of hierarchical knowledge. As future work we want to show that even general language semantic compound splitting can be conducted, e.g. by incorporating WordNet's hypernym/hyponym relations.

5   Conclusion

We introduce a technique for compound splitting in technical domains. Therefore, we uncover and resolve the main obstacle of currently used statistical approaches by integrating two further steps: consideration of unknown constituents in the splitting options generation and their semantic disambiguation. By integrating unknown constituents, we are able to split compounds that remained unsplit so far. I.e. we overcome the limitation of the lexical coverage in the corpus, which results in decomposition of only compounds, whose full set of constituents is included in the corpus. Further, we optimize the disambiguation of the correct split option. By resolving the semantic relation of the constituents, we satisfy the requirement to include semantic information into the decomposition. Compared to pure statistical compound splitting, our approach increases the accuracy by more than 74 % and shows that the integration of the previously mentioned steps are a necessity for successful domain-specific compound splitting.

Acknowledgments. This research has been supported in part by the KDI project, which is funded by the German Federal Ministry of Economics and Technology under grant number 01MT14001 and by the EU FP7 Diachron project (GA 601043).

References

1. Langer, S.: Zur morphologie und semantik von nominalkomposita. In: Proceedings of KONVENS (1998)
2. Monz, C., de Rijke, M.: Shallow morphological analysis in monolingual information retrieval for dutch, german, and italian. In: Peters, C., Braschler, M., Gonzalo, J., Kluck, M. (eds.) CLEF 2001. LNCS, vol. 2406, pp. 262–277. Springer, Heidelberg (2002)
3. Wilmanns, J.C., Schmitt, G.: Die Medizin und ihre Sprache. Ecomed, Landsberg (2002)
4. Brown, R.D.: Corpus-driven splitting of compound words. In: Proceedings of TMI (2002)
5. Braschler, M., Ripplinger, B.: How effective is stemming and decompounding for german text retrieval? Information Retrieval **7**, 291–316 (2004)
6. Koehn, P., Knight, K.: Empirical methods for compound splitting. In: Proceedings of the EACL (2003)
7. Stymne, S.: German compounds in factored statistical machine translation. In: Nordström, B., Ranta, A. (eds.) GoTAL 2008. LNCS (LNAI), vol. 5221, pp. 464–475. Springer, Heidelberg (2008)
8. Popović, M., Stein, D., Ney, H.: Statistical machine translation of german compound words. In: Salakoski, T., Ginter, F., Pyysalo, S., Pahikkala, T. (eds.) FinTAL 2006. LNCS (LNAI), vol. 4139, pp. 616–624. Springer, Heidelberg (2006)
9. Fritzinger, F., Fraser, A.: How to avoid burning ducks: combining linguistic analysis and corpus statistics for german compound processing. In: Proceedings of WMT and MetricsMATR, pp. 224–234 (2010)
10. Schmid, H.: Improvements in part-of-speech tagging with an application to german. In: Proceedings of the EACL-SIGDAT Workshop, Dublin, Ireland (1995)
11. Radiological Society of North America: Radlex (2012). (http://rsna.org/RadLex.aspx)
12. Bretschneider, C., Oberkampf, H., Zillner, S., Bauer, B., Hammon, M.: Corpus-based translation of ontologies for improved multilingual semantic annotation. In: Proceedings of the 3rd SWAIE Workshop (2014)

A Comparison of MT Methods for Closely Related Languages: A Case Study on Czech – Slovak and Croatian – Slovenian Language Pairs

Jernej Vičič[1]([✉]) and Vladislav Kuboň[2]

[1] University of Primorska, IAM, Muzejski trg 2, 6000 Koper, Slovenia
jernej.vicic@upr.si
[2] Faculty of Mathematics and Physics, UFAL,
Charles University in Prague, Prague, Czech Republic
vk@ufal.mff.cuni.cz

Abstract. This paper describes an experiment comparing results of machine translation between two pairs of related Slavic languages. Two language pairs on three different translation platforms were observed in the experiment. One pair represents really very close languages (Czech and Slovak), the other pair are slightly less similar languages (Slovenian and Croatian). The comparison is performed by means of three MT systems, one for each pair representing rule-based approach, the other one representing statistical (same system for both language pairs) approach to the task. Both sets of results are manually evaluated by native speakers of the target languages.

Keywords: Machine translation · Related languages · Comparison

1 Introduction

Machine translation (MT) for related languages presents a special field of MT where systems exploit lexical, morphological and syntactic similarity of related languages to (possibly) balance the lack of good quality language data. Two MT paradigms are mostly used in the production of MT systems for related languages: One of the methods, which guarantees relatively good results for the translation of closely related languages is the method of a rule-based shallow-transfer approach. It has a long tradition and it has been successfully used in a number of MT systems such as Apertium [1] for Romance languages, Česílko [2], for language pairs with Czech as a source, Guat [3] for Slavic languages, mostly language pairs with Slovene. Statistical Machine Translation (SMT) with the most known "players" Google Translate, Microsoft Bing and SMT toolkit Moses [4]. The existence of Google Translate which nowadays enables the automatic translation even between relatively small languages made it possible to investigate advantages and disadvantages of both approaches. This paper introduces

P. Král and V. Matoušek (Eds.): TSD 2015, LNAI 9302, pp. 216–224, 2015.
DOI: 10.1007/978-3-319-24033-6_25

the next step in this direction - the comparison of results of three different systems for two different language pairs - one pair of really very closely related languages(Czech (CES) and Slovak (SLK)) and the other pair of related, but slightly less similar languages (Slovenian (SLV) and Croatian (HRV)). The rule based systems used are different for each language pair, the representative of data-driven systems is Google Translate in both cases. There have been many debates as to which machine translation paradigm is most suitable for the MT for related languages, one of the first steps into answering this question is [5].

2   State of the Art

There has already been a lot of research in Machine Translation evaluation. There are quite a few conferences and shared tasks devoted entirely to this problem such as NIST Machine Translation Evaluation [6] or Workshop on Statistical Machine Translation [7]. [8] presents a research on how systems from two different MT paradigms cope with a new domain. [9] presents a research on how relatedness of languages influences the translation quality of an SMT system. The presented research is based on the methods presented in [5], but it introduces a new language pair of not-so closely related languages (Croatian – Slovenian) to the original pair (Czech – Slovak). The novelty of the presented paper is in the focus on machine translation for closely related languages and in the comparison of the two mostly used paradigms for this task: shallow parse and transfer RBMT and SMT paradigms.

3   Methodology

Two language pairs were selected for the experiment: one pair (Czech to Slovak) with very high degree of similarity at all levels (morphological, syntactic, semantic) and the other one (Croatian to Slovenian) exhibiting more differences but still closely related languages. This choice allows to compare not only the results of the basic paradigms (rule-based and stochastic), but also adds a kind of second dimension, aiming at the question whether the results are influenced by the similarity or not, whether any of the paradigms can naturally profit from the similarity more than the other. We are of course aware that for a complete answer to this question it would be necessary to test more systems and more language pairs, but in this phase of our experiments we do not aim at obtaining a complete answer, our main goal is to develop a methodology and to perform a pivot testing showing the possible directions of future research.

The second important decision concerned the method of evaluation. Our primary goal was to set up a method which would be relatively simple and fast, thus allowing to manually (the reasons for manual evaluation are given in Section 3.1) process reasonable volume of results. The second goal concerned the endeavor to estimate evaluator's confidence in their judgments. The third goal was to adapt or exploit state-of-the art methods of human evaluation of machine translation results.

3.1   Experiment Outline

The aim of the experiment was to test the applicability of two most used Machine
Translation paradigms to a task of machine translation of related languages. The
evaluation relied on the methodology similar to that used in the 2013 Workshop
on Statistical Machine Translation [7]. We conducted manual evaluation of all
system outputs for all language pairs consisting of ranking individual translated sentences according to the translation quality (the evaluators had access
to the original sentence). Unlike the ranking of the SMT Workshop which worked
always with 5 translations, our task was much simpler and the ranking naturally consisted of ranking translated sentences of both systems for each language
pair. The evaluator indicated which of the two systems is better, having also
the chance to indicate that the translation quality does not differ or that both
translations are identical (this was mostly used in the case of the Czech to Slovak
systems which produced relatively large number of identical results - see section
4). The evaluation was done by comparing the output of both systems using the
original sentence as reference. The evaluators were presented a list of 200 test
sentences with translations in random order and they selected the best translation (ties were allowed). The reason why we didn't use any automatic measure of
translation quality was quite natural. After a period of wide acceptance of automatic measures like BLEU or NIST, recent experiments seem to prefer manual
evaluation methods. Many papers such as [10] and authors of workshops such as
WMT 2013 [7] contend that automatic measures of machine translation quality
are an imperfect substitute for human assessments, especially when it is necessary to compare different systems (or, even worse, the systems based on different
paradigms). Three translation systems were selected for the experiment: *Google
Translate* was selected as the most used SMT translation system; *Cesílko* [2]
was used for the Czech – Slovak language pair, the translation direction was
defined by the system as this is the only direction this system supports. *HBS –
SLV* translation system [3] based on Apertium [1] was used for the Croatian language – Slovenian language language pair, the translation direction was selected
pragmatically, this direction yields better translation results as reported by the
system maintainers [11]. The on-line publicly available versions of the systems
were used in the experiment to ensure the reproducibility of the experiment.

3.2   Test Data

In this section, we describe how we collected test data and computed the results.
Our evaluation is based upon a small, yet relevant, test corpus. Because one
of the systems undergoing the evaluation has been developed by Google, the
creation of the test set required special attention. We could not use any already
existing online corpus as Google regularly enhances language models with new
language data. Any online available corpus could have already been included in
the training data of Google Translate, thus the results of the evaluation would
have been biased towards the SMT system. Therefore we have decided to use
fresh newspaper texts.

We have selected 200 sentences from the newspaper articles of the biggest daily newspapers for each source language. For Croatian we have used a single newspaper, namely "Jutarnji list" http://www.jutarnji.hr/, the test examples were collected on . Several headline news were selected in order to avoid author bias although the domain remained daily news. For Czech we have selected articles from a wider variety of online newspapers, namely "iDnes" http://www.idnes.cz/, "Lidovky" http://www.lidovky.cz/ and "Novinky" http://www.novinky.cz/. The test set has been created from randomly selected articles from the issues in the week between 14.7.2014 and 18.7.2014. The test data is publicly available for further experiments and to test the results of our experiment at the University of Primorska language technologies server[1].

4   Results

The results are divided into two sections each covering one language pair as each language pair was tested using a different translation system and also the relatedness between languages in each language pair is different (Czech and Slovak are much closer than Croatian and Slovenian). The results of the evaluations are presented in two forms: the count of the decisions for all the evaluators for every translation pair (the possible outcomes: all evaluators agree that one of the systems is better, some of the evaluators prefer one system over the other, it is impossible to determine the agreement of the evaluators). The evaluators were asked to mark which translation they consider to be better. Ties were not allowed, but the evaluators were also asked to mark identical sentences. This requirement served also as a kind of thoroughness check, too many unrecognized identical sentences could indicate that the evaluator lost concentration during the task; The proportions of the count of the clear wins for each evaluator for every translation pair are presented in separate tables showing another view on the same evaluation results. The Kruskal-Wallis [12] test was used because the evaluated values are presented as ordinal variable. The null hypothesis H_0: The samples come from populations with equal medians.

4.1   Czech – Slovak

The evaluation has been performed by 6 native speakers of Slovak, the sentences have been randomized so that no evaluator could know which of the two systems produced which translation. The results of the CES – SLK counting evaluation part of the experiment are summarized in the Table 1, the results show a big difference in the number of wins by the SMT system.

The rows of Table 1 marked as *Clear win* of one of the systems represent the sentences where none of the evaluators marked the other system as the *better one*. Win by voting does not distinguish how many evaluators were against the system marked by the majority as being the better of the two.

---

[1] http://jt.upr.si/research_projects/related_languages/

The results clearly indicate that the quality of Google Translate is better, although it clearly dominates in less than one third of translations. The large number of identical sentences also means that although Cesílko produced only 5% of translations which were clearly better than those of Google, it reached absolutely identical quality of translation in yet another 21.5%. This actually means that the top quality translations have been achieved in 26.5% by Cesílko and in 51% by Google Translate. According to our opinion, this ratio (approximately 2:1 in favor of the SMT approach) more realistically describes the difference in quality that the ratio of clear wins (approx. 6:1 for Google Translate).

Table 2 shows the same evaluations presented as proportions of wins by each system summarised for each rater. The p-value calculated using the Kruskal-Wallis test and the significance level $\alpha = 0.01$ is $p = 0.0008$, and since $p = 0.0008 < \alpha = 0.01$, the null hypothesis is rejected. Hence the translation quality of the SMT system is better.

Table 1. Evaluation of results for the CES – SLK counting evaluation.

	Sent. count	Percentage
Identical sentences	43	21.5%
Clear win of RBMT	10	5%
Clear win of SMT	59	29.5%
Win by voting - RBMT	23	11.5%
Win by voting - SMT	62	31%
Draw	3	1.5%
Total	200	100%

Table 2. The evaluations of each evaluator for the CES – SLK language pair. The proportion of wins for each system, the draws were eliminated. The proportion of wins of the SMT system is bigger than the RBMT, the Kruskal-Wallis test suggests that the systems are different – Null hypothesis(the systems are not different) was rejected (significance level $\alpha = 0.01$ and p-value $p = 0.0008$.

Value	Rater 1	Rater 2	Rater 3	Rater 4	Rater 5	Rater 6
COUNT (Google)	112,00	70,00	80,00	81,00	83,00	59,00
COUNT (Cesílko)	41,00	73,00	72,00	74,00	65,00	10,00
Proportion (Google)	73,20	48,95	52,63	52,26	56,08	85,51
Proportion (Cesílko)	26,80	51,05	47,37	47,74	43,92	14,49

4.2   Croatian – Slovenian

Two evaluation experiments were done for this language pair:

1. The results were evaluated by 4 evaluators. All evaluators were native speakers of the target language (Slovenian) with a good knowledge of the source language. Three of them came from the computer science field with a strong interest in language technologies (pragmatically), one of the evaluators was linguist.

2. The RBMT system's vocabulary was populated with the most frequent out-of-vocabulary words. The test data was re-translated and evaluated by a mixed set of evaluators (6 evaluators, 2 from the first experiment and 4 new evaluators, 3 linguists and 1 computer science student). The set of evaluators was chosen pragmatically, not all the original evaluators were available at the second evaluation.

At first the experiment involving 4 evaluators was done, but the results indicated that a new evaluation should be done to prove the findings of the first evaluation: all the evaluators of the first experiment suggested that the most frequent errors of the RBMT system were all connected to the the out-of-vocabulary errors. These errors further escalated with the impossibility of using the local agreement rules (unknown words). The evaluation methodology was the same in both experiments and it is described in Section 3.1. The evaluators were presented with a portion of the same test set (unfortunately we did not have time to populate the whole dictionary, we covered only the first 63 sentences) and graded the translations with 4 possible outcomes as in the experiment presented in Section 4.1. The results of the HRV – SLV counting evaluation part of the experiment for the original RBMT system are summarized in the Table 3, the results show a big difference in the number of wins by the RBMT system and an evene bigger for the enhanced RBMT system.

Table 4 shows the same evaluations presented as proportions of wins by each system summarised for each rater. The p-value calculated using the Kruskal-Wallis test and the significance level $\alpha = 0.01$ is $p = 0.0$, and since $p = 0.0021 < \alpha = 0.01$, the null hypothesis is rejected. Hence the translation quality of the RBMT system is better.

Table 3. Evaluation of results for the HRV – SLV, original and enhanced RBMT system counting evaluation.

System:	Original RBMT		Enhanced RBMT	
	Sent. count	Percentage	Sent. count	Percentage
Identical sentences	0	0%	0	0%
Clear win of RBMT	69	34.5%	32	51%
Clear win of SMT	47	23.5%	7	11%
Win by voting - RBMT	30	15%	11	17.5%
Win by voting - SMT	18	9%	7	11%
Draw	36	18%	6	9.5%
Total	200	100%	63	100%

4.3 Inter-rater Agreement

The inter-rater agreement was also calculated using the Fleiss' Kappa [13] which is an extension of the Cohen Kappa for multiple raters. The results are presented in Table 6 and show the κ values for all experiments, the column agreement shows the range the κ value belongs to according to the tables presented by [14].

Table 4. The evaluations of each evaluator for the HRV – SLV language pair, for the original RBMT system. The proportion of wins for each system, the draws were eliminated. The proportion of wins of the RBMT system is bigger than the SMT, the Kruskal-Wallis test suggests that the systems are different – Null hypothesis(the systems are not different) was rejected (significance level $\alpha = 0.01$ and p-value $p = 0.0021$.

Value	Rater 1	Rater 2	Rater 3	Rater 4
COUNT 1 (Google)	85	88	74	81
COUNT 2 (Apertium)	108	103	113	110
Proportion (Google)	44%	46%	39.5%	42.5%
Proportion (Apertium)	56%	54%	60.5%	57.5%

Table 5. The evaluations of each evaluator for the HRV – SLV language pair, for the enhanced RBMT system. The proportion of wins for each system, the draws were eliminated. The proportion of wins of the RBMT system is bigger than the SMT, the Kruskal-Wallis test suggests that the systems are different – Null hypothesis(the systems are not different) was rejected (significance level $\alpha = 0.01$ and p-value $p = 0.0$.

Value	Rater 1	Rater 2	Rater 3	Rater 4	Rater 5	Rater 6
COUNT (Google)	12	11	12	6	9	21
COUNT (Apertium)	38	34	43	35	31	32
Proportion (Google)	24%	24.5%	22%	14.5%	22.5%	39.5%
Proportion (Apertium)	76%	75.5%	78%	85.5%	77.5%	60.5%

Table 6. The inter-rater agreement for all experiments.

Experiment	Nr. of cases	Nr. of raters	Kappa value	Agreement
CES – SLK	200	5	0.6069	Substantial
HRV – SLV original	200	4	0.4040	Moderate
HRV – SLV enhanced	63	6	0.2766	Fair

5 Conclusions and Further Work

One of the main reasons for the experiment described in the paper was to fully confirm the preliminary findings presented in [5] where the clear winner of the head-to-head comparison between a representative of the SMT paradigm (Google translate) and a representative of the shallow transfer RBMT paradigm (Cesílko) was the SMT system. Parts of the comparison were redone, but the new results only further confirm the results of the first part of the experiment. The second part of the experiment involved a new language pair: Croatian – Slovenian, using a different toolkit with a similar architecture to the first part of the experiment. The RBMT system came out as a clear winner although the inter-rater agreement was quite low. A cursory examination of the results could rise doubts in the quality of implementation of the experiment, but closer inspection of errors and basic operation of the Google translation system still allows the interpretation of conflicting results. The Google translate system translates between Croatian and Slovenian language through a pivot language – English (first translating from Croatian to English and then from English to Slovenian). English language does not support all the features of the language pair and it is also not a related language to the pair. Hence a big difference in translation quality. The Google system translates directly between the Czech and Slovak languages. Some of the translated passages even suggest that Google uses a set of rules in the translation of this language pair although this fact still needs to be further examined and confirmed. Although our experiment represents only the first step in systematic evaluation of machine translation results between closely related languages, it has already brought very interesting results. It has shown that contrary to a popular belief that RBMT methods are more suitable for MT of closely related languages, Google Translate outperforms the RBMT system Cesílko. The similarity of source and target language apparently not only allows much simpler architecture of the RBMT system, it also improves the chances of SMT systems to generate good quality translation, although this results need further examination. The most surprising result of our experiment is the high number of identical translations produced by both systems not only for short simple sentences, but also for some of the long ones, as well as very similar results produced for the rest of the test corpus. In the future, we would like to develop phrase-based SMT system based on Moses [4] and to investigate its behavior on the same language pairs. Inclusion of a new language pair from a different language group with a good RBMT system already available, such as Catalan – Spanish and a direct translation in Google translate should further confirm our findings.

References

1. Corbi-Bellot, A.M., Forcada, M.L., Ortiz-Rojas, S.: An open-source shallow-transfer machine translation engine for the Romance languages of Spain. In: Proceedings of the EAMT Conference, HITEC e.V., pp. 79–86 (2005)
2. Hajič, J., Hric, J., Kuboň, V.: Machine translation of very close languages. In: Proceedings of ANLP, pp. 7–12 (2000)
3. Vičič, J.: A fast implementation of rules based machine translation systems for similar natural language. Informatica 37(4), 455–456 (2013)
4. Koehn, P., et al.: Moses: open source toolkit for statistical machine translation. In: Proceedings of ACL 2007, pp. 177–180 (2007)
5. Kuboň, V., Vičič, J.: A comparison of MT methods for closely related languages : a case study on Czech - Slovak language pair. In: EMNLP, pp. 92–98 (2014)
6. NIST: NIST 2009 Open Machine Translation Evaluation (MT09). Technical report, NIST (2009)
7. Bojar, O., et al.: Findings of the 2013 workshop on SMT. In: 2013 Workshop on SMT, pp. 1–44 (2013)
8. Weijnitz, P., Forsbom, E., Gustavii, E., Pettersson, E., Tiedemann, J.: MT goes farming: comparing two machine translation approaches on a new domain. In: LREC, pp. 1–4 (2004)
9. Kolovratnik, D., Klyueva, N., Bojar, O.: Statistical machine translation between related and unrelated languages. In: Proceedings of TPIT, pp. 31–36 (2009)

10. Callison-Burch, C., Osborne, M., Koehn, P.: Re-evaluating the role of BLEU in machine translation research. In: Proceedings of EACL, pp. 249–256 (2006)
11. Apertium: Apertium: machine translation toolbox (2014)
12. Kruskal, W.H., Wallis, W.A.: Use of Ranks in One-Criterion Variance Analysis. Journal of the American Statistical Association 47(260), 583–621 (1952)
13. Fleiss, J.L.: Measuring nominal scale agreement among many raters (1971)
14. Landis, J.R., Koch, G.G.: The measurement of observer agreement for categorical data. Biometrics 33(1), 159–174 (1977)

Ideas for Clustering of Similar Models of a Speaker in an Online Speaker Diarization System

Marie Kunešová[1,2](✉) and Vlasta Radová[1,2]

[1] Faculty of Applied Sciences, Department of Cybernetics,
University of West Bohemia, Univerzitní 8, 306 14 Plzeň, Czech Republic
{mkunes,radova}@kky.zcu.cz
[2] Faculty of Applied Sciences, New Technologies for the Information Society,
University of West Bohemia, Univerzitní 8, 306 14 Plzeň, Czech Republic

Abstract. During online speaker diarization, a situation may occur where a single speaker is being represented by several different models. Such situation leads to worsened diarization results, because the diarization system considers every change of a model to be a change of speakers. In the article we describe a method for detecting this situation and propose several ways of solving it. Experiments show that the most suitable option is treating multiple GMMs as belonging to a single speaker, i.e. updating all of them with the same data every time one of them is assigned a new segment. In that case, there was a relative improvement in Diarization Error Rate of 30.69% in comparison with the baseline system.

Keywords: Speaker clustering · Speaker diarization · Speaker segmentation

1 Introduction

In automatic speech processing, speaker diarization is the task of distinguishing between different speakers within an audio recording and identifying the intervals in which they are active. Or in other words, determining "Who spoke when?". This is generally done without any prior knowledge about the actual identities and number of speakers [1].

In a previous paper [2], we used our diarization system, which was itself based on ideas proposed in [3,4]. The main property of the system is that it creates models of new speakers online. During this process, a single speaker may be assigned two or more models. In this paper, we detect such situations and explore ideas for resolving them by creating a single model out of all the models that belong to a single speaker. We will refer to this process as *clustering*.

The paper is organized as follows. Section 2 describes the base diarization system. In Sect. 3, an offline algorithm for model clustering is described, and in Sect. 4 several approaches to online model clustering are proposed. They are evaluated in Sect. 5 and Sect. 6 gives some conclusions.

© Springer International Publishing Switzerland 2015
P. Král and V. Matoušek (Eds.): TSD 2015, LNAI 9302, pp. 225–233, 2015.
DOI: 10.1007/978-3-319-24033-6_26

2   The Diarization System

The online diarization system uses Gaussian Mixture Models (GMMs) to represent the individual speakers. The basic principle is as follows [2]:

The system starts with only two GMMs, one for each gender, which are trained in advance. The audio stream is divided into short segments and for each of them, the system decides if the segment corresponds to an already known speaker or a new one by comparing the likelihoods of the gender dependent and speaker models. In the case of a new speaker, a new model is created by copying one of the gender dependent models. Otherwise, one of the existing models is selected. The assigned model is then adapted using the data from the segment.

The online system for audio speaker diarization consists of several modules:

1. Feature extraction and voice activity detection
2. Speech segmentation
3. Speaker identification and novelty detection
4. Online GMM learning

Feature Extraction and Voice Activity Detection. For feature extraction, we used the LFCC with 25 filters in range from 50 Hz to 8 kHz based on 25 ms FFT window with 10 ms shift. 20 cepstral coefficients were computed without the energy coefficient. No cepstral normalization was performed.

This module also performs an energy-based voice activity detection (VAD), with every frame being labeled as *speech* or *silence* based on a threshold.

Speech Segmentation. Using the information obtained from the VAD and parameters such as the minimum and maximum segment length and the maximum pause length in a segment, the speech is divided into short segments. Of each segment, only the frames labeled as *speech* are used in the subsequent steps. In our experiments, this has lead to both reduced computation time and improved performance.

Speaker Identification and Novelty Detection. For each speech segment the system uses a maximum-likelihood classification to determine both the speaker's gender (using the gender dependent models) and their most likely identity out of the existing speaker-model candidates. Afterwards, a likelihood ratio test is used to decide whether the segment belongs to the chosen identity, or represents an entirely new speaker.

The likelihood ratio is as follows:

$$L(\mathbf{x}, \lambda_{sp}) = \frac{P_{\lambda_{sp}}}{P_{\lambda_{gen}}} ,$$ (1)

where $\mathbf{x}$ is a speech segment and $P_{\lambda_{sp}}$ and $P_{\lambda_{gen}}$ are the likelihoods of the winning speaker and the appropriate gender dependent model, respectively.

If $\log L(\mathbf{x}, \lambda_{sp}) \geq \theta$, the segment $\mathbf{x}$ belongs to the speaker represented by the model λ_{sp}. Otherwise it belongs to an entirely new speaker.

The optimal value of decision threshold θ was found experimentally.

Online GMM Learning. For the adaptation of GMMs we use an online variant of the Expectation-Maximization algorithm, as described in [6], with values of the parameters as proposed in [4].

One of the most problematic areas of the diarization system is the selection of the data-dependent decision threshold θ, which is used to decide whether a speech segment belongs to a new speaker or an already known one. If this threshold is set too low, multiple speakers may be assigned the same model. Conversely, if it is too high, speech belonging to a single real speaker may be divided between several different models.

Figure 1 illustrates a situation that commonly arises in the latter case: the similarity of two models which in reality represent the same speaker causes the system to frequently switch between the two, giving the illusion of an exchange between two speakers where there is in fact only one.

Both types of error can dramatically reduce the overall performance of the system, yet it may not be possible to completely eliminate them by the suitable selection of the threshold θ alone. The presence of background noise such as cough in the audio seems to be especially problematic in this matter, making it likely that both types of error will occur to some smaller degree no matter how well we choose θ.

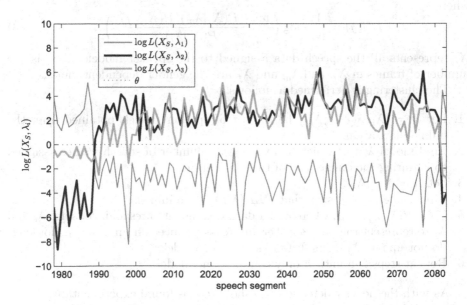

Fig. 1. Logarithm of the likelihood ratio $L(X_S, \lambda_i)$ from (1) calculated for three different speaker models GMMs in a part of one recording. Highest value of $\log L(X_S, \lambda_i)$ represents the winning speaker for the given speech segment. Models λ_2 and λ_3 represent the same speaker.

To combat this issue, we propose the following solution: we select a higher decision threshold, so that an excess number of speaker models is created, but then we implement an additional algorithm that identifies any models that are likely to correspond to the same speaker and cluster them all into one.

In [2], we simplified the task by performing an offline clustering after the whole audio recording had been processed. Here, we apply a similar method to perform the clustering online, as part of the main diarization system.

First, we will describe a modified version of the original offline variant, which serves for comparison. Then we will explain the online method.

3   Offline Clustering

For offline clustering, we first perform the basic diarization of the whole audio recording. Then we use the cross-likelihood ratio (CLR, [5]) to compute distances between all pairs of models and identify groups of models which likely correspond to the same real speakers.

The distance between two models λ_i and λ_j is defined as

$$CLR(\lambda_i, \lambda_j) = D(i,j) + D(j,i) \,, \tag{2}$$

where

$$D(i,j) = \frac{1}{N_i} \cdot \log \left(\frac{\max(P(X_i|\lambda_m), P(X_i|\lambda_f))}{P(X_i|\lambda_j)} \right) \,, \tag{3}$$

X_i represents all the speech data assigned to the speaker model λ_i, N_i is the number of frames in X_i, and λ_m and λ_f are the gender dependent models.

The clustering is performed as follows:

1. Let $\Lambda = \{\lambda_1, \lambda_2, \ldots, \lambda_N\}$ be the set of speaker models obtained from the online diarization.
2. Find model $\lambda_i \in \Lambda$ which had the lowest number of speech frames assigned to it during the online part of the diarization.
3. Set $\Lambda = \Lambda - \{\lambda_i\}$.
4. Find model $\lambda_j \in \Lambda$ such that $CLR(\lambda_i, \lambda_j)$ is minimal
5. If $CLR(\lambda_i, \lambda_j) < \phi$, where ϕ is a data-dependent threshold, consider λ_i and λ_j to represent the same speaker and reassign speech from λ_i to λ_j. However, do not update λ_j or its distances to other models.
6. Repeat steps 2–5 until Λ only contains one model.

As with the novelty detection threshold θ, ϕ is found experimentally.

4   Online Clustering

To identify similarities between models online, we used a method based on the offline variant. It has the benefit of requiring very little additional computation time, as most of the necessary calculations, namely the likelihoods of speaker models for each speech segment, are already being performed as part of the base system.

To find the distance between models λ_i and λ_j at time t, we use the following modification of the CLR distance:

$$d(\lambda_i, \lambda_j, t) = \min\left(D(i, j, t), D(j, i, t)\right),\qquad(4)$$

$$D(i, j, t) = \frac{1}{N_i(j, t)} \cdot \sum_{\mathbf{x} \in S_i(j,t)} \log\left(\frac{\max(P(\mathbf{x}|\lambda_m), P(\mathbf{x}|\lambda_f))}{P(\mathbf{x}|\lambda_j)}\right).\qquad(5)$$

Here, $S_i(j, t)$ represents a set of all the speech segments which were assigned to speaker model λ_i between the creation of λ_j and the current time t (and for which we thus have calculated the likelihood of λ_j), $N_i(j, t)$ is the total number of frames of the speech segments contained in $S_i(j, t)$.

A notable change from (2–3) is the replacement of the sum which was in (2) with a minimum of the two values in (4). This has shown in our experiments to lead to a slightly better performance of the online clustering.

Once the system decides that several of the models represent the same speaker, there are two possible approaches to clustering, apart from simply discarding all of the models except one. We can either use a suitable method to transform all of the similar models into a single GMM, or we can retain all of them while treating them as a single speaker.

4.1 Merging Multiple GMMs into a Single One

The simplest choice we have considered to merge several models into a single one is to obtain a *weighted sum of the original Gaussian mixtures*. This is computationally very simple and thus causes no immediate delay for the system. Yet the increased number of Gaussian components causes redundancy in the model and will also slow down future calculations.

As an alternative which preserves the original number of Gaussian components, we chose to replace all of the models to be merged with a *single new GMM trained using all the data* originally assigned to all. This causes a significant delay in the whole process, so it is not suitable for use in practical applications where online diarization is required. However, we can consider this approach to be the best way to merge the models together and we will use it for comparison in our experiments.

4.2 Treating Multiple GMMs as Belonging to a Single Speaker

In this approach to dealing with similarities of speaker models, the system retains all of the models. However, the models are treated as belonging to the same speaker. It means that all of them are being updated with the same data every time one of them is assigned a new segment.

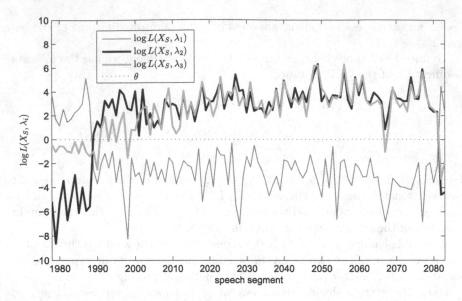

Fig. 2. Logarithm of the likelihood ratio $L(X_S, \lambda_i)$ from (1) calculated for three different speaker models GMMs in a part of one recording. Highest value of $\log L(X_S, \lambda_i)$ represents the winning speaker for the given speech segment. Models λ_2 and λ_3 represent the same speaker and both are being updated with the same date starting with speech segment 2008.

Because of this, after a certain number of updates these models should become nearly identical, as illustrated by Fig. 2. At that point, it may be possible to discard all of them except one. The choice of which of the models to keep is not a clear one, however. We will deal with this issue later in Sect. 5.

Similarly to the weighted sum of GMMs described in Section 4.1, this approach also causes no immediate delay for the system, but the additional GMM updates will slow down the calculations for future speech segments.

5   Experiments

Experiments were done on a set of 8 recordings from Czech parliament meetings with a total of 30 hours of labeled audio. Gender dependent models were trained using 30 seconds of speech from each of 16 women and 70 men. For audio segmentation, the minimum and maximum segment lengths were chosen as 1 and 5 seconds, respectively.

For performance evaluation, the Diarization Error Rate (DER) was used, as described in [7]. It is defined as the fraction of time that is not correctly assigned to a speaker or to non-speech and is the sum of three types of error rates: missed speech (*miss*, speech incorrectly labeled as silence), false alarm (*FA*, silence incorrectly labeled as speech) and speaker error (*SE*, speech labeled

Table 1. Diarization performance in terms of Speaker Error (SE) and DER [%] for different numbers of updates (N_{upd}) after which one of the original models is deleted.

N_{upd}	keep the more trained model		keep the less trained model	
	SE	DER	SE	DER
0	4.23	6.84	3.52	6.13
10	4.18	6.78	3.40	5.99
20	4.13	6.73	3.39	5.98
30	4.10	6.70	**3.35**	**5.94**
50	4.04	6.63	3.36	5.95
∞	3.46	6.05	3.46	6.05

as a wrong speaker). A forgiveness collar of 0.25 seconds around the reference speaker boundaries was used.

In our experiments, we have examined the three approaches to online speaker clustering which were described in Sect. 4. They were:

1. Weighted GMM summation
2. GMM retraining
3. Treating multiple GMMs as belonging to a single speaker

Before we evaluate the results of all approaches, we need to look at approach 3 and explore the option of discarding all but one of the original models after a set number of updates. Without loss of generality, we assume from this moment on that we are deciding between only two models.

As shown in Table 1, we have found that discarding the model which was trained using a greater amount of data while keeping the other one leads to a lower DER than the alternative. This may be caused by two factors: the "bigger" model could be more "polluted" by noise and misclassified speech or there is a change in the acoustics, such as a change in the speaker's voice during a longer speech or a change in the background noise.

The results in Table 1 show that the best results were obtained when discarding one of the models after 30 updates and therefore we will use this number of updates for the final comparison.

Having selected which of the original models in approach 3 to discard and when, we can compare the achieved results with the other approaches, as well as with the offline clustering and the base system without clustering. The results are shown in Table 2.

In addition to the evaluation of the immediate decisions, which are obtained after a segment is processed (column *online*), the table also contains the final values which can be achieved by retroactively relabeling previous speech whenever two models are found to represent the same speaker (column *final*).

Results in column *online* show that rather than attempting to create a new model by merging two similar ones, it is better to treat them as belonging to the same speaker and discard one of them after some time. Best results were obtained when discarding the model which was trained using the greater amount

Table 2. Comparison of the diarization performance on test data in terms of DER (%). The slight differences in missed speech rates and also in false alarm rates among different variants are caused by the removal of very short pauses within the speech of a single speaker.

	online				final			
	miss	FA	SE	DER	miss	FA	SE	DER
Without clustering	1.56	1.05	5.96	8.57	—	—	—	—
Offline clustering	—	—	—	—	1.51	1.05	**2.18**	**4.75**
Weighted GMM summation	1.55	1.07	4.94	7.55	1.50	1.06	3.90	6.46
GMM retraining	1.51	1.06	4.08	6.66	1.47	1.06	3.66	6.19
Deleting one GMM immediately ($N_{upd} = 0$)	1.55	1.06	3.52	6.13	1.50	1.06	3.00	5.57
Multiple GMMs for a speaker, $N_{upd} = 30$	1.52	1.07	**3.35**	**5.94**	1.49	1.06	2.99	5.54
Multiple GMMs for a speaker, $N_{upd} = \infty$	1.52	1.07	3.46	6.05	1.48	1.06	3.00	5.54

of speech after 30 updates. In this case, there was a relative improvement in DER of 30.69% in comparison with the base system.

Even better results can be achieved with the offline clustering, but this method cannot be used for online diarization.

6   Conclusion

In the paper we have dealt with a situation where during online speaker diarization a single speaker is being represented by several models. We have proposed several different methods of creating a single model out of them. Although the results of none of the online methods reach the results of the offline method, the relative improvement is more than 30% compared to the baseline system.

Acknowledgments. This research was supported by the Grant Agency of the Czech Republic, project No. GAČR GBP103/12/G084 and by an internal grant SGS-2013-032. Access to computing and storage facilities owned by parties and projects contributing to the National Grid Infrastructure MetaCentrum, provided under the programme "Projects of Large Infrastructure for Research, Development, and Innovations" (LM2010005), is greatly appreciated.

References

1. Anguera, X., Bozonnet, S., Evans, N., Fredouille, C., Friedland, G., Vinyals, O.: Speaker Diarization: A Review of Recent Research. IEEE Transactions on Audio, Speech, and Language Processing **20**, 356–370 (2012)
2. Campr, P., Kunešová, M., Vaněk, J., Čech, J., Psutka, J.: Audio-video speaker diarization for unsupervised speaker and face model creation. In: Sojka, P., Horák, A., Kopeček, I., Pala, K. (eds.) TSD 2014. LNCS, vol. 8655, pp. 465–472. Springer, Heidelberg (2014)
3. Geiger, J., Wallhoff, F., Rigoll, G.: GMM-UBM based open-set online speaker diarization. In: Proc. Interspeech, pp. 2330–2333 (2010)

4. Markov, K., Nakamura, S.: Never-ending learning system for on-line speaker diarization. In: IEEE Workshop on Automatic Speech Recognition & Understanding, ASRU 2007, pp. 699–704 (2007)
5. Reynolds, D., Singer, E., Carlson, B., O'Leary G., McLaughlin, J., Zissman, M.: Blind clustering of speech utterances based on speaker and language characteristics. In: Proceedings of the 5th International Conference on Spoken Language Processing, vol. 7, pp. 3193–3196 (1998)
6. Sato, M., Ishii, S.: On-line EM algorithm for the Normalized Gaussian Network. Neural Computation 12, 407–432 (2000)
7. National Institute of Standards and Technology. http://www.itl.nist.gov

Simultaneously Trained NN-Based Acoustic Model and NN-Based Feature Extractor

Jan Zelinka[1,2]([⊠]), Jan Vaněk[1,2], and Luděk Müller[1,2]

[1] Department of Cybernetics, Faculty of Applied Sciences,
University of West Bohemia, Univerzitní 8, 306 14 Plzeň, Czech Republic
{zelinka,vanekyj,muller}@kky.zcu.cz
[2] Faculty of Applied Sciences, New Technologies for the Information Society,
University of West Bohemia, Univerzitní 8, 306 14 Plzeň, Czech Republic

Abstract. This paper demonstrates how standard feature extraction methods such as PLP can be successfully replaced by a neural network and methods such as mean normalization, variance normalization and delta coefficients can be simultaneously utilized in a neural-network-based acoustic model. Our experiments show that this replacement is significantly beneficial. Moreover, in our experiments, also a neural-network-based voice activity detector was employed and trained simultaneously with a neural-network-based feature extraction and a neural-network-based acoustic model. The system performance was evaluated on the British English speech corpus WSJCAM0.

Keywords: Neural networks · Speech recognition · Feature extraction

1 Introduction

This paper describes an application of Neural Networks (NN) in miscellaneous parts of a speech recognition system. Namely, we have focused on acoustic models (AM), a feature extraction (FE) and a voice activity detection (VAD). Our main goal was to replace all these techniques with NNs. There are still only sparse attempts to apply NNs as the FE for speech recognition [1] [2] [3] [4] [5] [6] and there are even some papers which report a slightly higher error for the raw signal. [7].

This paper describes our experiments where a NN representing an AM and also a NN representing a feature extraction (also called front-end) were successfully applied. In our experiments, the standard PLP [8] was more or less directly approximated by a NN. Usually, convolutional neural networks are applied [9]. But, standard neural networks are sufficient for this task. Some simple techniques well known in image processing were applied. Our experiments demonstrate that NNs can be successfully used in all mentioned cases. Moreover, we have made efforts to build an efficient small (i.e. fast) NN. The novelty of this paper is the demonstration how methods such as mean or variance normalization, delta

© Springer International Publishing Switzerland 2015
P. Král and V. Matoušek (Eds.): TSD 2015, LNAI 9302, pp. 234–242, 2015.
DOI: 10.1007/978-3-319-24033-6_27

coefficients and even a NN-based voice activity detector can be utilized in a NN-based FE. The system performance was evaluated on the British English speech corpus.

This paper is organised as follows: Section 2 deals with NNs for approximating the PLP method. The role of a mean normalization and some other techniques is disputed in Section 3. Section 4 describes our NN-based AM. Results of our experiments where the NNs are combined are shown in Section 5.

2   Neural-Network-Based Feature Extraction

In this paper, as an input of each FE we took always the absolute spectrum |FFT|. The used Hamming window and FFT are linear transforms that a NN can perform without a hitch. Operations that are necessary for |FFT| computation (i.e. square and square root) can be provided by particular activation functions or can be approximated adequately by a relatively small NN. Hence, using |FFT| is very similar to using a "raw" signal.

Our preliminary experiments demonstrate that to start a training process with a NN trained as a PLP approximation is much more efficient than to initialize a NN randomly and train the NN-based FE together with a NN-based AM (also randomly initialized or even trained using PLP and then fixed as one complex NN). Especially, results of experiments with both NNs randomly initialized are poor [7]. For that reason, we decided to (less or more directly) initialize the NN-based FE as an approximation of PLP.

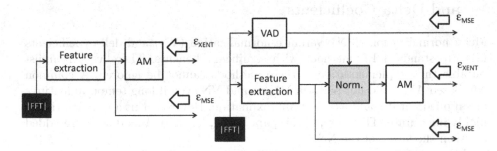

Fig. 1. The proposed alternative approach for a NN-based AM training.

Both parts (i.e. FE and AM) could be trained separately with satisfactory results but, in this paper, another approach was also investigated. Because the composition of a NN-based FE and a NN-based AM forms a deep neural network, well known approach for deep neural networks training was adopted. This approach is shown in Figure 1. The NN-based FE and the NN-based AM were connected together before the training. Two gradients were computed: gradient for the FE part (with parameters θ) is computed by means of Mean Square Error (MSE) criterion (ε_{MSE}) and gradient for the whole system is computed by means of Cross-Entropy (XENT) criterion (ε_{XENT}). Naturally, the resultant gradient is the sum $\frac{\partial \varepsilon}{\partial \theta} = \frac{\partial \varepsilon_{MSE}}{\partial \theta} + \frac{\partial \varepsilon_{XENT}}{\partial \theta}$. The motivation behind this approach is an assumption that even though the approximation of PLP is still not precise this approach leads to a more precise AM (in comparison with the approach where both parts are trained separately) for a posteriors estimate. When mean or variance normalization use an NN-based VAD, the third gradient is computed. In this case, the criterion is MSE. All these additional gradients were applied until the performance of an investigated system exceeded some chosen threshold. After that, only XENT gradient was applied on the complex NN. This approach could put the training in the following unwanted situation: PLP features might be so well designed that every minor change of the NN which simulates PLP can decrease the performance. Our experiments showed that this concern unnecessary.

In this paper, each NN-based FE has two layers. Activation functions are standard sigmoid functions in the hidden layer and linear functions in the output layer. Each NN-based FE has only 256 neurons in its hidden layer. |FFT| has 256 features and these NNs for FE have a dozen neurons in their output layers. Therefore, a structure of a NN for FE is always $256 \times 256 \times 12$.

In the input of each NN-based AM, eleven subsequent vectors were concatenated and time-shifted where time-shift was from -5 to 5. A NN can be copied eleven times and these NNs can be formed into one single NN which can be simply concatenated with a NN-based AM. The resultant NN would be still relatively small and thus usable for further training.

3   Mean Normalization, Variance Normalization and Delta Coefficients

Mean normalization (MN), variance normalization (VN) and delta coefficients (Δ) are standard FE methods with significant benefits. Thereby, it is indisputable that implementation of these methods cannot be ignored. A common NN-based AM could not perform MN or/and VN even if long temporal features are applied but it could use their approximation. However, this approximation is highly inaccurate. Therefore, in this paper, MN, VN and Δ modules were added between the layers of the NN.

MN transforms an input sequence $X = (x_1, \ldots, x_T)$ into an output sequence $Y = (y_1, \ldots, y_t)$ according to the formula

$$Y = \text{MN}(X), \ y_t = x_t - \frac{1}{T} \sum_{\tau=1}^{T} x_\tau. \tag{1}$$

During the backpropagation, it is necessary to compute $\frac{\partial \varepsilon}{\partial X}$, where ε is a criterion. MN is a linear transform therefore $\frac{\partial \varepsilon}{\partial X}$ can be computed by a linear transform too. After a short and simple derivation, the following surprisingly plain equation can be obtained: $\frac{\partial \varepsilon}{\partial X} = \text{MN}\left(\frac{\partial \varepsilon}{\partial Y}\right)$. A bias b in the layer just before MN can be omitted because obviously $\text{MN}(x + b) = \text{MN}(x)$. Delta coefficients computation is also a linear transform hence $\frac{\partial \varepsilon}{\partial X}$ computation is a very simple operation.

From the point of view of techniques such as cepstral normalization, usage of Weighted Mean Normalization (WMN) is more reasonable. The output Y is computed according to the equation

$$Y = \text{WMN}(X, W), \ y_t = x_t - \frac{\sum_{\tau=1}^{T} w_\tau x_\tau}{\sum_{\tau=1}^{T} w_\tau}, \tag{2}$$

where $W = (w_1, \ldots, w_T)$ is a sequence of weights, $w_t \geq 0$ for each $t = 1, \ldots, T$, and $\sum_{\tau=1}^{T} w_\tau > 0$. It is easy to deduce that $\frac{\partial \varepsilon}{\partial X} = \text{WMN}\left(\frac{\partial \varepsilon}{\partial Y}, W\right)$ holds. Usually, some VAD determines weights as 1 when a speaker is speaking and as 0 otherwise. In this paper, a NN was employed as a VAD. This NN has one single neuron in its output layer and the activation function of this neuron gives non-negative values. In this paper, standard sigmoid function was applied. For training, gradient $\frac{\partial \varepsilon}{\partial w_t}$ is computed according to the following formula

$$\frac{\partial \varepsilon}{\partial w_t} = -\frac{\sum_{\tau=1}^{T} \frac{\partial \varepsilon}{\partial y_\tau}}{\sum_{\tau=1}^{T} w_\tau} y_t. \tag{3}$$

VN is computed as follows:

$$y_t = \frac{x_t}{\sigma}, \ \sigma = \sqrt{\frac{1}{T} \sum_{\tau=1}^{T} (x_\tau - \mu)^2}, \ \mu = \frac{1}{T} \sum_{\tau=1}^{T} x_\tau. \tag{4}$$

Although gradient $\frac{\partial \varepsilon}{\partial x_t}$ is much less straightforward, with a little effort, one can see that the following equation holds:

$$\frac{\partial \varepsilon}{\partial x_t} = \frac{1}{\sigma} \frac{\partial \varepsilon}{\partial y_t} - \frac{1}{T} \frac{x_t - \mu}{\sigma^2} \sum_{\tau=1}^{T} \frac{\partial \varepsilon}{\partial y_\tau} y_\tau. \tag{5}$$

4 Neural-Network-Based Acoustic Model

A NN-based AM computes posteriors probabilities from its input features. Because there is a threat of overtraining due to a relatively small corpus, each NN has only two layers. In our experiments, standard sigma functions were activation functions in the hidden layers with 1024 neurons and the softmax function was the activation function in the output layers. Posteriors were computed for context-independent units. Used phonetic alphabet has 43 phonemes. Because each phoneme is modeled as a three state unit, the output layer provides estimates of 129 posterior probabilities.

In our previous works, a special long temporal spectral pattern feature processing technique was used in the hybrid NN/HMM systems. However in this paper, these features were replaced with a sequence of PLP vectors (or their NN substitutes) constituting one large vector. The sequence of PLP vectors needs several operations for time-shifting applied before the posteriors computation. Formally, the time-shift operation output is computed with regard to a recording border in the following way: $y_t = x_{t'}$ where $t' = \max\{\min\{t + s, T\}, 1\}$ where s is the time-shift. Obviously, the time-shift operation is a linear transform thus computing gradient $\frac{\partial \varepsilon}{\partial X}$ is also a linear transform and furthermore it is trivial: The time-shift only must be reversed except boundaries where relevant gradients must be added up.

In this paper, the XENT criterion was always a criterion of optimality during each AM training. The employed training algorithm was a momentum method. The NN training was stochastic, i.e. a subset was selected from a training set. Because we use the time-shifting, MN, VN, and delta operation, entire recording must be processed. A relatively high bunch size was chosen to prevent a biased gradient because in a short bunch all vectors would be probably from one recording.

5   Experiments and Results

The British English speech corpus WSJCAM0 [10] was used for the system performance evaluation in the following experiments. This corpus includes 7861 utterances (i.e. approximately 15 hours of speech) in the training set. Phonetic alphabet consists of 43 phones (including silence and inhale). The experiments were performed on the development sets si_dt5a and si_dt5b. In the corpus, A particular trigram language model for both sets is prescribed. The set si_dt5a was used to find the optimal word insertion penalty and language model weight. The set si_dt5b was used strictly as an evaluation set. Our proprietary real-time LVSCR decoder [11] [12] was applied.

Neither MN nor VN was performed in the first experiment. The results (i.e. word accuracies) are shown in Table 1. The results for standard PLP features (with delta and delta-delta coefficients) and a standard GMM-based AM (with 16 components per state) are in the first row. The results for a standard hybrid NN/HMM are in the second row. The third row contains the results for the NN-based FE trained separately by means of the MSE criterion as a PLP approximator. The training process was stopped when the accuracy reached almost the result from the previous row (on the development set si_dt5a). On the test set si_dt5b, accuracy significantly decreased but, naturally, we had to ignore this fact during the training process. The AM was the same model as the AM used in the second row and thus the AM was constant. In this experiment, the influence of the NN-based AM training must be distinguished from the influence of the NN-based FE training. Therefore, in the third and the fourth rows, the results for the fixed parameters of the NN-based AM are presented. The results of simple joining of both NNs are in the third row and the results of the NN-based FE training are in the fourth row. The fifth row contains the results where both NNs were properly trained together. The first experiment has proved that the standard PLP features can be beneficially replaced with the NN. Figure 2 shows this basic AMs with PLP or the NN-based FE.

Table 1. The Recognition accuracies without MN, VN and delta coefficients.

Feature Extraction	Ac. Model	si_dt5a (dev.)	si_dt5b (test)
PLP + Δ + $\Delta\Delta$	GMM	83.7%	82.2%
PLP	NN	87.1%	85.9%
init. NN	fixed NN	87.1%	85.1%
NN	fixed NN	88.0%	86.7%
NN	NN	**88.6%**	**87.6%**

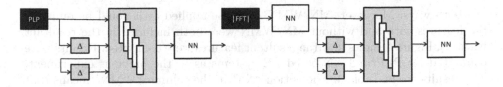

Fig. 2. An AM which uses PLP or NN-based FE.

In the second experiment, the benefit of MN/WMN was investigated. Figure 3 shows how AMs with PLP or NN-based FE use MN or WMN with the NN-based VAD. The baseline system was a NN-based AM that estimates posteriors from PLP with mean normalization, delta and delta-delta coefficients. This system is denoted as MN(PLP)+Δ+$\Delta\Delta$ in Table 2. The second system was the baseline system modified by WMN and by the NN-based VAD (see Figure 3). The NN-based VAD had 1024 neurons in its hidden layer. This system is denoted as WMN(PLP)+Δ+$\Delta\Delta$. In the next system, the NN-based FE and MN are used. This system is denoted as MN(NN)+Δ+$\Delta\Delta$. The fourth system is the third system modified by WMN and the NN-based VAD. This system is denoted as WMN(NN)+Δ+$\Delta\Delta$. The results are shown in Table 2. This experiments shown that methods such as MN can be applied beneficially. Also, strictly separated training of the NN-based FE is not necessary. No significant difference between MN and WMN was noticed.

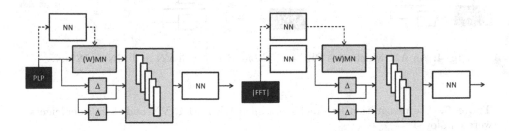

Fig. 3. An AM which uses PLP or NN-based FE with MN or WMN.

In the last experiment, a benefit of VN was investigated. In comparison with MN/WMN that does not change delta coefficients, there are two different possibilities how the delta coefficients could be computed. Because we did not want to chose one way or double the number of investigated AMs, we simply

Table 2. The recognition accuracies obtained when MN and delta coefficients were employed.

System	si_dt5a (dev.)	si_dt5b (test)
MN(PLP)+Δ+$\Delta\Delta$	87.5%	86.5%
WMN(PLP)+Δ+$\Delta\Delta$	87.9%	86.4%
MN(NN)+Δ+$\Delta\Delta$	89.6%	89.6%
WMN(NN)+Δ+$\Delta\Delta$	**89.9%**	**89.9%**

use both ways. Moreover, MN/WMN could be applied or it could be omitted. So, features with and without MN/WMN were both included in the resultant features. Figure 4 shows how the resultant features were computed. Results were computed for PLP and NN-based FE systems as in the previous experiment. The results are in Table 3. The option "No" in the column "WMN" means that MN was applied instead of WMN. These results significantly exceeded the results in both previous experiments. Again, no significant difference between MN and WMN was noticed. In contrary to this fact, using the NN-based FE increases accuracy significantly not only in this experiment but also in both previous experiments.

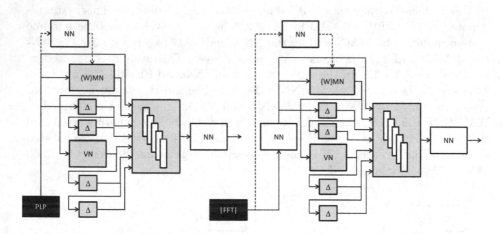

Fig. 4. An AM which uses PLP or NN-based FE with VN and MN or WMN.

Table 3. The recognition accuracies obtained when MN, VN and delta coefficients were employed.

Features Extraction	Mean norm.	si_dt5a (dev.)	si_dt5b (test)
PLP	MN	89.5%	88.7%
PLP	WMN	89.8%	88.8%
NN	MN	91.6%	**90.9%**
NN	WMN	**91.7%**	90.9%

6   Conclusion and Future Work

In this paper, the NN-based FEs for automatic speech recognition system were investigated. We focused on combination of the NN-based FE and the NN/HMM hybrids. Our experiments showed that the NN-based FE can sufficiently approximate standard method such as PLP and it can lead to more accurate results in speech recognition when a machine learning is simultaneously applied on the NN-based FE and the NN-based AM. Moreover, this paper demonstrates that methods mean or variance normalization and delta coefficients and even an NN-based VAD can be utilized successfully in a NN-based AM.

In the future, some experiments described in this paper will be done for context-dependent units. Also some NN-based speaker normalization or speaker adaptation techniques will be integrated in the described system. Beside this we are going to make some experiments with the NN-based FE in non-hybrid speech recognition systems.

Acknowledgments. This research was supported by the Ministry of Culture Czech Republic, project No.DF12P01OVV022.

References

1. Grézl, F., Karafiát, M.: Semi-supervised bootstrapping approach for neural network feature extractor training. In: ASRU, pp. 470–475. IEEE (2013)
2. Sainath, T.N., Kingsbury, B., Mohamed, A.R, Ramabhadran, B.: Learning filter banks within a deep neural network framework. In: ASRU, pp. 297–302. IEEE (2013)
3. Narayanan, A., Wang, D.: Ideal ratio mask estimation using deep neural networks for robust speech recognition. In: 2013 IEEE International Conference on Acoustics, Speech and Signal Processing (ICASSP), pp. 7092–7096, May 2013
4. Fernandez Astudillo, R., Abad, A., Trancoso, I.: Accounting for the residual uncertainty of multi-layer perceptron based features. In: 2014 IEEE International Conference on Acoustics, Speech and Signal Processing (ICASSP), pp. 6859–6863, May 2014
5. Seps, L., Málek, J., Cerva, P., Nouza, J.: Investigation of deep neural networks for robust recognition of nonlinearly distorted speech. In: INTERSPEECH, pp. 363–367 (2014)
6. Sainath, T.N., Peddinti, V., Kingsbury, B., Fousek, P., Ramabhadran, B., Nahamoo, D.: Deep scattering spectra with deep neural networks for LVCSR tasks. In: 15th Annual Conference of the International Speech Communication Association, INTERSPEECH 2014, September 14–18, 2014, Singapore, pp. 900–904 (2014)
7. Tüske, Z., Golik, P., Schlüter, R., Ney, H.: Acoustic modeling with deep neural networks using raw time signal for lvcsr. In: Interspeech, Singapore, pp. 890–894, September 2014
8. Heřmanský, H.: Perceptual linear predictive (PLP) analysis of speech. J. Acoust. Soc. Am. **57**(4), 1738–52 (1990)
9. Chang, S., Morgan, N.: Robust CNN-based speech recognition with Gabor filter kernels. In: 15th Annual Conference of the International Speech Communication Association, INTERSPEECH 2014, September 14–18, 2014, Singapore, pp. 905–909 (2014)

10. Robinson, T., Fransen, J., Pye, D., Foote, J., Renals, S.: Wsjcam0: a british english speech corpus for large vocabulary continuous speech recognition. In: Proc. ICASSP 1995, pp. 81–84. IEEE (1995)
11. Psutka, J., Švec, J., Psutka, J.V., Vaněk, J., Pražák, A., Šmídl, L.: Fast phonetic/lexical searching in the archives of the czech holocaust testimonies: advancing towards the MALACH project visions. In: Sojka, P., Horák, A., Kopeček, I., Pala, K. (eds.) TSD 2010. LNCS, vol. 6231, pp. 385–391. Springer, Heidelberg (2010)
12. Pražák, A., Psutka, J.V., Psutka, Loose, Z.: Towards live subtitling of TV ice-hockey commentary. In: Cabello, E., Virvou, M., Obaidat, M.S., Ji, H., Nicopolitidis, P., Vergados, D.D. (eds.) SIGMAP, pp. 151–155. SciTePress (2013)

Named Entity Recognition
for Mongolian Language

Zoljargal Munkhjargal[1]([✉]), Gabor Bella[2],
Altangerel Chagnaa[1], and Fausto Giunchiglia[2]

[1] DICS, National University of Mongolia, 14200 Ulaanbaatar, Mongolia
{zoljargal,altangerel}@num.edu.mn
[2] DISI, University of Trento, 38100 Trento, Italy
fausto@disi.unitn.it, gabor.bella@unitn.it

Abstract. This paper presents a pioneering work on building a Named
Entity Recognition system for the Mongolian language, with an agglutinative morphology and a subject-object-verb word order. Our work
explores the fittest feature set from a wide range of features and a
method that refines machine learning approach using gazetteers with
approximate string matching, in an effort for robust handling of out-of-
vocabulary words. As well as we tried to apply various existing machine
learning methods and find optimal ensemble of classifiers based on
genetic algorithm. The classifiers uses different feature representations.
The resulting system constitutes the first-ever usable software package
for Mongolian NER, while our experimental evaluation will also serve as
a much-needed basis of comparison for further research.

Keywords: Mongolian named entity recognition · Genetic algorithm ·
Machine learning · String matching

1 Introduction

The volume of textual information made available every day exceeds by far the
human ability to understand and process it. As a consequence, automated information extraction has become an essential and pervasive task in computing. One
particular component of such systems is Named Entity Recognition (NER) that
consists of identifying personal names, organization names, and location names
within sentences of natural language text. NER is an extensively researched
topic. However, In less-studied and resource lack languages such as Mongolian,
there is not enough research.

While the general problem of NER has been approached from widely varying perspectives, methods tend to follow 1) dictionary-based, 2) rule-based, 3)
stochastic machine learning-based approaches or 4) combinations of these. If
applied directly to text, these approaches are not considered robust as they
cannot tackle spelling mistakes, orthographic variations, or out-of-vocabulary
names, the latter being a very common phenomenon as the set of commonly used

© Springer International Publishing Switzerland 2015
P. Král and V. Matoušek (Eds.): TSD 2015, LNAI 9302, pp. 243–251, 2015.
DOI: 10.1007/978-3-319-24033-6_28

named entities (people and things) is open and constantly evolving. It is in great part for these reasons that statistical machine learning-based methods, based on manually pre-annotated text corpora, have become the basis of most NER systems. State-of-the-art results have been obtained using Maximum Entropy [1], Hidden Markov Models, Support Vector Machines [2], and Conditional Random Fields [3].

However, for agglutinative languages such as Mongolian, supervised learning methods tend to produce weaker results. This is due to the morphology, characterized by an almost unbound number of word forms. As a result, machine learning is hindered by frequent occurrences of word forms rarely or never seen during training.

Ensembling several NER classifiers that each one is based on different feature representation and different classification approach improves general performance accuracy [4] [5] [6]. This general improvement depends on a diversity of classifiers that see the NER task from different aspects. However, exploring the fittest-feature set and selecting appropriate classifier for constructing an ensemble classifier are a difficult problem.

This paper describes what we believe to be the first serious attempt at designing a supervised NER system for Mongolian. The system implements an ensemble approach consisting of supervised machine learners with a corresponding new annotated corpus, newly created gazetteers, as well as a simple rule-based matcher. A genetic algorithm is applied to find optimal classifier ensemble. Furthermore, to tackle the out-of-vocabulary word form problem, we took inspiration from studies showing how approximate string distance metrics can be used for robust name-matching tasks [7] [8], for taking into account inflectional variations of names.

2 Mongolian Names

Mongolian is an agglutinative language that a word is inflected by rich suffix chains in the verbal and nominal domains. It is often considered part of Altaic language family that includes Turkic languages, Korean and Japanese. As Hungarian is also agglutinative, from a computational linguistic point of view very similar problems need to be solved in the two languages [9].

Most personal names are compounds of two or more simple names or common words (that can themselves serve as names). For example, Ганболд (Ganbold) joins two common nouns: Ган (Gan-steel) and Болд (Bold-alloy).

The full personal name is written in the reverse order with respect to the Western convention: either a *patronymic* or a *matronymic* (roughly equivalent to the surname) comes first and a given name (equivalent to the first name) comes second. In general, the surname is inflected in a genitive case, depending on vowel harmony and on several other morphological factors. Because of the usage of patronymic/matronymic as surnames instead of distinct family names, the order of the surname and the given name is never inversed. The full name also consists of abbreviation of surname, which ends with period, and given name. This is commonly used in newswire domain.

In general, organization and location names are capitalized. For names of international organizations, countries, councils, ministries and associations consisting of several words, all of the words have to be capitalized. For other typical multi-word names—such as industry branches, provinces, districts, scientific and cultural organizations—the head word is capitalized and the others remain in lowercase.

If a proper name of any kind is a composite of two names and the second name starts with a vowel, a hyphen is placed between two names and the second word is capitalized: Баруун-Урт (Baruun-Urt: city name).

Another feature that makes NER more difficult for Mongolian is the subject-object-verb word order: boundaries between named entities are easy to miss when the subject and the object are both proper names.

3   The NER System

A range of machine learning algorithms are successfully applied to the task of NER. To effectively solve classification problem, we hypothesized that ensemble of a diverse set of classifiers would benefit the performance, assuming different methods lies in tackling the problem from different angles. Depending on the various feature combinations, a classifier produces various results, too. Thus we tried to find the most optimal combination of features sets using a brute force method. The building blocks of our system are shown in Fig 1.

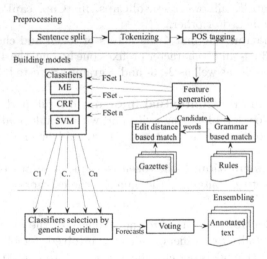

Fig. 1. Outline of building complex NER model. FSet: feature set that is a subset of 5 group feature sets; C: classifier that trained on particular subset.

3.1 Preprocessing

From the point of NER, feature of context/trigger words and sentence position is one of common clues to determine NEs. Thus we should involve a sentence detector. The Mongolian proper name writing rules such as surname abbreviation and hyphenation is leading to consider one word as two or more tokens. Further, we hope that tokenization before classification is more suitable for the sequence of instance (feature vector of token) tracking algorithms (ME, SVM, CRF, HMM). As well as part-of-speech (POS) tag is commonly used for a statistical NE classifier as a feature [10] [11].

3.2 Feature Generation

We experimented with a rich set of features, describing the characteristic of the token with its context (a moving window of size five), many of which are used in related works [5], [6], [12] and [9].

1. **Orthographic properties of the word form**: is first letter capitalized, is entire word uppercased, is entire word lowercased, does it contain any hyphen, does token only consist of punctuation marks, does token contain at least one punctuation mark except hyphen and word length.
2. **Word Shape**: long pattern (maps all uppercase characters to "X", all lowercase ones to "x", and the rest to "_"), and short pattern (consecutive character types are not repeated: "X_x") and a symbolic feature outputting one of following labels: allLowerCase, allCaps, firstCap, capPeriod, onlyDigit, onlyPunct, hyphened or other.
3. **Affix information**: the first 4 characters of token and character 3-grams. We validated 3-, 4- and 5-character prefixes one by one, the 4-character prefix reaches best result, as well as 2-, 3- and 4-grams are tested, 3-gram gives the best result.
4. **Morphological and Contextual information**: full part-of-speech (POS) tag, high and low-level POS tag (with and without information about inflection, resp.), position in the sentence (start, mid or end) and is the word between quotes.
5. **Gazetteer information**: if the token is included in one of the gazetteers, it receives a feature containing the name of the category.

Feature Set Selection. To measure the strength of the above five groups of features we trained all of classifiers, which we involved, for all possible sub set of the five groups (31 models per classifier). Between 15 and 20 best performing models achieved very similar results better than the others, in each classifier. We used the top 5 models of each classifier, and then recombined the models in a voting scheme.

Grammar- and Edit Distance-Based Matching. To complement the statistical classifier by making use of existing Mongolian name resources, we implemented a simple pattern-based and a gazetteer-based recognizer. The former is a regex-based matcher using simple grammatical rules while the latter uses fuzzy string matching on name lists.

A simple rule set representing an intentionally rough (and, by consequence, easy-to-implement and fast-to-apply) grammatical model of Mongolian proper names is created. The rules, given below, are optimized for recall rather than precision, since their main purpose is to extract candidate named entities that will be further verified using string distance metrics and gazetteers.

1. Personal name: a) capitalized word in genitive + capitalized word; b) capitalized word + period + capitalized word.
2. Organization name: a) sequence of capitalized words + organization designator (e.g., **XXK** "Co. Ltd.", холбоо "association", and компани "company"); b) capitalized word + sequence of lowercased words + organization designator; all-capital word (e.g., **МУИС-NUM**).
3. Location name: a) capitalized word + location designator (e.g., гудамж "street", талбай "square", хот "city").

Edit Distance Metrics. In order to robustly match inflectional form or naming variations, we validated using the main string similarity measures used in [7] and [8]. A token based metrics are Levenshtein, Smith-Waterman (SW) [13] (gap cost is *1*, the mismatch cost is *-1* and the match cost is *2*), Jaro and Jaro-Winkler (JW)[14]. The multi-token based metrics are Jaccard similarity, Fleggi-Shunter [15], Monge-Elkan [16] and SoftTFIDF [7].

In [8], considering the declension paradigm of Polish, approved a basic and time efficient metric based on the longest common prefix information, which would intuitively perform well in the case of single-token names. We have been inspired by this method and modified the declension paradigm to agglutinative feature. It is defined as $CP_b = (|lcp(s,t)| + b)^2/(|s| * |t|)$. b is set to 0 . If s ends in common suffix inflection such as "-ын" (genitive case), "-ынд" (genitive case + locative case), b is set to 1.

3.3   Classifiers

For the statistical classifer, we used the ME (OpenNLP v1.5.3), CRF (CRF++ v0.53) and SVM (Yamcha v0.33).

3.4   Ensembling

A Genetic algorithm (GA) [17] is a search method of an optimal solution, inspired by biological evaluation processes.

In our experiments, the genome for optimal ensembles from a set of $n = 15$ classifiers (5 models for each 3 classifiers), can be a binary string of length n, in which every bit represents a classifier. A bit value 1 means the classifier is selected and 0 is vice versa. For instance the chromosome 101010101010101 represents an ensemble in which every first classifier is used. Our implementation follows specific parameters and methods described in [6] and [5].

To combine the outputs of classifiers we use majority voting mechanism. We implemented and tested the following decision function: each classifier's output tag votes for a NE class for each token, and the tag that has highest score wins.

4   Resource Building

We manually annotated a Mongolian POS-tagged corpus [11] from the Newswire domain with NE tags. The corpus consists of 310 articles, about 277,000 tokens, 14,837 sentences, 4,382 personal names, 4,932 location names, and 3,366 organization names.

This corpus is equivalently created to English corpora used on the CoNLL02 and CoNLL03 conferences, in format, annotation style and the number of NEs.

To guarantee the accuracy of tagging we set up a three-stage annotation procedure: first, linguists manually labeled the corpus with NE tags using clearly established guidelines; secondly, a reviewer validated the annotations, and finally, the linguist and reviewer discussed borderline cases. The agreement of linguist and reviewer F-measure was 98.56% (using conlleval script).

For gazetteers, we had access to a free-to-use 170,000-entry list of Mongolian personal names and to a 80,000-entry list of Mongolian company names.

For locations, we used *Geonames.org* that contains 4,151 names of mountains, rivers, landmarks, and localities of Mongolia. However, these names were in various *Latin* transcriptions because of crowdsourcing; we therefore had to convert 2,410 names from *Latin* transcriptions to *Mongolian Cyrillic*. We also extracted about 3,500 location names from *Wikipedia*.

5   Experimental Results and Discussion

A randomly selected 10% segment of the NE corpus was used as a test set and rest of 90% was used as a training set. We used the CoNLL03 evaluation methodology [10].

5.1   Results in Edit Distance Functions

To validate efficiency of each string distance metrics, we compared 4,382 real-world names of the corpus to the non-inflected 176,343 person names (comparison pairs are 770 million). Table 1 shows results for each metrics that we involved. The SW based metrics achieved the best recall score, whereas Levenshtein was the worst metric. However, the SW based metrics matched irrelevant pairs. For example, "Лувсаншара-" (ваас (person name) + аас (ablative case)) is matched to "шарав" and scored those as 1.0. The JW and the combination of Monge-Elkan with JW achieved the best F1-score 94.1% and 91.6%. We therefore chose it as default string matching method.

Table 1. The experimental results for edit distance metrics.

Metrics	Pre	Re	F1	Metrics	Pre	Re	F1
On a token				On multiple tokens			
Levenshtein	73.4	71.9	72.6	MEl & SW	82.4	94.7	88.1
SmithWaterman (SW)	85.1	97.5	90.8	MEl & JW	89.3	93.8	91.6
Jaro	93.2	88.9	90.1	Jaccard & DM	79.3	67.5	72.9
JaroWinkler (JW)	95.5	92.8	94.1	Fleggi-Shunter	75.6	70.1	72.7
Common prefix	84.5	83.1	83.8	SoftTFIDF & JW	92.2	90.2	91.2
				SoftTFIDF & SW	86.8	94.8	90.6

5.2 Results in Feature Selection Experiments

We built 31 models for each 3 classifier methods (totally 93) from the available NE features. Table 2 presents the top 15 models that achieved the best F-measure for each classifier. The best individual classifier shows F-measure values 86.94%. One interesting thing is that ME reached its best performance without orthographic feature. The gazetteer feature was frequently involved in the best performing classifiers. The first observation that can be made about applying gazette feature with robust string matching method is its good impact to the fine grained classifier. It can be also observed that CRF achieves the best-performance for the Mongolian NER task.

5.3 Results in the Classifier Ensemble

We tested the ensemble of different classifiers that are trained using one kind of classification method and 5 different feature set. As shown in Table 3, the scores of the ensemble of the best 5 models of ME, the F-measure is decreased by 0.97% from best individual ME. For CRF ensemble; the F-measure is improved by 0.42%. Finally, for it also decreased from best individual performance. We gathered the optimal genome 00001 10011 11111. The first 5 bits represents the ME classifiers, ordered per the best F1-accuracy rank (according to Table 2), followed by 5 CRFs and 5 SVMs. The optimal ensemble reached to 90.59% precision, 85.88% recall and 88.17% F1 score.

Table 2. The top five feature sets for each classifier.

ME		CRF		SVM	
Fgroups	F1	Fgroups	F1	Fgroups	F1
2,4,5	83.69	1,2,3,4,5	86.94	1,2,3,4,5	86.66
2,4	83.04	1,3,4,5	86.91	1,2,3,4	86.57
1,2,4	82.92	1,2,3,4	86.75	1,3,4	86.56
1,2,4,5	82.82	2,3,4,5	86.75	2,3,4	86.16
1,2,3,4	82.81	2,3,4	86.62	1,2,4,5	86.11

Table 3. Result of ensemble on feature set.

Classifier	Pre	Re	F1
Ensemble of MEs	88.86	77.38	82.72
Ensemble of CRFs	90.83	84.14	87.36
Ensemble of SVMs	88.19	84.75	86.43

We seriously tried to apply existing machine learning and string matching methods into NER, as well as created the NER corpus and gazettee for Mongolian language. The experimental results confirm that genetic algorithm can be successfully applied to the task of finding a classifier ensemble that outperforms the best individual classifier and simple ensemble method. However, the performance improvement measured in our experiments is not satisfactory, since running many classifiers that are trained on different feature sets took more CPU time than one classifier.

We nevertheless consider these as promising first results, especially taking into account the difficulties of Mongolian grammar. They will also be useful as a quantitative basis for comparison for further research in Mongolian NER.

References

1. Bender, O., Och, F.J., Ney, H.: Maximum entropy models for named entity recognition. In: Proceedings of CoNLL-2003, pp. 148–151 (2003)
2. Isozaki, H., Kazawa, H.: Efficient support vector classifiers for named entity recognition. In: Proceedings of the 19th International Conference on Computational Linguistics (COLING-2002), Taipei, Taiwan (2002)
3. Lafferty, J., McCallum, A., Pereira, F.: Conditional random fields: Probabilistic modelsfor segmenting and labeling sequence data. In: Machine Learning International Workshop (2001)
4. Florian, R., Ittycheriah, A., Jing, H., Zhang, T.: Named entity recognition through classifier combination. In: Proceedings of the Seventh Conference on Natural Language Learning at HLT-NAACL (2003)
5. Desmet, B., Hoste, V.: Dutch named entity recognition using classifier ensembles. In: Preceedings of the 20th Meeting of Computational Linguistics, Netherlands (2010)
6. Ekbal, A., Saha, S.: Maximum entropy classifier ensembling using genetic algorithm for ner in bengali. In: Proceedings of the International Conference on Language Resource and Evaluation (LERC) (2010)
7. Cohen, W., Ravikumar, P., Fienberg, S.: A comparison of string distance metrics for name-matching tasks. In: IJCAI 2003 Workshop on Information Integration on the Web (IIWeb 2003), Acapulco, Mexico, pp. 73–78 (2003)
8. Piskorski, J., Wieloch, K., Pikula, M., Sydow, M.: Towards person name matching for inflective languages. In: NLPIX, Beijing, China (2008)
9. Szarvas, G., Farkas, R., Kocsor, A.: A multilingual named entity recognition system using boosting c4.5 decision tree learning algorithms. In: Todorovski, L., Lavrač, N., Jantke, K.P. (eds.) DS 2006. LNCS (LNAI), vol. 4265, pp. 267–278. Springer, Heidelberg (2006)

10. Sang, E.F.T.K., De Meulder, F.: Introduction to the conll-2003 shared task: Language-independent named entity recognition. In: CoNLL-2003, Canada (2003)
11. Purev, J., Odbayar, C.: Part of speech tagging for mongolian corpus. In: The 7th Workshop on Asian Language Resources, Singapore (2009)
12. Simon, E., Kornai, A.: Approaches to hungarian named entity recognition. In: PhD thesis, Budapest University of Technology and Economics. (2013)
13. Smith, T., Waterman, M.: Identification of common molecular subsequences. Journal of Molecular Biology **147**, 195–197 (1981)
14. Winkler, W.E.: The state of record linkage and current research problems. In: Technical report, Statistical Research Division, U.S. Bureau of the Census, Washington, DC (1999)
15. Fleggi, I.P., Sunter, A.B.: A theory for record linkage. Journal of the American Statistical Society **64**, 1183–1210 (1969)
16. Monge, A., Elkan, C.: The field matching problem: Algorithms and applications. In: Proceedings of Knowledge Discovery and Data Mining, pp. 267–270 (1996)
17. Goldberg, D.E.: Genetic algorithm in search, optimization, and machine learning. Addison-Wesley Publishing Company, Boston (1989)

Comparing Semantic Models for Evaluating Automatic Document Summarization

Michal Campr[✉] and Karel Ježek

Department of Computer Science and Engineering, FAS,
University of West Bohemia, Univerzitn 8, 306 14 Pilsen, Czech Republic
{mcampr,jezek_ka}@kiv.zcu.cz

Abstract. The main focus of this paper is the examination of semantic modelling in the context of automatic document summarization and its evaluation. The main area of our research is extractive summarization, more specifically, contrastive opinion summarization. And as it is with all summarization tasks, the evaluation of their performance is a challenging problem on its own. Nowadays, the most commonly used evaluation technique is ROUGE (Recall-Oriented Understudy for Gisting Evaluation). It includes measures (such as the count of overlapping n-grams or word sequences) for automatically determining the quality of summaries by comparing them to ideal human-made summaries. However, these measures do not take into account the semantics of words and thus, for example, synonyms are not treated as equal. We explore this issue by experimenting with various language models, examining their performance in the task of computing document similarity. In particular, we chose four semantic models (LSA, LDA, Word2Vec and Doc2Vec) and one frequency-based model (TfIdf), for extracting document features. The experiments were then performed on our custom dataset and the results of each model are then compared to the similarity values assessed by human annotators. We also compare these values with the ROUGE scores and observe the correlations between them. The aim of our experiments is to find a model, which can best imitate a human estimate of document similarity.

Keywords: Contrastive opinion summarization · Summarization evaluation · ROUGE · Semantic models · Tfidf · lsa · lda · Word2vec · Doc2vec · Document similarity

1 Introduction

In recent years, with rapid growth of information available online, the research area of automatic summarization has been attracting very much attention. Automatic document summarization aims to transform an input text into a condensed form, in order to present the most important information to the user. Summarization is a very challenging problem, because the algorithm needs to understand the text and this requires some form of semantic analysis and grouping of the

© Springer International Publishing Switzerland 2015
P. Král and V. Matoušek (Eds.): TSD 2015, LNAI 9302, pp. 252–260, 2015.
DOI: 10.1007/978-3-319-24033-6_29

content using world knowledge. Therefore, attempts at performing true abstraction (generating the summary from scratch) have not been very successful so far. Fortunately, an approximation called extraction exists and is more feasible for the vast majority of current summarization systems, which simply need to identify the most important passages of the text to produce an extract. The output text is often not coherent but the reader can still form an opinion of the original content.

A very challenging problem, which arises, is the evaluation of summarization quality. There are dozens of possible ways for the evaluation of summarization systems, and these methods can be classified basically into two categories [1]. *Extrinsic* techniques judge the summary quality on the basis of how helpful summaries are for a given task, such as classification or searching. On the other hand, *intrinsic* evaluation is directly based on analysis of the summary, which can involve a comparison with the source document, measuring how many main ideas from it are covered by the summary, or with an abstract written by a human.

Recently, one particular method has become very popular for the evaluation of automatic summarization. ROUGE [2] (Recall-Oriented Understudy for Gisting Evaluation) includes measures for automatically determining the quality of system summaries by comparing them to ideal human-made summaries. These measures count the number of overlapping units such as n-grams, word sequences, or word pairs between the system summary and the ideal summaries created by humans.

Since the evaluation of automatic summarization is based on a comparison between the system summary and a human-made one, we wondered if it is possible to utilize other NLP (Natural Language Processing) methods for evaluating the system summaries. In this paper, we examine some of the most popular NLP models and their performances in the task of assessing document similarity.

This paper firstly describes, what data we used for evaluating these models and how we annotated them. Then, we describe the models which we used in our experiments, each in its own section. Lastly, we provide the results and performances of chosen models in computing document similarities, and their comparison to human annotators.

2   Dataset

Our current research is focused on a specific variation of automatic summarization: contrastive opinion summarization. The main goal is to analyze the input documents, in our case restaurant reviews, and construct two summaries, one depicting the most important positive information and the other providing negative information. For this task, we constructed a collection from czech restaurant reviews downloaded from www.fajnsmekr.cz, in total of 6008 reviews for 1242 restaurants. For human annotation, however, this is too much, so we manually selected 50 restaurants, each with several reviews, so that their combined length is at least 1000 words. Three annotators then independently created two summaries for each restaurant, each with approximately 100 words.

This collection of gold summaries can already be used for evaluating our system summaries using the ROUGE metrics. However, we wondered, whether any other method could be utilized for this task, so we enhanced our collection in such a way, that it can be used for experimenting with document similarity. The main idea is to utilize the manually created summaries for finding the best algorithm for summarization evaluation. We will be investigating the similarity between those gold summaries the same way as if we were comparing the system summaries with the human-made ones. The process of additional annotation of our collection is described in the following subsection.

2.1 Data Annotation

We presented three annotators with 150 pairs of summaries, three positive summaries for each of the 50 restaurants. Each annotator was asked to assign a similarity score between them. The most obvious problem here is, how a human can come up with such a value after reading the texts. We devised a process of acquiring this score based on SCUs (Summary Content Units) used in the so called Pyramid evaluation [3] and combined it with a technique of annotation used in [4].

We asked our annotators to find pairs of facts which can be assigned the highest possible similarity score, according to our scale. Some can be pretty straightforward, such as those that praise the quality of service or food (assigned a value between 4 to 2), but other facts, that are missing or redundant in any of the summaries would be assigned with 0 or 1. These values are then averaged and thus the final score is assigned to the summary pair.

4 - Completely equivalent
3 - Mostly equivalent, differs in unimportant details
2 - Roughly equivalent, discussing the same topic, but important information differs
1 - Not equivalent, but roughly discussing a similar topic
0 - Different topics

3   Examined Language Models

In order to algorithmically perform a comparison of two documents, it is necessary to transform the original documents (plain text) into a representation, which a computer can understand, i.e a vector of features. The main problem is, what type of features to use. It is worth noting, that a common step for all models mentioned here is a preprocessing step, where the input string is tokenized into words and each word is lemmatized. The result is a set of terms corresponding to the input document. These terms can be used in several ways for constructing a model of the document. Among some of the more basic models are:

- Boolean model - equals to 1 if a term t occurs in document d and 0 otherwise
- Term frequency $tf(t, d)$ - raw number of times that term t is in the document
- Logarithmically scaled term frequency $\log(tf(t, d) + 1)$
- Augmented frequency - to prevent a bias towards longer documents, e.g. raw frequency divided by the maximum raw frequency of any term in the document

Besides those, there are more, however we decided to utilize only the most widely used and recognized models. In addition, we added to our experiments two relatively new ones, which are lately gaining much popularity (Word2Vec and Doc2Vec). Models used in our experiments are TfIdf, LSA, LDA, Word2Vec, Doc2Vec, and each one is briefly described in the following sections.

There is also the possibility of utilizing external linguistic resources, such as WordNet [5], for processing synonyms or other semantic information. However, our intention is to minimize the dependency of our methods on any language-specific tool. Also, we found a variety of colloquial expressions, which are not present in WordNet, and thus we decided not to use any such tool.

We should note beforehand, that (if not specified otherwise) the similarity score for each pair of documents is computed as the cosine similarity between two feature vectors v_1 and v_2:

$$sim(v_1, v_2) = \frac{\sum_{i=1}^{m} v_1[i] * v_2[i]}{\sqrt{\sum_{i=1}^{n} v_1[i]^2} * \sqrt{\sum_{i=1}^{n} v_2[i]^2}}, \tag{1}$$

3.1 TfIdf

The TfIdf (Term frequency - Inverse document frequency) weight of a term is a statistical measure used to evaluate how important a word is to a document in a collection or a corpus. Its importance increases proportionally to the number of times the term appears in the document, but is offset by its frequency in the corpus. The TfIdf weight of a term is a product of its frequency $tf(t, s)$ and inverse document frequency:

$$idf(t, D) = \log \frac{N}{|d \in D : t \in d|}, \tag{2}$$

where D is the document set, N is the size of set D and $|d \in D : t \in s|$ is the number of documents from D where the term t appears. The final value is then computed as:

$$tfidf(t, d, D) = tf(t, d) \cdot idf(t, D). \tag{3}$$

3.2 LSA

Latent Semantic Analysis (LSA) [6], also known as Latent Semantic Indexing (LSI), is a method for extracting and representing the contextual meaning of words by statistical computations performed on a corpus of documents. The underlying idea is that the totality of information about all the word contexts, in which a given word does and does not appear, provides a set of mutual constraints that largely determines the similarity of meaning of words. The adequacy of LSA's reflection of human knowledge has been established in a variety of ways.

The creation of an LSA model starts with building a $m \times n$ matrix A, where n is the number of documents in the corpus and m is the total number of terms that appear in all documents. Each column of A represents a document d and each row represents term t.

There are several methods on how to compute the elements a_{td} of matrix A representing term frequencies, and among the most common are: Term frequency, TfIdf or Entropy. In our experiments, we present only models based on TfIdf, because they provided the best results.

With the matrix A built, LSA applies Singular Value Decomposition (SVD), which is defined as $A = U\Sigma V^T$, where $U = [u_{ij}]$ is an $m \times n$ matrix and its column vectors are called left singular vectors. Σ is a square diagonal $n \times n$ matrix and contains the singular values. $V^T = [v_{ij}]$ is an $n \times n$ matrix and its columns are called right singular vectors. This decomposition provides latent semantic structure of the input documents, which means, that it provides a decomposition of documents into n linearly independent vectors, which represent the main topics of the documents. If a specific combination of terms is often present within the document set, it is represented by one of the singular vectors.

3.3 LDA

Latent Dirichlet Allocation (LDA) [7] can be basically viewed as a model which breaks down the collection of documents into topics by representing the document as a mixture of topics with their probability distributions. The topics are represented as a mixture of words with a probability representing the importance of the word for each topic.

Since LDA provides probability distributions of topics, it is possible to use statistical measures for quantifying the similarity between two documents. There are many such measures, like KL-divergence, Hellinger distance or Wasserstein metric to name just a few. In our experiments, we chose to use the Hellinger distance (further denoted as LDA-h in the results) along with the cosine similarity to see, how those methods differ.

3.4 Word2Vec

Briefly, Word2vec is a two-layer neural net published by Google in 2013. It implements continuous bag-of-words and skip-gram architectures for computing vector representations of words, including their context. The skip-gram representation popularized by Mikolov [8], [9], [10] has proven to be more accurate than other models due to the more generalizable contexts generated. The output of Word2Vec is a vocabulary of words, which appear in the original document, along with their vector representations in an n-dimensional vector space. Related words and/or groups of words appear next to each other in this space.

Since Word2Vec provides vector representations only for words, we need to combine them in some way to get a representation of the whole document. This can be done by averaging all the word vectors for the given document, and thus creating just one document vector, which can be compared to another by cosine similarity (model designated as 'Word2Vec'). We also experimented with an n-gram analogy (denoted as 'W2V-pn'), i.e. combining word vectors for phrases with n words and then computing similarities between these phrase-vectors from both input documents.

In our experiments, we utilized the Word2Vec implementation (as well as Doc2Vec) in Python, called gensim, by Radim Řehůřek [11].

3.5 Doc2Vec

Googles Word2Vec project has created lots of interests in the text mining community. It provides high quality word vectors, however there is still no clear way to combine them into a high quality document vector. Doc2vec (Paragraph2Vec) modifies the Word2Vec model into unsupervised learning of continuous representations for larger blocks of text, such as sentences, paragraphs or entire documents. In [12], an algorithm called Paragraph Vector is used on the IMDB dataset to produce some of the most state-of-the-art results to date. In part, it performs better than other approaches, because vector averaging or clustering lose the order of words, whereas Paragraph Vectors preserve this information. Because of this, we also experimented with an n-gram analogy, i.e. computing Paragraph Vectors for phrases of length n and then comparing those phrases from both input document. The original model, which computes vectors for the whole documents is denoted as 'Doc2Vec' and the n-gram analogies are denoted as 'D2V-pn'.

4   Evaluation

As was described in section 2.1, we manually annotated 150 pairs of summaries with their similarity score, resulting in a total 450 human-made assumptions. Our main goal is to find such a model, that would provide the best correlation with human intuition. The annotated scores are based on a score scale with values ranging between 0 to 4, however in all the following texts and figures, they are converted into a 0-1 scale in order to be comparable with the models' scores. The average Pearson correlation coefficient between annotators is 0.8988.

Our results regarding the performance on document similarity assessment show, that the tested models can be basically divided into two groups:

1. models with no apparent correlation with human ratings
2. models showing significant correlation

The first group contains models: TfIdf, LSA, LDA and LDA-h. Figure 1 shows all these models in comparison to the averaged human-made values (dashed line). It is clear, that these models do not show any significant correlation, see Table 1 for exact values. This behaviour is most likely caused by the models' tendency to over-generalize the features, and by the fact, that they all work on the bag-of-words basis, effectively disregarding the word order. The Figure 1 shows, that these models compute very high values in all cases and do not provide scores from the full scale between 0 to 1, as are the human-made scores.

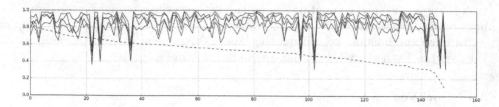

Fig. 1. Correlations of TfIdf, LSA, LDA and LDA-h models with average annotator scores (dashed line).

The second group of models contains: Word2Vec, Doc2Vec and their variations. These models show higher values of Pearson correlation with annotated data, see Table 1. Although Word2Vec does not show a very high correlation value (0.4009), it is apparent that it is able to capture a more sophisticated document structure. The same applies to Doc2Vec. Its base correlation (0.5523) with average annotated data shows, that its ability to take the word order into account provides a better document latent structure.

An interesting observation is, that methods comparing phrase-vectors (3.4) show a significant improvement over the base models. The best being W2V-p2 (0.4626) and D2V-p2 (0.6614).

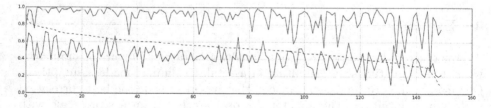

Fig. 2. Correlations of Word2Vec (higher values) and Doc2Vec models with average annotator scores (dashed line).

The last set of results was obtained using the ROUGE measures: ROUGE-1, ROUGE-2 and ROUGE-SU4. These measures are nowadays frequently used for summarization evaluation. From Figure 3 and Table 1 is apparent that these metrics provide the best overall correlations with annotated data, where the best one is 0.7276 for ROUGE-1.

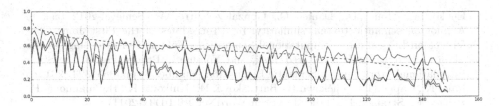

Fig. 3. Correlations of ROUGE metrics with average annotator scores.

Table 1. Pearson correlations between models and annotators. The 'avg' model score is computed against averaged scores from annotators.

	TfIdf	LSA	LDA	LDA-h	Word2Vec	Doc2Vec	ROUGE-1	ROUGE-2	ROUGE-SU4
a1	0.0430	0.0249	0.0890	0.0601	0.4111	0.4289	0.6376	0.5503	0.5237
a2	0.1152	0.0582	0.1651	0.1133	0.3777	0.5380	0.7248	0.6083	0.5814
a3	0.1287	0.0675	0.1678	0.1091	0.2742	0.5031	0.5633	0.5460	0.5271
avg	0.1103	0.0580	0.1613	0.1077	0.4009	0.5523	0.7276	0.6481	0.6209

	W2V-p1	W2V-p2	W2V-p3	W2V-p4	W2V-p5	D2V-p1	D2V-p2	D2V-p3	D2V-p4	D2V-p5
a1	0.3823	0.4042	0.3875	0.3912	0.3925	0.5491	0.5760	0.5330	0.5027	0.4759
a2	0.4623	0.4821	0.4707	0.4736	0.4785	0.6285	0.6585	0.6122	0.5724	0.5474
a3	0.3276	0.3430	0.3357	0.3361	0.3523	0.4368	0.5155	0.5195	0.4943	0.4859
avg	0.4410	0.4626	0.4493	0.4519	0.4420	0.6071	0.6614	0.6311	0.5954	0.5735

5 Conclusion

We explored performances of five language models (plus their variants) for computing document similarity. We aimed to find a model, which would best imitate the human estimates and, if successful, we could use this model for evaluation of automatic summarization along with the ROUGE measures. The best model proved to be the Doc2Vec (specifically D2V-p2) with score 0.6614. However, the best overall score was provided by the ROUGE-1 metric (0.7276), showing us, that there is still more research needed for semantic models to be able to outperform today's standard measures. Nevertheless, the Doc2Vec model shows promising results and we intend to conduct more experiments in this area.

Acknowledgments. This project was supported by grant SGS-2013-029 Advanced computing and information systems.

References

1. Steinberger, J., Ježek, K.: Evaluation measures for text summarization. Computing and Informatics **25**, 1001–1025 (2012)
2. Lin, C.: Rouge: A package for automatic evaluation of summaries. In: Text Summarization Branches Out: Proceedings of the ACL 2004 Workshop, vol. (1), pp. 74–81 (2004)
3. Nenkova, A., Passonneau, R.: Evaluating content selection in summarization: The pyramid method. In: HLT-NAACL, pp. 145–152 (2004)

4. Eneko, A., Mona, D., Daniel, C., Gonzalez-Agirre, A.: Semeval-2012 task 6: A pilot on semantic textual similarity. In: Proceedings of the First Joint Conference on Lexical and Computational Semantics - Volume 1: Proceedings of the Main Conference and the Shared Task, and Volume 2: Proceedings of the Sixth International Workshop on Semantic Evaluation, Number 3, pp. 385–393 (2012)
5. Pala, K., Čapek, T., Zajíčková, B., Bartůšková, D., Kulková, K., Hoffmannová, P., Bejček, E., Straňák, P., Hajič, J.: Czech WordNet 1.9 PDT (2011)
6. Deerwester, S., Dumais, S., Landauer, T.: Indexing by latent semantic analysis. Journal of the American Society for Information Science **41**(6), 391–407 (1990)
7. Blei, D., Ng, A., Jordan, M.: Latent dirichlet allocation. The Journal of Machine Learning Research **3**, 993–1022 (2003)
8. Mikolov, T., Chen, K., Corrado, G., Dean, J.: Efficient estimation of word representations in vector space. In: Proceedings of Workshop at ICLR, January 2013
9. Mikolov, T., View, M., Sutskever, I., Chen, K., Corrado, G., Dean, J.: Distributed representations of words and phrases and their compositionality. In: Proceedings of NIPS (2013)
10. Mikolov, T., Yih, W.T., Zweig, G.: Linguistic regularities in continuous space word representations. In: Proceedings of NAACL HLT (2013)
11. Řehůřek, R., Sojka, P.: Software framework for topic modelling with large corpora. In: Proceedings of the LREC 2010 Workshop on New Challenges for NLP Frameworks, Valletta, Malta, pp. 45–50. ELRA, May 2010. http://is.muni.cz/publication/884893/en
12. Le, Q., Mikolov, T.: Distributed representations of sentences and documents. In: Proceedings of The 31st International Conference on Machine Learning, vol. 32, pp. 1188–1196 (2014)

Automatic Labeling of Semantic Roles with a Dependency Parser in Hungarian Economic Texts

Zoltán Subecz[✉]

College of Szolnok, Technical, Agricultural and Economic Analysis Department,
Szolnok, Hungary
subecz@szolf.hu

Abstract. The events in natural language texts and the detection and analysis of the semantic relationships or semantic roles of these events play an important role in several natural language processing (NLP) applications such as summarization and question answering. In this study we introduce a machine learning-based approach that can automatically label semantic roles in Hungarian texts by applying a dependency parser. In our study we dealt with the areas of purchases of companies and news from stock markets. For the tasks we applied binary classifiers based on rich feature sets. In this study we introduce new methods for this application area. According to our best knowledge, this is the first result for automatic labeling of semantic roles with a dependency parser in Hungarian texts, for domain specific roles. Having evaluated them on test databases, our algorithms achieve competitive results as compared to the current English results.

Keywords: Information extraction · Event detection · Semantic role labeling

1 Introduction

One of the main tasks of Information extraction besides named entity detection is *event detection* [8]. Finding and analyzing events in a text, the way they relate to each other in time, is crucial to extracting a more complete picture of the contents of a text. Besides event detection, the labeling of the semantic relations of these events is an important task (*Semantic Role Labeling*, SRL). The detection of the events and their semantic roles can be utilized in several fields of natural language processing, for example, in the fields of summarization, machine translation and question answering.

In this paper we deal with *semantic role labeling*. This means the identification of semantic relations within a *semantic frame*. A *frame* is a schematic representation of situations involving various participants, props, and other

© Springer International Publishing Switzerland 2015
P. Král and V. Matoušek (Eds.): TSD 2015, LNAI 9302, pp. 261–272, 2015.
DOI: 10.1007/978-3-319-24033-6_30

conceptual roles. In our study we dealt with the frames of *purchases of companies* and *news from stock markets*.

Semantic role labeling is one of the most progressive areas in natural language processing (NLP) nowadays. For English texts *constituency-based syntactic parsers* are used for preprocessing, because English language is strongly configurational, where most of the sentence level syntactical information is expressed by word order. Conversely, Hungarian is a morphologically rich language with free word order. *Dependency parsers* are very useful for morphologically rich languages with free word order, like Hungarian, because they facilitate the connection of non-neighboring but coherent words. Therefore, we used a *dependency parser* for our Hungarian texts.

In the simplest cases but not always, the *roles* were the *syntactic relationships* of the target word. Often, the sought-after role was located far from the target word in the dependency tree, many times in the other part of the sentence. In some other cases, at the position expected on the basis of the syntactic relationship we did not find the sought-after role. This was, quite often, due to the fault of the syntactic parser, so we had to find the roles farther away from the target word in the tree and to filter the closer, false positive candidates.

2   Related Work

Initially, SRL research focused solely on verbs, the verbs were examined independently, and general roles were being looked for (e.g. Agent, Patient, and Instrument). The texts of the PropBank corpus [11] were used, in which selected verbs and associated semantic roles are annotated for English texts. CoNLL-2004 and CoNLL-2005 Shared Tasks dealt with this subject [2].

Later, the verbs were not examined independently any more but were grouped into thematic categories (frames). Besides the general roles, domain-dependent roles were investigated. For this tasks the texts of the FrameNet corpus were used, in which semantic roles are annotated for English texts. These also deal with verbs primarily but look for nonverbal target words too. An important basic study was made by D. Gildea and D. Jurafsky [7] in the SRL theme. The Senseval-3 task [9] and the ACE program [1], besides other NLP tasks, were also concerned with the SRL theme.

Some studies for semantic role labeling for Hungarian language have already been created. Farkas et al. [5] used a *rule-based* method for semantic role labeling. We applied a *machine learning* method for the same problem. In contrast with the rule based method, the machine learning method doesn't require so many resources and preprocessing, and can be applied automatically also for other domains. Ehmann et al. [4] look for only two general roles (Agent, Patient) in Hungarian psychological texts for semantic role labeling. We have labeled not only the two most frequent general roles but more domain-dependent roles too. We investigated five roles for the company purchase frame and eight roles for the news from stock markets frame. We looked for roles of only verbal and infiniti-val target words. In our system we applied the *Hungarian WordNet* [10] for the semantic characterization of the examined words. Since several meanings may belong to one word form in the WordNet, we performed *word sense disambigua-tion (WSD)* between the particular senses with the *Lesk algorithm.* [8]

Subecz et al. [13] also analyzed texts with a dependency parser for semantic role labeling for Hungarian texts but their methods and results were simpler, not very well worked-out, and were published only in Hungarian. For English texts *constituency-based syntactic parsers* are used usually. For Hungarian texts we used a *dependency parser.* According to our best knowledge, ours is the first result for *semantic role labeling with a dependency parser for domain-specific rules for Hungarian texts* which is published in English.

3 Semantic Frames and Semantic Roles

Nowadays many information extraction systems work with *domain specific frames.* It is practical to investigate the events of one domain within one *frame* since the same *roles* belong to each event that we grouped into one frame. If target words are processed independently, we can only work with fewer train-ing data. *Grouping the target words* within frames significantly decreases this problem since the training data of more target words are cumulative.

We looked for roles for *verbal and infinitival* target words. In the first part of our study we dealt with the frame of *purchases of companies.* The following target word lemmas were examined in the given frame: *ad (give), árul (sell), bekebelez (incorporate), beruház (invest), elad (sale), értékesít (market), gyarapít (thrive), kap (receive), kereskedik (trade), szerez (get), vásárol (purchase), vesz (buy).* For the target words the next roles were looked for: *Vevő (Customer), Eladó (Seller), Árucikk (Item), Ár (Price), Dátum (Date).*

Next we tested our model on the *news from the stock markets* domain. The follow-ing target word lemmas were examined: *befejez (end), csökken (slip), csúszik (slide), emelkedik (advance), esik (drop), gyengül (decline), kezd (begin), növel (increase), nő (grow), nyer (gain), nyit (open), szerez (earn), ugrik (jump), változik (change), veszít (lose), zár (close).* For the target words the following roles were looked for: *Instrumentum (Instrument), Ár (Price), Elmozdulás-irány (Shift-direction), Elmozduls-érték (Shift-value), Piac (Market), Dátum (Date), Idő (Time), Mennyiség (Volume).*

We performed the most measurements on the *purchases of companies* domain. Only the basic configurations were used to test our model on the *news from the stock markets* domain.

Examples for the roles in the two domains: In the examples *target words* are highlighted in **bold** and *roles* can be found in [square bracket]. Types of the given roles are indicated in subscripts.

1. *[A svéd Ericsson]*$_{Eladó}$ *bejelentette, hogy [a német Infineonnak]*$_{Vevő}$ *adja el [chip-gyártó részlegét]*$_{Áru}$, *[400 millió euróért]*$_{Ár}$. *(The Swedish Ericsson announced that it had sold its chip manufacturing department to the German Infineon for 400 million Euros.)*

2. *[A japán Hitachi Ltd.]*$_{Vevő}$ *[2,05 milliárd dollárért]*$_{Ár}$ *megveszi [az IBM merevlemez-meghajtókat gyártó üzletágát]*$_{Áru}$ *- jelentette a Bloomberg. (The Japanese Hitachi Ltd. purchases IBM's hard disk drive business for $2.05 billion - Bloomberg announced.)*

3. *[A Budapesti Értéktőzsde]*$_{Piac}$ *[hivatalos részvény indexe, a BUX]*$_{Instrumentum}$ *[36,54 pontos]*$_{Elmozdulás-érték}$ *[csökkenéssel]*$_{Elmozduls-irány}$, *[7.376,30 ponton]*$_{Ár}$ *nyitott*$_{Idő}$ *[csütörtökön]*$_{Dátum}$. *(The BUX index, the official index of the Budapest Stock Exchange, started 36.54 points lower, at 7.376,30 on Thursday.)*

4. *[Hétfőn]*$_{Dátum}$ *[5871,3 ponton]*$_{Ár}$ *zárt*$_{Idő}$ *[az FTSE-100]*$_{Instrumentum}$, *[8,5 ponttal, azaz 0,1 százalékkal]*$_{Elmozdulás-érték}$ *[alacsonyabban]*$_{Elmozdulás-irány}$ *a pénteki zárónál. (The FTSE 100 ended at 5871.3 on Monday, 8.5 points lower, 0.1 % below the Friday closing price.)*

According to the examples, a role often consists of several words and the sentences don't usually contain all the roles.

4   Corpus, Programs

To test our application we applied that version of the short news section of Szeged Corpus [3], where semantic roles are annotated. 1000-1000 sentences of this corpus were used for the two domains. We used 10-fold cross-validation for training and evaluation.

We applied *binary classification* for the problems. For this the C4.5 Decision Tree classifier of the Mallet machine learning package[1] was used. For linguistic preprocessing segmentation, morphological analysis, POS-tagging and dependency parsing we applied the *Magyarlanc 2.0* program package [14].

4.1   The Syntactic Parse Tree

Sentences make up a *dependency tree* on the basis of the syntactic relations between words. The tree's topmost element is the *Root*. The sentence's words are in the tree's *nodes*, the *syntactic relationships* between the words are represented by the *branches*. If the role consists of several words, then these words compose a sub-tree within the main tree. The sub-tree is attached to the main tree by its head word. The closer the headword is to the target word in the parse-tree or the sentence, the greater the possibility of identifying the role is.

The semantic role doesn't follow from the type of syntactic relationship. So, be-sides the syntactic relations several other features should be investigated in the sentence. The task is complicated by the fact that the syntactic parser also makes mistakes, so these mistakes and the false decisions made on the basis of these mistakes also appear in our results. We would have got better results if our texts had been hand-annotated.

---

[1] http://mallet.cs.umass.edu/

5   Classification

The *target word* was given for each input sentence. The task was to lookup the given role. The *candidates* in classification were the nodes of the dependency tree. One node in the sentence is the head word of the requested role. These are the *true* cases in classification; the other nodes are the *false* cases. The classifier labels the requested role in the given sentence. *We didn't give* the classifier the information whether the sentence contains the particular role or not. We applied *rigid rules* in evaluation: only that decision was accepted which labels exactly the annotated node. Neither the trees that contain this node, nor the sub-trees of this node were accepted as positive decision. We would get better results if we applied more flexible rules.

5.1   Feature Set

We assigned features for the candidates on the training and evaluation set. We used the common features of the SRL tasks [7] too. Besides them our feature set was extended with new ones. The *new features* were selected according to the characteristics of the *Hungarian language*. To do this, we also applied the dependency tree, the relationship of the candidate and the target word in the dependency tree because it's often an important attribute of the roles.

The *following features* were defined for each event candidate:

Surface features: *Bigrams and Trigrams*: The character bigrams and tri-grams of the end of the examined words. *Position*: simply indicates whether the constituent to be labeled occurs before or after the predicate. *Distance-in-sentence*: The distance between the candidate and target in the sentence.

Morphological features: Since the Hungarian language has rich morphology, several morphology-based features were defined. We defined the MSD codes (morpho-logical coding system) of the event candidates using the following morphological features: *type, mood, case, tense, person of possessor, number, definiteness*. The following features were also defined: *POS-code, Lemma*: the POS code and lemma of the candidate and the target.

Dependency tree features-1: These features are usually applied for SRL tasks [7]. The relationship of the candidate and the target in the dependency tree was investigated. *POS-code-path*: the POS codes of the nodes between the candidate and the target were recorded sequentially. We also indicated the up or down direction of the particular connection in the dependency tree. Example: C↑S↑V↑C↑V↑V↓V↓N↓N↓A. *The-POS-code-of-the governing-category*: The POS code of the topmost node in the path between the candidate and the target. *Direct-syntactic-relationship*: The type of the syntactic relationship between the candidate and target, if exist.

Dependency tree features-2: These *new features* were designed on the basis of the dependency tree and were fine-tuned in accordance with the characteristics of Hungarian language. *Distance-in-dependency-tree*: The number of nodes between the candidate and target. *Lemma-path*: Just as in the case of the POS-code-path feature but the lemmas of the node between the candidate and the target were recorded sequentially. Example: Budapesti↑Értéktőzsde↑ honlap↑közöl↓megvásárol   (Budapest↑Exchange↑website↑announce↓purchase). *Syntactic-relationship-path*: similarly to the previous one, the syntactic relationships between the nodes were recorded on the path from the candidate to the target. Example: ↑COORD↑SUBJ↓ATT↓INF↓OBJ↓ATT. *Named-entity-under-candidate*: indicates whether the candidate or the candidate's sub-tree contains a named entity. Some roles usually contain named entities. *Named-entity-distance*: the distance between the named entity and the candidate in the candidate's sub-tree, if exist. *Lemma-above-candidate*: the lemma of the node above the candidate.

Bag of words features: We applied the following methods, *which are new com-pared to the previous studies*. These features were also designed on the basis of the dependency tree and were fine-tuned in accordance with the characteristics of Hungarian language. *Candidate-sub-tree-lemmas-average*: Each candidate designates a sub-tree. This feature characterizes not only the headword but also the other words of the sub-tree. The lemmas of the sub-tree were investigated with the *bag of words* model. First, the probability of the particular lemma that belongs to the sub-tree of a positive candidate was calculated for each lemma in the training set. Second, the average of the previous probabilities that belong to the lemmas of candidate's sub-tree was calculated for the particular candidate. With this average, the lemmas of the candidate's sub-tree are well characterized as bag of words. *Candidate-sub-tree-lemmas-max*: Similar to the previous feature but here, instead of the average, the largest probability was selected from the probabilities that belong to the lemmas of candidate's sub-tree. This feature helps to recognize the most significant word of the candidate's sub-tree since the sub-trees of the positive candidates often have a significant word that usually belongs to the sub-trees of other positive candidates. *Lemma-path-BagOfWords-average* and *Lemma-path-BagOfWords-max*: Similar to the *Lemma-path* feature, but the Lemma-path feature characterizes the lemmas between the candidate and the target with a fixed lemma sequence, so it strongly depends on the sequence of the lemmas. It treats two sequences of lemmas as different even if they differ only in one element. It's only suitable for recognizing identical Lemma-paths. In the case of the feature defined actually the lemmas were characterized with bag of words, which is independent of the lemma sequence so it also indicates the similarities of the paths. Similarly to the Candidate-sub-tree-lemmas-average and the Candidate-sub-tree-lemmas-max features we created the bag-of-words-average and the bag-of-words-max features for the lemmas of the Lemma-path. *InSentence-environment-N-lemmas-BagOfWords-average* and *InSentence-environment-N-lemmas-BagOfWords-max*: In the sentence, the N-distance lemma environment of the candidate was characterized with bag of words in the cases of N=3 and N=5. Similarly to the previous features, we created the bag-of-words-average and the bag-of-words-max features for this N-distance lemma environment.

WordNet features: These new features were also designed on the basis of the dependency tree and the *Hungarian WordNet* [10], and were fine-tuned in accordance with the characteristics of Hungarian language. The semantic relations of the Word-Net hypernym hierarchy were used. We applied the following method, *which is new compared to the previous studies. Candidate-WordNet-BagOfWords-average* and *Candidate-WordNet-BagOfWords-max*: With bag of words model we collected the synsets above the lemma in the WordNet hierarchy for all lemmas of the candidate's sub-tree in the training set. Since several meanings may belong to a word form in the WordNet, we performed *word sense disambiguation (WSD)* between the particular senses with the *Lesk algorithm.* [8]: Synsets in the WordNet contain definition and illustrative sentences. In the case of polysemic event candidates, we counted how many words from the syntactic environment of the event candidate can be found in the definition and illustrative sentences of the particular WordNet synset (neglecting stop words). That sense was chosen which contained the highest number of common words. After disambiguation we counted the probability of the particular synset to belong to a positive candidate for all the synsets collected from the training set. Based on this, all these synsets were characterized with a probability. Similarly to the Bag of words features: all candidates were characterized with the BagOfWords-average and the BagOfWords-max features.

5.2   Using the Probabilities of the Base Features

The features for the candidates were selected using *two main methods. In the first method* we used the base features introduced in the previous section. *In the second method*, instead of the base features, we applied probabilities calculated from the base features. In the case of each feature instance, we calculated on the basis of the training set how many times it occurred and how many times the candidate was positive. For example, if the *Candidate-lemma = Corp.* instance occurs in 11 cases, and in 7 cases it belongs to a positive candidate, then the probability is 0.64. In this case, we didn't give the base feature to the classifier, but instead we used its probability. In the previous example: *Candidate-lemma-probability* = 0.64. With this method the *size of the classifier's vector space* and thus the *running time* were *significantly reduced.* It was useful in the development phase. *In the third method*, the features of the previous two methods were applied together.

Having compared the three methods we can see that, in most cases, the *second (probability) method* used independently produces the best results.

5.3   Reducing the Number of Feature-occurrences

The *size of the vector space was reduced* with the following technique: only those *feature-occurrences* were treated that appeared at least three times in the training set. With this method the *running time was significantly reduced,* and only the unimportant feature-occurrences from the point of view of the classification were left out.

5.4 Grouping Target Words

In this section we wanted to find the answer for the following question: can we get better results by grouping target words within a domain or by treating them together? In the *purchases of companies* domain, the selection of the *customer* and the *seller* roles is facilitated by the fact that the subject of the particular target word is usually customer or seller. Therefore, we sorted out the targets into two groups with the following simple technique: Those targets were put in the *customer-centric* group, where the subject is usually *customer*. Example: vesz (buy), vásárol (purchase), szerez (get). Those targets were put in the *seller-centric* group, where the subject is usually *seller*. Example: elad (sell), értékesít (market). In a third case the targets were not put in categories. We have found that grouping only helps in detecting the *Customer* role but produces worse results in all other cases. This is illustrated in Table 2-4.

5.5 Baseline Methods

Baseline methods were investigated on the *purchases of companies* domain. The following candidates were judged positive:

In the *Item role*: the objects of the target.

In the *Price role*: which were included in a previously made currency list.

In the *Date role*: which were included in the following list: years 1990-2014, month names, days of the week, numbers from 1 to 31.

In the *customer-centric targets at the Customer role* and in the *seller-centric targets at the Seller role*: the subjects of the target.

In the *customer-centric targets at the Seller role*: which ends with the following character trigrams:-tól, -től, -ból, -ből (Hungarian suffixes).

In the *seller-centric targets at the Customer role*: dative case relation to the target verb.

5.6 Statistical Data

On the *purchases of companies* domain:
Number of sentences: 1000
Sentences containing the given role:
Customer: 783, Seller: 579, Item: 1025, Price: 299, Date: 312
(The number of Item role is greater than the number of sentences because there are sentences, which contain more than one item.)
On the *news from stock markets* domain:
Number of sentences: 1000
Sentences containing the given role: Instrument: 787, Price: 530, Shift-direction: 431, Shift-value: 683, Market: 485: Date: 436, Time: 109, Volume: 302

Table 1. Results for baseline methods

	Role	Precision	Recall	F-measure
Customer-centric targets	Customer	48.24	59.73	**53.37**
	Seller	54.77	72.13	**62.26**
	Item	73.25	73.25	**73.25**
	Price	67.33	96.02	**79.16**
	Date	34.74	57.89	**43.42**
Seller-centric targets	Customer	78.18	44.10	**56.39**
	Seller	42.63	47.50	**44.93**
	Item	77.47	72.97	**75.15**
	Price	62.64	93.44	**75.00**
	Date	23.95	46.51	**31.62**

6   Results

6.1   Results for Baseline Methods

In the course of evaluation the precision (P), recall (R) and F-measure (F) metrics were used. According to the following results, *our machine learning model by far outperformed the Baseline methods.*

6.2   Results for Grouping the Target Words

The model in the case of Customer-centric targets achieved the best results in the Price and Item roles and the worst results in the Customer role. It in the case of Seller-centric targets achieved the best results in the Item and Price roles and the worst results in the Seller role. The model without grouping the targets achieved the best results in the Price and Item roles, and the worst results in the Seller role.

According to the results in Tables 2, grouping of the targets helped only in labeling the Customer role and produced worse results in the other roles.

Table 2. Results for Grouping the Target Words (F-measure)

Role	Customer-centric targets	Seller-centric targets	Without grouping the targets
Customer	65.55	67.89	63.73
Seller	68.65	56.36	59.10
Item	77.59	79.69	81.77
Price	83.94	76.69	84.83
Date	70.26	60.20	71.72
Average	**73.20**	**68.17**	**72.23**

6.3   Results for Ablation Analysis

We examined the efficiency of the particular *feature groups* with ablation analysis for all of the five roles when the target words were not grouped. In this case the particular feature groups were left out from the whole feature set, and we trained on the basis of the features that had remained. The results can be found in Table 3. The figures of the table show how the results changed after leaving out the particular feature group. The negative figure indicates that the investigated feature group has positive influence on the labeling of the particular role.

Table 3. Results for ablation analysis

		Role				
		Customer	Seller	Item	Price	Date
Ablated features	Surface	-1.99	-4.31	-2.3	0,64	-0.13
	Morpho-logical	0.16	-4.54	-0.66	-2.12	-0.51
	Dependency tree-1	-0.83	-1.25	-2.71	-0.17	-0.34
	Dependency tree-2	-2.15	-2.72	-0.38	-1.25	-1.12
	Bag of words	0.39	-1.06	-0.45	1.07	-3.19
	WordNet	0.17	-0.53	-0.03	-1.63	-0.58

According to the results, the best performers were the *Dependency tree-1* and the *Dependency tree-2* feature groups. Their impact was positive in all role cases. The following feature groups have positive impact on four roles: Surface, Morphological and WordNet. The Bag of words feature group had positive impact on three roles only.

6.4   Results for the News from Stock Markets Domain

The model, introduced in the previous chapters, was tested on labeling the eight roles of the *news from stock markets* domain. The results can be found in Table 4. According to the results, *the model performed well on this domain too*.

Table 4. Results for the news from stock markets domain

Role	Precision	Recall	F-measure
Instrument	74.28	68.54	71.03
Price	88.38	84.82	86.47
Shift-direction	82.04	81.9	81.7
Shift-value	72.71	67.28	69.56
Market	77.92	69.25	71.99
Date	80.46	69.25	72.39
Time	82.78	75.09	78.06
Volume	82.08	72.24	76.07
Average	**80.08**	**73.55**	**75.91**

6.5   The Comparison of the Results with the Related Works

For English texts D. Gildea and D. Jurafsky [7] performed the task for many frames and many roles within the frames. They primarily dealt with verbal targets but also labeled roles for nonverbal targets. As an average they achieved an F-measure of 63. Our algorithms achieved competitive results (F-measure of 72.23 and 75.91) even though we investigated only two frames and five and eight roles for each, and dealt with only verbal and infinitival target words.

7   Discussion, Conclusions

In this paper, we introduced our machine learning approach based upon a rich feature set, which can label semantic roles in Hungarian texts automatically with a dependency parser. We dealt with the *purchases of companies* and *news from stock markets* frames. Within these frames 1000-1000 annotated sentences were processed. In the *purchases of companies* domain five, in the *news from stock markets* eight domain specific roles were labeled. In our rich feature set the Surface, Morphological, Dependency tree, Bag of words and WordNet features were applied. The efficiency of the feature groups was tested with an ablation analysis. The base features were expanded with *probabilities* calculated from the base features. Although the texts we examined cover fewer topics than the works presented for English texts, our algorithms achieved competitive results as compared to the current results for English.

References

1. Ahn, D.: The stages of event extraction. In: Proceedings of the ARTE 2006 Proceedings of the Workshop on Annotating and Reasoning about Time and Events, pp. 1–8 (2006)
2. Carreras, X., Màrquez, L.: Introduction to the CoNLL-2005 shared task: semantic role labeling. In: CONLL 2005 Proceedings of the Ninth Conference on Computational Natural Language Learning, p. 8997 (2005)
3. Csendes, D., Csirik, J.A., Gyimóthy, T.: The szeged corpus: A POS tagged and syntactically annotated hungarian natural language corpus. In: Sojka, P., Kopeček, I., Pala, K. (eds.) TSD 2004. LNCS (LNAI), vol. 3206, pp. 41–47. Springer, Heidelberg (2004)
4. Ehmann, B., Lendvai, P., Miháltz, M., Vincze, O., László, J.: Szemantikus szerepek a narrative kategoriális elemzés (NARRCAT) rendszerében. IX. Magyar Számítógépes Nyelvészeti Konferencia, Szeged, pp. 121–123 (2013)
5. Farkas, R., Konczer, K., Szarvas, G.Y.: Szemantikus keretillesztés és az IE-rendszer automatikus kiértékelése. II. Magyar Számítógépes Nyelvészeti Konferencia, pp. 49–53. Szegedi Tudományegyetem, Szeged (2004)
6. Fillmore, C.L., Ruppenhofer, J., Baker, C.F.: Framenet and representing the link between semantic and syntactic relations. In: Churen, H., Winfried, L. (eds.) Frontiers in Linguistics. Language and Linguistics Monograph Series B, vol. I, p. 1959. Institute of Linguistics, Academia Sinica, Taipei (2004)

7. Gildea, D., Jurafsky, D.: Automatic labeling of semantic roles. Computational Linguistics Journal **28**(3), 245–288 (2002)
8. Jurafsky, D., Martin, J.H.: Speech and Language Processing: An Introduction to Natural Language Processing, Computational Linguistics, and Speech Recognition. Prentice-Hall, Upper Saddle River (2000)
9. Litkowski, K.C.: SENSEVAL-3 TASK: Automatic labeling of semantic roles. In: Proceedings of SENSEVAL-3, the Third International Workshop on the Evaluation of Systems for the Semantic Analysis of Text (2004)
10. Miháltz, M., Hatvani, C.S., Kuti, J., Szarvas, G.Y., Csirik, J., Prószéky, G., Váradi, T.: Methods and results of the hungarian wordnet project. In: Tanács, A., Csendes, D., Vincze, V., Fellbaum, C., Vossen, P. (eds.) Proceedings of the Fourth Global WordNet Conference (GWC 2008), pp. 311–320. University of Szeged, Szeged (2008)
11. Palmer, M., Gildea, D., Kingsbury, P.: The Proposition Bank: An annotated corpus of semantic roles. Computational Linguistics **31**(1), 71105 (2005)
12. Subecz, Z.: Detection and classification of events in hungarian natural language texts. In: Sojka, P., Horák, A., Kopeček, I., Pala, K. (eds.) TSD 2014. LNCS (LNAI), vol. 8655, pp. 68–75. Springer, Heidelberg (2014)
13. Subecz, Z.: Szemantikus szerepek automatikus cmkézése függőségi elemző alkalmazásával Magyar nyelvű gazdasági szövegeken. XI. Magyar Számítógépes Nyelvészeti Konferencia, Szeged, pp. 95–108 (2014)
14. Zsibrita, J., Vincze, V., Farkas, R.: magyarlanc: A toolkit for morphological and dependency parsing of hungarian. In: Proceedings of RANLP 2013, pp. 763–771 (2013)

Significance of Unvoiced Segments and Fundamental Frequency in Infant Cry Analysis

Anshu Chittora[✉] and Hemant A. Patil

Dhirubhai Ambani Institute of Information and Communication Technology,
Gandhinagar, Gujarat, India
{anshu_chittora,hemant_patil}@daiict.ac.in
http://www.daiict.ac.in

Abstract. In this paper, significance of unvoiced segments and fundamental frequency (F_0) in infant cry analysis is investigated. To find out the unvoiced segments from the infant cry F_0 contour is used. For extraction of F_0 contour, Teager Energy operator (TEO) based pitch extraction algorithm is used. TEO gives the running estimate of the signal energy in terms of its amplitude and instantaneous frequency. To quantify the importance of proposed features in infant cry analysis of variance (ANOVA) method is applied. It has been found that quantification of unvoiced segments and fundamental frequency in the cry, deliver information about the maturation of cry production system. In infant cry analysis, presence of high unvoicing ratio in a cry cannot be attributed to presence of pathology, like adult vocal fold pathological sounds.

Keywords: Fundamental frequency · Phonation · Dysphonation · Unvoicing · ANOVA

1 Introduction

Research in infant cry analysis was started with spectrographic analysis. Spectrographic analysis is a time- frequency analysis method [1,2]. In spectrogram, various cry modes are defined, viz., falling, rising, flat melody, phonation, hyperphonation, etc. These modes have been used to identify the pathologies [2]. Infant cry is also studied from infant developmental perspective and for understanding the cry production mechanism and related neurological developments. In last two decades, this area of research has been explored and a little work is done towards cry classification, pathology identification and analysis of infant cry.

Cries that are rhythmic are easier for parents to understand and make infant more predictable. Shorter periodicities are evident within the cry itself, as in the regularity of phonation. More accurately, cry phonation occurs during the expiratory phase of the respiratory cycle. We (especially parents) detect the pain cry because of the unusually long respiratory phase that upsets the normal cycle. Even shorter periodicities are found in the pitch of the cry. The vocal

© Springer International Publishing Switzerland 2015
P. Král and V. Matoušek (Eds.): TSD 2015, LNAI 9302, pp. 273–281, 2015.
DOI: 10.1007/978-3-319-24033-6_31

folds oscillate to produce the cry, and the frequency of vibration of vocal folds is called the fundamental frequency (F_0) as the neural control of vocal folds changes. Periodicities in F_0 over time can be expected under normal conditions of vagal input. A constantly varying tension on the laryngeal musculature should produce inherent cycles in the pitch of an infants cry (as vocal folds oscillate around the mean position), again describing relaxation or oscillation system [3].

Recent work on infant cry analysis includes classification of normal $vs.$ hearing impaired infants cry classification using Mel Frequency Cepstral Coefficients (MFCC), classification of pain $vs.$ hunger cries using MFCC as feature with neural networks, classification of normal vs. pathological cries using MFCC and extraction of fundamental frequency of infant cry [4–7].

This paper serves two purposes: 1.Algorithm for fundamental frequency (F_0) estimation of infant cry is proposed. Because of large range of F_0 values (250 Hz-1 kHz), conventional F_0 extraction algorithms do not perform well for infant cry signals. The estimated fundamental frequency contour and its harmonic structure can be used to extract prosodic features for cry analysis. 2. From this F_0 contour, unvoiced regions are extracted. The ratio of unvoicing frames to total number of frames in a cry is verified for its importance using ANOVA (ANalysis Of VAriance) method. Results show that unvoicing percentage in a cry is an important parameter for analysis of infant cry.

2   Teager Energy Operator (TEO) Based F_0 Estimation

In this section, F_0 estimation from speech signals [8] is explained. In the pre-processing stage, signal is passed through a 4th order lowpass Butterworth filter with cutoff frequency of $1\ kHz$ (because highest F_0 in infant is $1\ kHz$). From the filtered signal, TEO profile of the signal is calculated. Computation of TEO requires three consecutive samples of speech. TEO of a signal is defined as:

$$x_{teo}(n) = x^2(n) - x(n-1) * x(n+1) \tag{1}$$

The TEO profile of a signal gives the running estimate of its energy. From the TEO profile of the signal, the peaks of TEO signal are identified. The difference in peak location gives a rough estimate of pitch values. The pitch for a speech frame from peak differences is calculated as follows:

- Calculate the number of peaks in a frame. If the number of peaks is more than three, then take median of the value as pitch of the frame, else take the frame as unvoiced segment. Median of the pitch values is considered to ignore effect of outlier pitch values (due to unvoicing effect) instead of mean as proposed in [8].
- After getting the pitch values for each of the frame, calculate fundamental frequency (F_0) by dividing the pitch value from sampling frequency.

From Fig. 1, it can be seen that TEO signal is able to pick the peaks in the voiced signals. The unvoiced segments are considered as the segments for which the F_0

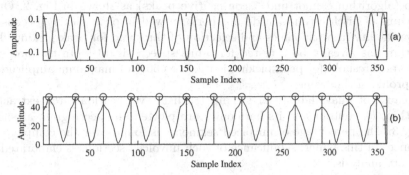

Fig. 1. Extraction of pitch values from TEO profile, (a) cry signal and (b) TEO profile of (a).

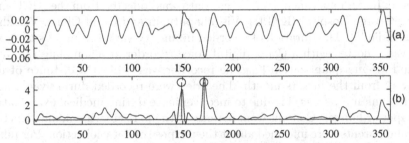

Fig. 2. Extraction of pitch values from TEO profile (unvoiced signal), (a) cry signal and (b)

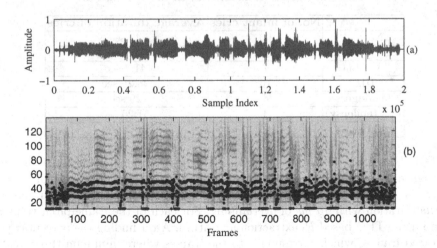

Fig. 3. Spectrogram of infant cry. (a) infant cry signal and (b) spectrogram and F_0 contour extracted from (a), Y-axis represents NFFT bins which corresponds to frequency in Hz).

is zero (algorithm cannot find *3* consecutive peaks) as shown in Fig. 2. Other algorithms of pitch estimation do not perform well in peak picking because of high variability in pitch values and selection of threshold for peak picking. This algorithm has already been tested for adult speech signal [8].In the proposed algorithm threshold for peak picking is taken as 50% of maximum amplitude of TEO profile of the frame.

Because of unavailability of ground truth for verification pitch extraction algorithm, F_0 contour and its harmonics are plotted on spectrogram as shown in Fig. 3. The percentage of unvoiced segments is found over the complete cry utterance. In this paper, significance of such unvoiced sections is elaborated for infant cry analysis.

3 Experimental Results

Database: In our experiments, infant cry data was collected from the NICU units of King George Hospital (K.G.H), Visakhapatnam and Child Clinic, Visakhapatnam, India [9]. The duration of recorded infant cry varied from *30* s to *2* min. Data was collected with a Cenix digital voice recorder at a sampling rate of *12 kHz* and *12-bits* quantization [9]. The recorder was kept at a distance of *6-10 cms* away from the infants mouth. The cries were recorded during vaccination and while infants were crying due to inconvenience during medical examination, wet diaper or hunger. Consent from parents of the participating infants has been taken and parents were informed about the purpose of data collection. Per infant, one cry was collected and the details of number of cries for normal, newborns and pathological cries are given in Table 1.

Table 1. Number of infant cries for different classes

	No. of infant cries	Average duration of cries
Normal	90	43 sec
New born	40	42.6 sec
Asthma	7	41.7 sec
HIE	16	28.25 sec
Meningitis	4	29 sec
Fits (epilepsy)	4	51.25 sec
Misc.	10	43.8 sec

Data Analysis: In the pre-processing stage, the infant cry signal is passed through a 4th order Butterworth lowpass filter with cutoff frequency of *1 kHz*. After the signal is being filtered, the filtered signal is divided in the frames of *30 ms* duration with *15 ms* of overlapping. For each of the frame pitch value is found using TEO based F_0 extraction algorithm. After finding the percentage of unvoiced frames, which corresponds to the frames where minimum three peaks are not available.

The unvoiced segments are tested for their significance in infant cry analysis using ANOVA (Analysis of Variance) analysis [10]. Results are shown below in

Table 2. ANOVA analysis of infant cry signal for significance of unvoiced frames

S. NO.	Case	F-Ratio	Probability
1	Normal vs. HIE	5.64	0.0194
2	Normal vs. fits	11.67	0.001
3.	Normal vs. pathology HIE/meningitis/fits	7.39	0.0076
4	Normal vs. Newborn	84.04	1.22E-15
5	Normal vs. Asthma	4.78	0.0313
6	All	22.81	4.89E-17

Table 3. Mean values of unvoiced frames in a cry episode

S. NO.	Infant class	Mean UV	Standard Deviation
1	HIE	0.3467	0.0848
2	meningitis	0.2518	0.0443
3	Fits	0.4396	0.081
4	Normal	0.292	0.0846
5	New born	0.4224	0.0444
6	Asthma	0.2212	0.039

Table 2. It can be seen that the value of *F-ratio* is much higher than *1* hence it suggests the rejection of null hypothesis that all infants came from the same group. This implies that relative amount of unvoiced regions is a significant feature in analysis of infant cry. From Table 2, we can see that the F ratio of unvoiced frames in a cry is higher in pathological infants cries compared to normal infant cries. It shows the change in the unvoicing-to-voicing ratio in infant due to presence of pathology. Same results are obtained for normal infants and newborn (birth cry) analysis. In neonates as well, percentage of the unvoiced segments are different than the normal infants. Further details of unvoiced frames to total cry frames ratio is given in Table 3.

The mean values and standard deviation of the unvoiced ratio given in Table 3, indicates that fits and HIE cries have higher number of unvoiced frames. Asthma and meningitis have comparatively higher voiced frames. This observation suggests that asthma cry is similar to a normal cry. Normal infant cries

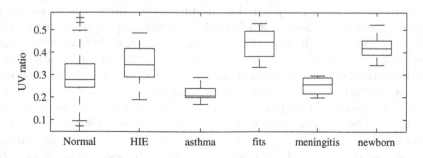

Fig. 4. Boxplot of unvoicing ratio in infant cries

have a wide range of unvoicing ratio because it consists of all cries of infants of all age groups from neonatal to *1* year old. Moreover, it has many pain cries also. In pain cries researchers have reported higher values of unvoiced regions in spectrograms of the cry. In neonates, the percentage of unvoiced frames is much higher than normal ones and it is even higher than pathological infant cries.

It has already been observed that in adult voice pathological speech, the unvoiced segments are higher compared to healthy adult speech. A feature which can capture this unvoicing information may perform better in pathological and healthy speech classification.

Table 4. ANOVA analysis of infant cry signal for significance of F_0

S. NO.	Case	F-Ratio	Probability
1	Normal vs. asthma	15.82	0.0001
2	Normal vs. fits	2.93	0.09
3	Normal vs. pathology HIE/meningitis/fits	0.08	0.782
4	Normal vs. Newborn	21.96	7.22E-06

Table 4 shows the F-ratio values for ANOVA analysis carried out for finding significance of F_0 in infant cry analysis. It can be observed that for *95%* confidence interval the mean fundamental frequency feature does not work well for normal and pathological infant cry analysis. The F-ratio values are higher for normal and newborn birth cries and normal and asthma infant cries, *i.e.*, asthma infants and newborns have different values of F_0 or different functioning/ anatomy of vocal folds.

4   Discussion

For an infant, crying is a way to express his emotions and needs. Production of cry requires coordinated functioning of respiratory, laryngeal and supralaryngeal network and neurophysiological coordination. Early research work on infant cry analysis is based on spectrographic analysis of infant cry. Researchers have reported many cry modes from these spectrographic modes. The primarily used modes are phonation, hyperphonation, dysphonation, double harmonic breaks and glide. Identifying these modes from the spectrogram requires expertise and experience. To identify the phonation and dysphonation in the spectrogram, proposed work can be used. It can be seen from Fig. 3 that the proposed method coincides with the harmonic present in the spectrogram and also able to identify dysphonation regions. These regions are addressed in this paper as unvoiced frames. The presence of dysphonation or unvoiced frames indicates presence of noise and non-linearity in the cry production system. In adult voice pathology, the presence of this noise represents dysfunction of vocal folds. Similarly, in infants higher percentage of unvoicing in cry represents less coordination among cry production organs or neural integration.

Table 5. Mean values of F_0 in a cry episode

S. NO.	Infant class	Mean F_0	Std
1	HIE	331.04	53.15
2	Meningitis	397.42	19.54
3	Fits	333.33	0
4	Normal	354.07	65.58
5	New born	280.8	24.8
6	Asthma	443.75	46.1

On the other side, higher unvoicing in newborn birth cry indicates less integration of cry production mechanism. In newborns, vocal folds are not fully developed hence the newborn birth cry lack phonation in its spectrogram as shown in Fig. 5. Integration of vocal system and development of vocal folds in the first three months of infant life has been reported in [11]. The paper, quantifies the noise reduction in cry production with infant age in initial *3* months of age.

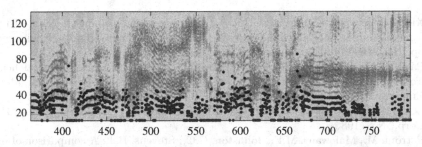

Fig. 5. Spectrogram of newborn infant cry

Apart from percentage of unvoicing in the cry, analysis of F_0 using ANOVA is shown in Table 4. It can be seen that mean F_0 does not carry significant role in normal and pathological cry classification except in classifying normal and fits (epilepsy) infant cries. The mean values and standard deviation of the F_0 vales are shown in Table 5. From Table 4 and Table 5, we can observe that the mean F_0 of normal and newborn cries are significantly different. This observation also suggest the same result mentioned earlier, that with growing age infant vocal folds are developed and cry production mechanism becomes comparatively mature. In fits which is a neurological disorder, can be a cause of change in F_0 compared to normal infants and presence of high unvoiced regions in entire cry episode.

5 Summary and Conclusions

In this paper, F_0 extraction algorithm using TEO is proposed for infant cry pitch extraction. This algorithm works well for infant cries. This can also be used for defining spectrographic modes from pitch contours, automatically. It was shown that higher unvoicing percentage (dysphonation) cannot guarantee pathological state of an infant, as it is also visible in newborn birth cries and neonatal cries. The F_0 values also changes with the maturation of cry production system as evidenced from normal and newborn cry analysis. Significant changes in pitch values from normal infant cries are indicators of some pathological state of an infant. It is observed in some genetic diseases, e.g., cri-du-chat, where F_0 is different from normal infant cries.

In future, we would like to define various spectrographic modes from F_0 contours. We would like to analyze infant cry from developmental perspective of infant with growing age.

Acknowledgments. Authors would like to thank Department of Electronics and Information Technology (DeitY), Government of India, New Delhi,India and authorities of DA-IICT for carrying out this research work.

References

1. Xie, Q., Ward, R.K., Laszlo, C.A.: Determining normal infants' level-of-distress from cry sounds. In: Canadian Conference on Electrical and Computer Engineering, pp. 1094–1096 (1993)
2. Patil, H.A.: Cry Baby: Using spectrographic analysis to assess neonatal health from an infants cry. In: Neustein, A. (ed.) Advances in Speech Recognition, Mobile Environments, Call Centers and Clinics, pp. 323–348. Springer (2010)
3. Lester, B.M.: Infant Crying: Theoretical and Research Perspective. Plenum Pub. Corp., NY (1985)
4. Petroni, M., Malowany, M.E., Johnston, C.C., Stevens, B.J.: A comparison of neural network architectures for the classification of three types of infant cry vocalizations. In: IEEE 17th Annual Conference on Engineering in Medicine and Biology Society (EMBS), vol. 1, pp. 821–822 (1995)
5. Garcia, J.O., Reyes Garcia, C.A.: Mel-frequency cepstrum coefficients extraction from infant cry for classification of normal and pathological cry with feed-forward neural networks. In: Proc. of the Int. Joint Conf. on Neural Networks, vol. 4, pp. 3140–3145 (2003)
6. Lederman, D., Cohen, A., Zmora, E., Wermke, K., Hauschildt, S., Stellzig-Eisenhauer, A.: On the use of hidden Markov models in infants' cry classification. In: 22nd Convention of Electrical and Electronics Engineers, pp. 350–352 (2002) (in Israel)
7. Patil, H.A.: Infant identification from their cry. In: 7th Int. Conf. on Advances in Pattern Recognition, ICAPR 2009, vol. 4, pp. 107–110 (2009)
8. Patil, H.A., Srikant, V.: Effectiveness of Teager energy operator for epoch detection from speech signals. Inter. J. of Speech Technology **14**(4), 321–337 (2011)

9. Buddha, N., Patil, H.A.: Corpora for analysis of infant cry. In: Int. Conf. on Speech Databases and Assessments, Oriental COCOSDA, Hanoi, Vietnam (2007)

10. Details of ANOVA. http://www.psychstat.missouristate.edu/introbook/sbk27.html (last Accessed on January 18th, 2014)

11. Fuamenya, N.A., Robb, M.P., Wermke, K: Noisy but Effective: Crying Across the First 3 Months of Life. J. Voice (2014) (in Press)

Speech Corpus Preparation for Voice Banking of Laryngectomised Patients

Markéta Jůzová[1,3](✉), Jan Romportl[2,3], and Daniel Tihelka[3]

[1] Department of Cybernetics, University of West Bohemia, Pilsen, Czech Republic
juzova@kky.zcu.cz
[2] Department of Interdisciplinary Activities, New Technologies – Research Centre,
University of West Bohemia, Pilsen, Czech Republic
rompi@ntc.zcu.cz
[3] New Technologies for the Information Society,
University of West Bohemia, Pilsen, Czech Republic
dtihelka@ntis.zcu.cz

Abstract. This paper focuses on voice banking and creating person-alised speech synthesis of laryngectomised patients who lose their voice after this radical surgery. Specific aspects of voice banking are discussed in the paper, including a description of the adjustments of the generic methods. The main attention is paid to the speech corpus building since the quality of synthesised speech depends a lot on the speech units variability and the number of their occurrences. Also some statistics and characteristics of the first experimental voices are presented and the possibility of using different speech synthesis methods depending on the voice quality and speech corpus size is pointed out.

Keywords: Voice banking · Personalised speech synthesis · Speech unit · Greedy algorithm · Text corpus building · Corpus recording

1 Introduction

Total laryngectomy, i.e. surgical removal of a larynx with all its parts including vocal cords, is a radical treatment procedure which is often unavoidable to safe life of patients who were diagnosed with severe laryngeal cancer. In spite of being very effective with respect to the primary treatment, it significantly handicaps the patients due to the permanent loss of their ability to use voice and produce speech.

Methods and applications of computer text-to-speech (TTS) synthesis can be utilized to improve the patients quality of life after he/she loses the ability

The research leading to these results has received funding from the Norwegian Financial Mechanism 2009-2014 and the Ministry of Education, Youth and Sports under Project Contract no. MSMT-28477/2014, Project no. 7F14236 (HCENAT). This work was also supported by the grant of the University of West Bohemia, project No. SGS-2013-032.

© Springer International Publishing Switzerland 2015
P. Král and V. Matoušek (Eds.): TSD 2015, LNAI 9302, pp. 282–290, 2015.
DOI: 10.1007/978-3-319-24033-6_32

to speak naturally using his/her own voice. We have discussed and analysed in more detail various aspects of TTS for laryngectomised patients (including usage scenarios) in [1] and [2] where we emphasize that it is very important to endorse effective mutual interaction and cooperation between primary medical treatment processes and ongoing advances in TTS research and development, because this offers a possibility for "digital conservation" of patients original voice for his/her future speech communication – a procedure hereafter called *voice banking* (or voice conservation).

Voice banking in this sense means that the patient records his/her voice before the planned laryngectomy and a personalized TTS system modeling this voice is then created. Our feedback from several laryngectomised patients who underwent voice banking before the surgery shows that it brings them significant improvement to their quality of life. However, as with any clinical practice, formal empirical evidence is needed, and therefore we are now in the process of Quality of Life Study (QoLS) with the patients of Department of Otorhinolaryngology and Head and Neck Surgery, University Hospital Motol in Prague (within the HCENAT project).

This paper thus focuses on one particular technical aspect essential for QoLS and voice banking in general: how to design and organize the speech recording process so that the patients prior to total laryngectomy are able to record speech data suitable enough for a personalized TTS system with a reasonable level of quality.

A typical interval between diagnosis of laryngeal cancer, its staging and indication for total laryngectomy and surgery itself is approximately 2 weeks and because the surgery is live-saving, any delay might have severe consequences. Therefore, the speech recording task here is very challenging, giving us very little time and hard limits for the size of the corpus, not to mention the quality of the speaker who is most likely a very distressed person freshly diagnosed with a potentially fatal disease, facing permanent loss of voice in couple of days and being absolutely non-trained in speech recording (and quite often even in using a computer).

2   Building of the Speech Corpus

Given the aforementioned, our task is to prepare such a (smaller) text corpus for the recording which would achieve the quality and naturalness of the final synthesis as high as possible. Since we do not know in advance how many sentences each patient will be able to record, the corpus must be prepared so that it ensures maximum possible coverage of speech units no matter the number of recorded sentences.

When building a phonetically and prosodically rich corpus for this purpose, we have adapted the process of sentence selection introduced in [3,4], described further.

2.1 Greedy Algorithm for Sentences Selection

The algorithm ensures that the selected sentences contain a desired number of occurrences of every speech unit k. At the beginning of every step j, we have a set of already selected sentences $\bar{R}$ and a set R containing sentences from which to select. D, the desired number of occurrences of each unit is also set. Finally, let us denote by N_k^j a number of occurrences of the unit k in the set $\bar{R}$. All sentences s from R are evaluated by a score r_s^j according to the following formula 1:

$$r_s^j = \sum_{k=1}^{K} \min[\,\max\,[0, D - N_k^j],\ S_k^s] \tag{1}$$

where S_k^s represents the number of speech units k in the sentence s. The sentence s^* with the highest score r_s^j is chosen in this step j and transferred from R to the set of chosen sentences $\bar{R}$. The process can be stopped by a maximum number of chosen sentences or if all speech units are contained in the selected set at least D-times.

Since the list of all speech units used in the greedy algorithm is obtained from the available texts, it is not guaranteed that all theoretically possible units are included. Nevertheless, a probability of such a unit appearance in the process of speech synthesis is very small.

2.2 Optimal Sentence Length Preselection

As in a general synthesizer, a large collection of Czech newspapers texts covering various domains like news, culture, economy, sport, etc. was used. These texts contain 524,472 sentences of various word length. Both indicative and interrogative sentences were included to ensure a sufficient diphone coverage in the two fundamental terminating prosodic contexts.

First, only the sentences of the length from 5 to 8 words were taken into account for a few following steps, since longer sentences would be difficult for the patients to concentrate on (in comparison, a generic corpus recorded by a professional speaker contains sentences of up to 15 or 18 words). This should be a good compromise between the number of words patients are able to read fluently and the need to keep the usually longer natural sentence structure (an average sentence length in the source texts is 14 words). Also sentences containing some very long, usually foreign-like words difficult for reading, were excluded. The number of available sentences thus, after described preselection, decreases to $|R| = 97{,}033$, approx. one fifth of which are questions. And since the algorithm below works on phonetic level, grapheme-to-phoneme rules (like in [5]) were employed to convert textual sentences to their phonetic form.

2.3 Step-by-Step Sentence Selection Procedure

At the beginning of each step, the numbers of units' occurrences in the sentences selected so far ($|\bar{R}|$) are evaluated. After that, greedy algorithm from 2.1 is used to choose sentences to be moved from R to $\bar{R}$ until the required number of units is achieved. Once a sentence is stored to $\bar{R}$, it will remain there forever.

1. In the first step of our proposed algorithm, we consider a raw phoneme as a speech unit and we require at least 15 occurrences of every phoneme in the selected sentences. The algorithm ends with $|\bar{R}| = 96$ sentences.

2. However, a requirement on a prosodically (not only phonetically) rich speech corpus was also given, so the phonetic transcription was enriched with one of the four types of the *prosodeme*, a symbolic prosodic feature (0 - "null" prosodeme, 1 - declarative, 2 - interrogative, 3 - non-terminating, [6]).
   Therefore, a speech unit is now a phone-in-prosodeme. We demanded once again the number of occurrences equal to 15. The algorithm selected 380 additional sentences to meet the requirement, $|\bar{R}| = 476$ in total.

3. Since the unit-selection synthesizer $ARTIC$ ([7]) developed at the authors' department works with diphones, the third step is focused on diphones' coverage; the requirement of at least 2 occurrences of each diphone in $\bar{R}$ was set. The total number of selected sentences stopped at $|\bar{R}| = 1{,}040$.

4. Having selected more than 1,000 sentences so far, we have changed the paradigm of the corpus preparation. Until this point, the input texts were chosen with respect to the patients, however when a patient is able and willing to record this number of sentences, we can start to optimize the corpus with respect to the speech synthesis method we use. To the input set of sentences, additional almost 1,500 sentences, manually created to cover rare Czech diphones, were added and used together.
   Therefore, still the same algorithm but with extended R was used to select sentences with at least 5 occurrences of each diphone in $\bar{R}$, after that $|\bar{R}| = 2{,}317$.

5. The last step applied the greedy algorithm to diphones-in-prosodemes with the requirement of at least 3 occurrences. However, the meeting of the requirement will enlarge the set to almost 7,000 sentences which is impossible to record in this case. Therefore, the process was stopped when the number 2,500 was achieved.

6. Finally, the algorithm balanced the distribution of diphones-in-prosodemes ([3], formula 2) until reaching 3,500 selected sentences. When balancing, a sentence with the highest score S is moved from R to $\bar{R}$,

$$S = -\sum_{i=1}^{I} \frac{n_i + n_i'}{n} \log_2 \frac{n_i + n_i'}{n} , \qquad (2)$$

where I is the number of different diphones-in-prosodemes, n_i is the frequency of i-th diphone in $\bar{R}$, n_i' is the frequency of i-th diphone in R, n is a sum of all diphones in both sets and $0 \log_2 0$ is set to 0.

Based on our experiences, no patient have managed to record the whole prepared corpus yet, nevertheless, the number 3,500 was chosen to be sure that the text corpus is well prepared and large enough even for a very capable speaker with more time.

2.4　Text Corpus Statistics

The selected text corpus contains 3,500 sentences, both declarative and inter-rogative (approx. 6 : 1). The sequential selection process described above should guarantee the widest possible coverage of speech units (phonemes, diphones in different prosodemes) in the group of sentences selected in the particular step. Some more detailed statistics of the corpus are shown in Table 1, as well as the comparison with the original sentence set and one of the generic text corpora recorded by a professional speaker.[1]

Table 1. Text corpus statistics (P_x means a prosodeme X, #-symbol means "number of"). The second part of the table contains information about percentages of units appearing at least k-times in the text.

statistics		text collection	created corpus	generic corpus
# sentences		98,498	3,500	12,241
# phones		3,662,031	132,161	631,719
# diphones		3,563,533	128,661	605,663
units	k			
phones	15x	100.00 %	100.00 %	100.00 %
	1000x	92.86 %	80.95 %	80.95 %
diphones	2x	96.10 %	96.10 %	97.20 %
	10x	80.09 %	61.10 %	79.69 %
phones in P_0	15x	100.00 %	100.00 %	100.00 %
	100x	100.00 %	97.62 %	95.24 %
phones in P_1	15x	100.00 %	100.00 %	97.62 %
	100x	92.86 %	88.10 %	95.24 %
phones in P_2	15x	88.10 %	88.10 %	88.10 %
	100x	83.33 %	33.33 %	57.14 %
phones in P_3	15x	92.86 %	92.86 %	97.62 %
	100x	88.10 %	85.71 %	95.24 %
diphones in P_0	1x	99.01 %	98.94 %	95.88 %
	3x	89.16 %	87.10 %	86.72 %
diphones in P_1	1x	85.61 %	79.59 %	86.26 %
	3x	76.90 %	56.77 %	72.44 %
diphones in P_2	1x	58.82 %	45.36 %	61.07 %
	3x	44.86 %	24.73 %	41.98 %
diphones in P_3	1x	73.21 %	64.78 %	90.23 %
	3x	61.87 %	46.78 %	76.49 %

　　As follows from the table, not all required minimum occurrences were met in the process of sentence selection. E.g. the requirement of at least 15 occurrences of all phones-in-prosodemes was set in the step 2 but the collection of sentences used for the selection does not contain enough units in prosodemes 2 and 3

---

[1] Generic corpora were selected from a broad collection of Czech newspapers texts containing more than 500,000 sentences.

(interrogative and non-terminating prosodic features). So the greedy algorithm selected such sentences to have as many units as available. A similar situation occurred in the next steps because the texts do not contain even 2 occurrences of all diphones.

2.5 Recording

The process of corpus recording for the purposes of speech synthesis is analogous to the process of usual "big" corpus recording. The recording management and controlling policy is described in [8,9].

The best way is to record the sentences in a soundproof studio; nevertheless, due to little time available, we have often accepted an alternative way – a recording in a quiet room in a hospital which is usually more comfortable for the patients. In both cases, a special recording tool is used for displaying the sentences and checking the loudness of the audio and sufficient lengths of pauses at the beginning/end of every sentence. The recording runs in a sentence-by-sentence manner in order to prevent speakers from reading any sentence in the context of the previous. The speech corpus should be read naturally but with no emotions and expressiveness, so all the speakers are instructed to try to keep consistent news-reading-like speech rate, pitch and prosody style during the entire recording while keeping their normal, innate style of speaking. While professional speakers have no problems with that, it was sometimes difficult for the patients to meet our requirements because of their inexperience and health problems.

Although the patient can read the whole sentence before pressing the record button and the sentence length was intentionally lowered a lot, relatively significant number of mistakes have often been noticed. Also many breaths were recorded in the middle of sentences in spite of instructing the speakers to take always a breath before the recording itself. As written in Section 1, the patients are often absolutely non-trained in speech recording and speaking aloud longer time, so the speech corpus is usually inconsistent. Besides the problem with extra breaths and mistakes in reading, they sometimes stammer when reading a longer, usually foreign-like word, or even repeat a syllable. The recording process enables a rerecording when a speaker makes a mistake, but we have recommended the patients to continue with another sentence so that more sentences are recorded. Also little practice at using computers makes the recording more difficult. Despite frequent pauses for relaxing, the patients' voice becomes more and more tired and the differences between the first and the last sentences recorded in one day are evident.

Because of many non-speech events like breathing and clicking, mispronunciations and unintelligible pronunciations, the careful orthographic annotation of all recorded sentences is necessary. It is very important to know the right correspondence between the speech signals and their representations on orthographic level (which is later converted to phonetic level). Due to a high number of non-speech events and misannotations, the annotators's work was much more difficult compared to the generic corpus. However, the precise corpus annotation is, in this case, all the more important since much fewer units are available during the synthesis.

3   First Observations

The very first (pilot) voice was recorded based on the generic text corpus (a special text corpus described in the previous section had not been prepared yet at the time of the recording). That corpus contains longer sentences and it was sometimes difficult for the patient to read the whole sentence fluently. Furthermore, the recording was not done in a professional recording studio and the patient could spend only one day recording due to the imminent surgery. He managed to read almost 700 sentences but his voice was tired a lot at the end of the day which resulted in an audible difference between the first and the last sentences. Although the speech corpus contains quite a lot of phonemes and diphones, the variability of speech units is far from being complete and many units for the unit selection method, being considered the most natural, are missing (the corpus was not prepared with respect to this issue). Nevertheless, the other two approaches of speech synthesis – single-unit-instance syntehsis (SUI) and HMM synthesis [10] – were tested, the better one for this patient seems to be the first.

Other patients have already recorded sentences from the new text corpus described above. Several patients were recorded up to now (i.e. before the deadline of this paper) and the personalised TTS system was created and tested for all of them. According to our observations, the patients can be divided into two groups by the number of sentences and the overall speech quality. Table 2 shows some corpora statistics for one representative patient per each group, the pilot voice and the representative professional speaker.

Table 2. Speech corpora statistics

speaker	pilot voice	group 1	group 2	semi-professional	professional
# sentences	694	469	1,038	2,014	12,241
# recording hours	6	6	8	10	60
# recording days	1	2	2	2	20
corpus length (min)	73.3	46.2	62.8	116.6	886.2
# phones in total	36,686	18,649	39,731	80,639	631,719
# different diphones	1,008	898	1,189	1,324	1,310
percentage of diphones	71.44 %	63.64 %	84.27 %	93.83 %	95.84 %
speech synthesis method	SUI	SUI/HMM	unit selection	unit selection	unit selection

The patients in the first group usually have (almost) all the problems described in Section 2.5 like extra breaths, mispronunciations, unintelligible pronunciations, inconsistent recording, problems with using a computer etc. They record only a few hundred sentences, so utilization of the unit selection method is not possible because there is generally an insufficient number of units for selection. The other two approaches of speech synthesis are applied and the TTS systems are created and offered to patients (the quality of each of them depends a lot on a specific patient's voice).

The second group comprises the patients who record relatively quite a lot of sentences given the non-typical conditions of the recording. Those people have usually no (or just a little) problems with the recording itself and controlling the recording process on a computer. Since more than 1,000 sentences are recorded, the speech corpus contains all the sentences selected in the first three steps of the selection algorithm and thus the units' coverage is quite high, so a synthesis utilising the unit selection method can be used. For any missing diphone in the corpora, the replacements were manually defined as in the generic TTS.

The patient denoted as "semi-professional" in Table 2 is a special case of the second group. This patient had been used to speaking aloud and had also had experience with speech recording before, so he managed to record more than 2,000 sentences in two days with not many stumbles and breaths. Despite his voice tiredness at the end of both days, the unit selection synthesis is fluent and not much worse than the generic voices, in the authors' subjective view.

Out of the total number of 1,411 diphones in the prepared text corpus, not all of them are contained in the speech corpora. First, the patients (group 1 especially) did not record that many sentences. Second, some diphones are missing due to pauses and breaths in the middle of sentences or an incorrect reading of some foreign-like words.

4   Conclusions

We have shown that we are able to create a reasonable personalised TTS for almost any patient prior to total laryngectomy, even in cases when the patient's voice quality is already very low, he/she has no prior experience with voice recording and there are quite hard time constraints. Such a flexibility is crucial for future implementation of this procedure to the clinical practice.

Indeed, the resulting TTS speech quality very strongly depends on the quality of the recorded speech corpus, but there is almost always a backup solution in a form of a SUI or HMM synthesis for worst case scenarios (too few sentences recorded, dramatic problems in the patient's voice quality etc.). We have also shown that for the unit selection method (generally considered as delivering the most natural synthetic speech), it is usually enough to record just slightly more than 1,000 sentences from a patient in order to produce a personalised TTS of a reasonable quality.

The most important aspect of the situation described by this paper is to make the recording process as smooth and easy as possible for the patients in order to help them in their extremely complicated situation and not to discourage them from entering the voice banking procedure. We have successfully achieved the first stage of this and now we can move on to a full-blown Quality of Life Study involving more patients and giving us valuable feedback for further improvements.

References

1. Hanzlíček, Z., Romportl, J., Matoušek, J.: Voice conservation: towards creating a speech-aid system for total laryngectomees. In: Kelemen, J., Romportl, J., Zackova, E. (eds.) Beyond Artificial Intelligence. TIEI, vol. 4, pp. 203–212. Springer, Heidelberg (2013)
2. Romportl, J., Řepová, B., Betka, J.: Vocal rehabilitation of laryngectomised patients by personalised computer speech synthesis. In: Zehnhoff-Dinnesen, A., Schindler, A., Wiskirska-Woznica, B., Zorowka, P., Nawka, T., Sopko, J. (eds.) Phoniatrics. European Manual of Medicine. Springer (2015 in press)
3. Matoušek, J., Romportl, J.: On building phonetically and prosodically rich speech corpus for text-to-speech synthesis. In: Proceedings of the Second IASTED International Conference on Computational Intelligence. ACTA Press, San Francisco, pp. 442–447 (2006)
4. Matoušek, J., Psutka, J., Krůta, J.: Design of speech corpus for text-to-speech synthesis. In: Eurospeech 2001 - Interspeech, Proceedings of the 7th European Conference on Speech Communication and Technology, Aalborg, Denmark, pp. 2047–2050 (2001)
5. Matoušek, J., Tihelka, D., Psutka, J.: New slovak unit-selection speech synthesis in ARTIC TTS system. In: Proceedings of the World Congress on Engineering and Computer Science 2011, San Francisco, USA, pp. 485–490 (2011)
6. Romportl, J.: Structural data-driven prosody model for TTS synthesis. In: Proceedings of the Speech Prosody 2006 Conference, pp. 549–552. TUDpress, Dresden (2006)
7. Matoušek, J., Tihelka, D., Romportl, J.: Current state of czech text-to-speech system ARTIC. In: Sojka, P., Kopeček, I., Pala, K. (eds.) TSD 2006. LNCS (LNAI), vol. 4188, pp. 439–446. Springer, Heidelberg (2006)
8. Matoušek, J., Romportl, J.: Recording and annotation of speech corpus for czech unit selection speech synthesis. In: Matoušek, V., Mautner, P. (eds.) TSD 2007. LNCS (LNAI), vol. 4629, pp. 326–333. Springer, Heidelberg (2007)
9. Matoušek, J., Tihelka, D., Romportl, J.: Building of a speech corpus optimised for unit selection TTS synthesis. In: Proceedings of 6th International Conference on Language Resources and Evaluation, LREC 2008. ELRA (2008)
10. Hanzlíček, Z.: Czech HMM-based speech synthesis. In: Sojka, P., Horák, A., Kopeček, I., Pala, K. (eds.) TSD 2010. LNCS, vol. 6231, pp. 291–298. Springer, Heidelberg (2010)

An Open Source Speech Synthesis Frontend for HTS

Markus Toman$^{(\boxtimes)}$ and Michael Pucher

FTW Telecommunications Research Center Vienna,
Donau-City-Straße 1, 1220 Vienna, Austria
{toman,pucher}@ftw.at
http://www.ftw.at

Abstract. This paper describes a software framework for HMM-based speech synthesis that we have developed and released to the public. The framework is compatible to the well-known HTS toolkit by incorporating hts_engine and Flite. It enables HTS voices to be used as Microsoft Windows system voices and to be integrated into Android and iOS apps. Non-English languages are supported through the capability to load Festival format pronunciation dictionaries and letter to sound rules. The release also includes an Austrian German voice model of a male, professional speaker recorded in studio quality as well as pronunciation dictionary, letter to sound rules and basic text preprocessing procedures for Austrian German. The framework is available under an MIT-style license.

Keywords: Speech synthesis · HTS · Hidden Markov Model · Frontend · Software

1 Introduction

Hidden-Markov Model (HMM) based speech synthesis provides a methodology for flexible speech synthesis while keeping a low memory footprint [1]. It also enables speaker adaptation from average voice models, allowing the creation of new voice models from sparse voice data [2], as well as techniques like interpolation [3][4] and transformation [5][6] of voices. A well-known toolkit for creating HMM-based voice models is HTS [7]. Separate software toolkits are available to actually synthesize speech waveforms from HTS models. A popular, freely available framework is hts_engine [8]. Speech synthesis frontends on the other hand provide means for analyzing and processing text, producing the necessary input for the synthesizer. In HTS this input is a set of labels where usually each label represents a single phone and contextual information, including surrounding phones, position in utterance, prosodic information etc. While not exclusively being frontends and not specifically targeted for HTS, popular choices are Festival [9] or Flite [10]. Festival is a complex software framework for building speech synthesis systems focusing Unix-based operating systems. Flite was built as a lightweight alternative to Festival with low memory footprint and fast runtime in mind.

© Springer International Publishing Switzerland 2015
P. Král and V. Matoušek (Eds.): TSD 2015, LNAI 9302, pp. 291–298, 2015.
DOI: 10.1007/978-3-319-24033-6_33

Our main goal when creating the presented frontend framework was to easily allow HTS voices to be used with the Speech Application Programming Interface 5 (SAPI5). This allows the framework to be installed on different versions of the Microsoft Windows operation system as speech synthesis engine, making HTS voice models available as system voices to applications like screen readers, e-book creators etc. The second goal was simple integration of new languages and phone sets. The third goal was portability to mobile devices. The framework is available under an MIT-style license at http://m-toman.github.io/SALB/.

Flite has been adapted for HTS in the Flite+hts_engine software [8] and due to its small and portable nature it seemed like a good fit to our requirements. The structure of Flite makes integrating new languages rather cumbersome.[1] Therefore our framework integrates Flite for text analysis of English while additionally providing a second text analysis module that can utilize Festival style pronunciation dictionaries and letter to sound trees. Text preprocessing tasks (e.g. number and date conversion) can be added to the module in C++. Adding a completely new text processing module is also possible. The framework includes hts_engine for speech waveform synthesis and can be extended by other synthesizers. The framework also includes a free voice model of a male, professional speaker for Standard Austrian German.

2   Voice Model "Leo"

With the framework we provide a free voice model of a male, professional speaker for Standard Austrian German called "Leo". The model is built from 3,700 utterances recorded in studio quality using a phonetically balanced corpus. The phone set used in the voice can be seen in Table 1. A pronunciation dictionary with 14,000 entries, letter to sound rules and procedures for number conversion are also included.

3   Framework Architecture

The general architecture of the SALB framework is shown in Figure 1. Frontend modules provide means to communicate with the user or other applications through different channels. For example the user can directly trigger speech synthesis tasks by the Command-Line-Interface (CLI), other applications can use the SAPI5 interface to use the framework in a uniformly manner together with other synthesis engines that implement SAPI5. The frontend modules use the C++ Application Programming Interface (API) of the core module `manager`, which in turn coordinates the backend modules, performing text processing and the actual speech synthesis task. The C++ API can be used directly to embed the framework in other applications.

---

[1] We have published instructions on adding a new language to Flite: http://sourceforge.net/p/at-flite/wiki/AddingNewLanguage/

Table 1. Phone set used for Austrian German voice "Leo".

Category	Phones (IPA)
Vowels (monoph.)	ɑ ɑː ɒ ɒː ɐ e ɛ ɛː i ɪ iː ɔ o oː ø ɶː ɐ œ œː ə u ʊ uː ʏ y yː
Vowels (monoph.) nasalized	ɒ̃ː ɔ̃ː æ̃ː œ̃ː
Diphthongs	aɪ̯ ɒːɐ̯ ɑːɐ̯ ɒɪ̯ aʊ̯ ɛɐ̯ ɛːɐ̯ ɪɐ̯ iɐ̯ iːɐ̯ ɔɐ̯ ɔːɐ̯ oːɐ̯ ɔʏ øːɐ̯ œɐ̯ ʊɐ̯ uːɐ̯ ʊːɐ̯ ʏɐ̯ yːɐ̯
Plosives (stops)	b̥ d̥ ɡ̊ k ʔ p t
Nasal stops	m n ŋ
Fricatives	ç x f h s ʃ v z ʒ
Approximants	j
Trill	r
Lateral approx.	l

3.1   Manager Module and API

The core of the framework is the manager module which provides a uniform API for frontend modules or other applications. It provides abstractions for different elements of the speech synthesis process. This API is provided by the `TTSManager` class. A `TextFragment` is a piece of text in a given language. Each `TextFragment` has `FragmentProperties` associated which control synthesis parameters (e.g. voice, speaking rate) for this text fragment. Multiple `TextFragment` objects can form a `Text` object. This can be used to synthesize a text consisting of fragments with different synthesis parameters (e.g. a text read by different voices). A `Text` or `TextFragment` object with associated `FragmentProperties` can be passed to a `TTSManager` object which executes the speech synthesis process and returns a `TTSResult` object. This process is depicted in Figure 2. The `TTSManager` object first selects an adequate `TextAnalyzer` object based on the value in `FragmentProperties` specifying the text analyzer to use or a value specifying the language of the text. The `TextFragment` is then passed to the `TextAnalyzer` object which returns a series of `Label` objects in a container called `Labels`. A `Label` represents the basic unit for synthesis, usually being a single phone with contextual information. The `TTSManager` then selects an adequate `Synthesizer` implementation, again based on `FragmentProperties`, and passes the `Label` to it. The `Label` class is responsible for providing the desired format. Currently the only available format is the HTS label format, which can easily be stored as special character delimited string. The `Synthesizer` returns a `TTSResult` object containing the synthesized waveform as `vector` of discrete samples as well as meta information.

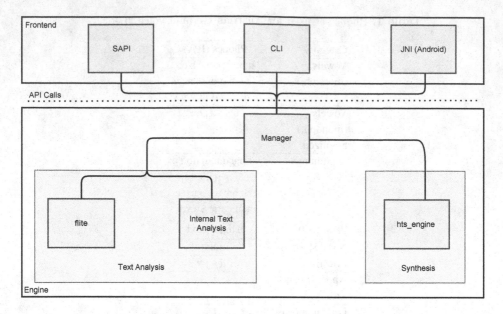

Fig. 1. General framework architecture

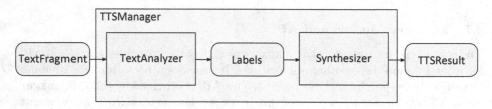

Fig. 2. Data flow in synthesis process

3.2   Frontend Interfaces

The following sections describe the frontend interfaces currently included in the framework.

Speech Application Programming Interface (SAPI). The framework provides a frontend implementing the SAPI5. This allows the registration of the framework as speech synthesis engine on Microsoft Windows platforms, therefore enabling HTS voices to be registered as system voices. We provide a Microsoft Visual Studio project [11] to compile a SAPI-enabled dynamic link library for 32-bit and 64-bit systems that can be registered with the operating system. Subsequently, voices can be added using the `register_voice` tool that comes with the framework. Lastly a script, that allows to create installer packages that install or uninstall the engine and associated voice models, is bundled with the framework.

Command Line Interface. The distribution also contains a simple command line tool which, given all necessary input for the text analysis and synthesis modules, produces RIFF wav output files from textual input.

Android. Integration in Android apps is possible through the Java Native Interface (JNI) and the Android Native Development Kit (NDK) [12]. The framework comes with Android make files, a JNI wrapper and a Java class for demonstration purposes.

3.3 Text Analysis Modules

The following sections describe the text analysis modules currently included in the framework.

Flite. For converting English text to a series of labels for synthesis, we integrated Flite. The class `FliteTextAnalyzer` (derived from `TextAnalyzer`) is a wrapper converting input and output data for and from Flite.

Internal Text Analyzer. The distribution also comes with an internal text analysis module `InternalTextAnalyzer` (derived from `TextAnalyzer`). This module reads a specific rules file consisting of a pronunciation dictionary and letter to sound rules in Festival style format. Preprocessing of text (e.g. for numbers and dates) can be added by extending the `Normalizer` class which delegates to different implementations based on the chosen language. We provide a simple `Normalizer` for Austrian German (`AustrianGermanNormalizer`) as well as a comprehensive rules file which can be used as an example to integrate new languages. This module uses an utterance structure which consists of the classes `Phrase`, `Word`, `Syllable` and `Phone`. The `PhraseIterator` class is used to navigate in this structure to build the resulting `Label` object.

3.4 Synthesis Modules

The following section describes the currently available synthesis module in the framework.

hts_engine. This module provides a wrapper around hts_engine and is implemented by the class `HTSEngineSynthesizer` (derived from `Synthesizer`). The `Label` objects provide strings in HTS label format which are the input to hts_engine. The resulting waveform is then converted and encapsulated in a `TTSResult` object and returned to the `TTSManager` and subsequently to the caller. We have changed the algorithm for changing the speaking rate to linear scaling due to the results in [4].

4   Adding New Languages

One main goal when developing the framework was the possibility to easily integrate voice models of other languages than English. As literature aiding this process is scarce, we present some basic guidelines in the following sections.

4.1   Gathering Data

Before creating a recording script, a defined phone set is needed. These already exist for many languages. If not, the inventory of all relevant phones of a language should be defined in cooperation with phoneticians. One possibility is to gather conversational speech data in the target language and then produce manual transcriptions. The granularity of the transcription is very important and has a direct impact on the quality of the final voice models. For example, diphthongs can be modeled as separate phone symbols or split into two symbols. When a phone set is defined, recording scripts can be generated either through manual transcription or by using orthographic text (e.g. from newspapers) and a letter to sound system. The recording script should contain all phones in as many triphone contexts, better quinphone contexts, as possible. In any case each phone should occur multiple times (preferably at least 10 times, considering that a set set will also be split off the training data). Given a data set of phonetically transcribed sentences, algorithms to solve the set cover problem can be employed to produce a final recording script. From this data set, a test set can be selected by the same procedure (e.g. select the smallest set that contains each phone at least 2 times). Speakers should be recorded in studio setting and with neutral prosody. The output of this step is a phone set, recording scripts and a corpus of recorded utterances in waveforms and transcriptions.

4.2   Integration

The internal text analyzer reads rules for text processing from an input stream. These rules consist of a lexicon and a letter to sound tree. A word is first looked up in the lexicon. If it can not be found there, the letter to sound rules are used to create the phonetic transcription. A voice model and the text processing rules are sufficient for bassic speech synthesis and building SAPI5 voice packages using the framework. Extending the source code is necessary if more sophisticated text processing is required, the C++ class `AustrianGermanNormalizer` can be used as an example for this. The method `InternalTextAnalyzer::TextFragmentToPhrase` can be adapted to implement alternatives to the Festival lexica and letter to sound rules.

Lexicon. The lexicon (or pronunciation dictionary) part of the text rules is a set of mappings from orthography to phonetics. The framework uses Festival style lexica in Scheme syntax.[2] A lexicon can be derived from the data corpus used for the recording or from publicly available pronunciation dictionaries.

---

[2] Festival lexicon definition: http://www.cstr.ed.ac.uk/projects/festival/manual/festival_13.html

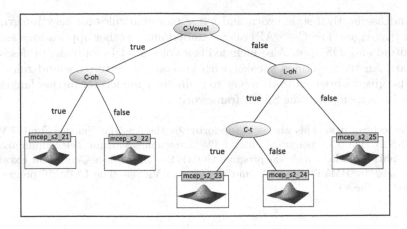

Fig. 3. Questions in decision tree based clustering

Letter-to-sound Rules. The internal text analyzer is able to read Festival style letter to sound rules.[3] A method for building letter to sound rules from an existing lexicon can be found in [13]. For languages with an orthography very close to the phonetics, the letter to sound rules can also be hand-crafted.

4.3   Training Voice Models

Voice models to be used with the SALB framework can be trained using the HTS toolkit. Available demonstration scripts for speaker dependent voices can be adapted for this purpose. When using the demonstration scripts, it is necessary to replace raw sound files, labels and question files. Label files can be produced using the SALB framework once at least a lexicon containing all words from the training set is available. The question files contain questions used in the decision tree based clustering [14]. An illustrative example of the usage of the phone questions in a decision tree can be seen in Figure 3. A minimal question file should at least contain all phone identity questions (e.g. for a phone "oh" and quinphone models, at least the following questions should be defined: "LL-oh", "L-oh", "C-oh", "R-oh", "RR-oh"). When all parameters (e.g. sampling rate) have been set, the training process can be started to produce an "htsvoice" model file that can be used with the SALB framework.

5   Conclusion

We have presented an open source speech synthesis framework, a software that bridges existing tools for HTS-based synthesis like hts_engine and Flite with SAPI5 to enable HTS voices to be used as Windows system voices. It allows to

---

[3] Festival LTS rules: http://www.cstr.ed.ac.uk/projects/festival/manual/festival_13. html#SEC43

load and use Festival style lexica and letter to sound rules for easy integration of new languages. The C++ API allows embedding in other applications as well as Android and iOS apps. Also included is a voice model of a male, professional Standard Austrian German speaker with lexicon and letter to sound rules. We have also given a brief tutorial on how to train voice models for further languages and add use them with the SALB framework.

Acknowledgments. This work was supported by the Austrian Science Fund (FWF): P22890-N23. The Competence Center FTW Forschungszentrum Telekommunikation Wien GmbH is funded within the program COMET - Competence Centers for Excellent Technologies by BMVIT, BMWA, and the City of Vienna. The COMET program is managed by the FFG.

References

1. Zen, H., Tokuda, K., Black, A.W.: Statistical parametric speech synthesis. Speech Communication **51**(11), 1039–1064 (2009)
2. Yamagishi, J., Kobayashi, T.: Average-voice-based speech synthesis using HSMM-based speaker adaptation and adaptive training. IEICE Transactions on Information and Systems **E90–D**(2), 533–543 (2007)
3. Pucher, M., Schabus, D., Yamagishi, J., Neubarth, F., Strom, V.: Modeling and interpolation of Austrian German and Viennese dialect in HMM-based speech synthesis. Speech Communication **52**(2), 164–179 (2010)
4. Valentini-Botinhao, C., Toman, M., Pucher, M., Schabus, D., Yamagishi, J.: Intelligibility analysis of fast synthesized speech. In: Proc. Interspeech, Singapore, pp. 2922–2926, September 2014
5. Wu, Y.J., Nankaku, Y., Tokuda, K.: State mapping based method for cross-lingual speaker adaptation in HMM-based speech synthesis. In: Proceedings of the 10th Annual Conference of the International Speech Communication Association (INTERSPEECH), Brighton, United Kingdom, pp. 528–531 (2009)
6. Toman, M., Pucher, M., Schabus, D.: Cross-variety speaker transformation in HSMM-based speech synthesis. In: Proceedings of the 8th ISCA Workshop on Speech Synthesis (SSW), Barcelona, Spain, pp. 77–81, August 2013
7. Zen, H., Nose, T., Yamagishi, J., Sako, S., Masuko, T., Black, A.W., Tokuda, K.: The HMM-based speech synthesis system (HTS) version 2.0. In: Proceedings of the 6th ISCA Workshop on Speech Synthesis (SSW), Bonn, Germany, pp. 294–299, August 2007
8. HTS working group: hts-engine. http://hts-engine.sourceforge.net/
9. University of Edinburgh: Festival. http://www.cstr.ed.ac.uk/projects/festival/
10. Carnegie Mellon University: Flite. http://www.festvox.org/flite/
11. Microsoft Corporation: Visual Studio. https://www.visualstudio.com/
12. Google Inc: Android NDK. https://developer.android.com/tools/sdk/ndk/
13. Black, A.W., Lenzo, K., Pagel, V.: Issues in building general letter to sound rules. In: The Third ESCA Workshop in Speech Synthesis, pp. 77–80 (1998)
14. Yoshimura, T., Tokuda, K., Masuko, T., Kobayashi, T., Kitamura, T.: Simultaneous modeling of spectrum, pitch and duration in HMM-based speech synthesis. In: Proceedings of the 6th European Conference on Speech Communication and Technology (EUROSPEECH), Budapest, Hungary, pp. 2374–2350, September 1999

Tibetan Linguistic Terminology on the Base of the Tibetan Traditional Grammar Treatises Corpus

Pavel Grokhovskiy(✉), Maria Khokhlova,
Maria Smirnova, and Victor Zakharov

Saint-Petersburg State University, Saint-Petersburg, Russia
{plgr,2321781}@mail.ru, khokhlova.marie@gmail.com, vz1311@yandex.ru

Abstract. The paper is devoted to Tibetan grammatical terminology. For this purpose Tibetan grammatical works corpus was created. At the same time Russian translations of the works were added to the corpus, so it is factually a parallel Tibetan-Russian corpus. The corpus represents the collection of grammar treatises of the Tibetan grammatical tradition formed in VII-VIII c. The corpus is useful to researchers of the Tibetan linguistic tradition as well as to those specialized in linguistic studies of classical and modern Tibetan and its teaching. On the basis of corpus a specific grammatical lexical database is created. The database will be useful both to tibetologists and general linguistics specialists.

Keywords: Linguistical terminology · Tibetan language · Corpus linguistics · Parallel corpus · Morphology · Tagging · Lexical database

1 Introduction

The project focuses on the creation of the Tibetan grammatical terminology database and Tibetan traditional grammar treatises corpus. The origin of linguistic tradition in Tibet is dated back to the creation of the first grammatical treatises "Sum cu pa" and "Rtags kyi 'jug pa" (VII-VIII c.).

The Tibetan linguistics is mainly based on grammars created by Buddhist scholars and is thus highly connected with Indian tradition. Methods of language characterization and analysis are greatly different from those of Western linguistics. Modern Tibetan scholars continue to follow and develop the Tibetan grammatical tradition. The corpus includes the basic grammar treatises and commentaries, which are considered to be the most important grammatical works within the Tibetan grammatical tradition. On the basis of the corpus a specific grammatical lexical database is created which will be useful both to tibetologists and general linguistics specialists.

© Springer International Publishing Switzerland 2015
P. Král and V. Matoušek (Eds.): TSD 2015, LNAI 9302, pp. 299–306, 2015.
DOI: 10.1007/978-3-319-24033-6_34

2   Modern Corpus Linguistics of the Tibetan Language

Despite the fact that scholars in different countries (Germany, Great Britain, People's Republic of China, USA and Japan) are engaged in working out of Tibetan texts corpus presentation, still there is no common standard for it.

Last four conferences of the International Association for Tibetan Studies (IATS Seminar) included section «Tibetan Information Technology», where computer technology projects in the field of Tibetan studies were represented. They also include projects focusing on the creation of Tibetan corpus.

The creation of Tibetan language corpora abroad has just begun. The cooperative research project 441 under the guidance of B. Zeissler in Eberhard Karls University (Tübingen, Germany) included the subproject B11 «Semantic roles, case relations, and cross-clausal reference in Tibetan» (2002-2008) [1]. In 2012 U. Pagel from Department of the Study of Religions and N. Hill from Department of China and Inner Asia and Departments of Linguistics began development of the Tibetan corpus to contain 1 million syllables. Its texts cover three historical periods of the Tibetan language: preclassical, classical and modern[1].

The first difference between the corpus of the Tibetan traditional grammar treatises and the projects mentioned above is the development of special system of linguistic tags [2][3]. The second difference is related to the involved materials. All texts represent one of the traditional Tibetan sciences – linguistics.

3   Corpus Structure

The project has two main tasks: creation of a parallel corpus of the Tibetan grammatical treatises with Russian translation and creation of a specific grammatical lexical database with frequency characteristics and semantic relations.

Tibetan texts and Russian translations in the corpus are aligned by sentence breaks of the Tibetan part. Words of the Tibetan part are tagged morphologically. It should be noted that segmentation of Tibetan texts is a sophisticated problem because according to the traditional Tibetan orthography only syllable borders are marked (Tibetan syllable coincides with a word form approximately in 95 cases out of 100).

There are special programs for automatic segmentation, for example, Hunalign, Vanilla, etc. However, the process of their implementation for the needs of the Tibetan language is quite difficult and not included in the project tasks. Comparatively small corpus volume and the necessity of manual tokenisation argue for manual segmentation and tagging.

4   Corpus Annotation

Tibetan text undergoes morphological tagging (lemmatization, part-of-speech tagging, grammatical annotation of verb forms, eliminating of grammatical

Table 1. Tags for function words in the Tibetan language

No.	Tag	Function word	Example
1	Cj	Conjunction	dang
2	Pp	Postposition	drung du
3	Erg	Ergative	allomorphs kyis, gyis, gis, s, yis
4	Com	Comitative	dang
5	Dat	Dative	la
6	Loc	Locative	na
7	Dest	Destinative	allomorphs tu, du, ra, ru, su
8	Abl	Ablative	las
9	El	Elative	nas
10	Comp	Comparative	allomorphs pas, bas
11	Gen	Genitive	allomorphs kyi, gyi, gi, 'i, yi
12	Fin	Final particle	allomorphs go, ngo, do, no, bo, mo, 'o, ro, lo, so, to
13	Top	Topicalizing particle	ni
14	Ind	Indefinite particle	allomorphs cig, zhig, shig
15	Emph	Emphatic particle	allomorphs kyang, yang, 'ang
16	Quant	Quantifier ('that much', etc.)	tsam, kho na, 'ba' zhig, snyed
17	Pl	Plural marker	rnams
18	Quot	Quotation marker	allomorphs ces, zhes

homonymy). The Tibetan language tag system is developed. List of tags for the most widely used function words is given in Table 1.

Corpus texts are also provided with metadata about genre, date of creation, author.

Every word is represented by the following data: word form in Tibetan script, word form in Latin transliteration, lemma in Tibetan script, lemma in Latin transliteration, part-of-speech tag, terminological tag (Table 2). The program tool for automatic tagging TreeTagger is supposed to be adapted to the Tibetan language. In this regard the manually tagged corpus will be used as training corpus.

5   Corpus Search and Usage

The system Sketch Engine (as well as Nosketch Engine) is used as a corpus manager. The corpus can be accessed at http://corpora.spbu.ru. The corpus interface allows searching and filtration by all elements of annotation as well as brief and broad concordance and word lists creation. To create a frequency list all the above mentioned elements and their combinations could be used as attributes.

Currently the corpus contains 33913 tokens. Using corpus it is possible to search word forms, lemmas in the Tibetan script and Latin transliteration as well as collocations for given words and phrases, translations in parallel texts, operate with statistic data, etc. For an extended search the regular expression language is used.

---

[1] http://www.soas.ac.uk/news/newsitem73472.html as accessed on 29. 03. 2015.

Table 2. Fragment of aligned text

Word form (Tibetan sript)	Word form (transliteration)	Lemma (Tibetan script)	Lemma (transliteration)	Part-of-speech tag	Terminological tag
<s>					
<align>					
དེ་	de	དེ་	de	P	
ནི་	ni	ནི་	ni	Top	
སྡུད་	sdud	སྡུད་	sdud	VN	Gram L TGrMark
དང་	dang	དང་	dang	Cj	
འབྱེད་པ་	'byed pa	བྱེད་	byed	VN	Gram L TGrMark
དང	dang	དང	dang	Cj	
‖	//	‖	//	Punct	
རྒྱུ་མཚན་	rgyu mtshan	རྒྱུ་མཚན་	rgyu mtshan	N	Gram GenLex TGrMark
ཚེ་སྐབས་	tshe skabs	ཚེ་སྐབས་	tshe skabs	N	Gram L TGrMark
གདམས་ངག་	gdams ngag	གདམས་ངག་	gdams ngag	N	Gram GenLex TGrMark
ལྔ	lnga	ལྔ	lnga	Num	
འོ	'o	འོ	'o	Fin	
‖	//	‖	//	Punct	

The corpus of the Tibetan traditional grammar treatises could be used both for the purposes of theoretical and applied linguistics. The corpus is a source for lexicography, semantics and grammar investigations. Such corpus opportunities as statement comparison, search of lexical units equivalents allow to reduce time of working with teaching and information materials (e.g. dictionaries). Thus the corpus is a useful tool for translation and language teaching.

6   Lexical Database of the Tibetan Grammatical Treatises Corpus

6.1   Special Tagging of Grammatical Terminology

It is not typical for the Tibetan linguistics to emphasize such traditional subdisciplines of Western linguistics as phonology, morphology and syntax. Basic terms of the Tibetan grammatical tradition denote basic units of different language levels [4].

Most Tibetan authors begin their grammatical works with the description of the Tibetan alphabet, different types of graphemes and corresponding phonemes, rules of syllable composition and grapheme/phoneme combination as well as morphonological rules. Tibetan grammars also contain the description of function words and morphemes.

The Tibetan linguistic tradition borrowed Indian idea of seven cases. In Indian linguistics cases are connected with the kāraka category which represents an intermediate level between semantics and morphology. This system of kāraka categories was also borrowed by Tibetans.

6.2   Tags for Grammatical Terminology

Elements of traditional grammatical metadescription such as terminological categories (phonological, syntax terms), Sanskrit equivalents for loans, links to synonyms, hyperonyms, hyponyms are added to the lexical database as well as scientific commentaries.

The use of special grammatical tags given in Table 3 makes it possible to divide different terminological fields: grammatical terminology (tag Gram) and terms of traditional sciences (tag GenScien).

Certain tags stand for models of terms origin: by terminologisation of common words (tag GenLex) or through borrowing (tag L).

The tag TBas is used for basic grammatical terminology. Polysemy is the main feature of the Tibetan terminology in general and basic grammatical terms in particular. Therefore one of the main tasks was to separate phonological terms (tag TPhon) and terms for different types of graphemes (tag TGra).

Grammatical terms imported from the corpus through the use of special grammatical tags form the lexical database of Tibetan grammatical terminology, which contains additional information about the origin language for loans, foreign equivalents, way of borrowing (phonetic or semantic borrowing, calquing, hybrid terms), etc. There is more than 2000 occurrences of grammatical terms in the current corpus database.

Table 3. Tags for grammatical terminology

Characteristic of classification	Tag	Meaning
Terminological	Gram	term of the Tibetan grammatical tradition
field	GenScien	general scientific term
Origin	GenLex	term of Tibetan origin
	L	borrowed term
Type of	TBas	basic grammatical term
terminology	TPhon	phonological term
	TGra	grapheme type
	TGrMark	name of auxiliary morphemes and lexemes
	TCGr	case grammar term

6.3   Structure of Lexical Database

Lexical database of the Tibetan grammatical treatises corpus contains lexical units selected from the Tibetan part of the corpus by appropriate tags.

TEI recommendations (Text Encoding Initiative) are taken as a methodological basis for database exchange format [5]. It is important that TEI has tags for relation links to create network data representation in linear XML files.

Let's describe lexical database representation template in XML format according to database structure. This template has several divisions and representation levels.

Lexical unit level Lexical unit of the database in XML begins with record:

```
1  <entry n="1" type="lex">
2  <term>ཨུ་ལི་</term>
3  <pron>A li</pron>
```

where index number of a lexical unit (n="1"), its type (type="lex" – lexical unit) and entry word ཨུ་ལི་ in Tibetan script (tag <term>) and transcription (tag <pron>) are given.

It is followed by the block with grammatical information (tag <gramGrp>):

```
<gramGrp>
    <pos> N </pos>
</gramGrp>
```

which contains part-of-speech tag (tag <pos>) and additional grammatical information (tags <gen>, <flex>, etc.).

The same level includes one or several etymological information blocks (tag <etym>):

```
<etym n="1">
    <lbl> Sumcupa </lbl>
    <date> 8|$^{\rm th}$| c.</date>
</etym>
```

which contains sequence number of the etymology block, etymological tag (tag <etym>), source of information (tag <lbl>), date of the first usage (tag <date>). Also other tags, like <lang> (language, from which a word was presumably borrowed), could be used.

Link level and example level are represented in the same way. In the end data are converted from exchange format into Microsoft SQL on the Microsoft.NET platform.

6.4   User Interface

The database interface is a window application powered by Microsoft.NET and closely integrated with a system core. During the interface development the following tasks were set:

1. searching all lexemes in the database;
2. displaying lexemes in a convenient form;
3. manual adding of new lexemes;
4. editing of available lexemes;
5. loading (importing) lexical unit records from XML to TEI format;
6. saving (exporting) database records into TEI format.

Functions of the grammatical lexical database are as follows:

1. statistical data where appropriate;
2. to retrieve content of an entry for a given word;
3. to retrieve all synonyms for a given word;
4. to retrieve all hyponyms for a given word;
5. to retrieve all hyperonyms for a given word;
6. to show frequency in the database for a given terminological tag;
7. to show frequency in the corpus for a given terminological tag;
8. to retrieve all lexemes marked by a given terminological tag.

7   Future Works

In the future the corpus is supposed to be provided with syntactic annotation, extended and developed in a more extensive corpus of Tibetan texts including those dedicated to other traditional Tibetan sciences: Buddhist religious doctrine, logic, medicine, craft, poetics, synonymics, prosody, astrology and drama.

8   Conclusion

Pilot version of the Tibetan grammatical treatises corpus will be useful to all researchers of Tibetan grammatical written texts. Nowadays there are no available Tibetan language corpora aligned and translated into Russian. Therefore the corpus also could be useful for linguistic research, Tibetan language study and teaching.

Frequency dictionary of Tibetan lexical units (grammar terms) and semantic analysis of the lexical database will form a sort of linguistic ontology that includes hyponyms and hyperonyms, polysemic words and synonyms.

All this will allow to analyze the development of language in general and structure of terminological fields in particular, and to estimate the terminologisation degree of common words.

Acknowledgement. This corpus and database development was supported by the grant No. 13-06-00621 «The Pilot Version of Tibetan Grammar Texts' Electronic Corpus» by Russian Foundation for Basic Research. Authors also acknowledge Saint-Petersburg State University for a research grant 2.38.293.2014 «Modernizing the Tibetan Literary Tradition» which enabled the conceptual study of the original Tibetan texts.

References

1. Wagner, A., Zeisler, B.: A syntactically annotated corpus of Tibetan. In: Proceedings of the 4th International Conference on Language Resources and Evaluation, Lisboa, May 2004

2. Grokhovskiy, P.L.: Kategorii skazuemosti i nominalizatsii deystviya (substantivno-ad"ektivnye formy) v klassicheskom tibetskom yazyke. In: Guzev, V.G., (ed.) Ocherki po Teoreticheskoy Grammatike Vostochnykh Yazykov: Sushchestvitel'noe i Glagol, Izdatel'skiy dom SPbGU, pp. 269–288 (2001)
3. Grokhovskiy, P.L.: Grammatika imeni sushchestvitel'nogo v klassicheskom tibetskom yazyke. In: Guzev, V.G. (ed.) Ocherki po Teoreticheskoy Grammatike Vostochnykh Yazykov: Sushchestvitel'noe i Glagol, Izdatel'skiy Dom SPbGU, pp. 76–91 (2001)
4. Smirnova, M.O.: Bazovye terminy tibetskoy grammaticheskoy traditsii. In: Vestnik Sankt-Peterburgskogo universiteta. Seriya 13. Vostokovedenie. Afrikanistika. Vypusk 1, SPb, pp. 23–34 (2014)
5. Burnard, L., Bauman, S. (eds.) TEI P5: Guidelines for electronic text encoding and interchange (2010)

Improving Multi-label Document Classification of Czech News Articles

Jan Lehečka$^{(\boxtimes)}$ and Jan Švec

Department of Cybernetics, Faculty of Applied Sciences,
University of West Bohemia, Univerzitní 8, 306 14 Plzeň, Czech Republic
{jlehecka,honzas}@kky.zcu.cz
http://www.kky.zcu.cz

Abstract. In this paper, we present our improvement of a multi-label document classifier for text filtering in a corpus containing Czech news articles, where relevant topics of an arbitrary document are to be assigned automatically. Different vector space models, different classifiers and different thresholding strategies were investigated and the performance was measured in terms of sample-wise average F_1 score. Results of this paper show that we can improve the performance of our baseline naive Bayes classifier by 25% relatively when using linear SVC classifier with sublinear *tf-idf* vector space model, and another 6.1% relatively when using regressor-based sample-wise thresholding strategy.

Keywords: Multi-label classification · Topic identification · Threshold selection · Thresholding strategy

1 Introduction

With growing volume of electronic text documents available online, text filtering systems are increasingly required, especially in large text corpora. For purpose of text filtering, each document in the corpus can be associated with one or more label describing the topic of the document (e.g. "football", "politics" etc.). The natural requirement for modern text filtering systems is to assign these labels for an arbitrary document automatically. When the process of revealing topics from a text is based on a supervised learning (e.g. a classifier trained from manually labeled data), it is called multi-label document classification task.

Almost any binary classifier can be used for multi-label document classification using *one-vs-the-rest* strategy, which trains one binary classifier per label. Such multi-label classifier outputs a vector of scores for each document (*soft prediction*), which has to be processed by a thresholding strategy to obtain a binary vector (*hard prediction*), that is "true" or "false" for each label.

As for the thresholding strategy, there is a large volume of published studies reporting various attempts. Older attempts are using a fixed threshold to produce hard predictions, like *RCut*, *PCut*, *SCut* [1,2], also dynamic threshold *MCut*, which sets the threshold for each document in the highest difference

© Springer International Publishing Switzerland 2015
P. Král and V. Matoušek (Eds.): TSD 2015, LNAI 9302, pp. 307–315, 2015.
DOI: 10.1007/978-3-319-24033-6_35

between successive scores and thus doesn't need any training, has been presented [3]. In recent years, many other various thresholding strategies have been published in [4–6] and in many other papers.

2 Current System and Baseline Classifier

Our current system is a language modelling corpus [7] containing large, constantly growing, volume of text documents from various sources, mainly web-mined news articles. With increasing volume of documents in the corpus, the need of automatic document classification for purpose of data filtering became essential [8].

Currently in this system, a multi-label naive Bayes classifier is used and trained from documents incoming from one selected news server, which we believe to be labeled thoroughly. For each new document, the classifier produces a probability distribution over all topics and N most probable labels are assigned to the document. The number of assigned labels is currently fixed on the average number of labels in the training data, which is $N = 3$, although a threshold selection method has been recently reported to perform better [5].

3 Training and Testing Data

For the purpose of comparing different classifiers and thresholding strategies, we exported from the corpus all labeled data from a selected news server and split them into years of publishing. Then, from each year, we added documents from a 2-months epoch (different for each year) into the testing data set while keeping documents from the rest of the year as the training data set. Using this division technique, all documents were split roughly in ratio 5:1 into training and testing data sets while avoiding the absence of older topics in the testing data set.

Our training data set consists of $205k$ documents ($70M$ words total) with vocabulary size $700k$ and $21k$ different labels. Because of the lack of training data assigned to low-frequency labels, we decided to use only labels assigned to at least 30 documents, which decreased the number of labels to 1843. Our testing data consists of $44k$ documents.

4 Evaluation Metric

For binary classification tasks, where the retrieved output for each tested document is either 1 or 0 (i.e. the document has the label or not), the standard and widely used evaluation metric of a test outcome is the F_1 score

$$F_1 = \frac{2PR}{P + R}, \tag{1}$$

which is the harmonic mean of precision P and recall R, where

$$P = \frac{tp}{tp + fp} = \frac{|\{\text{relevant documents}\} \cap \{\text{retrieved documents}\}|}{|\{\text{retrieved documents}\}|}, \tag{2}$$

$$R = \frac{tp}{tp + fn} = \frac{|\{\text{relevant documents}\} \cap \{\text{retrieved documents}\}|}{|\{\text{relevant documents}\}|}, \quad (3)$$

where tp are true positives, fp false positives and fn false negatives.

However, to evaluate outcome of multi-label classification test, where the retrieved output for each tested document is binary vector (1 or 0 for each label), some average value of F_1 is needed. There are several options:

- *micro-average*: total true positives, false negatives and false positives are counted across all the documents and used to compute P and R for (1);
- *label-wise average*, also known as *macro-average*: F_1 is computed for each single label and average value is obtained;
- *sample-wise average*: F_1 is computed for each single document and average value is obtained.

We decided to use *sample-wise average* F_1, because it gives us good image, of how well an unknown document would be labeled in average, although it produces slightly lower scores then the other metrics. In the following sections, we use short notation F_1 for *sample-wise average* F_1.

We should also define, how to deal with *out-of-training-data* labels. The standard approach is to use the same set of labels for training and evaluating phase, and thus ignore all labels in the testing data set, which were not seen during the training phase. This approach can be tricky, because one can artificially increase scores from measured metrics by simply selecting fewer labels for training phase and thus ignore more labels in evaluation phase and thus ignore more errors. However, we believe that the better approach is to consider all relevant labels, that were not retrieved by classifier, as a mistake. Of course, the later approach produces lower F_1 scores, because when evaluating documents labeled with *out-of-training-data* labels, it is impossible to obtain high recall, so the upper bound of F_1 score is not 1, but substantially lower. In the case of our data, where we used only labels assigned to 30+ documents, the upper bound of (*sample-wise average*) recall was $R^{ceil} = 0.906$ and $F_1^{ceil} = 0.943$, which is the highest value we could possibly reach.

We believe the later approach better reflects the true performance of a document classifier, therefore we decided to use it as an evaluation metric for our experiments.

5   Vector Space Models

When dealing with a text, direct use of text representation (i.e. a sequence of words) in classification tasks is impractical, because of a variable length of each document. Therefore, it is desirable to convert raw text data to a vector space model, where each instance of data (i.e. each document in a corpus) is represented as a vector of predefined features.

Choosing the right vector space model and the right set of features suitable for particular task, is very important and can have strong impact on document classification performance. In this section, we present most popular vector space models and show their performance in our problem of multi-label document classification of Czech news articles.

Very popular vector space models based on bag-of-words model (i.e. a document-term matrix), which are widely used in document classification tasks, are *tf* (term frequency), *idf* (inverse document frequency) and it's powerful combination *tf-idf*. The *tf* reflects within document frequency of words (terms), whereas *idf* holds information about term frequencies across all documents.

Pure *tf* uses only raw frequencies, $tf_{td} = N(t,d)$, where $N(t,d)$ denotes how many times term t occurs in document d. However in practice, some normalization is usually applied on *tf* model:

- *cosine normalization* is a length-normalization used to scale all values to a $[0,1]$ while penalizing documents with high individual term frequencies or with many different terms:

$$tfc_{td} = \frac{tf_{td}}{\sqrt{tf_{t_1 d}^2 + tf_{t_2 d}^2 + \dots + tf_{t_V d}^2}} \tag{4}$$

- *sublinear tf* claims that if a term occurs twenty times in a document, it doesn't carry twenty times the significance of a single occurrence, but the significance has rather logarithmic scale:

$$tfs_{td} = \begin{cases} 1 + \log tf_{td} & \text{if } tf_{td} > 0 \\ 0 & \text{else} \end{cases} \tag{5}$$

On the other hand, *idf* reflects how common or rare the term is across all documents. It is the logarithmically scaled inverse proportion of documents containing the term:

$$idf_t = \log \frac{N}{N(t)}, \tag{6}$$

where N is total number of documents and $N(t)$ denotes number of documents containing the term t. If a term occurs in almost every document, it's *idf* will be very low (almost zero), and rare terms that occur few times in whole corpus, will gain high *idf*.

Combination of *tf* and *idf* leads to very powerful vector space models for text classification task. The simple product $tf_{td} \cdot idf_t$ is not commonly used, because it completely ignores terms with $idf = 0$, i.e. terms which occur in every document. Instead, formula

$$tf\text{-}idf_{td} = tf_{td}[idf_t + 1] = tf_{td} + tf_{td} \cdot idf_t \tag{7}$$

is usually used. In this paper, different normalizations of *tf* vector space model were used as both stand-alone model and in combination with *idf* using (7).

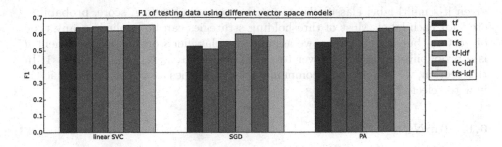

Fig. 1. F_1 score using different vector space models and different multi-label classifiers measured on testing data

5.1 Performance on Our Data

To see which vector space model is suitable for our data, we run following experiment. We converted raw text data to several different vector space models described in this section. For each vector space model, we trained three widely used linear classifiers using all the training data:

- *linear SVC* (linear Support Vector Classifier): SVC with linear kernel function, implemented in terms of *liblinear* [9];
- *SGD* (Stochastic Gradient Descend classifier): linear classifier with stochastic gradient descend learning;
- *PA* (Passive-Aggressive classifier) [10].

We used classifier implementations from *scikit-learn* [11] while leaving all parameters set on default values. To train a multi-label classifier, we used *one-vs-the-rest* strategy, which trains one binary classifier per label (document has the label or not). Then, all the testing data (converted to corresponding vector space model) have been labeled using the trained classifier, and the F_1 score has been measured. Results of this experiment are shown in Figure 1, where *tfc* denotes term frequency with cosine normalization (4) and *tfs* denotes sublinear term frequency (5). As a thresholding strategy, we used *top3* (see section 6.1).

We can see from the graph, that applying normalization on the term frequency, have mostly positive effect on the document classifier performance (in terms of F_1 score). The highest F_1 score has been obtained from *linear SVC* when using *tfs-idf* ($F_1 = 0.6554$), but the difference from *tfc-idf* ($F_1 = 0.6550$) or from the score obtained from *PA* when using *tfs-idf* ($F_1 = 0.6408$), is very small.

For experiments in the next chapter, we will use *linear SVC* classifier trained with data converted to *tfs-idf* vector space model.

6 Thresholding Strategy

A thresholding strategy describes the way how to select a set of relevant labels $\mathscr{L}_d^{rel}$ from a set of all possible labels $\mathscr{L} = \{l_k\}_{k=1}^K$ for an arbitrary document d

given it's multi-label classifier's output, i.e. one number (e.g. score, probability etc.) for each label. Most of thresholding strategies can be applied on any type of vectors, but in this paper, we assume the classifier's output for document d is a probability distribution over topics $P_d(k), k = 1, ..., K, \sum_{k=1}^{K} P_d(k) = 1$. In this section, we present some commonly used strategies as well as some our ideas how to select $\mathscr{L}_d^{rel}$.

6.1   TopN

TopN thresholding strategy, also known as $RCut$ [1] selects N most probable labels. It is very simple strategy with obvious drawback: the same number of labels is assigned to every document ignoring how probable the topics are. N can be set for example as an average number of labels in the training data set.

6.2   Threshold Selection

Next very simple thresholding strategy is to assign only labels of topics with the probability higher then some defined threshold t:

$$\mathscr{L}_d^{rel} = \{l_k \in \mathscr{L} : P_d(k) \geq t\}. \tag{8}$$

However, defining one fixed threshold for all documents can be impractical, because the higher number of relevant labels the document has, the lower corresponding probabilities are. There are many possibilities, how to set the threshold (or more thresholds) in more general way [1–5].

Our approaches are mainly based on learning thresholds from the probability distribution over topics $P_d^{train}(k), d \in \mathscr{D}^{train}, k = 1, ..., K$ of documents from training data set $\mathscr{D}^{train}$, which has been obtained by classifying the training data set after training the classifier. For $\mathscr{D}^{train}$, we also know document's true labels, which can be described by function

$$true(d, l_k) = \begin{cases} 1 & \text{if label } l_k \text{ is the true label of the document } d, \\ 0 & \text{else.} \end{cases} \tag{9}$$

Here, we describe thresholds used in this paper.

Label-wise thresholding: for each label $l_k \in \mathscr{L}$, one threshold t_k was set. First, two sets of probabilities were created for each l_k:

$$\mathscr{P}_k^{true} = \{P_d^{train}(k) : true(d, l_k) = 1, d \in \mathscr{D}^{train}\}, \tag{10}$$

$$\mathscr{P}_k^{others} = \{P_d^{train}(k) : true(d, l_k) = 0, d \in \mathscr{D}^{train}\}. \tag{11}$$

Then, performance of the following label-wise thresholds were investigated:

$$t_k^{true} = \min(\mathscr{P}_k^{true}), \tag{12}$$

$$t_k^{others} = \max(\mathscr{P}_k^{others}), \tag{13}$$

Table 1. Performance of different multi-label document classifiers

	strategy	P	R	F_1
naive Bayes classifier (*baseline*)	top3	0.486	0.607	0.524
SGD	top3	0.687	0.548	0.592
PA classifier	top3	0.742	0.594	0.641
linear SVC	top3	0.759	0.607	**0.655**
linear SVC	t_k^{true}	0.269	0.755	0.376
	t_k^{others}	0.784	0.468	0.554
	t_k^{mean1}	0.765	0.639	0.668
	t_k^{mean2}	0.758	0.672	**0.685**
linear SVC	$t_d^{\mathscr{R}}$	0.811	0.635	0.684
	$t_d^{\mathscr{R}sort}$	0.772	0.683	**0.695**

$$t_k^{mean1} = \frac{\min(\mathscr{P}_k^{true}) + \max(\mathscr{P}_k^{others})}{2}, \tag{14}$$

$$t_k^{mean2} = \frac{\text{mean}(\mathscr{P}_k^{true}) + \text{mean}(\mathscr{P}_k^{others})}{2}, \tag{15}$$

where $\text{mean}(x)$ denotes the mean value of set x, $\min(x)$ minimal value and $\max(x)$ maximal value of x. Now, (8) can be modified for *label-wise thresholding*:

$$\mathscr{L}_d^{rel} = \{l_k \in \mathscr{L} : P_d(k) \geq t_k\}, \tag{16}$$

where t_k can be obtained by any of (12), (13), (14), (15).

Sample-wise thresholding: the threshold t_d for each document d is obtained from a linear regressor $\mathscr{R}$ trained from $P_d^{train}(k)$. Target values for each document $d \in \mathscr{D}^{train}$ were set in the middle of mean probability of document's true labels and mean probability belonging to an irrelevant labels (i.e. in the spirit of (15) but in a sample-wise manner).

After the regressor $\mathscr{R}$ is trained, the probability distribution over topics $P_d(k)$, $k = 1, ..., K$ for document d can be used as an input of $\mathscr{R}$, then threshold $t_d^{\mathscr{R}}$ is returned and (8) can be modified for *sample-wise thresholding*:

$$\mathscr{L}_d^{rel} = \{l_k \in \mathscr{L} : P_d(k) \geq t_d^{\mathscr{R}}\}. \tag{17}$$

We also tried sorting $P_d^{train}(k)$ for each $d \in \mathscr{D}^{train}$ before training the regressor, i.e. we didn't care which label is relevant for the document d , but we rather trained the regressor from differences between successive probabilities. We denote these thresholds as $t_d^{\mathscr{R}sort}$.

7 Conclusion and Future Work

The results of this paper are summarized in Tab. 1. From label-wise thresholds best performed t_k^{mean2} (15), however, the best F_1 score was achieved when generating threshold for each document using regressor with sorted probabilities on input.

As we can see from the table, we can improve (in terms of F_1 score) the performance of our baseline multi-label document classifier by 25% relatively when using linear SVC classifier with *tfs-idf* vector space model, and another 6.1% relatively when using linear regressor to obtain thresholds, which together makes the improvement over baseline 32.6% relatively.

As it seems we are reaching the upper bound where we can go with classical linear classifiers, we'd like to try also neural networks, especially convolutional neural networks, which have recently became very popular in the field of image categorization. It would also be isteresting to compare our document classifier with winning classifiers of recent WISE 2014 challenge [6], where a lot of novel approaches have been introduced.

Acknowledgments. This research was supported by the Grant Agency of the Czech Republic, project No. GAČR GBP103/12/G084. Access to computing and storage facilities owned by parties and projects contributing to the National Grid Infrastructure MetaCentrum, provided under the programme "Projects of Large Infrastructure for Research, Development, and Innovations" (LM2010005), is greatly appreciated.

References

1. Yang, Y.: A study of thresholding strategies for text categorization. In: Proceedings of the 24th Annual International ACM SIGIR Conference on Research and Development in Information Retrieval, SIGIR 2001, pp. 137–145. ACM, New York (2001)
2. Montejo-Ráez, A., Ureña-López, L.A.: Selection strategies for multi-label text categorization. In: Salakoski, T., Ginter, F., Pyysalo, S., Pahikkala, T. (eds.) FinTAL 2006. LNCS (LNAI), vol. 4139, pp. 585–592. Springer, Heidelberg (2006)
3. Largeron, C., Moulin, C., Géry, M.: MCut: a thresholding strategy for multi-label classification. In: Hollmén, J., Klawonn, F., Tucker, A. (eds.) IDA 2012. LNCS, vol. 7619, pp. 172–183. Springer, Heidelberg (2012)
4. Fan, R.E., Lin, C.J.: A study on threshold selection for multi-label classification. National Taiwan University, Department of Computer Science, pp. 1–23 (2007)
5. Skorkovská, L.: Dynamic threshold selection method for multi-label newspaper topic identification. In: Habernal, I. (ed.) TSD 2013. LNCS, vol. 8082, pp. 209–216. Springer, Heidelberg (2013)
6. Tsoumakas, G., Papadopoulos, A., Qian, W., Vologiannidis, S., D'yakonov, A., Puurula, A., Read, J., Švec, J., Semenov, S.: WISE 2014 challenge: multi-label classification of print media articles to topics. In: Benatallah, B., Bestavros, A., Manolopoulos, Y., Vakali, A., Zhang, Y. (eds.) WISE 2014, Part II. LNCS, vol. 8787, pp. 541–548. Springer, Heidelberg (2014)
7. Švec, J., Hoidekr, J., Soutner, D., Vavruška, J.: Web text data mining for building large scale language modelling corpus. In: Habernal, I., Matoušek, V. (eds.) TSD 2011. LNCS, vol. 6836, pp. 356–363. Springer, Heidelberg (2011)
8. Skorkovská, L., Ircing, P., Pražák, A., Lehečka, J.: Automatic topic identification for large scale language modeling data filtering. In: Habernal, I., Matoušek, V. (eds.) TSD 2011. LNCS, vol. 6836, pp. 64–71. Springer, Heidelberg (2011)
9. Fan, R.E., Chang, K.W., Hsieh, C.J., Wang, X.R., Lin, C.J.: Liblinear: A library for large linear classification. J. Mach. Learn. Res. 9, 1871–1874 (2008)

10. Crammer, K., Dekel, O., Keshet, J., Shalev-Shwartz, S., Singer, Y.: Online passive-aggressive algorithms. J. Mach. Learn. Res. **7**, 551–585 (2006)
11. Pedregosa, F., Varoquaux, G., Gramfort, A., Michel, V., Thirion, B., Grisel, O., Blondel, M., Prettenhofer, P., Weiss, R., Dubourg, V., Vanderplas, J., Passos, A., Cournapeau, D., Brucher, M., Perrot, M., Duchesnay, E.: Scikit-learn: Machine learning in Python. Journal of Machine Learning Research **12**, 2825–2830 (2011)

Score Normalization Methods for Relevant Documents Selection for Blind Relevance Feedback in Speech Information Retrieval

Lucie Skorkovská[(✉)]

Faculty of Applied Sciences, New Technologies for the Information
Society and Department of Cybernetics, University of West Bohemia,
Univerzitní 8, 306 14 Plzeň, Czech Republic
lskorkov@ntis.zcu.cz

Abstract. This paper aims at the automatic selection of the relevant
documents for the blind relevance feedback method in speech information retrieval. Usually the relevant documents are selected only by simply determining the first N documents to be relevant. On the contrary,
the previous first experiments with the automatic selection of the relevant documents for the blind relevance feedback method has shown
the possibilities of the dynamical selection of the relevant documents
for each query depending on the content of the retrieved documents
instead of just blindly defining the number of the relevant documents to
be used in advance. In the first experiments, the World Model Normalization method was used. Based on the promising results, the experiments
presented in this paper try to thoroughly examine the possibilities of
the application of different score normalization techniques used in the
speaker identification task, which was successfully used in the related
task of multi-label classification for finding the "correct" topics of a newspaper article in the output of a generative classifier.

Keywords: Query expansion · Blind relevance feedback · Spoken
document retrieval · Score normalization

1 Introduction

The field of information retrieval (IR) has received a significant attention in the
past years, especially since large audio-visual databases are available on-line, the
research in the field of information retrieval extends to the retrieval of speech
content. Experiments performed on the speech retrieval collections containing
conversational speech [1] suggest that classic information retrieval methods alone
are not sufficient enough for successful speech retrieval. The biggest issue here is
that the query words are often not found in the documents from the collection.
One cause of this problem is the high word error rate of the automatic speech
recognition (ASR) causing the query words to be misrecognized. The second
cause is that the query words was actually not spoken in the recordings and

© Springer International Publishing Switzerland 2015
P. Král and V. Matoušek (Eds.): TSD 2015, LNAI 9302, pp. 316–324, 2015.
DOI: 10.1007/978-3-319-24033-6_36

thus are not contained in the documents. To deal with this issue the query expansion techniques are often used.

One of the favorite query expansion methods often used in the IR field is the relevance feedback method. The idea is to take the information from the relevant documents retrieved in the first run of the retrieval and use it to enhance the query with some new terms for the second run of the retrieval. The selection of the relevant documents can be done either by the user of the system or automatically without the human interaction - the method is then usually called blind relevance feedback. The automatic selection is usually handled only by selecting the first N retrieved documents, which are considered to be relevant.

In this paper we will present the thorough experiments aimed at the better automatic selection of the relevant documents for the blind relevance feedback method. Our idea is to apply the score normalization techniques used in the speaker identification task [2][3] to dynamically select the relevant documents for each query depending on the content of the retrieved documents instead of just experimentally defining the number of the relevant documents to be used for the blind relevance feedback in advance.

2 Information Retrieval System

For all experiments the language modeling approach [4] was used as the information retrieval method. The collection is described in Section 4.1.

2.1 Query Likelihood Model

The query likelihood model with a linear interpolation of the unigram language model of the document with an unigram language model of the whole collection was used. The idea of this method is to create a language model M_d from each document d and then for each query q to find the model which most likely generated that query, that means to rank the documents according to the probability $P(d|q)$. The Bayes rule is used:

$$P(d|q) = P(q|d)P(d)/P(q), \tag{1}$$

where $P(q)$ is the same for all documents and the prior document probability $P(d)$ is uniform across all documents, so both can be ignored. We have left the probability of the query been generated by a document model $P(q|M_d)$, which can be estimated using the maximum likelihood estimate (MLE):

$$\hat{P}(q|M_d) = \prod_{t \in q} \frac{tf_{t,d}}{L_d}, \tag{2}$$

where $tf_{t,d}$ is the frequency of the term t in d and L_d is the total number of tokens in d. To deal with the sparse data for the generation of the M_d we have used the mixture model between the document-specific multinomial distribution and the multinomial distribution of the whole collection M_c with interpolation parameter λ. So the final equation for ranking the documents according to the query is:

$$P(d|q) \propto \prod_{t \in q}(\lambda P(t|M_d) + (1 - \lambda)P(t|M_c)). \qquad (3)$$

The retrieval performance of this IR model can differ for various levels of interpolation, therefore the λ parameter was set according to the experiments presented in [5] to the best results yielding value - $\lambda = 0.1$.

2.2 Blind Relevance Feedback

Query expansion techniques based on the blind relevance feedback (BRF) method has been shown to improve the results of the information retrieval. The idea behind the blind relevance feedback is that amongst the top retrieved documents most of them are relevant to the query and the information contained in them can be used to enhance the query for acquiring better retrieval results.

First, the initial retrieval run is performed, documents are ranked according to the query likelihood computed by (3). Then the top N documents are selected as relevant and the top k terms (according to some importance weight L_t, for example *tf-idf*) from them is extracted and used to enhance the query. The second retrieval run is then performed with the expanded query.

Since we are using the language modeling approach to the information retrieval, for the terms selection we have used the importance weight defined in [4]:

$$L_t = \sum_{d \in R} \log \frac{P(t|M_d)}{P(t|M_c)}, \qquad (4)$$

where R is the set of relevant documents. In the standard approach to the blind relevance feedback the number of documents and terms is defined experimentally in advance the same for all queries. In our experiments we would like to find the number of relevant documents for each query automatically by selecting the "true" relevant documents for each query to dynamically set the number of top retrieved documents to be used in BRF.

3   Score Normalization for Relevant Documents Selection

The score normalization methods from the open-set text-independent speaker identification (OSTI-SI) problem were successfully used in the task of the multi-label classification to select the relevant topics for each newspaper article [6] in the output of a generative classifier. This is the same problem as in the information retrieval task, where as the result only the ranked list of documents with their likelihoods is returned. When the search is done the user of the retrieval system will look though the top N documents and therefore the specific selection of which document is relevant and which not is not needed. On the contrary, when the blind relevance feedback is used, the selection of the true relevant documents can be very useful.

This problem is quite similar to the OSTI-SI problem. The speaker identification can be described as a twofold problem: First, the speaker model best matching the utterance has to be found and secondly, it has to be decided, if the utterance has really been produced by this best-matching model - find out if the speaker is truly "relevant". The relevant documents selection can be described in the same way: First, we need to retrieve the documents which have the best likelihood scores for the query and second, we have to choose only the relevant documents which really generated the query. The only difference is that we try to find more than one relevant document. The normalization methods from OSTI-SI can be used in the same way, but have to be applied to all documents likelihoods.

3.1   Score Normalization Methods

After the initial run of the retrieval system described in Section 2 we have the ranked list of documents with their likelihoods $p(d|q)$ computed by equation (3). We have to find the threshold for the selection of the relevant documents. A score normalization methods have been used to tackle the problem of the compensation for the distortions in the utterances in the second phase of the open-set text-independent speaker identification problem [2]. In the IR task, the likelihood score of a document is dependent on the content of the query, therefore the beforehand set number of relevant documents is not suitable. Similarly as in the OSTI-SI [2] we can define the decision formula:

$$p(d_R|q) > p(d_I|q) \rightarrow q \in d_R \quad \text{else} \quad q \in d_I, \tag{5}$$

where $p(d_R|q)$ is the score given by the relevant document model d_R and $p(d_I|q)$ is the score given by the irrelevant document model d_I. By the application of the Bayes' theorem, formula (5) can be rewritten as:

$$\frac{p(q|d_R)}{p(q|d_I)} > \frac{P(d_I)}{P(d_R)} \rightarrow q \in d_R \quad \text{else} \quad q \in d_I, \tag{6}$$

where $l(q) = \frac{p(q|d_R)}{p(q|d_I)}$ is the normalized likelihood score and $\theta = \frac{P(d_I)}{P(d_R)}$ is a threshold that has to be determined. Setting this threshold θ a priori is a difficult task, since we do not know the prior probabilities $P(d_I)$ and $P(d_R)$. Similarly as in the OSTI-SI task the document set can be open - a query belonging to a document not contained in our set can easily occur. A frequently used form to represent the normalization process [2] can therefore be modified for the IR task:

$$L(q) = \log p(q|d_R) - \log p(q|d_I), \tag{7}$$

where $p(q|d_R)$ is the score given by the relevant document and $p(q|d_I)$ is the score given by the irrelevant document. Since the normalization score $\log p(q|d_I)$ of an irrelevant document is not known, there are several possibilities how to approximate it:

World Model Normalization (WMN). The unknown model d_I can be approximated by the collection model M_c which was created as a language model from all documents in the retrieval collection. This technique was inspired by the World Model normalization [7]. The WMN method was used in the first experiments with the score normalization methods application in the blind relevance feedback method [8]. The normalization score of a model d_I is defined as:

$$\log p(q|d_I) = \log p(q|M_c). \tag{8}$$

Unconstrained Cohort Normalization (UCN). For every document model a set (cohort) of N similar models $C = \{d_1, ..., d_N\}$ is chosen [9]. These models in the set C are the most competitive models with the reference document model, i.e. models which yield the next N highest likelihood scores. The normalization score is given by:

$$\log p(q|d_I) = \log p(q|d_{UCN}) = \frac{1}{N} \sum_{n=1}^{N} \log p(q|d_n). \tag{9}$$

Standardizing a Score Distribution. Another solution called Test normalization (T-norm) stated in [9] is to transform a score distribution, resulting from a different test conditions, into a standard form. The formula (7) now has the form:

$$L(q) = (\log p(q|d_R) - \mu(q))/\sigma(q), \tag{10}$$

where $\mu(q)$ and $\sigma(q)$ are the mean and standard deviation of the whole document likelihood distribution. This approach has similarities to WMN, the main difference here is the use of the standard deviation of the distribution.

Threshold Selection. Even when we have the likelihood scores normalized, we still have to set the threshold for verifying the relevance of each document in the list. Selecting a threshold defining the boundary between the relevant and the irrelevant documents in a list of normalized likelihood is more robust, because the normalization removes the influence of the various query characteristics. Since in the former experiments [8] the threshold was successfully defined as a percentage of the normalized score of the best scoring document, the threshold θ will be similarly defined as the ratio k of the best normalized score. A thorough analysis of different parameters settings is presented in Section 4.3 and the dependency between the threshold ratio k setting and the size of the set C for the UCN method is examined.

4   Experiments

Since the first experiments [8] with the use of the World model score normalization method for the selection of relevant documents for the blind relevance feedback has shown promising results, in-depth experiments have been performed.

First, we have done thorough experiments with the setting of the standard blind relevance feedback method - the selection of the number of documents and the number of terms. We have found the best parameters settings and selected it for our baseline. Then detailed experiments with the WMN method were performed. Since in the first experiments only few parameters settings have been tried out, in this paper we present thorough experiments aimed to find out the best WMN parameters settings. Since the WMN method achieved very promising result, we have also experimented with other methods from the speaker identification area - UCN and T-norm method. The UCN method has been shown to achieve even better results than the WMN method in the application on the multi-label classification task [10]. The results of all the experiments are presented in Section 4.3.

4.1 Information Retrieval Collection

The experiments were performed on the spoken document retrieval collection used in the Czech task of the Cross-Language Speech Retrieval track organized in the CLEF 2007 evaluation campaign [1]. The collection contains automatically transcribed spontaneous interviews (segmented by a fixed-size window over the transcribed text into 22 581 "documents") and two sets of TREC-like topics - 29 training and 42 evaluation topics. Each topic consists of 3 fields - <title> (T), <desc> (D) and <narr> (N).

The training topic set was used for our experiments and the queries were created from all terms from the fields T, D and N. Stop words were omitted from all sets of query terms. All the terms were also lemmatized, since lemmatization was shown to improve the effectiveness of information retrieval in highly inflected languages (as is the Czech language) [11][12][13]. For the lemmatization an automatically trained lemmatizer described in [5][14] was used.

4.2 Evaluation Metrics

The mean Generalized Average Precision (mGAP) measure that was used in the CLEF 2007 Czech task was used as an evaluation measure. The measure (described in detail in [15]) is designed for the evaluation of the retrieval performance on the conversational speech data, where the topic shifts in the conversation are not separated as documents. The mGAP measure is based on the evaluation of the precision of finding the correct beginning of the relevant part of the data.

4.3 Results

Number of Terms. In the first experiments [8] the number of terms was selected, according to the settings used for BRF in the paper dealing with the experiments on this collection [13], to 5 terms. Later experiments in the paper [8] has shown that the more terms used the better the retrieval score. We have done experiments with the number of terms to select with all the described methods in this paper. The number of terms was selected from 5 to 45 terms, with 5 term interval (5, 10, 15...). The premise that more terms are always better has shown not to be true. All methods have shown best results when selecting around 20-30 terms. For comparison of the following methods all of them are presented when selecting the 30 best terms (according to the weight (4)).

Standard BRF. For the standard blind relevance feedback the number of documents is defined beforehand, it is the same for each query in the set. We have experimented with the number of documents to select equal to 5, 10, 20, 30, 40, 50 and 100. Best results were achieved with 20 documents, almost the same with 30 documents. When selecting more documents the mGAP score dropped down, only when selecting also more terms the score was almost as with 20 documents. The less documents selected the lower was the score. For the baseline the selection of 20 documents was used.

Score Normalization. In score normalization methods, the number of documents to select for the BRF is dependent on the threshold θ defined as the ratio k of the best normalized score. The final number of documents selected this way is different for each query in the set. The experiments with the different ratio setting (from 0.1 to 0.95 with 0.05 distance) were done for all the score normalization methods presented.

For the **WMN method** this is the only parameter to chose. The best results were obtained with $k = 0.5$. The setting is not very sensitive, the results for $k = 0.55$ or $k = 0.45$ was almost the same. The promising results of the application of the WMN method led us to the idea to try other score normalization methods, especially the UCN, which was shown to achieve even better results either in the area of speaker identification [9] and multi-label classification [10].

In the **UCN method** apart from the ratio k also the size C of the cohort has to be set. Experiments with C from 5 to 125 with distance 10 were performed. The ratio k and the cohort size C depends on each other directly, because the normalization score in (9) is bigger (an average from the higher likelihoods) for a smaller cohort size. The best setting was $k = 0.25$ with $C = 85$.

The **T-norm method** was also experimented with, the best results were obtained with $k = 0.55$. Again as with the WMN method the setting is not very sensitive, the results for $k = 0.6$ or $k = 0.5$ was almost the same.

The final comparison can be seen in Table 1. As can be seen from the table the BRF methods in all cases achieved better score than without BRF. All the score normalization methods achieved better mGAP score than the standard BRF, the best score achieved the UCN method but it was very close to the WMN method.

Statistical relevance tests across the queries in the set was done for the verification of our claims. First we claim that the use of the BRF method achieves better results than without it. The difference has shown to be statistically significant (with the significance level $\alpha = 0.05$ for standard BRF and $\alpha = 0.01$ for

Table 1. IR results (mGAP score) for no blind relevance feedback, with standard BRF and BRF with score normalization. 30 terms were used to enhance each query in all cases.

query set / method parameters	no BRF -	standard BRF # of doc.= 20	BRF - WMN k=0.5	BRF - UCN k=0.25, C=85	BRF - T-norm k=0.55
train TDN	0.0392	0.0513	0.0568	0.0570	0.0564

score normalization BRF). Then we claim that with UCN method the results are better than with standard BRF. The difference has shown to be statistically significant with the significance level $\alpha = 0.05$. The Wilcoxon Matched-Pairs Signed-Ranks Test was used for all tests.

5   Conclusions

This article has shown the experiments with the use of the score normalization methods for selection of the relevant documents for the blind relevance feedback in speech information retrieval. The results are showing that with the score normalization better retrieval results can be achieved than with the standard blind relevance feedback with the number of relevant documents set beforehand. We have also confirmed that the blind relevance feedback in any form is very useful in the speech information retrieval.

The retrieval results are for each query the best with different number of documents used (because the number of truly relevant documents is different for each query). In the standard BRF the number of relevant documents is set the same for all the queries, therefore the mean results for the set of queries can not be the best which can be achieved. The use of score normalization methods for the automatic dynamic selection of relevant documents for each query independently solves this problem.

The number of terms to be selected was chosen the same for all queries. We have experimented with different number of terms to be selected, the number of terms significantly affects the retrieval results. The experiments on how to select this number automatically - different for each query will be the subject of our future research.

Acknowledgments. The work was supported by the Ministry of Education, Youth and Sports of the Czech Republic project No. LM2010013 and by the University of West Bohemia, project No. SGS-2013-032.

References

1. Ircing, P., Pecina, P., Oard, D.W., Wang, J., White, R.W., Hoidekr, J.: Information retrieval test collection for searching spontaneous Czech speech. In: Matoušek, V., Mautner, P. (eds.) TSD 2007. LNCS (LNAI), vol. 4629, pp. 439–446. Springer, Heidelberg (2007)

2. Sivakumaran, P., Fortuna, J., Ariyaeeinia, M.A.: Score normalisation applied to open-set, text-independent speaker identification. In: Proceedings of Eurospeech, Geneva, pp. 2669–2672 (2003)
3. Zajíc, Z., Machlica, L., Padrta, A., Vaněk, J., Radová, V.: An expert system in speaker verification task. In: Proceedings of Interspeech, vol. 9, pp. 355–358. International Speech Communication Association, Brisbane (2008)
4. Ponte, J.M., Croft, W.B.: A language modeling approach to information retrieval. In: Proceedings of SIGIR 1998, pp. 275–281. ACM, New York (1998)
5. Kanis, J., Skorkovská, L.: Comparison of different lemmatization approaches through the means of information retrieval performance. In: Sojka, P., Horák, A., Kopeček, I., Pala, K. (eds.) TSD 2010. LNCS, vol. 6231, pp. 93–100. Springer, Heidelberg (2010)
6. Skorkovská, L.: Dynamic threshold selection method for multi-label newspaper topic identification. In: Habernal, I. (ed.) TSD 2013. LNCS, vol. 8082, pp. 209–216. Springer, Heidelberg (2013)
7. Reynolds, D.A., Quatieri, T.F., Dunn, R.B.: Speaker verification using adapted gaussian mixture models. In: Digital Signal Processing 2000 (2000)
8. Skorkovská, L.: First experiments with relevant documents selection for blind relevance feedback in spoken document retrieval. In: Ronzhin, A., Potapova, R., Delic, V. (eds.) SPECOM 2014. LNCS, vol. 8773, pp. 235–242. Springer, Heidelberg (2014)
9. Auckenthaler, R., Carey, M., Lloyd-Thomas, H.: Score normalization for text-independent speaker verification systems. Digital Signal Processing 10(1–3), 42–54 (2000)
10. Skorkovská, L., Zajíc, Z.: Score normalization methods applied to topic identification. In: Sojka, P., Horák, A., Kopeček, I., Pala, K. (eds.) TSD 2014. LNCS, vol. 8655, pp. 133–140. Springer, Heidelberg (2014)
11. Ircing, P., Müller, L.: Benefit of proper language processing for Czech speech retrieval in the CL-SR Task at CLEF 2006. In: Peters, C., Clough, P., Gey, F.C., Karlgren, J., Magnini, B., Oard, D.W., de Rijke, M., Stempfhuber, M. (eds.) CLEF 2006. LNCS, vol. 4730, pp. 759–765. Springer, Heidelberg (2007)
12. Psutka, J., Švec, J., Psutka, J.V., Vaněk, J., Pražák, A., Šmídl, L., Ircing, P.: System for fast lexical and phonetic spoken term detection in a czech cultural heritage archive. EURASIP J. Audio, Speech and Music Processing 2011 (2011)
13. Ircing, P., Psutka, J.V., Vavruška, J.: What can and cannot be found in Czech spontaneous speech using document-oriented IR methods — UWB at CLEF 2007 CL-SR track. In: Peters, C., Jijkoun, V., Mandl, T., Müller, H., Oard, D.W., Peñas, A., Petras, V., Santos, D. (eds.) CLEF 2007. LNCS, vol. 5152, pp. 712–718. Springer, Heidelberg (2008)
14. Kanis, J., Müller, L.: Automatic lemmatizer construction with focus on OOV words lemmatization. In: Matoušek, V., Mautner, P., Pavelka, T. (eds.) TSD 2005. LNCS (LNAI), vol. 3658, pp. 132–139. Springer, Heidelberg (2005)
15. Liu, B., Oard, D.W.: One-sided measures for evaluating ranked retrieval effectiveness with spontaneous conversational speech. In: Proceedings of ACM SIGIR 2006, SIGIR 2006, pp. 673–674. ACM, New York (2006)

Imbalanced Text Categorization Based on Positive and Negative Term Weighting Approach

Behzad Naderalvojoud(✉), Ebru Akcapinar Sezer, and Alaettin Ucan

Computer Engineering Department, Hacettepe University, Ankara, Turkey
{n.behzad,ebru,aucan}@hacettepe.edu.tr,
http://humir.cs.hacettepe.edu.tr

Abstract. Although term weighting approach is typically used to improve the performance of text classification, this approach may not provide consistent results while imbalanced data distribution is available. This paper presents a probability based term weighting approach which addresses the different aspects of class imbalance problem in text classification. In this approach, we proposed two term evaluation functions called as PNF and PNF^2 which can produce more influential weights by relying on the imbalanced data sets. These functions can determine the significance of a term in association with a particular category. This is a crucial point because in one hand a frequent term is more important than a rare term in a particular category according to feature selection approach, and on the other hand a rare term is no less important than a frequent term based on *idf* assumption of traditional term weighting approach. Incorporation of these two approaches at the same time is the main idea that make them superior to other weighting methods. The achieved results from experiments which were carried out on two popular benchmarks (Reuters-21578 and WebKB) demonstrate that the probability based term weighting approach yields more consistent results than the other methods on the imbalanced data sets.

Keywords: Text classification · Class imbalance problem · Term weighting approach · Machine learning

1 Introduction

In text classification, class imbalance problem typically occurs when the number of documents of some classes is higher than the numbers of the others. In the imbalanced datasets, classes containing more number of instances are known as major classes while the ones having relatively less number of instances are called as minor classes. At this point, most of standard classifiers tend towards major classes and consequently show poorly performance on the minor classes. In other words, there should be as many examples belonging to major classes as examples belonging to minor ones [1,2]. This fundamental requirement cannot be always

© Springer International Publishing Switzerland 2015
P. Král and V. Matoušek (Eds.): TSD 2015, LNAI 9302, pp. 325–333, 2015.
DOI: 10.1007/978-3-319-24033-6_37

met and standard applications of machine learning algorithms may not provide satisfactory results. One of the effective approaches to resolve this problem which is also useful in text mining, is *term weighting strategy* via *tfidf* method [3]. *Tfidf* weighting is used to express how much a term can be important in a certain document while documents are represented in the Vector Space Model (VSM). In text classification, VSM is used to represent documents as term vectors and *tfidf* as a traditional term weighting scheme provides an influential solution for classification of imbalanced texts in many studies [4,5]. Debole and Sebastiani [6] proposed a number of supervised variant of *tfidf* weighting by replacing *idf* with feature selection metrics and presented a category based weighting scheme for classification task. In the other study [7] the supervised term weighting, *tf.rf*, was proposed based on distribution of relevant documents. The *rf* metric indicates the relevance level of a term with respect to a category. They evaluated *tf.rf* weighting scheme using SVM and kNN algorithms over different corpora and showed it consistently preforms well. In [8] a probability based term weighting scheme which can better distinguish documents in a minor category was introduced. In another one, [4] addressed the feature selection process for solving the class imbalance problem and took into consideration the abilities and characteristics of various metrics for feature selection. They asserted that negative features make a positive influence on the classification performance. In a more recent study, [5] explored the feature selection policies in text categorization by using SVM classifier.

In this study, we tackle the class imbalance problem using a probability based weighting scheme for a better multi-class classification task. Actually, two category based functions named as PNF and PNF^2 are proposed as a global component of term weighting scheme. These functions are based on two probabilities of relevant documents distribution. PNF^2 is designed as a two-sided function which takes into account either positive or negative terms. By this way, it can indicate either the type of term relevancy or the strength of relevancy (or irrelevancy) with respect to a specific category. Conversely, PNF is known as one-sided version of PNF^2 which can only determine the power of relevancy. In fact, we can distinguish documents better either in minor or major categories by replacing *idf* with the proposed category based metrics.

2   Term Weighting Approach

To better distinguish documents in the VSM, the term weighting approach is applied to represent documents. At first, traditional methods inspired by information retrieval are used for the purpose of term weighting. Their basic assumptions can be listed as follows: (1) "multiple appearances of a term in a document are no less important than single appearance" (*tf* assumption); (2) "rare terms are no less important than frequent terms" (*idf* assumption); (3) "for the same quantity of term matching, long documents are no more important than short documents" (*normalization* assumption) [6]. *Tfidf* as a standard weighting scheme has been used in many studies [4,7–9], because it provides an effective solution for the classification of imbalanced texts by relying on these assumptions. It has been formulated in form of multiplying term frequency (tf) by inverse document frequency (idf). The common and normalized form of $tfidf$ weighting are shown in Eq. 1 [3,10]:

$$tfidf(t_i, d_j) = tf(t_i, d_j) \times \log(\tfrac{N}{N_{t_i}}) \qquad w_{i,j} = \frac{tfidf(t_i,d_j)}{\sqrt{\sum_{k=1}^{|T|} tfidf(t_k,d_j)^2}} \qquad (1)$$

where $tf(t_i, d_j)$ denotes the number of times that term t_i occurs in document d_j, N is the number of documents in the training set, N_{t_i} denotes the number of documents in which term t_i occurs at least once and $|T|$ denotes the number of unique terms. Actually, $tfidf$ method is constituted from local and global principles. The frequency of a term within a specific document (tf) provides the local principle in the term weighting scheme and inverse document frequency (idf) supplies the global principle. Even if tf is used as a term weighting scheme alone, it can perform well [3,7,10]. On the other hand, idf is considered as an unsupervised function since it does not take into account the category membership in documents.

In text classification, if labeled documents are available, the term weighting approach which uses the prior known information can be applicable and named as supervised term weighting [6]. In this approach, metrics used in the term selection phase are replaced by the idf function, because the aim of term selection phase is to associate terms with each category. In fact, supervised approach uses category based term selection metrics as global component of term weighting scheme. In this study, we use the popular term selection metrics employed in [7] for supervised term weighting scheme. These metrics are represented by information elements in Table 1 (please see Table 2 for 'a', 'b', 'c' and 'd').

Table 1. Employed metrics as the global component of term weighting scheme in the experiments

Metric name	Formula
Chi square (X^2)	$N \frac{(ad-bc)^2}{(a+c)(b+d)(a+b)(c+d)}$
Information gain (ig)	$\frac{a}{N} log \frac{aN}{(a+c)(a+b)} + \frac{b}{N} log \frac{bN}{(b+d)(a+b)} + \frac{c}{N} log \frac{cN}{(a+c)(c+d)} + \frac{d}{N} log \frac{dN}{(b+d)(c+d)}$
Odds ratio (or)	$log \frac{ad}{bc}$
Relevance frequency (rf)	$log(2 + \frac{a}{max(1,c)})$

3 Proposed Positive and Negative Based Term Weighting Scheme

In the supervised functions, a one-sided function like rf or or only takes relevant terms that appear mostly in the given category into consideration, whereas two-sided function like X^2 or ig takes into account the irrelevant terms that do not mostly appear in the given category, as well as relevant ones. In this study, a two-sided function (Eq. 2) is proposed for global component of term weighting scheme based on two probabilities of relevant documents; i.e. $P(t_i|C_j)$ which is known as the probability of documents from category C_j where term t_i occurs at least once and $P(t_i|\bar{C}_j)$ which is considered as the probability of documents not from category C_j where term t_i occurs at least once. The main idea is to specify the degree of being relevant or non-relevant for a term with respect to each category where the negative documents outnumber the positive ones. To achieve this, the difference between two probabilities is computed as shown in Eq. 2. In fact, if $P(t_i|C_j)$ is bigger than $P(t_i|\bar{C}_j)$, which basically indicates that term t_i is relevant to category C_j, then the term is labeled as a positive term associated with category C_j and otherwise is assumed as negative. By dividing the difference into the summation of two probabilities, the normalized values of weights are obtained and the weights are transformed to [-1, 1] interval.

In imbalanced cases, use of conditional probabilities plays an important role in the weighting process, because it creates a balanced situation between categories. It means that the document frequency for a certain term is computed based on its distribution over classes. Moreover, PNF^2 function can assign high weights to the terms, which are rare or frequent, according to their distributions in different classes. This is a crucial point because in one hand a frequent term is more important than a rare term in a particular category according to feature selection approaches, and on the other hand, a rare term is no less important than a frequent term based on idf assumption. We named the proposed function as PNF^2 which is the abbreviation of *Positive Negative Features* and power of 2 symbolizes that equation is designed as two-sided.

$$PNF^2(t_i, C_j) = \frac{P(t_i|C_j) - P(t_i|\bar{C}_j)}{P(t_i|C_j) + P(t_i|\bar{C}_j)} \tag{2}$$

To estimate the probabilities of Eq. 2, four information elements shown in Table 2 are used. In Table 2, C_j denotes the class corresponding to the jth category in the dataset; t_i is the ith term in the vocabulary set; $a_{i,j}$, $b_{i,j}$, $c_{i,j}$ and $d_{i,j}$ denote the document frequencies associated with the corresponding conditions. Therefore, the probabilities are calculated by using Eq. 3:

$$P(t_i|C_j) = \frac{a_{i,j}}{a_{i,j}+b_{i,j}} \qquad P(t_i|\bar{C}_j) = \frac{c_{i,j}}{c_{i,j}+d_{i,j}} \tag{3}$$

If PNF^2 is used as a global component of term weighting scheme, either positive or negative values are assigned to terms. When PNF^2 computes a

Table 2. Fundamental information elements which are used in feature selection functions

	Containing term t_i	Not containing term t_i
Belonging to class C_j	$a_{i,j}$	$b_{i,j}$
Not belonging to class C_j	$c_{i,j}$	$d_{i,j}$

negative value for a term, it shows not only the term is irrelevant for given category but also it has a negative effect for that category as much as its absolute value. To eliminate the negative effect, the one-sided form of PNF^2 (Eq. 4) is defined as another alternative for the global component of term weighting scheme. In fact, we transform PNF^2 to one-sided function abbreviated as PNF and compare it with the performance of PNF^2.

$$PNF = 1 + PNF^2 \tag{4}$$

PNF function does not produce any negative weights for terms and it assigns just low positive values to non-relevant terms instead of negative. Thus, PNF function does not transform the trend of weighting to the negative space. Since the weighting scheme is employed for only training data, this approach becomes plausible.

4   Empirical Observation of Term Weighting and Feature Selection Approaches

In this part, we try to make a comparative explanation by using a realistic example. First, the scores of terms in the *grain* category of Reuters dataset are calculated by using two popular feature selection metrics i.e. *ig*, X^2 and proposed PNF metrics; then the scores of terms are sorted in descending order to select top 4 terms of each metric. Actually, *grain* is a minor category with 41 documents and Table 3 lists a, c and *idf* values of the selected top 4 terms. At this point, we want to emphasize the differences between feature selection and term weighting approaches. Feature selection means the identification of more representative terms, and selected features should represent the most number of documents. As a result, they ignore rare terms. On the other hand, a term weighting scheme which uses *idf* as a global component, gives higher score to terms with low document frequency. As can be seen from Table 3, *idf* values of terms selected by PNF are higher than the *idf* values of other terms. The difference between term weighting and feature selection approaches can be obviously proven with c values. Although, most of terms selected by *ig* and X^2 metrics have high document frequency in non-*grain* categories (i.e. high c values), terms selected by PNF metric have 0 values for the c parameter. Since use of feature selection metrics for category based weighting purposes has been preferred in the previous studies [6–8], we have to evaluate our proposed metrics by comparing with them. The last point is that, proposed PNF metric has closer approach to *idf* than the others, but unlike *idf*, PNF is proposed for category based weighting.

5   Experiments

In this study, all experiments were conducted on two different benchmarks such as Reuters-21578 and WebKB. The Reuters-21578 dataset has been widely used in text classification researches as an imbalanced collection [6,8,9,11,12]. The R8

Table 3. The characteristics of top 4 terms selected by different manners for *grain* category in Reuters-21578 dataset

Terms	X^2			IG			PNF		
	a	c	idf	a	c	idf	a	c	idf
t_1	36	15	6.75	36	15	6.75	14	0	8.61
t_2	14	0	8.61	24	52	6.17	3	0	10.84
t_3	11	5	8.42	14	0	8.61	3	0	10.84
t_4	24	52	6.17	11	5	8.42	3	0	10.84

version of Reuters dataset which was used in the experiments [13], consists of two major categories called as *earn* and *acq* with almost 52% and 30% class distributions respectively and 6 minor categories with almost 3% class distributions. WebKB dataset consists of four categories of web pages collected from computer science departments of four universities[13]. This dataset contains two minor categories called as *project* and *course* with almost 10% and 20% class distributions respectively and two major categories with 30% and 40% class distributions. For both datasets, experiments were performed on the original training and test sets obtained from benchmarks [13]. Standard text preprocessing steps were applied and all features were used in classification.

To analyze the effect of different weighting methods on the imbalanced data classification problem, we need a simple classifier which makes term weighting scheme as the most effective factor in the learning process. In the proposed classifier algorithm which is inspired by Rocchio, after representing documents in VSM and applying a term weighting scheme, learning process is realized by combining training document vectors $\vec{d}$ into a vector $\vec{c}_j$ for each category. The vector $\vec{c}_j$ is computed for category C_j by dot dividing of two vectors as Eq. 5:

$$\vec{c}_j = \frac{1}{\vec{a}_j} \sum_{d \in C_j} \vec{d} \tag{5}$$

In Eq. 5 the $\vec{a}_j$ is the vector yielded from the document frequency of terms with respect to category C_j (as shown in Table 2) and $\sum_{d \in C_j} \vec{d}$ yields the summation of document vectors which belong to category C_j. Consequently, the set of $\vec{c}_j$ vectors which are computed for each category, represent the learned model. This model is used to classify document d^t which has never seen before. This test document is represented by the vector $\vec{d}^t$ which has only tf values as weights. In order to classify the test document, cosine similarity is computed between two vectors such as $\vec{d}^t$ and each of $\vec{c}_j$. Finally, the vector $\vec{d}^t$ is assigned to the category which has the highest similarity with $\vec{d}^t$ as indicated in Eq. 6.

$$F(\vec{d}^t) = \arg\max_{c_j \in C} \frac{\vec{c}_j}{||\vec{c}_j||} \cdot \frac{\vec{d}^t}{||\vec{d}^t||} \tag{6}$$

Precision (P), Recall (R) and F-measure were used to evaluate the performance of classification.

Table 4. The F-measure values of different term weighting schemes for Reuters-21578 dataset

Categories	The term weighting schemes						
	$tf.idf$	$tf.X^2$	$tf.ig$	$tf.or$	$tf.rf$	$tf.PNF^2$	$tf.PNF$
earn	0.771	0.512	0.845	0.945	**0.981**	0.950	**0.981**
acq	0.450	0.654	0.831	0.921	0.957	0.952	**0.961**
crude	0.698	0.896	0.887	0.867	0.902	0.835	**0.945**
trade	0.542	0.867	0.886	0.771	**0.906**	0.802	0.898
money-fix	0.646	0.789	0.781	0.798	0.719	0.834	**0.868**
interest	0.754	0.792	0.779	0.852	0.776	0.838	**0.881**
ship	0.539	0.831	0.679	0.781	0.806	0.794	**0.845**
grain	0.667	0.889	0.889	**0.900**	0.800	0.750	**0.900**
Macro Average	0.633	0.779	0.822	0.854	0.856	0.844	**0.910**
Micro average	0.687	0.639	0.836	0.912	0.945	0.925	**0.958**

Achieved F-measure values for the different weighting methods employed in Reuters-21578 benchmark are listed in Table 4. As can be seen, $tf.PNF$ term weighting method consistently outperforms all other methods for all categories except one case. The results obtained from $tf.PNF^2$ can be competitive with the other methods. The $tf.PNF$ weighting scheme, which eliminates the negative impact considered in $tf.PNF^2$, significantly improves the performance of the classification. The superiority of $tf.PNF$ scheme can also be seen by micro and macro averaged F-measure values. Another point is that the $tfidf$ weighting scheme cannot provide a good distinction between categories and consequently performs weakly on the whole categories.

In WebKB benchmark, the superiority of $tf.PNF^2$ and $tf.PNF$ can be observed among the other methods as shown in Table 5. Although the $tf.PNF$ is known as the best weighting scheme by possessing the highest micro and macro averaged F-measure values, $tf.PNF^2$ gives better results for minor categories. It can be also observed that the performance $tf.ig$, $tf.X^2$ and $tf.rf$ are degraded in contrast with their previous results on the Reuters benchmark and cannot keep their relative goodness. At this point, it can be said that they cannot perform well on different imbalanced circumstances and may not yield consistent results. Conversely, $tf.PNF$, $tf.PNF^2$ and $tf.or$ can provide more reliable results since they can make a relative minimum range of fluctuation in their results.

According to the achieved results from two benchmarks (Tables 4 and 5), the proposed two functions as a global component of term weighting scheme yield better results than the others. Moreover, the category based term weighting schemes outperform the traditional $tfidf$ in most cases. In other words, $tfidf$ cannot make any clear distinction between documents of the different classes in multi-class classification task. As mentioned in section 4, ig and X^2 are successful for feature selection task [5] but they cannot consistently perform well as a global component of term weighting scheme in imbalanced text classification.

Table 5. The F-measure values of different term weighting schemes for WebKB dataset

Categories	The term weighting schemes						
	$tf.idf$	$tf.X^2$	$tf.ig$	$tf.or$	$tf.rf$	$tf.PNF^2$	$tf.PNF$
student	0.636	0.587	0.588	0.636	0.735	0.705	**0.852**
faculty	0.372	0.236	0.224	0.688	0.673	0.750	**0.757**
course	0.608	0.014	0.006	0.859	0.662	**0.887**	0.860
project	0.088	0.403	0.424	**0.649**	0.443	**0.649**	0.617
Macro Average	0.426	0.310	0.311	0.708	0.628	0.747	**0.772**
Micro average	0.549	0.452	0.454	0.703	0.683	0.749	**0.805**

To determine the statistical significance of the results, we performed ANOVA test on the F-measure values gained by the methods for categories. According to results of ANOVA test for Reuters-21578 and WebKB benchmarks, since the P-value of the test is less than 0.05 for each case (P-value equals 0.0000 for Reuters and 0.0028 for WebKB), there are statistically significant differences between the macro-averaged F-measure values of $tf.PNF$ with the other different schemes at the 95.0% confidence level.

6   Conclusion

In this study, we tackled the class imbalance problem by category based term weighting approach and PNF^2 and PNF were proposed as a global component of term weighting scheme based on the probabilities of relevant documents frequency. Experiments were made with several methods on two different benchmarks. According to our findings, the $tf.PNF$ term weighting scheme is the best in all experiments and can provide the best tradeoff between precision and recall. Despite the wide range of fluctuation in the results of $tf.ig$ and $tf.X^2$, $tf.PNF^2$ as a two-sided method achieves more expectable results with high F-measure values. Additionally, one-sided functions (i.e. or, rf and PNF) consistently perform better than the two-sided ones (i.e. ig and X^2), however, PNF^2 presents competitive results in contrast with or, rf functions. As a result, the PNF and PNF^2 functions as a global component of term weighting scheme are recommended for imbalanced classification task.

References

1. Japkowicz, N., Stephen, S.: The class imbalance problem: A systematic study. Intelligent Data Analysis **6**(5), 429–449 (2002)
2. Chawla, N.V., Japkowicz, N., Kotcz, A.: Editorial: special issue on learning from imbalanced data sets. ACM Sigkdd Explorations Newsletter **6**(1), 1–6 (2004)
3. Salton, G., Buckley, C.: Term-weighting approaches in automatic text retrieval. Information Processing & Management **24**(5), 513–523 (1988)
4. Ogura, H., Amano, H., Kondo, M.: Comparison of metrics for feature selection in imbalanced text classification. Expert Systems with Applications **38**(5), 4978–4989 (2011)

5. Taşcı, Ş., Güngör, T.: Comparison of text feature selection policies and using an adaptive framework. Expert Systems with Applications **40**(12), 4871–4886 (2013)
6. Debole, F., Sebastiani, F.: Supervised term weighting for automated text categorization. In: Sirmakessis, S. (ed.) Text Mining and its Applications. STUDFUZZ, vol. 138, pp. 81–97. Springer, Heidelberg (2004)
7. Lan, M., Tan, C.L., Su, J., Lu, Y.: Supervised and traditional term weighting methods for automatic text categorization. IEEE Transactions on Pattern Analysis and Machine Intelligence **31**(4), 721–735 (2009)
8. Liu, Y., Loh, H.T., Sun, A.: Imbalanced text classification: A term weighting approach. Expert Systems with Applications **36**(1), 690–701 (2009)
9. Ren, F., Sohrab, M.G.: Class-indexing-based term weighting for automatic text classification. Information Sciences **236**, 109–125 (2013)
10. Sebastiani, F.: Machine learning in automated text categorization. ACM Computing Surveys (CSUR) **34**(1), 1–47 (2002)
11. Sun, A., Lim, E.P., Liu, Y.: On strategies for imbalanced text classification using svm: A comparative study. Decision Support Systems **48**(1), 191–201 (2009)
12. Erenel, Z., Altınçay, H.: Nonlinear transformation of term frequencies for term weighting in text categorization. Engineering Applications of Artificial Intelligence **25**(7), 1505–1514 (2012)
13. Cachopo, A.M.d.J.C.: Improving Methods for Single-label Text Categorization. PhD thesis, Universidade Técnica de Lisboa (2007)

CloudASR: Platform and Service

Ondřej Klejch, Ondřej Plátek, Lukáš Žilka$^{(\boxtimes)}$, and Filip Jurčíček

Faculty of Mathematics and Physics, Institute of Formal and Applied Linguistics,
Charles University in Prague, Malostranské náměstí 25,
118 00 Prague, Czech Republic
lukas@zilka.me

Abstract. CloudASR is a software platform and a public ASR web-service. Its three strong features are state-of-the-art online speech recognition performance, easy deployment, and scalability. Furthermore, it contains an annotation interface for the addition of transcriptions for the recordings. The platform API supports both batch and online speech recognition. The batch version is compatible with Google Speech API. New ASR engines can be added onto the platform and can work simultaneously.

Keywords: Automatic speech recognition · Cloud · Cloud service · Kaldi

1 Introduction

Recent advances in automatic speech recognition popularized voice interaction as a way of interacting with computers, which in turn increased demand for easy-to-use voice recognition systems that developers can use in their applications. Some examples of such applications are voice-controlled web browsers, email and SMS readers for car drivers, or voice assistants. Developers typically integrate voice recognizers into their application in two ways: Either they run voice recognition directly on the user's device, or they use an off-device voice recognizer to which the device connects through the Internet. In this paper, we are interested in the latter setup, in particular a voice recognition web-service in the cloud.

We present an open-source platform called CloudASR that gives everyone the possibility to run a customizable and scalable voice recognition service in her own cloud. By customizability, we primarily mean the ability to use custom recognition models[1]. Kaldi ASR engine [1] is supported out-of-the-box, but it is possible to plug in other recognizers like CMUSphinx [2] or RWTH [3] by implementing appropriate wrappers. In addition, we operate a public web service that runs on the presented platform with ready-to-use recognizers for English and Czech using the Kaldi ASR system modified for online recognition [4].

---

[1] A recognition model represents the parameters of an ASR system, and roughly said it determines its performance given a language and a domain.

P. Král and V. Matoušek (Eds.): TSD 2015, LNAI 9302, pp. 334–341, 2015.
DOI: 10.1007/978-3-319-24033-6_38

There are ASR web-services available on the Internet that provide some of the features described above, but none of them offer online recognition and full customization. Google Speech API[2] is very fast and accurate for general speech and supports a lot of languages, but is not customizable and has unclear licensing conditions. Nuance's[3] commercial recognizers are more flexible but they are paid and closed-source. ISpeech ASR API[4] is also customizable (users can select expected speech type: text messages, voice mail, dictation) but does not allow full customization and is also paid. AT&T Watson[5] allows users to provide a custom grammar (language model) but its free use is limited. In [5], Otosense-Kaldi, a promising software was mentioned, which would offer similar features as our platform, but its announced release date is already a year past due.

In the following, we give the general overview of the platform architecture (in Sect. 2), show how the platform scales with increased workload (Sect. 3), briefly describe the public ASR web-service (Sect. 4), and conlude the paper (in Sect. 5).

2   CloudASR Platform

The CloudASR platform consists of a Frontend server,[6] a Master server, and Worker servers that collaborate together to provide the ASR web service. The platform architecture was designed with stress on scalability, high availability, customizability, and maintainability, and adheres to the common cloud industry best practices. All of the components are described and their roles are illustrated in Figure 1 on a simple scenario. Each of the servers runs inside a Docker container[7]. Docker provides an isolated environment for the execution of the servers and an API for their deployment.

Clients communicate with the platform using either HTTP POST requests in the batch recognition mode, or WebSockets in the online recognition mode. Developers using Google Speech API can change their ASR backend to CloudASR just by modifying the web service URL because the APIs are compatible. Additionally, we implemented JavaScript library for the online recognition mode. An example how to use this library is shown in Figure 2.

The whole platform is covered by unit tests, which guarantees robustness with respect to future development.

2.1   Platform Scalability

The most computationally intensive part of the platform are the Workers, which perform the actual speech recognition, and the Frontends, which serve as a proxy

---

[2] https://www.google.com/intl/en/chrome/demos/speech.html
[3] http://dragonmobile.nuancemobiledeveloper.com/public/
[4] http://www.ispeech.org/api
[5] http://developer.att.com/apis/speech
[6] We use the term "server" here to denote a running process on a physical/virtual machine.
[7] https://docs.docker.com/

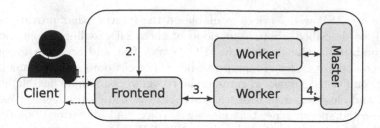

Fig. 1. The CloudASR service in a simple ASR scenario. The Client request is served in the following way: When a Client sends a request to CloudASR service (no. 1 on the picture), the request comes to the Frontend server, which asks Master for a particular Worker according to the requested recognition model (no. 2). Frontend then sends data to the Worker and waits for the recognition results (no. 3). Workers periodically send heartbeats to Master with information about their state (no. 4).

between the Client and the Workers and monitor the connection. For scalability, both of them can run in multiple instances on multiple machines. The only bottleneck may be the Master server, which manages the Workers. However, in our experiments it had no problem handling up to 1,000 Workers[8].

2.2 Platform High Availability

We implemented heart-beat protocols for mitigating consequences of any worker faults. Workers send periodic heart-beat signals to Master with their status. When Master does not receive an update from a Worker for 15 seconds, it assumes the Worker has failed and does not schedule any future tasks for it. The signle point of failure of the platform is the Master server. When the Master server dies the platform cannot process any requests, but as soon as it starts working again everything goes back to normal.

2.3 Platform Maintenance and Customizability

The platform can be run either on a single machine or in a Mesos cluster [6], using the provided set of scripts that start, restart, and shutdown the service. In the single-machine scenario, the only system requirement is a working Docker installation. For a Mesos cluster deployment, the cluster needs to support deployment of Docker containers.

A new type of ASR engine can be added easily, as illustrated in Figure 3, by implementing an ASR class and adding dependencies into the Docker configuration file.

---

[8] Looking at http://zeromq.org/results:10gbe-tests-v031, which benchmarks the performance of the ZeroMQ library we use for communication, we expect that it should be able to handle much more Workers.

```
var speechRecognition = new SpeechRecognition();
speechRecognition.onStart = function() {
    console.log("Recognition started");
}

speechRecognition.onEnd = function() {
    console.log("Recognition ended");
}

speechRecognition.onError = function(error) {
    console.log("Error occured: " + error);
}

speechRecognition.onResult = function(result) {
    console.log(result);
}

var lang = "en-wiki";
$("#button_start").click(function() {
    speechRecognition.start(lang);
});

$("#button_stop").click(function() {
    speechRecognition.stop()
});
```

Fig. 2. JavaScript code that can be used for speech recognition in Google Chrome.

2.4 Annotation Interface

All of the processed recordings are stored in a network-shared file system and all users can access them from the annotation interface shown in Figure 4. They can confirm the correctness of the recognition or provide their own transcriptions, along with additional information such as whether the recording contains speech or whether it was recorded by a native speaker. We also provide scripts for creating CrowdFlower[9] jobs to obtain the transcriptions from paid workers. We plan to publish all recordings from the public CloudASR web-service along with their transcriptions in a data set that can be used for training ASR systems.

3   Evaluation

We performed a set of experiments to show that the CloudASR platform is ready for production use. First, we compared the word error rate (WER) and real time factor (RTF) of CloudASR batch mode with Google Speech API on a specific

---

[9] http://www.crowdflower.com/

```
class DummyASR:
    def recognize_chunk(self, chunk):
        return (1.0, 'Dummy interim result')

    def get_final_hypothesis(self):
        return [(1.0, 'Dummy final result')]

    def reset(self):
        pass
```

Fig. 3. Illustration of the API for an ASR engine on CloudASR platform. Each ASR engine needs to be wrapped into a Python class with the listed methods. In this example, the recognizer does not perform any recognition, just provides dummy values as results for any input.

dataset to show that domain-specific models, which are CloudASR's primary designation, outperform the general ones and are faster. Second, we compared the latency of the CloudASR online mode with the batch mode to demonstrate the superior performance of online ASR for interactive speech applications. Finally, we stress-tested CloudASR in the batch mode to see how the platform handles high load.

3.1   Real Time Factor and Word Error Rate

We compare WER and RTF of CloudASR batch mode with Google Speech API to illustrate that we achieve similar speed and that indeed there is a big difference when we are given the capability to adapt the recognizer to a particular domain. As the CloudASR recognizer we used the test set from the Czech Public Transportation Information Domain [7], with the recognition models described in [4]. Google Speech API achieves 64% WER and an RTF of 0.33, while CloudASR scores 22% WER and an RTF of 0.17.

3.2   Online vs. Batch Mode Latency

Figure 5 shows the delay between the utterance end and the availability of the recognition results for CloudASR online and batch mode. We used the same dataset as in Sect. 3.1. Leveraging the time while the recording is still being made for computing the recognition hypotheses brings shorter waiting time for the final hypothesis. It can be seen that for online recognition, the recording length does not make a difference in the delay whereas for batch recognition latency increases proportionaly to the recording length.

3.3   Parallel Requests Benchmark

We tested how CloudASR behaves under the load of up to 1,000 parallel requests. Ideally we would run the tests on a cluster with thousands of nodes, but due

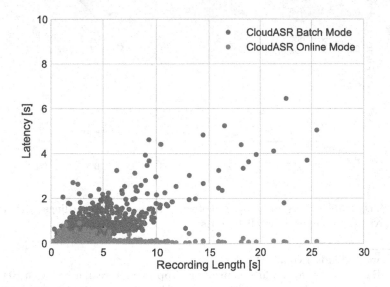

CloudASR   Demo   Transcribe   Documentation   View on GitHub   Login with Google+

CloudASR Transcribe Recordings

Instructions

Numerals

Please, do not use numbers to transcribe numerals; spell them out:

| 53 | NO |
| fifty three | YES |

Orthography

Please, try to avoid spelling errors. Punctuation and capitalisation generally do not matter, but please donot capitalise sentence-initial letters (unless they belong to an acronym, see the next sentence). The exception is with acronyms – please transcribe acronyms in uppercase letters.

Transcribe recording

▶ ●——————— 0:01 ◀)) ●——●

☐ It's not a speech
☐ Speaker is native
☐ Language is offensive

Save

Fig. 4. The CloudASR annotation interface where the users can provide transcription and other information about the recordings.

to lack of resources we resort to a simulation. We use dummy Workers which do not perform real recognition but only accept the input recording, wait half length of the recording, and send back a dummy result. This enables us to run multiple Worker instances on a single machine, and scale the number of Workers to 1,000. This approach is still a valid test of the communication capabilities of the platform and provides a meaningful picture of its performance in practice.

Fig. 5. A latency comparison for CloudASR online and batch mode.

We run 1,000 dummy Workers on a Mesos cluster with 5 physical machines (4 CPUs, 16 GB RAM each), and HAProxy [8] load-balancer which spread the workload across 5 Frontend servers. We tested the platform with a different number of parallel requests and recordings of different lengths. The results are plotted on Figure 6.

The results show that high load adds just a very little overhead until a point of 500 requests per second is reached. At that point, the Frontend servers need to handle more than 100 requests per second each, which corresponds to a transfer rate of 140 Mbps. From that point on, the Frontends are overloaded and the latency increases noticeably with the number of parallel requests. If more Frontend servers are used, the platform should be able to handle even more requests.

4   CloudASR Web-Service

CloudASR web service is publicly available on the Internet[10], running on the infrastructure provided by CESNET's Metacentrum project[11]. The service provides unlimited speech recognition in English and Czech with state-of-the-art KALDI recognition toolkit in return for the speech data sent to it. We plan to expand the service to support more languages if there is a demand for it.

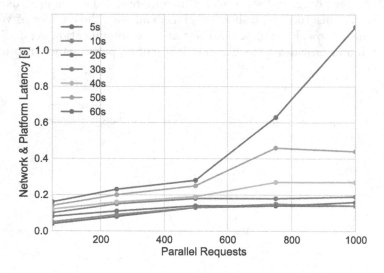

Fig. 6. The graph shows platform & network latency for recordings with various lengths given the number of parallel requests.

---

[10] http://www.cloudasr.com/

[11] CESNET is an non-for-profit organization supporting research in Czech Republic, and Metacentrum is its project with the aim of building a computational infrastructure for research purposes.

5   Conclusion and Future Work

We described the CloudASR platform and the public web service that it powers. We demonstrated how it achieves scalability, high availability, and customizability.

In the future, we would like to add an easy way for the users to upload their own language models. Also, we would like to use the transcribed recordings to adapt the acoustic models for specific recording conditions, if the users will provide enough of them, to improve the recognition performance.

Acknowledgments. This research was partly funded by the Ministry of Education, Youth and Sports of the Czech Republic under the grant agreement LK11221, core research funding, grant GAUK 2076214 and GAUK 1915/2015 of Charles University in Prague. This research was (partially) supported by SVV project number 260 224. This work has been using language resources distributed by the LINDAT/CLARIN project of the Ministry of Education, Youth and Sports of the Czech Republic (project LM2010013). Cloud computational resources were provided by the MetaCentrum under the program LM2010005 and the CERIT-SC under the program Centre CERIT Scientific Cloud, part of the Operational Program Research and Development for Innovations, Reg. no. CZ.1.05/3.2.00/ 08.0144.

References

1. Povey, D., Ghoshal, A., Boulianne, G., Burget, L., Glembek, O., Goel, N., Hannemann, M., Motlicek, P., Qian, Y., Schwarz, P., et al.: The Kaldi speech recognition toolkit. In: ASRU (2011)
2. Lamere, P., Kwok, P., Gouvea, E., Raj, B., Singh, R., Walker, W., Warmuth, M., Wolf, P.: The CMU SPHINX-4 speech recognition system. In: ICASSP (2003)
3. Rybach, D., Gollan, C., Heigold, G., Hoffmeister, B., Lööf, J., Schlüter, R., Ney, H.: The RWTH aachen university open source speech recognition system. In: Interspeech (2009)
4. Plátek, O., Jurčíček, F.: Free on-line speech recogniser based on Kaldi ASR toolkit producing word posterior lattices. In: SIGDIAL (2014)
5. Morbini, F., Audhkhasi, K., Sagae, K., Artstein, R., Can, D., Georgiou, P., Narayanan, S., Leuski, A., Traum, D.: Which ASR should I choose for my dialogue system? In: SIGDIAL (2013)
6. Hindman, B., Konwinski, A., Zaharia, M., Ghodsi, A., Joseph, A.D., Katz, R.H., Shenker, S., Stoica, I.: Mesos: a platform for fine-grained resource sharing in the data center. In: NSDI, vol. 11, pp. 22–22 (2011)
7. Korvas, M., Plátek, O., Dušek, O., Žilka, L., Jurčíček, F.: Vystadial 2013-czech data (2014). http://hdl.handle.net/11858/00-097C-0000-0023-4670-6
8. Tarreau, W.: Haproxy-the reliable, high-performance tcp/http load balancer (2012). http://haproxy.lwt.eu

Experimental Tagging of the ORAL Series Corpora: Insights on Using a Stochastic Tagger

David Lukeš[✉], Petra Klimešová, Zuzana Komrsková, and Marie Kopřivová

ICNC, Faculty of Arts, Charles University in Prague, Prague, Czech Republic
{lukes,klimesova,komrskova,kopriva}@korpus.cz,
http://www.korpus.cz

Abstract. The ORAL series corpora of spontaneous spoken Czech currently contain neither lemmatization nor part of speech tagging. The main reason for this is that readily available NLP tools, designed primarily with written texts in mind, underperform when applied directly to speech transcripts, due to various morpohological and syntactic specificities of informal spoken language and the ways these are captured in transcription. Recently, the highly optimized open-source MorphoDiTa toolchain for training and applying stochastic tagging models was released; MorphoDiTa makes it easy and fast to experiment with incremental changes in the training procedure. The article discusses modifications to the morphological dictionary and training data used by the models which are necessary in order to improve their performance on the ORAL series corpora, as well as challenges which remain to be solved.

Keywords: Spoken corpus · Lemmatization · POS-tagging

1 Introduction

For some types of linguistic research, having a corpus that is morphologically tagged and lemmatized is paramount, especially in the case of such a highly inflected language as Czech in which target lexemes take on many different forms according to context. This applies even more to corpora such as the ORAL series corpora of informal spoken Czech, whose transcription guidelines reflect the even greater regional and register-based variety of spontaneous spoken language. However, at present, no such additional annotation is available in either of the three aforementioned corpora, because the differences between spoken and written language are such that the existing tools for automatic linguistic annotation of texts, which are developed primarily with written language in mind, yield suboptimal results when directly applied to speech transcripts.

This paper resulted from the implementation of the Czech National Corpus project (LM2011023) funded by the Ministry of Education, Youth and Sports of the Czech Republic within the framework of Large Research, Development and Innovation Infrastructures.

© Springer International Publishing Switzerland 2015
P. Král and V. Matoušek (Eds.): TSD 2015, LNAI 9302, pp. 342–350, 2015.
DOI: 10.1007/978-3-319-24033-6_39

Recently, the MorphoDiTa (Morphological Dictionary and Tagger) framework was released as an end-to-end solution for training and applying POS-tagging models [1]. Compared to previous resources of the kind (Morce [2], Featurama,[1] COMPOST [3]), MorphoDiTa's aim is to be fully openly available and extremely resource-efficient while retaining excellent performance on the large tagsets typical of highly-inflected languages such as Czech. MorphoDiTa's speed and memory consumption optimizations in particular allow for more flexible, iterative workflows, which favour trial and error because the whole cycle (from modifying the setup through training to testing) takes several hours at most instead of days. These characteristics make it ideal for exploring new territory, such as addressing the issues of training a tagger which would perform reasonably well on spoken language transcripts in the absence of a hand-annotated gold standard to train on. Our goal was therefore to explore the MorphoDiTa toolchain and find ways in which it can be adapted to perform better when lemmatizing and tagging the ORAL series corpora. This paper reports our approach to the problem and assesses preliminary results based on ongoing work.

1.1 Speech Transcripts vs. Written-Text-Based NLP Tools

Two possible approaches come to mind when trying to apply NLP tools devised for written data to speech transcripts (irrespective of whether the tools are rule-based, stochastic or mixed; see e.g. [4,5] for a review of these methods):

1. adapt the transcript (see e.g. [6])
2. adapt the analysis tools (our approach)[2]

The first option might seem more alluring, because it allows the user to buy into the full stack of advanced NLP tools available for processing written language (syntactic parsing, semantic analysis, machine translation etc.). For some types of data, corresponding to the more formal and preferrably monologic region of the spectrum, this might constitute a useful technique, because they lend themselves more easily to such normalization. This approach can be particularly valuable in settings where a processing pipeline is already set up whose ulterior elements depend on such normalized input.

However, in general, perceiving spoken language utterances as "broken" surface forms which should be repaired and consequently mapped back onto the underlying "correct" syntax and morphology is not a helpful analogy. Such a perspective is especially misguided when the data is destined to be primarily used in linguistic research and consists of spontaneous, unscripted, informal speech. Aposiopeses, restarts, contaminations etc. are integral *features* of spoken

---

[1] See http://sourceforge.net/projects/featurama/.

[2] A third, stopgap solution is to use the analysis tools as they are in full awareness of their partial inappropriateness (see e.g. the DIALOG corpus [7], which consists of TV broadcast transcripts, i.e. speech that is considerably more "well-behaved" than the ORAL corpora data because of the formal or semi-formal setting).

language, not flaws which can be unambiguously eradicated, and should be analyzed and described as such. In particular, elaborate rule-based tagging systems
often rely on sentence and clause boundaries to infer relationships among words
and thereby disambiguate homonymous forms [8]. Yet these are notoriously hard
to establish in spontaneous speech, with syntax and prosody often giving conflicting cues, and may ultimately not even be a useful concept. Indeed, the bulk
of research to date suggests that the sentence is not an appropriate unit of analysis for spoken language; instead, an analysis in terms of individual clauses loosely
coupled into clause complexes is much more fruitful [9, Chap. 2].

In contrast, stochastic tagging tools do not rely on sentence units in any
fundamental way, they focus primarily on local windows of varying lengths[3] to
infer grammatical properties, which means they are significantly less sensitive
to whether the entire structure is syntactically "well-formed" or not. At least
in theory then, it should suffice to train a tagger on a representative sample
of manually tagged gold standard speech transcripts. Unfortunately, this would
require widespread prior agreement on the particulars of the transcription approach in order for the resulting tagger to be useful for more than one project,
or else it would be hard to justify the costs of this labour-intensive task. Additionally, spontaneous speech may not contain as many statistically salient purely
formal patterns of the kind that allow a stochastic tagger to operate, because
it exhibits higher syntactic irregularity and unpredictability (cf. [11] for a discussion of online performance/processing and the irreversibility of time as the
factors which decisively shape spoken language). To clarify: while spontaneous
speech is generally less lexically rich than writing and often uses ready-made
chunks as building blocks, the way these chunks are combined and interspersed
is much less formally constrained, because the speaker can rely on the hearers'
ability for contextual disambiguation and their understanding of the situation
as a whole, i.e. semantics and pragmatics.

1.2 ORAL Series Corpora

The ORAL series is a family of informal spontaneous spoken Czech corpora
devised and compiled at the Institute of the Czech National Corpus. It currently comprises the following installments: ORAL2006 [12], ORAL2008 [13]
and ORAL2013 [14], totalling about 4.8M tokens spanning the entire territory
of the Czech Republic.

Their transcription guidelines have been optimized for fluent reading while
providing a good approximation of the text as actually uttered by the given
speaker even to the non-expert user (see e.g. [15,16] on why it is a good idea
to carefully negotiate a tradeoff between readability and fidelity when devising
a spoken language transcription scheme). As a result, the transcripts contain

---

[3] The MorphoDiTa tagger is an implementation of the averaged perceptron algorithm
(see [10]), which in theory allows referring to a context of arbitrary length. In practice
however, this context is limited to a pre-set n-gram value (usually trigrams) so that
the Viterbi algorithm can be used for efficient decoding.

a substantial amount of forms which either do not exist at all in written language, or even worse, do exist, but with a different interpretation, which leads to increased homonymy; see Tab. 1 for a few examples.

Table 1. Some variants contained in the ORAL series corpora which are unknown to MorfFlex CZ [17], a standard morphological dictionary resource for Czech (for completeness sake: there are many non-standard forms the dictionary already accounts for). The forms *bej* and *prče* are homonymous with entries already contained in the dictionary. For a full rationale behind the transcription guidelines employed in the ORAL series corpora see [16].

lemma	gloss	variants unknown to MorfFlex CZ
dělat	to do	*dělajú, dělaš* (dialectal), *dělál, děléj* (final/emphatic lengthening)
být	to be	*budó, budú, býl, byzme, só, som, sú, zme* (dialectal), *bej* (careless)
protože	because	*poče, potože, prče, proe, protoe, protže, prože, prtoe, prtože, prtže, přže, ptože, pže* (careless variants), *protožes* (contraction with *jsi*)

In addition, syntax is often severely dislocated overall. Whereas ORAL2006 and ORAL2008 attempt to reconstruct syntactic relationships by using conventional punctuation, no sentence segmentation is available in ORAL2013, which uses pausal punctuation instead in order to minimize the amount of interpretation imposed on the material.

As can be seen from the examples provided, any model designed for tagging spoken language transcripts will always need to cater for the peculiarities of the particular transcription system in use. Whereas written language tagging systems have to cater for what may occur "naturally" in the texts of a given language, spoken language transcription systems are devised by linguists, which means they are much more arbitrary and less constrained in terms of the conventions they might implement. While extending the morphological dictionary to include new variants is a relatively easy task, annotating a set of training data by hand is an endeavour too laborious to undergo each time a spoken corpus project with a new or revised set of transcription guidelines is initiated.[4]

2   Method

Our goal was to try and improve the results (lemmatization and tagging) of a state-of-the-art stochastic tagger (MorphoDiTa) when fed the ORAL series corpora as input, compared to its performance using the standard tagging models for Czech [18] which are distributed along with it. MorphoDiTa adopts the usual scheme for tagging Czech, which consists in first generating all morphologically

---

[4] That the guidelines do in fact change is easy to demonstrate on the differences between the ORAL series corpora, e.g. with the syntactic vs. pausal punctuation approach cited above.

plausible tags for every token in a sentence using a morphological dictionary and then disambiguating them by picking the ones that yield the highest overall score for the sentence according to the model. Scores are based on a set of features whose values depend on the word form being disambiguated and its context (surrounding word forms and their lemmas and/or tags if they have already been successfully disambiguated). Considering this, the following areas presented themselves as open to adjustments:

- the morphological dictionary (MorfFlex CZ [17]) lacks some of the word forms contained in the transcripts (or some of their morphological interpretations in the case of homonymous forms) because they rarely if ever appear in print (see e.g. [19] for a recent treatment of Czech diglossia)
- the hand-annotated training set (the Prague Dependency Treebank 3.0 [20]) consists solely of written language, but the frequency of homonymous spoken language forms can in some cases be artifically increased in order to bias the models in their favour (see below for *sem* ∼ *to be*.1SG.PRES.IND / *here* as a typical example)
- it might also be appropriate to adjust the features used to score the sentences

We proceeded as follows: (1) run the MorphoDiTa tagger with the default models; (2) assess errors, identify systematic problems; (3) train a modified tagger (the types of possible modifications are listed above); (4) repeat from point (2).

Since there does not exist a gold standard tagging and lemmatization of the ORAL series corpora, the evaluation of the effect of our adjustments, and of the validity and usefulness of our further suggestions, can unfortunately be only fragmentary and mostly qualitative in nature at this point in time.

3   Results and Discussion

3.1   Morphological Dictionary Modifications

The goal of these modifications is to selectively fix under- and over-generation of the morphological dictionary, i.e. provide more accurate input to the disambiguation phase (tagging proper). This involves both additions to and removals from the dictionary.

First, we identified candidate non-standard forms in our data; our heuristic for this was based on the decisions of the aspell (http://aspell.net/) spell-checker. Using aspell over MorfFlex CZ was an attempt to maximize recall: the latter is a much more comprehensive resource and some of its more obscure entries may be homonymous with some non-standard transcription forms (e.g. *bej* from Tab. 1 has an entry as a noun). 7076 word forms were flagged as suspicious by aspell, compared to only 2400 by MorfFlex (58 of which were unique to that list and were therefore examined as well). In a one-off task of this kind, it was deemed preferable to take a more conservative approach and examine more entries manually. We then annotated each form with its standard counterpart (where available) and generated additional morphological dictionary entries based on this information by giving the form all lemma + tag combinations which already existed in the dictionary for its standard counterpart. For instance, for the form *budó*, the entry budó být VB-P---3F-AA---[5] was generated based on the entry for its counterpart *budou*, budou být VB-P---3F-AA---. Forms without standard counterparts need to have their entries created from scratch. Note that in the case of regular alternations, these entries may also be rule-generated: e.g. any 3PL.PRES.IND verb entry may spawn a variant with the dialectal ending *ó*.

As for removals, some are fairly straightforward, as in cases where homonymy is introduced through an obscure reading of a form which is unlikely ever to be correct given the nature of the data (e.g. *sou* lemmatized as *sou [an old French coin]* instead of *to be*; or *i*, a common conjunction, lemmatized as a graphical separator, which is only possibly appropriate for written data). In other cases however, such an approach is contentious: *pudem* might appear much more frequently as a colloquial variant of *půjdeme* ∼ *to go*.1PL.FUT.IND, but entirely excluding the possibility of it being an instance of *instinct*.M.SG.INSTR seems wrong. These cases might better be addressed through adjustments in the training data (see Sec. 3.2).

3.2 Training Set Modifications

The purpose here is to sway the bias of the tagger in favour of the most common reading of a homonymous word form, which is different in spoken language transcripts compared to the prevailing interpretation in written language. For instance, in texts, the form *sem* will almost exclusively be interpreted as the adverb *here*; conversely, in speech transcripts, the overwhelming majority of instances will correspond to the usual pronunciation of *jsem* ∼ *to be*.1SG.PRES.IND. Being based on written language, the hand annotated PDT 3.0 corpus, used as training data, does not reflect this. However, it is easy to tip this balance by replacing instances of *jsem* by *sem* in the original data and thus forcing the tagger to learn patterns of contextual disambiguation for this form during training instead of relying on trivial formal disambiguation (*jsem* is not homonymous). Similar adjustments would be warranted in the case of other (partially) overlapping paradigms where some of the readings are rare or ruled out in written language, such as that of the form *pudem* mentioned in Sec. 3.1.

This seems relatively straigthforward; in practice however, the results obtained with *sem* were only partially satisfactory. Using default MorphoDiTa models, 63,070 instances were tagged as *here* against 1,271 as *to be*. We then replaced roughly 95% of *jsem* instances in the training data by *sem* (the ratio of adverbial to verbal readings of the form in this modified PDT 3.0 was 108 : 2,931) and trained a custom model. After applying this new model in tagging,

---

[5] Format: form lemma tag; see [21] for a description of the tagset.

the ratio in the ORAL series corpora changed to 52,357 : 11,984, which is admittedly an improvement, but still wildly off the mark after but a cursory glance at a few of the relevant concordance pages.

Clearly, either the contexts of *(j)sem* differ so substantially between written and spoken language that the patterns learned in training do not generalize well to speech transcripts, or a modification of the feature template set used to infer these patterns is in order so as to achieve the desired effect. This will require further experimentation; more importantly, given the complex influence that feature templates exert over the overall performance of the model, a hand-annotated test set will be needed to gauge precisely the positive and negative effects of our interventions, else more is lost than gained by them.

3.3 Linguistic Concerns

Apart from the technical concerns listed above involved in selecting the correct tag + lemma combination for a given form in a given context, attempting to tag the ORAL series corpora has brought to the fore some linguistic concerns of a more fundamental nature: what even *is* the correct tag for such and such form? For instance, expletives such as *kurva, kurňa, vole, pičo* are identified as the appropriate forms of the corresponding nouns. Given the fact that they are not syntactically integrated and play instead the role of lexical fillers or pragmatic markers, should they be classified e.g. as particles instead of nouns?

Consider the case of *ty vole* (roughly an equivalent of *dude* in both its contact-establishing and surprise-expressing function): the slot of *vole* is open to a wide variety of options, many of which are morphologically frozen (*vado, vago, vogo, brd'o, bláho*), which would speak in favour of the particle interpretation. On the other hand, the construction is often creatively modified by filling in the slot with a newly-selected form which is always a noun in the vocative singular, suggesting that the underlying pattern in the speaker's mind is nominal after all. Perhaps it seems pointless trying to solve this issue at the POS level, because the semantics of a multi-word unit are clearly at play here, and the interesting piece of information to recover is the identity of the phraseme. FRANTA [22] is a tool currently under active development which strives to address precisely this kind of issue; however, it relies crucially on POS annotations to function.

Another quite specific concern is that MorfFlex CZ uses a separate lemma *von* instead of *on* for 3rd person personal pronoun occurrences exhibiting [v]-prothesis, which are common across the Western half of the Czech Republic. While the concern to set apart forms which correspond to a lower register is understandable, this seems redundant, as the [v]-prothesis information is already encoded in the word form, and the role of the lemma is on the contrary to provide some intuitive normalization, which should correspond to standard equivalents where they exist.

Finally, on a similar note, some word forms cannot be reasonably grouped under the same lemma, but are related enough that it would be a good idea to capture this relatedness explicitly. This could pertain to the myriad closely related colloquial variations on deictic pronouns (*todleto, tohlecto, tohlencto, tohlensto, tohlento, tohleto, toleto,* to name but a few), to various renderings of alphabetisms such as *DVD* or *ODS* (*ódées, ódéeska; dývko, dývýdýčko*) and similar items. For this purpose, a notion akin to the concept of hyperlemma proposed for diachronic corpora [23] would be useful. While the role of a hyperlemma is specifically to "iron out" diachronic variation by collecting various historical stages of the evolution of a word under one umbrella term, the analogy in spoken corpora would be a grouping based on looser semantic criteria (perhaps called "multilemma" for lack of a better term) whose purpose would be to maximize the recall of a query, on the premise that unwanted hits can always be filtered out given the manageable size of spoken corpora. In order to purposely avoid any claims as to the linguistic status of such groupings, multilemmas could simply list the forms they encompass instead of picking one of them as a label.

4 Conclusion

Though fraught with many errors that it was impossible to discuss in such a short amount of space, the experimental tagging of the ORAL series corpora is already useful in many cases where specifying a query using a regex on the word form is impractical at best, but frequently outright impossible (e.g. categories of verbal and nominal inflection across lemmas). A simple variant search tool based on this preliminary lemmatization has already been made available to make it easier to retrieve full paradigms including all pronunciation variants in the ORAL series corpora as currently released (see https://github.com/dlukes/achsynku for details).

Ideally, the next steps would consist of improving the annotation incrementally based on user feedback, so as to focus on areas which users themselves consider key. The Czech National Corpus has recently changed its policy with respect to published corpora, introducing versioning: original/older versions will always remain accessible for reference, but they may be superseded by newer versions which incorporate error corrections and various other improvements. This change in policy should easily enable such an iterative improvement cycle for the tagging and lemmatization of the ORAL series corpora.

References

1. Straková, J., Straka, M., Hajič, J.: Open-source tools for morphology, lemmatization, POS tagging and named entity recognition. In: Proceedings of 52nd Annual Meeting of the ACL: System Demonstrations, pp. 13–18. ACL (2014)
2. Votrubec, J.: Morphological tagging based on averaged perceptron. In: WDS 2006 Proceedings of Contributed Papers, pp. 191–195. Matfyzpress, Charles University, Prague (2006)
3. Spoustová, D., Hajič, J., et al.: Semi-supervised training for the averaged perceptron POS tagger. In: Proceedings of the 12th Conference of the European Chapter of the ACL, pp. 763–771. ACL, Athens (2009)
4. Petkevič, V.: Problémy automatické morfologické disambiguace češtiny. Naše řeč (4–5), 194–207 (2014)

5. Spoustová, D., Hajič, J., et al.: The best of two worlds: Cooperation of statistical and rule-based taggers for Czech. In: Proceedings of the Workshop on Balto-Slavonic Natural Language Processing 2007, pp. 67–74. ACL, Prague (2007)

6. Mikulová, M., Urešová, Z.: Rekonstrukce standardizovanáho textu z mluvené řeči. In: Kopřivová, M., Waclawičová, M. (eds.) Čeština v Mluveném Korpusu. NLN, Prague (2008)

7. Čmejrková, S., Jílková, L., Kaderka, P.: Mluvená čeština v televizních debatách: korpus DIALOG. Slovo a Slovesnost 65, 243–269 (2004)

8. Hnátková, M., Petkevič, V.: Pomocné sloveso být a automatická identifikace jeho hlavních funkcí. In: Čermák, F., Blatná, R. (eds.): Korpusová Lingvistika: Stav a Modelové Přístupy, pp. 131–152. NLN (2006)

9. Miller, J., Weinert, R.: Spontaneous Spoken Language: Syntax and Discourse. Clarendon Press, Oxford (1998)

10. Collins, M.: Discriminative training methods for Hidden Markov Models: Theory and experiments with perceptron algorithms. In: Proceedings of the 2002 Conference on Empirical Methods in Natural Language Processing, pp.1–8 (2002)

11. Auer, P.: On-line syntax: Thoughts on the temporality of spoken language. Language Sciences 31, 1–13 (2009)

12. Kopřivová, M., Waclawičová, M.: ORAL2006: korpus neformální mluvené češtiny. Ústav českého národního korpusu FF, Prague (2006)

13. Waclawičová, M., Křen, M.: ORAL2008: New balanced corpus of spoken Czech. In: Proceedings of the International Conference "Corpus Linguistics–2008", pp. 105–12. Saint Petersburg University Press, Saint Petersburg (2008)

14. Benešová, L., Křen, M., Waclawičová, M.: ORAL2013: Reprezentativní korpus neformální mluvené češtiny [ORAL2013: A representative corpus of informal spoken Czech]. Ústav českého národního korpusu FF, Prague (2013)

15. Ehlich, K., Rehbein, J.: Halbinterpretative arbeitstranskriptionen (HIAT). Linguistische Berichte 45, 21–41 (1976)

16. Waclawičová, M., Křen, M., Válková, L.: Balanced corpus of informal spoken Czech: compilation, design and findings. In: Proceedings of Interspeech, Brighton, pp. 1819–1822 (2009)

17. Hajič, J., Hlaváčová, J.: MorfFlex CZ (1990)

18. Straka, M., Straková, J.: Czech models (MorfFlex CZ + PDT) for MorphoDiTa (2013)

19. Bermel, N.: Czech diglossia: dismantling or dissolution? In: Árokay, J., et al. (eds.) Divided Languages? Springer, Switzerland (2014)

20. Bejček, E., Hajičová, E., et al.: Prague Dependency Treebank 3.0 (2013)

21. Hajič, J.: Disambiguation of Rich Inflection: Computational Morphology of Czech. Charles University Press, Prague (2004)

22. Hnátková, M., Kopřivová, M.: Identification of idioms in spoken corpora. In: Gajdošová, K., Žáková, A. (eds.) NLP, Corpus Linguistics, E-learning, pp. 92–99. RAM-Verlag (2013)

23. Kučera, K.: Hyperlemma: A concept emerging from lemmatizing diachronic corpora. In: Levická, J., Garabík, R. (eds.) Computer Treatment of Slavic and East European Languages, Tribun, pp. 121–125 (2007)

Development and Evaluation of the Emotional Slovenian Speech Database - EmoLUKS

Tadej Justin[1], Vitomir Štruc[1], Janez Žibert[2], and France Mihelič[1(✉)]

[1] Faculty of Electrical Engineering, University of Ljubljana, Tržaška 25,
1000 Ljubljana, Slovenia
{tadej.justin,vitomir.struc,france.mihelic}@fe.uni-lj.si
[2] Faculty of Mathematics, Natural Sciences and Information Technologies,
University of Primorska, Glagoljaška 8, 6000 Koper, Slovenia
janez.zibert@upr.si

Abstract. This paper describes a speech database built from 17
Slovenian radio dramas. The dramas were obtained from the national
radio-and-television station (RTV Slovenia) and were given at the universities disposal with an academic license for processing and annotating
the audio material. The utterances of one male and one female speaker
were transcribed, segmented and then annotated with emotional states
of the speakers. The annotation of the emotional states was conducted
in two stages with our own web-based application for crowd sourcing.
The final (emotional) speech database consists of 1385 recordings of one
male (975 recordings) and one female (410 recordings) speaker and contains labeled emotional speech with a total duration of around 1 hour
and 15 minutes. The paper presents the two-stage annotation process
used to label the data and demonstrates the usefulness of the employed
annotation methodology. Baseline emotion recognition experiments are
also presented. The reported results are presented with the un-weighted
as well as weighted average recalls and precisions for 2-class and 7-class
recognition experiments.

Keywords: Emotional speech database · Emotion recognition ·
Database development

1 Introduction

Paralinguistic information is commonly defined as speaker- or speech-related
information that cannot be described properly with phonetic or linguistic labels.
Examples of paralinguistic information include the level of intoxication of a
speaker, the speaker's mood, his/her interests, or the emotional state of the
speaker among others.

The design and development of speech databases, which include paralinguistics labels, demands interdisciplinary cooperation [1], [2], [3]. One of the most
demanding parts during the design stage is to properly define the paralinguistic labels. For example, when the selected paralinguistic labels are the emotional

© Springer International Publishing Switzerland 2015
P. Král and V. Matoušek (Eds.): TSD 2015, LNAI 9302, pp. 351–359, 2015.
DOI: 10.1007/978-3-319-24033-6_40

states of the speakers (such as in our case), discrete states need to be defined that can be associated with the paralinguistic labels. However, this represents a difficult task, since there is no generally established definition of what an emotional state is. In such cases, developers commonly resort to examples of good-practice from the literature, e.g., [4], [5],[6] which define examples of emotional states, or rely on expert knowledge and guidelines for building emotional databases, e.g.,[7].

The development of an emotional database is, in general, guided by the research goals. Clearly, different datasets are needed when studying human emotions from a theoretical perspective (where the goal is to understand how emotions are related to the psychological or biological processes in humans) or when developing applications that try to take the expressed emotions into account during their operation. The latter is also the case with speech technologies, such as emotional speech synthesis or speech recognition from emotionally colored speech, where the goal is to either make the synthesized speech sound more natural or improve the performance of speech recognition systems by accounting for the variability in the speech signal induced by changes in the emotional state of the speaker. With speech technologies, emotional databases typically consist of recordings and transcriptions of speech with additional paralinguistic labels [8].

Over the last few years the field of speech technologies has seen increased interest in modeling techniques and approaches that make use of paralinguistic speaker information [9]. Increased popularity in applications capable of natural human-computer interaction (HCI) using speech can also be observed over the last decade. Since speech technologies are for the most part language dependent, it is important to have suitable resources at ones disposal. Building a speech database with paralinguistic labels (in the form of annotated emotional states of the speakers) is the first steps when trying to improve the naturalness of the existing Slovenian speech synthesis systems or to develop emotion recognition systems for Slovenian speakers.

In this paper we focus on the development of an emotional speech database for the Slovenian language for applicative use - primarily in emotionally colored speech synthesis. We evaluate the annotated emotional speech material and present a use-case for the database, i.e., automatic emotion recognition from speech. We also elaborate on the importance of the two-stage annotation process of the emotional utterances and finally present the Slovenian emotional speech database - EmoLUKS.

The rest of the paper is structured as follows: in Section 2 we describe the preparation and annotation of the database and present a brief analysis of the annotated speech material. We evaluate the annotation of the EmoLUKS database and present emotion recognition experiments in Section 3. Finally, we conclude the paper in Section 4 with some final comments and directions for future work.

2   Emotional Speech Database - EmoLUKS

The development of emotional speech databases typically follows one of two different approaches in terms of design: *i)* With the first approach utterances of professional actors, who are capable of correctly articulating natural speech and imitate different emotional states, are read and then recorded. In general, such databases are typically comprised of predefined sets of sentences, which are commonly extracted from a large, language-dependent text corpus and selected in a way that balances the distribution of the base phonetic units in one of the target languages. The emotional labels of the sentences are commonly defined before the recording stage. Such an approach results in databases of simulated or acted emotions. *ii)* The second approach relies on the transcription, segmentation and annotation of pre-recorded speech material. In case the pre-recorded speech material corresponds to natural speech, the annotated database represent a database of spontaneous emotional states of speakers, while it may again represent a database of acted emotions if the pre-recorded data is read, such as in the case of radio shows or dramas. The main difference between the first and second approaches is in the way the emotional states of the speakers are annotated. The first one has predefined emotional labels, which are commonly evaluated with perception tests. With the second approach, annotators are commonly employed to annotate the utterances (again with perception tests), but the final labels of the speech utterances are decided based on a majority vote.

In our case we adopted the second approach, and used radio dramas to build a database of acted emotional speech. We selected sentences as the basic annotation units and decided on the final emotion labels based on a majority vote over the labels assigned to the sentences by the annotators.

2.1 Database Description and Preparation

With the help of the national radio-and-television station of Slovenia, i.e., RTV Slovenia, we obtained radio-drama recordings that were produced in a professional studio. We manually transcribed and segmented the radio dramas and also annotated the non-lexical data, which are commonly used in radio dramas, such as background music, various background noises, and various added audio effects. Additionally, we took extra caution when marking and segmenting the speaker's non-lexical sounds, such as crying, breathing, laughter, etc. For this process, the Transcriber annotation tool [10] was used. Once we obtained the transcribed and segmented audio, we extracted the utterances of each speaker based on full-sentence units that included only clear speech. In this way we obtained speech segments that are not too long and with enough contexts for annotating the emotional state of the speakers. Such utterances were prepared for further processing and annotation of the speaker's emotional state.

We transcribed 17 radio dramas with an approximate total time of 12 hours and 50 minutes. The transcribed material includes the segmentation of 16 speaker identities (5 female and 11 male) that produced at least 3 minutes of clear speech. From the transcribed and segmented material we extracted the utterances of one male (01m_av) and one female speaker (01f_lb) (clear speech only) for a total duration of 62 minutes for the male and 15 minutes the female speaker.

2.2 Defining the Emotional States

Each research field that interacts or deals with human emotions needs a proper definition of what an emotional state is. The differences in the theoretical models, on which the theory of emotions is based [11], clearly show how emotions can be subject to different interpretations. In the literature we found four different perspectives on emotional states [12]. The use of each perspective also determines the different relations between the emotional categories. The main assumption for modeling the delimitations of the emotional categories is that the differences between the observed emotional experiences of one category are smaller than the differences between the emotions from different categories. Various aspects to modeling the emotional relations are presented in [11].

We focus on the discretization of the emotional states based on Darwin's perspective [11]. Hence, we assume that there are some basic emotional states from which the basic discrete models of emotional categories were developed. Such a representation is one of the most popular approaches for presenting the emotional space. In this way we determined the discrete basic emotions in different emotional categories for annotating the recordings of radio drama: sadness, joy, gust, anger, fear, surprise and neutral.

2.3 Database Annotation Through Crowd-Sourcing

Annotating the emotional states from speech can be achieved with expert knowledge. While an expert on emotional states can annotate each speech signal, more objective labels can be obtained if several experts provide their expertise and annotate the data. Since there is no generally accepted definition of emotional states and due to the fact that all people are able to interpret human emotion to at least some degree, there is no need to look for experts in the field. Instead ordinary people can be asked to label the data and decide which emotions are expressed in each of the speech utterances.

The annotating procedure for speech material is a relatively time-consuming process. Therefore, it is advisable to provide a software application that allows to annotate utterances in a fast and reliable way. It is also desirable to allow the annotators to pick the annotating time by themselves. In recent years the most attractive way to approach such problems is to allow multiple annotators to annotate from any location thought the web. Such an approach is called crowd-sourcing [13] and has became the most popular solution for annotating different databases in the past decade, especially those meant for further processing or data mining.

Inspecting the literature and the available crowd-sourcing applications, we decided that none of these applications suited our needs - to allow volunteers to perform fast and easy annotation of the speech database. Therefore, we decided to develop our own crowd-sourcing application for audio or video resources.

Our web-based crowd-sourcing application was developed with the help of the available open-source software. It was designed as an add-on for the well-established Content Management System (CMS) Plone (er. 4.3.3) [1] based on

---
[1] http://plone.org

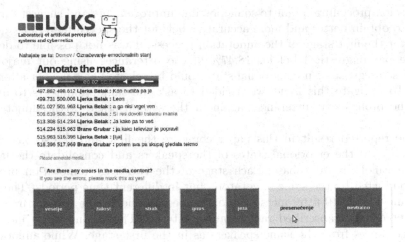

Fig. 1. A screen shoot of an annotator submitting decision "surprise" for the current speech utterance. An annotator is able to read the full context dialog, where the transcription of the current utterance is colored red. The annotator can also mark if the utterance contains an error in the transcription or segmentation.

the Zope [2] application server. The developed crowd-sourcing application consists of two different types of access. The first is designed for editors/administrators who can prepare the annotation procedure through the web. The second tries to offer the annotator a user-friendly annotation experience. Anonymous annotating is not permitted due to academic license associated with the database. The annotators can annotate the resource over a longer time period by simply logging-in/logging-out and starting at the last annotated audio or video file. The demonstration of the application for crowd-sourcing the audio or video signals developed at LUKS is available at http://emo.luks.fe.uni-lj.si. A snapshot of the user interface of the application is shown in Fig. 1.

The developed application works as follows. The speech utterance is randomly picked from the predefined media files that need to be annotated. It is automatically played when the submission form is loaded and the annotator can easily pass his/her decision with a simple mouse click. When the annotator makes the decision, the next media for the annotation is loaded.

We asked 5 volunteers (2 female and 3 male) to annotate the speech material. The annotation labels were picked as discrete basic emotions with an added normal state. During the process of annotation, all the evaluators wore headphones.

2.4   Analysis of the Annotated EmoLUKS Database

The development stage for the EmoLUKS emotional database was already described in [14]. Here, we try to present the difficulties of the first-stage

---

[2] http://zope.org

annotation procedure, point to some possible improvements and suggest further steps to obtain better and more accurate labels for the speech material.

After the first stage of the annotation process, it was not possible to decide on the paralinguistic label for 18.47% of the utterances using our majority-voting strategy, since no emotional state could be selected as a winner after the vote. To mitigate this issue, we decided to ask the same volunteers to annotate the problematic utterances again in the second stage of the annotation process.

The reported result in this paper compare the first and the second-stage annotation of the emotional states of the speakers and consolidate the labels for the EmoLUKS database. Each stage of the two-stage annotation process was conducted by the same annotators, but in different time periods. The first stage included 1010 utterances from one male and one female speaker from the segmented and transcribed radio-drama material. The second stage included 988 utterances from the same speakers as in the first stage. While annotating the first stage, the annotators reported some errors in the transcription and/or segmentation. All reported errors were corrected and the files were once again annotated in the second stage. Also, some utterances, which were too long, but were part of the first annotation stage, were segmented again from full sentences into shorter, but still meaningful parts.

Overall, the second stage included 421 of the corrected utterances from the first stage due to segmentation and/or transcriptions errors. The other 542 utterances were the same as in the first stage of the annotation process. In the second annotation stage we also included 25 utterances that were not included in the first stage. As can be seen from Table 1, where a brief summary of the annotation stages is presented, both annotation stages resulted in approximately the same fraction of utterances, where no winning label could be selected after the majority vote and, hence, the utterances had to be assigned to the "Undecided" category (18.47% vs. 19.53%). Since many of the utterances annotated in the second stage overlapped with the (problematic) utterances that were already annotated in the first stage, ten labels were available for these utterances to decide on a winning label. Clearly, this fact resulted in a reduction of the percentage of utterances in the "Undecided" category and consequently improved the percentage of utterances in all other categories (except for Disgust).

Other information that can be seen from Table 1 is: *i)* the number of utterances per speaker and annotation stage (marked as Utt. #), *ii)* the total duration of the annotated speech material per speaker and annotation stage (marked as Dur.), and *iii)* the fraction of annotated speech material (in terms of duration) per emotional category/label for each speaker and each annotation stage.

3   Baseline Emotion Recognition Experiments

In this section we present one possible use-case for the EmoLUKS database, i.e., developing and testing emotion recognition systems for affective computing

Table 1. Overview of the annotated material in the two-stage annotation process for the Slovenian emotional speech database EmoLUKS based on the annotators majority vote. Here the abbreviations Surp., Disg., Ang.. Neut., Sad., Undec. stand for Surprise, Disgust, Anger, Neutral, Sadness, and Undecided, respectively.

Eval.	Spk.	Utt. #	Dur.	Fraction of annotated speech material [% od total duration]							
				Joy	Surp.	Disg.	Ang.	Neut.	Fear	Sad.	Undec.
1.stage	01m_av	762	1:01:29	8.53	11.02	1.18	14.44	36.48	5.38	4.86	18.11
	01f_lj	348	14:56	11.78	14.94	4.02	28.45	11.21	8.05	2.30	19.25
	sum	1110	1:16:24	9.55	12.25	2.07	18.83	28.56	6.22	4.05	18.47
2.stage	01m_av	733	49:12	8.59	9.69	1.36	12.55	35.06	9.41	3.14	20.19
	01f_lj	255	8:27	10.59	17.25	6.27	21.96	12.94	10.98	2.35	17.65
	sum	988	57:39	9.11	11.64	2.63	14.98	29.35	9.82	2.94	19.53
EmoLUKS	01m_av	975	1:02:31	8.72	11.59	1.03	15.69	39.69	7.79	4.10	11.38
	01f_lj	410	14:44	12.44	16.83	4.15	27.56	13.17	10.98	3.90	10.98
	sum	1385	1:17:16	9.82	13.14	1.95	19.21	31.84	8.74	4.04	11.26

Table 2. Speaker-dependent emotion-recognition results performed on the EmoLUKS speech material. UA (un-weighted average) and WA (weighted average). The results are presented in the form of the mean values and standard deviation of the performance metrics over all five folds.

Speaker label	Problem	Recall		Precision	
		UA	WA	UA	WA
01m_av	2 class	0.70 ± 0.04	0.70 ± 0.04	0.70 ± 0.04	0.71 ± 0.04
	7 class	0.27 ± 0.04	0.50 ± 0.05	0.30 ± 0.06	0.47 ± 0.05
01f_lb	2 class	0.54 ± 0.04	0.79 ± 0.03	0.56 ± 0.05	0.77 ± 0.02
	7 class	0.32 ± 0.04	0.44 ± 0.05	0.31 ± 0.04	0.42 ± 0.05

applications. Towards this end, we use the OpenSMILE feature extractor [15] and its predefined configuration to extract the same features as were used in the baseline experiments during the Interspeech 2009 Emotion Challenge [16]. The idea behind this experiments is to use a state-of-the-art emotion recognition approach to demonstrate the difficulty of the database for the problem of emotion recognition.

Since the obtained radio-drama recordings were professionally recorded, but during different time periods with different producers and directors, we first normalized all the annotated utterances. The normalization and equalization of the gain was conducted with SoX (Sound eXchange)[3], using its default values. We used a 5-fold stratified cross-validation scheme to evaluate the SVM classifier [17] integrated into the WEKA Data Mining Toolkit [18], with default parameters. Since currently only speech material from one male and one female speaker is annotated, we performed speaker-dependent experiments. Thus, separate SVM classifiers ware trained for each speaker. The classification results are presented in Table 2. They represent two-class classification results for the neutral and the non-neutral emotional states and the seven-class emotional state classification

---

[3] http://sox.sourceforge.net

problem, where the reference label represents the majority vote of the annotated speech utterances.

Note that the performance of the recognition experiments is quite low. These results may partially be ascribed to the fact that for some emotional classes (such as "Disgust") only a small amount of data is available in the current version of the database, but also show how challenging this database is.

4   Conclusion and Feature Work

In this paper we presented our current efforts towards building a speech database with paralinguistic information. Specific attention was given to the two-stage annotation process for the emotional utterances from one male and one female speaker. Our analysis suggests that the efforts involved in setting up a second annotation stage were repaid. With the consolidated labels from the first- and second-stage of the annotation we were able to significantly decrease the amount of utterances that could not be labeled with a majority vote (undecided category). As indicated in Subsection 2.1, we still have some speech material from different speakers available for the annotation of the emotional states of the speaker. Therefore, we plan to expand the current EmoLUKS database as part of our future work.

References

1. Gajšek, R., Štruc, V., Mihelič, F., Podlesek, A., Komidar, L., Sočan, G., Bajec, B.: Multi-modal emotional database: AvID. Informatica (Slovenia) **33**(1), 101–106 (2009)
2. Batliner, A., Biersack, S., Steidl, S.: The prosody of pet robot directed speech: evidence from children. In: Proc. of Speech Prosody, pp. 1–4 (2006)
3. Koolagudi, S., Rao, K.: Emotion recognition from speech: a review. International Journal of Speech Technology **15**(2), 99–117 (2012)
4. Gajšek, R., Štruc, V., Dobrišek, S., Mihelič, F.: Emotion recognition using linear transformations in combination with video. In: 10th INTERSPEECH (2009)
5. Gajšek, R., Žibert, J., Justin, T., Štruc, V., Vesnicer, B., Mihelič, F.: Gender and affect recognition based on GMM and GMM-UBM modeling with relevance MAP estimation. In: 11th INTERSPEECH (2010)
6. Dobrišek, S., Gajšek, R., Mihelic, F., Pavešić, N., Štruc, V.: Towards efficient multi-modal emotion recognition. International Journal of Advanced Robotic Systems **10**(53), 1–10 (2013)
7. Cowie, R., Cornelius, R.R.: Describing the emotional states that are expressed in speech. Speech Communication **40**(1–2), 5–32 (2003)
8. Schuller, B., Steidl, S., Batliner, A., Burkhardt, F., Devillers, L., Müller, C., Narayanan, S.: Paralinguistics in speech and language – state-of-the-art and the challenge. Computer Speech & Language **27**(1), 4–39 (2013)
9. Yamashita, Y.: A review of paralinguistic information processing for natural speech communication. Acoustical Science and Technology **34**(2), 73–79 (2013)

10. Barras, C., Geoffrois, E., Wu, Z., Liberman, M.: Transcriber: Development and use of a tool for assisting speech corpora production. Speech Communication **33**(1–2), 5–22 (2001)
11. Cornelius, R.R.: The science of emotion: Research and tradition in the psychology of emotions. Prentice-Hall, Inc. (1996)
12. Cornelius, R.R.: Theoretical approaches to emotion. In: ISCA Tutorial and Research Workshop (ITRW) on Speech and Emotion (2000)
13. Howe, J.: The rise of crowdsourcing. Wired Magazine **14**(6), 1–4 (2006)
14. Justin, T., Mihelic, F., Žibert, J.: Development of emotional Slovenian speech database based on radio drama – EmoLUKS. In: Language Technologies. Proceedings of the 17th International Multiconference INFORMATION SOCIETY - IS 2014, vol. G, Institut "Jožef Stefan" Ljubljana, pp. 157–162 (2014)
15. Eyben, F., Weninger, F., Groß, F., Schuller, B.: Recent developments in openSMILE, the Munich open-source multimedia feature extractor. In: Proceedings of the 21st ACM International Conference on Multimedia, pp. 835–838. ACM (2013)
16. Schuller, B., Steidl, S., Batliner, A.: The Interspeech 2009 emotion challenge. In: 10th INTERSPEECH, pp. 312–315 (2009)
17. Keerthi, S., Shevade, S., Bhattacharyya, C., Murthy, K.: Improvements to Platt's SMO algorithm for SVM classifier design. Neural Computation **13**(3), 637–649 (2001)
18. Hall, M., Frank, E., Holmes, G., Pfahringer, B., Reutemann, P., Witten, I.H.: The WEKA data mining software: an update. ACM SIGKDD Explorations Newsletter **11**(1), 10–18 (2009)

A Semi-automatic Adjective Mapping Between plWordNet and Princeton WordNet

Ewa Rudnicka[1,2]([✉]), Wojciech Witkowski[1,2], and Michał Kaliński[1,2]

[1] Department of Computational intelligence, Wrocław University of Technology,
Wybrzeże Wyspiańskiego 27, Wrocław, Poland
{ewa.rudnicka,michal.kalinski}@pwr.edu.pl, wojciech62@gmail.com
[2] Department of English Studies, University of Wrocław, Kuźnicza 22,
Wrocław, Poland

Abstract. The paper presents the methodology and results of a pilot stage of semi-automatic adjective mapping between plWordNet and Princeton WordNet. Two types of rule-based algorithms aimed at generation of automatic prompts are proposed. Both capitalise on the existing network of intra and inter-lingual relations as well as on lemma filtering by a cascade dictionary. The results of their implementation are juxtaposed with the results of manual mapping. The highest precision is achieved in a hybrid approach relying on both synset and lexical unit relations.

Keywords: Bilingual wordnet alignment · Adjective mapping · Rule-based algorithms · Automatic prompts

1 Introduction

Originally, wordnets were constructed as single language databases (e.g. Princeton WordNet (henceforth, PWN) [1], or GermaNet [2]), yet soon efforts were made to link the existing monolingual wordnets into multilingual resources (e.g. EuroWordNet [3], or MultiWordNet [4]. The vast majority of those projects capitalised on the so called "transfer and merge" approach which essentially consists in the translation of Princeton WordNet's content and structure onto one of "national" languages. It is certainly fast and cost-effective, yet it does not allow to capture the actual lexico-semantic structure of a given language.

A rare example of a wordnet built fairly independently of Princeton WordNet is plWordNet (Slowosiec - Polish WordNet (henceforth, plWN) [5]). This makes any kind of linking of these two resources a non-trivial task. A large part of plWN noun synsets have been already mapped to the corresponding PWN synsets in the course of a manual mapping process partly enhanced by an automatic prompt system relying on the Relaxation Labeling (RL) algorithm paired with a filtering by a cascade dictionary [6,7]. The implementation of the RL algorithm was possible due to the fact that noun synset relation structures are vertical in both PWN and plWN, and most of PWN's relation types appear in plWN as well [8].

© Springer International Publishing Switzerland 2015
P. Král and V. Matoušek (Eds.): TSD 2015, LNAI 9302, pp. 360–368, 2015.
DOI: 10.1007/978-3-319-24033-6_41

Unfortunately, the same cannot be said about the adjective synset structure, horizontal in PWN and vertical in plWN, which renders the application of the RL algorithm impossible [9]. Therefore, mapping of adjective synsets constitutes a real challenge.

In the paper, we propose two types of automatic prompt generating algorithms dedicated to the enhancement of manual mapping of plWordNet and Princeton WordNet adjective synsets. They rely on intra-lingual synset and lexical unit relations as well as on the network of inter-lingual links between plWN and PWN synsets. The latter are possible to be applied, because certain adjective relations are relations between adjectives and nouns. The algorithms produce corresponding synset "nests" in both wordnets. Next, the lemmas of the adjectives synsets are filtered by a cascade dictionary in order to obtain semantically corresponding pairs of Polish-English synsets. These are presented to lexicographers in the WordNetLoom editing tool in the form of special links. Finally, results of the manual mapping are juxtaposed with the number and location of automatic prompts.

2   Adjective Relation Structure in plWordNet and Princeton WordNet

Before we proceed to an analysis of adjective relation structures, let us consider basic adjective counts in Princeton WordNet 3.1 and in plWordNet 2.2, provided in Table 1 below:

Table 1. Basic adjective data in PWN 3.1 and in plWN 2.2

	Princeton WordNet 3.1	plWN 2.2
Number of synsets	18185	17452
Number of lexical units	30072	21146
Number of lemmas	21808	13293
Number of synset relations	23491	23160
Number of lexical unit relations	21634	22467

The number of synsets and their relations is comparable between plWN and PWN. The number of lemmas and lexical units is higher by about one third in PWN, but the number of lexical unit relations is again comparable. Such counts point to a bigger polysemy of Polish adjective lemmas[1]. Still, in terms of key adjective numbers the resources are comparable. If the choice of the coded meanings also turned out comparable, it would create the potential for establishing a good number of inter-lingual synonymy links between plWN and PWN adjective synsets, but this can only be verified in the course of mapping.

---

[1] In fact, this is motivated by richer, and thus more precisely sense-distinguishing synset structure in plWN, see Table 2

Now, let us focus on adjective relation types and their respective frequencies within both wordnets in question. The relation structure of Princeton Word-Net was strongly motivated by the results of psycholinguistic experiments on the organization of mental lexicon carried out in the 60-ties of the XX. century. The results pointed to a hierarchical organization of nouns, troponymy-oriented verb structure and strongly opposition-based structure of adjectives [10,11]. Therefore, WordNet creators based the organization of the adjective domain on the relation of antonymy [1]. Antonymy is a lexico-semantic relation that holds between pairs of specific words, therefore it is coded as a relation between lexical units, not synsets. The main synset relation is a rather imprecise *Similar to* relation, coded in a horizontal, "dumbbell" model [12]. In sharp contrast, plWN adjective synset relation structure is vertical, similarly to noun and verb synset relation structures. Instead of one general relation, it employs a few more fine grained relations such as *Hyponymy*, *Gradability*, *Modifier* and *Value of the attribute* (the last two linking adjectives to nouns). In Table 2 below, we juxtapose PWN and plWN adjective synset relations and provide the counts of their instances in PWN 3.1 and in plWN 2.2[2].

Table 2. Adjective synset relations of PWN 3.1 and plWN 2.2

PWN relation name	number of instances	plWN relation name	number of instances
Similar to (Adj ↔ Adj)	**21434**		
Member of this domain (Adj → N)	1418		
Attribute (Adj → N)	**639**	**Value of the attribute (Adj → N)**	**4756**
		Hyponymy (Adj → Adj)	**5889**
		Modifier (Adj → N)	3030
		Gradability (Adj → Adj)	397

Evidently, there is only one pair of directly corresponding adjective synset relations between PWN and plWN: *Attribute - Value of the attribute*, yet their respective frequencies are very different (639 instances of *Attribute* vs. 4756 instances of *Value of the attribute*). Still, this correlation may be exploited for test purposes. Moreover, an analysis of (the semantic import) and specific instances of *Similar to* relation showed that often it encompasses cases of hyponymy and gradability.

---

[2]   The data are derived from the database of Princeton WordNet 3.1 (available from https://wordnet.princeton.edu/wordnet/download/current-version/) and from the database of plWordNet 2.2.

The structures of lexical unit relations of PWN and plWN adjectives bear much more resemblance. There are three pairs of directly corresponding relations: *Derivationally related form - Derivativity, Antonym - Antonymy,* and *Pertainym (Pertains to noun) - Cross-categorial synonymy.* The relations and their respective counts are given in Table 3 below:

Table 3. Adjective lexical unit relations of PWN 3.1 and plWN 2.2

PWN relation name	number of instances	plWN relation name	number of instances
Derivationally related form (Adj → N/V/Adj)	**14317**	**Derivativity (Adj → N/V/Adj)**	**2623**
Antonym (Adj ↔ Adj)	4024	Antonymy (Adj → Adj)	4026
Pertainym (pertains to noun) (Adj → N)	**3293**	**Cross-categorial synonymy (Adj → N)**	**11970**
		Characteristic (Adj → N)	1505
		Similarity (Adj → N)	596

Although the counts of the corresponding relations seem divergent again, it must be noted that plWN's *Cross-categorial synonymy* usually entails the derivational relatedness as well. The number of its instances added to *Derivativity* instances makes 14 593, and is very close to the number of *Derivationally related form* instances, 14317, which may be a good prognosis for their use in mapping.

3   Automatic Prompt Algorithms

The comparative analysis of adjective relation structures of plWN and PWN presented in the previous section has shown profound differences in the synset relation structure and much more correspondence in lexical unit relation structure. The divergent relation structure constitutes a real challenge for mapping. Still, some directly corresponding relations have been tracked in both networks and, what is particularly interesting, some of them are relations between adjectives and nouns, with the latter being almost wholly mapped. This creates the potential for exploiting the corresponding adjective relations and the already existing inter-lingual links between nouns for the purposes of the design of rule-based algorithms enabling the generation of prompt links and, in the effect, building the basis for mapping of adjectives. In the present section, we will propose two types of such algorithms. The first one is based on synset relations exclusively, the second one capitalises on both lexical unit and synset relations.

Mapping between plWordNet and Princeton WordNet takes place at the level of synsets, therefore synset relations are naturally the first to consider at the moment of starting the mapping of a new category. Within adjective synset relation structure, plWN and PWN have only one pair of directly corresponding relations: plWN *Wartosc cechy* - "Value of attribute" and PWN *Attribute*, both holding between adjectives and nouns. Thus, the first proposed algorithm used these relations plus the inter-lingual synonymy links between plWN and PWN nouns being their targets. However, as shown in Table 2 in Section 2, the number of PWN *Attribute* links is not big, so we decided to go deeper into relation nests and enriched Algorithm 1 (creating Algorithm 3.1) with additional relations, such as *Modifier*, *Hyponymy* and *Gradability* on the plWN side and with *Member of this domain* and *Similar to* on the PWN side. Its path is illustrated in Figure 1 below:

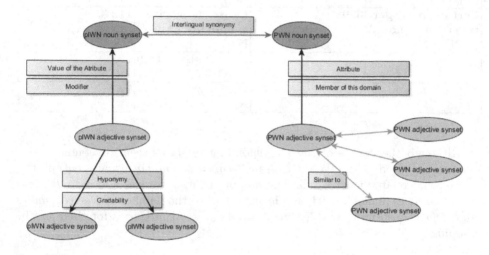

Fig. 1. The scheme of synset relation based Algorithm 1

Such design of Algorithm 1 was also motivated by the observation of a non-trivial conceptual correlation between some of these relations, especially between *Hyponymy* and *Gradability* and *Similar to* (mentioned in Section 2). In the last step lemmas of adjective synsets from the corresponding plWN and PWN synset "nests" were filtered by a large Polish-English cascade dictionary[3]. Synsets for whose lemmas the dictionary found a connection were linked by a special automatic prompt relation. An example of the implementation of Algorithm 1 is given in Figure 2 below (the red dashed lines signal the potential matches between plWN and PWN adjective synsets):

---

[3]   The cascade dictionary is composed of a few dictionaries with their data ordered in the hierarchy of importance. The topmost ones are [13] and a large proprietary dictionary of TiP company.

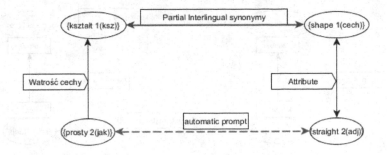

Fig. 2. An example of an automatically prompted pair of synsets resulting from the application of Algorithm 1

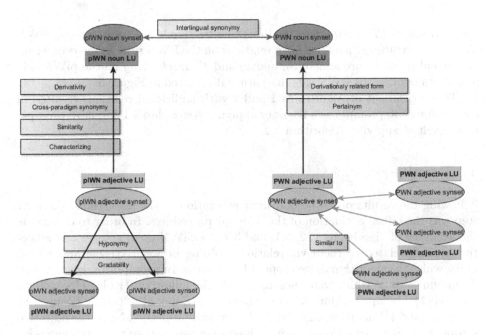

Fig. 3. The scheme of synset and lexical unit relation based Algorithm 2

As in the case of Algorithm 1, the last step is lemma filtering by a cascade dictionary. An example of the implementation of Algorithm 2 is given in Figure 4 below:

The second type of the proposed automatic prompt algorithm goes beyond synset relations and makes use of lexical unit relation structure which, as shown in Section 2, displays more correspondence than synset relation structure. Similarly to Algorithm 1, it starts with inter-lingual synonymy relation between plWN and PWN noun synsets. Then, it singles out their component lexical units which bear

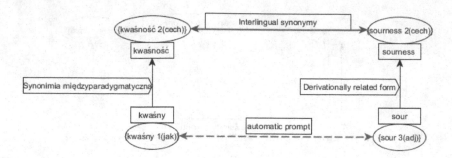

Fig. 4. An example of automatically prompted pair of synset resulting from the application of Algorithm 2

Derivativity or *Cross-paradigm* relations to adjective lexical units on the plWN side and *Derivationally related form* relations on the PWN side. In its second run, additional relations are added: *Similarity* and *Characterizing* on the plWN side and Pertainym on the PWN side. Its path is illustrated in Figure 3:

The enrichment of Algorithms 1 and 2 with additional relations resulted in establishing 530 prompts as a result of applying Algorithm 3.1 and 3873 prompts as a result of applying Algorithm 3.2.

4 Results

Following the results of the second implementation of the algorithms, the next step was the initial evaluation of the adopted procedures. In order to determine the accuracy of algorithms 1, 2, 3.1, and 3.2, six plWN noun synsets and adjective synsets related to them via relations holding between synsets and lexical units, which the algorithms have proposed the automatic prompts for, were chosen. The noun synsets, which were chosen, include the following: {Kolor 1} - "color", {Ksztalt 1} - "shape", {Material 1} - "material", {Smak 1} - "taste", {Zapach 1} - "smell", and {Wlasciwosc fizyczna 1} - "physical property". Next, the adjective synsets were given to lexicographers, whose task was to map their hyponyms on their PWN counterparts. After the editors have completed their task, the results of their decisions were compared to the results generated by algorithms 1, 2, 3.1 and 3.2.. The aim of the comparison was to test the degree to which the plWN adjective synsets related to synsets {Kolor 1} {Ksztalt 1} {Material 1} {Smak 1} {Zapach 1} {Wlasciwosc fizyczna 1} will be mapped onto PWN adjective synsets related to synsets {Color 1}, {Shape 1}, {Figure 6}, {Material 1}, {Taste 1}, {Olfactory property 1} and {Physical property 1}, respectively. The hypothesis that was tested was that Algorithms 2 and 3.2, which employ relations between the lexical units, will produce significantly better results than Algorithms 1 and 3.1., which

---

[3] Figures presented in Table 4 refer only to the six chosen synsets which were the focus of the evaluation

Table 4. Comparison of matching between automatic prompts and editors' choices[4]

Algorithm	All manually established relations	Relations established within corresponding algorithm-set nests	Relations established outside corresponding algorithm-set nests	%match
1	76	35	41	46%
2	561	458	103	82%
3.1	92	35	57	38%
3.2	784	589	195	75%

are based on relations between synsets. The relevant figures for the raw results are presented in Table 4 below.

As can be seen in Table 4, Algorithms 2 and 3.2 produce far better results than Algorithms 1 and 3.1. Accordingly, these results show that although the current mapping procedure applies to the level of synsets, it should ultimately be carried out at the level of lexical units.

5 Conclusions

An analysis of adjective relation structures in plWordNet and Princeton Word-Net revealed a number of contrasts: vertical vs. horizontal synset relation structure, different types and frequencies of relations. Thus, the mapping of adjectives has turned out to be a real challenge. To create a kind of mapping basis we have proposed two types of the automatic prompt generating algorithms and tested their performance on selected wordnet nests from the widely understood domain of physical property. The results of this pilot adjective mapping point to the greater precision of the second type of the proposed algorithm relying on both synset and lexical unit relations. This is probably motivated by the stronger correspondence between the types and frequencies of lexical unit relations of plWordNet and Princeton Wordnet. Still, the adjective part of plWordNet is under development and the manual adjective mapping is also in progress, which creates the chance for further verification of the proposed algorithms.

References

1. Fellbaum, C.: Wordnet: An electronic lexical database. MIT Press (1998)
2. Hamp, B., Feldweg, H., et al.: Germanet: a lexical-semantic net for German. In: Proceedings of ACL Workshop Automatic Information Extraction and Building of Lexical Semantic Resources for NLP Applications, pp. 9–15 (1997)
3. Vossen, P.: Eurowordnet general document version 3 final (2002)
4. Pianta, E., Bentivogli, L., Girardi, C.: Multiwordnet: developing an aligned multilingual database. In: Proceedings of the First International Conference on Global WordNet (January 2002)
5. Piasecki, M., Szpakowicz, S., Broda, B.: A Wordnet from the Ground Up. Oficyna Wydawnicza Politechniki Wroclawskiej, Wroclaw (2009)

6. Rudnicka, E., Maziarz, M., Piasecki, M., Szpakowicz, S.: A strategy of mapping polish wordnet onto princeton wordnet. In: COLING (Posters), pp. 1039–1048 (2012)
7. Kedzia, P., Piasecki, M., Przybycien, E.R.K.: Automatic prompt system in the process of mapping plwordnet on princeton wordnet. Cognitive Studies 13, 123–141 (2013)
8. Maziarz, M., Piasecki, M., Szpakowicz, S., Rabiega-Wiśniewska, J.: Semantic Relations among Nouns in Polish WordNet Grounded in Lexicographic and Semantic Tradition. Cognitive Studies 11 (2011)
9. Maziarz, M., Szpakowicz, S., Piasecki, M.: Semantic Relations among Adjectives in Polish WordNet 2.0: A New Relation Set, Discussion and Evaluation. Cognitive Studies/Études Cognitives 12, 149–179 (2012)
10. Collins, A.M., Quillian, M.R.: Retrieval time from semantic memory. Journal of Verbal Learning and Verbal Behavior 8(2), 240–247 (1969)
11. Deese, J.: The associative structure of some common english adjectives. Journal of Verbal Learning and Verbal Behavior 3(5), 347–357 (1964)
12. Sheinman, V., Fellbaum, C., Julien, I., Schulam, P., Tokunaga, T.: Large, huge or gigantic? identifying and encoding intensity relations among adjectives in wordnet. Language Resources and Evaluation 47(3), 797–816 (2013)
13. Piotrowski, T., Saloni, Z.: Slownik angielsko-polski i polsko-angielski. Wydawnictwo Naukowe PWN, Warszawa (2002)

Toward Exploring the Role of Disfluencies from an Acoustic Point of View: A New Aspect of (Dis)continuous Speech Prosody Modelling

György Szaszák[1]([✉]) and András Beke[2]

[1] Department of Telecommunications and Media Informatics,
Budapest University of Technology and Economics, Budapest, Hungary
szaszak@tmit.bme.hu
[2] Research Institute for Linguistics, Hungarian Academy of Sciences,
Budapest, Hungary
beke.andras@mta.nytud.hu

Abstract. Several studies use idealized, fluent utterances to comprehend spoken language. Disfluencies are often regarded to be just a noise in the speech flow. Other works argue that fragmented structures (disfluencies, silent and filled pauses) are important and can help better understanding. By extending the original concept of speech disfluency, the current paper involves the acoustic level and places the discontinuity of F0 in parallel with speech disfluencies. An exhaustive analysis of the advantages and disadvantages of using a continuous F0 estimate in prosodic event detection tasks is performed for formal and informal speaking styles. Results suggest that unlike in read (formal) speech, using a continuous, overall interpolated F0 curve is counterproductive in spontaneous (informal) speech. Comparing the behaviour of speech disfluencies and the effect of discontinuity of the F0 contour, results raise more general modelling philosophy considerations, as they suggest that disfluencies in informal speech may be by themselves informative entities, reflected also in the acoustic level organization of speech, which suggests that disfluencies in general are an important perceptual cue in human speech understanding.

Keywords: Disfluency · Interpolation · Prosody · Spontaneous speech

1 Introduction

Spontaneous speech tends to have incomplete syntactic, and even ungrammatical structure and is characterized by disfluencies, repairs and other non-linguistic vocalizations, etc. In general, spontaneous speech is hard to treat with conventional methods developed primarily for the formal or read speaking style, i.e. simple rule-based pattern learning and also data-driven approaches raise several difficulties. One of the most critical challenges is simply determining a broad segmentation of spontaneous speech, such as segmenting speaker turns into utterances.

© Springer International Publishing Switzerland 2015
P. Král and V. Matoušek (Eds.): TSD 2015, LNAI 9302, pp. 369–377, 2015.
DOI: 10.1007/978-3-319-24033-6_42

Sometimes agreement is also missing on the definition of prosodic categories [1]. In current work we rely on the prosodic hierarchy described in [2] and use the phono-logical phrase level for modelling. We define phonological phrases as prosodic units characterized by an own stress and followed with a less or more complete intonation contour.

Several phrasing or boundary detection approaches have been developed and analysed for read and slightly spontaneous speech (such as semi-formal speech used in information retrieval systems) [3], [4]. The automatic phrasing imple-mented for read speech in [5] was able to yield a reliable (accuracies ranging between 70-80%) phrasing down to the phonological phrase level, and also to separate intonational phrase level from the underlying phonological phrase level. This approach required clustering of a number of phonological phrase prototypes, which were then modelled by Hidden Markov Models/Gaussian Mixture Models (HMM/GMM) based on acoustic-prosodic features. Clustering of such charac-teristic prototypes in spontaneous speech was less successful: an effort to try to identify and cluster characteristic prosodic entities or phrase types in Hungarian spontaneous speech by using an unsupervised approach has lead only to partial success [6].

Beside recognizing spontaneous speech as a standalone phenomenon, requir-ing quite a different approach as compared to read speech, it has been a com-mon practice to try to trace back the processing of spontaneous speech to that of read speech. In other words, use and adapt algorithms or tools developed for read speech in tasks involving processing of spontaneous speech. A characteristic phenomenon, which is often in the focus of this "de-spontaneisation" is speech disfluency, heavily present in spontaneous speech. However, detecting and elim-inating or "repairing" disfluency might be counterproductive on some levels of speech processing. Several studies [7] [8] argue that if disfluencies are available in the speech transcription, those can play an important role in disambiguation between sentence-like units. The same can be the case on the acoustic level: pre-serving disfluency may be sometimes useful. In the current paper, the authors would like to focus on this aspect and compare read and spontaneous speech processing in a prosodic event detection related task.

A frequent disfluency type is filled pause in spontaneous speech. Cook and Lallijee suggested [9] that filled pauses may have something to do with the lis-tener's perception of disfluent speech. They showed that speech may be more comprehensible when it contains filler material during hesitations by preserving continuity and that filled pauses may serve as a signal to draw the listeners attention to the next utterance in order to help the listener not to be able to perceive the onset of the following utterance. Similarly, Swerts and Ostendorf found [10] that in human-machine interactions, turns introducing a new topic tended to have more disfluencies than other turns, showing that a speech rec-ognizer may exploit these disfluencies to detect discourse structure. Swerts et al. analysed the role of filled pauses in discourse structure [11]. They found that phrases following major discourse boundaries contain filled pauses more often.

The silent or filled pauses are generally located at syntactic or prosodic boundaries [12]. Hirst and Cristo demostrated that silent and filled pauses constitute the acoustic markers that enclose the prosodic units [13]. Silent pauses always involve the disruption of the F0 contour. Filled pauses on the other hand are often schwa-like hesitations which are hence voiced. Hereafter, a pitch reset may call the listeners attention and signal the prosodic boundary.

This paper focuses on exploring the advantages and disadvantages of continuity vs discontinuity in speech processing. Recently, several pitch trackers providing stable and overall continuous F0 estimate have been released [14]. Using a continuous, overall interpolated F0 estimation is compared to a case where F0 is kept fragmented or interpolated only partially. The effect of continuous vs discontinuous F0 is evaluated in a prosodic event detection task, separately for read and spontaneous speaking styles. The authors believe that beside yielding some basic, but practically important results these experiments may contribute to a better understanding of spontaneous speech.

The paper is organized as follows: first material (read and spontaneous speech corpora) are presented shortly, followed by the description of the phonological phrase segmentation algorithm used for evaluation in the experiments. Thereafter, experiments are run with original (undefined F0 for unvoiced frames), partially and overall interpolated F0 processing are presented and evaluated for read and spontaneous speech. Finally, conclusions are drawn.

2   Material and Methods

This section describes speech databases and basic processing tools used for the experiments later.

2.1   Speech Databases

BABEL is a Hungarian read speech database, designed for research [15]. A subset of BABEL is used in current experiments, which is labelled for intonational (IP) and phonological phrases (PP). The used subset contains 300 sentences uttered by 22 speakers, containing 2067 PPs, labelled according to 7 types as described in [5].

BEA is a spoken language database [16]. It is the first Hungarian database of its kind in the sense that it involves many speakers, very large amount of spontaneous, informal speech material. The recording conditions of the database were kept permanent and of studio quality. 8 spontaneous narratives were selected (4 male and 4 female) from the database. The sub-corpus was manually annotated by two different phoneticians. The annotation contained three levels: intonational phrases (IP), phonological phrases (PP) and also involved a word level transcription [2]. The IP can be thought of being a part of speech forming a unity in terms of stress and intonation contour, and is found often between two pauses. The database, in most cases the IP boundaries are bound to pauses. A number of filled pauses were perceived as separate IPs by the annotators.

Therefore, filled pauses (FP) were annotated separately in the transcription. As already explained, the IPs can be further divided into PPs based on intonation and stress pattern. A PP is a unity characterized by its own stress and intonation contour, but this latter can be unterminated (continued in next PP). The corpus contained 398 IPs and 751 PPs in total.

2.2 Automatic Segmentation for Phonological Phrases

This section describes the automatic PP segmentation algorithm used in the experiments. Being a prosodic event detection task, its behaviour is analysed with continuous and fragmented F0 patterns. Experiments are run in Hungarian language for read and spontaneous utterances separately.

PPs constitute a prosodic unit, characterized by an own stress and some preceding/following intonation contour. As this contour is specific, PPs can be classified and hence modelled separately, in a data-driven machine learning approach. The distinction between PPs consists of two components: the strength of stress the PP carries and the PPs? intonation contour. In this way 7 different types are distinguished. Modelling is done with HMM/GMM models, and PP segmentation is carried out as a Viterbi alignment of PPs for the utterances requiring segmentation. During this Viterbi alignment, all PPs are allowed to occur with equal probability. A parameter influencing insertion likelihood for PPs can be tuned to force or prevent a more dense segmentation for PPs. The more dense alignment we require, the higher the probability is the insertion of false PP boundaries, resulting often from a confusion between microprosodic variations and accents/prominences resulting from stress. The overall approach is documented in details in [5].

As acoustic-prosodic features, fundamental frequency (F0) and wide-band energy (E) are used [17]. Syllable duration is not used for Hungarian as it was not found to be a distinctive cue in this task [5]. Post-processing alternatives for F0 are described in the respective section later. For energy computation a standard integrating approach is applied with a window span of 150 ms. Frame rate is 10 ms. First and second order deltas are appended to both F0 and E streams.

Evaluation of PP segmentation is done with a 10-fold cross-validation. The PP alignment is generated with models trained on utterances different from the one under segmentation. The generated PP alignment is then compared to the reference obtained by hand-labelling. Detection is regarded to be correct if the boundary is detected within the $TOL=100$ ms vicinity of the reference. Once all utterances have the automatic PP segmentation ready, the following performance indicators are evaluated:

- recall (RCL) of PP boundaries,
- precision (PRC) of PP boundaries and
- the average time deviation (ATD) between the detected and the reference PP boundary.

3   Overall vs Partial Interpolation of F0

In this section, several scenarios are evaluated using globally or partially continuous $F0$ contours per utterance in the PP alignment task.

All extracted F0 contours are subject to error correction resulting from octal halving/doubling, done by signal processing tool described in [5].

Further post-processing of $F0$ varies according to the scenario, whereas energy is always kept unchanged (by nature continuous, energy is extracted by a 25 ms window each 10 ms, however smoothed with a mean filter of a span of 150 ms afterwards).

3.1   Post Processing Alternatives for F0

The basic interest is to see whether using a continuous contour (all vacancies interpolated) outperforms a non interpolated or partially interpolated contour in the PP detection task. Tested scenarios cover 3 cases as follows:

- Use the F0 contour as produced by a conventional pitch tracker (Snack V2.2.10 in our case [17], doubling/halving errors corrected automatically);
- Use a continuous F0 contour, interpolated at all unvoiced parts;
- Use a partially F0 interpolated contour, where interpolation is omitted if the length of the unvoiced interval exceeds a limit (250 ms in the experiments) or if $F0$ starts significantly higher (suspected pitch reset) than it was before the unvoiced segment (criterion applied in the experiments: $F0_{former} * 1.1 < F0_{current}$).

The motivation to constrain the disruption of the F0 contour in the partial interpolation scenario comes from the following considerations:

- The silence limit is set because a silent period longer than 250 ms can hardly be considered as fluent speech. In such cases the speaker may not employ a pitch reset as the silence in itself can be a clear acoustic marker of PP (and IP) boundary. If this happens interpolation may mask the PP boundary, although energy features are still likely to signal it.
- Medium strength pitch resets may often be smoothed by the F0 interpolation, which makes further detection more difficult. However, a factor of 1.1 is preferred in order to avoid that microprosodic disturbances give false PP (IP) boundary detection, which may happen in the vicinity of long plosives for example.

3.2   Results

Results are shown in Table 1. Regarding precision (PRC) and recall (RCL), they highly depend on a parameter influencing PP insertion likelihoods during the PP segmentation done with Viterbi alignment (see the PRC-RCL curve in Fig. 1). Therefore, segmentation results are shown for operating points where precision and recall are equal ($PRC=RCL$). The settings of the tolerance interval TOL

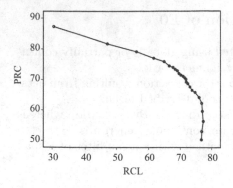

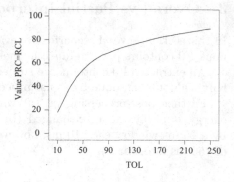

Fig. 1. Precision [%] and recall [%] as influenced by the PP insertion likelihood in the automatic PP segmentation. Read speech, total F0 interpolation, $TOL = 100ms$.

Fig. 2. Precision and recall in operating points defined by $PRC = RCL$ [%] depending on TOL [ms]. Read speech, total F0 interpolation.

Table 1. Precision (PRC) and recall (RCL) in operating points defined by $PRC = RCL$ and ATD for the 3 evaluated scenarios and for read and spontaneous speech ($TOL = 200\ ms$).

Style	F0 interpolation	$PRC = RCL$ [%]	ATD [ms]
	None	62.9	93.0
Read speech	Partial	68.2	80.1
	Total	81.2	55.9
	None	57.7	52.9
Spontaneous	Partial	69.7	44.1
	Total	66.3	44.8

also influence results (see Fig. 2). Unless written else explicitly for the experiments, $TOL = 200ms$ will be used, corresponding roughly to the length of a syllable on average. We believe this deviation is admissible in a supra-segmental detection task, since average PP length is of 659.0 ms in the read speech BABEL corpus and of 782.9 ms in the BEA spontaneous corpus.

As it can be seen from results in Table 1 read and spontaneous speech styles show different behaviour. Whereas in read speech, the more continuous the contour is the better the PP detection results are, this is not the case for spontaneous speech, where the best performing approach constitutes a compromise between do not interpolate at all and interpolate everything: the partially interpolated contour yields the best results for spontaneous speech, where interpolation is omitted if the length of the unvoiced interval exceeds 250 ms or if F0 starts by $F0_{former} * 1.1 < F0_{current}$ higher (suspected pitch reset) than it was before the unvoiced segment.

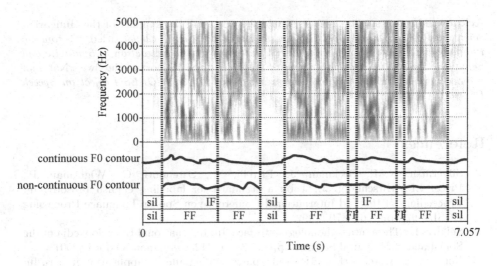

Fig. 3. An example with continuous and partially interpolated F0 contours with IP and PP labelling.

These results suggest that the discontinuity of F0 plays an important cue is spontaneous speech in human perception as well once automatic approaches can exploit it. On the other hand, these results may also make doubtful any attempt to try to "de-spontanize" spontaneous speech in order to transform it into "read" style, and treat it with tools developed for read speech.

4   Conclusions

In the present paper we investigated the effects of using continuous (overall interpolated) vs a fragmented, eventually partially interpolated F0 estimate. Although a noticeable tendency of nowadays is to favour pitch trackers yielding a totally continuous F0 estimate [14], results have shown that this is useful only in read, formal speaking styles, where the overall continuous F0 contour outperformed the partially interpolated one by 19.1% relative in the precision of a phonological phrase segmentation task. According to the results, a partially interpolated F0, where interpolation leaves intact places with longer unvoiced periods or pitch reset suspect F0 increase from one voiced segment to the other, yields by 5.1% relative better results over total interpolation in spontaneous speech in the same PP segmentation task. Beside speech technology applications where spontaneous speech seems to be better treated with only a piecewise interpolation of F0, results also suggest some other considerations regarding human speech perception, however, these latter remain to be confirmed with targeted experiments.

Acknowledgments. The authors would like to thank the support of the Hungarian Scientific Research Fund (OTKA) under contract ID *PD 112598*, titled *"Automatic Phonological Phrase and Prosodic Event Detection for the Extraction of Syntactic and Semantic/Pragmatic Information from Speech"* and the support of the Swiss National Science Foundation via the joint research project *"SP2: SCOPES Project on Speech Prosody"* (No. CRSII2-147611 / 1).

References

1. Silverman, K.M., Beckman, J., Pitrelli, M., Ostendorf, C., Wightman, P., Price, J.P., Hirschberg, J.: Tobi: a standard for labelling english prosody. In: Proceedings of the 2nd International Conference on Spoken Language Processing (ICSLP-92), pp. 867–870 (1992)
2. Selkirk, E.: The syntax-phonology interface. In: International Encyclopaedia of the Social and Behavioural Sciences, pp. 15407–15412. Pergamon, Oxford (2001)
3. Veilleux, N., Ostendorf, M.: Prosody/parse scoring and its application in atis. In: Proceedings of the Workshop on Human Language Technology, pp. 335–340 (1993)
4. Gallwitz, F., Niemann, H., Nöth, E., Warnke, W.: Integrated recognition of words and prosodic phrase boundaries. Speech Communication **36**(1–2), 81–95 (2002)
5. Szaszák, G., Beke, A.: Exploiting prosody for automatic syntactic phrase boundary detection in speech. Journal of Language Modeling **0**(1), 143–172 (2012)
6. Beke, A., Szaszák, G.: Unsupervised clustering of prosodic patterns in spontaneous speech. In: Sojka, P., Horák, A., Kopeček, I., Pala, K. (eds.) TSD 2012. LNCS, vol. 7499, pp. 648–655. Springer, Heidelberg (2012)
7. Medeiros, H., Batista, F., Moniz, H., Trancoso, I., Meinedo, H.: Experiments on automatic detection of filled pauses using prosodic features. Actas de Inforum **2013**, 335–345 (2013)
8. Swerts, M.: Filled pauses as markers of discourse structure. Journal of Pragmatics **30**, 485–946 (1998)
9. Cook, H., Lallijee, M.: The interpretation of pauses by the listener. Brit. J. Soc. Clin. Psy. **9**, 375–376 (1970)
10. Swerts, M., Ostendorf, M.: Prosodic and lexical indications of discourse structure in human-machine interactions. Speech Communication **22**(1), 25–41 (1997)
11. Swerts, A., Wichmann, A., Beun, R.J.: Filled pauses as markers of discourse structure. In: Proceedings ICSLP96, Fourth International Conference on Spoken Language Processing, pp. 1033–1036 (1996)
12. Zellner, B.: Pauses and the temporal structure of speech. In: Fundamentals of Speech Synthesis and Speech Recognition, pp. 41–62. John Wiley, Chichester (1994)
13. Hirst, D., Cristo, A.D.: Intonation Systems: A Survey of Twenty Languages. Cambridge University Press, New York (1989)
14. Ghahremani, P., BabaAli, B., Povey, D., Riedhammer, K., Trmal, J., Khudanpur, S.: A pitch extraction algorithm tuned for automatic speech recognition. In: Proceedings of the IEEE International Conference on Acoustics, Speech and Signal Processing, pp. 2494–2498 (2014)
15. Roach, P.S., Amfield, S., Bany, W., Baltova, J., Boldea, M., Fourcin, A., Goner, W., Gubrynowicz, R., Hallum, E., Lamep, L., Marasek, K., Marchal, A., Meiste, E., Vicsi, K.: Babel: an eastern european multi-language database. In: International Conf. on Speech and Language, pp. 1033–1036 (1996)

16. Neuberger, T., Gyarmathy, D., Gráczi, T.E., Horváth, V., Gósy, M., Beke, A.: Development of a large spontaneous speech database of agglutinative Hungarian language. In: Sojka, P., Horák, A., Kopeček, I., Pala, K. (eds.) TSD 2014. LNCS, vol. 8655, pp. 424–431. Springer, Heidelberg (2014)
17. Sjölander, K., Beskow, A.: Wavesurfer - an open source speech tool. In: Proceedings of the 6th International Conference of Spoken Language Processing, vol. 4, pp. 464–467 (2000)

Heuristic Algorithm for Zero Subject Detection in Polish

Adam Kaczmarek[1,2](✉) and Michał Marcińczuk[2]

[1] Computational Intelligence Reserach Group, Institute of Computer Science,
University of Wrocław, Wrocław, Poland
akaczmarek@cs.uni.wroc.pl
[2] G4.19 Research Group: Computational Linguistics and Language Technology,
Department of Computational Intelligence, Wrocław University of Technology,
Wrocław, Poland
michal.marcinczuk@pwr.edu.pl

Abstract. This article describes a heuristic approach to zero subject detection in Polish. It focuses on the zero subject detection as a crucial step in end-to-end coreference resolution. The zero subject verbs are recognized using a set of manually created rules utilizing information from different sources, including: a dependency parser, a shallow relational parser and a valence dictionary. The rules were developed and evaluated on the Polish Coreference Corpus. The experimental results show that the presented method significantly outperforms the only machine learning-based alternative for Polish, i.e., MentionDetector. We also discuss and evaluate the importance of zero subject detection for existing coreference resolution tools for Polish.

Keywords: Zero subject · Anaphora detection · Coreference resolution · Polish

1 Introduction

Zero subject is a linguistic phenomenon that occurs in certain languages, i.e., Balto-Slavic, Romanian, Spanish, Portuguese, Chinese and Japanese. It appears when an independent phrase, with a non-subject-blocking verb, lacks an explicit subject. In Polish the subject agreement is reflected in the verb morphology. We can see this problem as a binary classification of verbs whether they have or not a zero subject. The example below[1] shows a coreference between the zero-subject verb *consists* and the proper name *The Toronto Dominion Centre*:

Toronto Dominion Centre - kompleks handlowo-kulturalny (...). ϕ-**Sklada sie** z 3 czarnych budynków (...).

The Toronto-Dominion Centre - is a cluster of buildings (...) of commercial and cultural function. ϕ-***consists**_{sg:masc:ter}$ of three black buildings (...)

---

[1] The fragment comes from the KPWr corpus [1].

© Springer International Publishing Switzerland 2015
P. Král and V. Matoušek (Eds.): TSD 2015, LNAI 9302, pp. 378–386, 2015.
DOI: 10.1007/978-3-319-24033-6_43

This is very important issue for coreference resolution and has very significant impact on the results. Here we treat the end-to-end coreference resolution as a sequential process in which the determination of coreferntial relations is preceded by a mention detection. In this study we describe an algorithm developed for the problem of zero subject identification called *Minos* (Mention IdentificatioN for Omitted Subjects).

2 Related Work

Polish is one of the languages classified as a *pro-drop* language (Chomsky, [2]) meaning that omitted subjects are commonly seen phenomenon. There have been studies for a few other *pro-drop* languages on the problem of zero subject detection. The verbs lacking explicit subjects do not necessarily have elided subject, but they can also be impersonal and do not have subject at all, by their nature. Most researchers consider the problem of zero subject detection as a crucial preprocessing step in coreference resolution. There are also some other applications, for example in machine translation to improve the quality of translation of sentence with null verbs [3]. The existing researches show that the problem of zero subject detection is not trivial. The performance for different languages varies from 57.9% of F-measure for Portuguese [4] to 74.2% for Romanian [5]. The problem of zero subject detection is not very widely studied for Polish. According to our best knowledge there is only one system for Polish which handles zero subject detection, i.e. MentionDetector [6]. It uses a machine-learning technique for rule induction called RIPPER.

3 Problem Definition

As zero subjects in Polish are reflected in verbs' properties, in our work we will identify zero-subjects with corresponding verbs called further *zero-subject verbs* and we will basically consider the problem of zero subject detection as a binary classification of verbs. However, the free word order nature of Polish and rich properties of verbs make it a non-trivial task. We have made some observations on the nature of verbs in Polish resulting in a conclusion that the zero-subject verbs cannot be subject-blocking and are only a subclass of verbs not having explicitly given subject in text.

3.1 Functional Verb Classification

Basically, we employ two kinds of partition among verbs: depending on whether there is an explicit subject for given verb, and depending on whether the verb is subject-blocking or not. The *subject-blocking verbs* are verbs that by definition cannot have any subject explicit or implicit. Obviously there are no verbs that are subject-blocking and have explicit subject, what gives us three classes of verbs. While it is relatively simple how to distinguish verbs having explicit subject, we need to make some additional observations on subject-blocking verbs to distinguish them properly from zero subjects. The subject-blocking verbs can be shown in several subcategories describing their properties that make them distinguishable from zero subjects:

Predicatives (*pora, sposób, etc.*) — words act as verbs, but do not have inflexional endings and do not connect with subjects in nominative case.

Jak to osiagnac — doprawdy **nie wiadomo**.
*How to achieve it — this indeed **is not known**.*

Improper verbs — verbs that are inflected only by tense and mode, but not by person, gender or number:

Mnie juz w rzezni **mdli**!
I_{dative} *feel sick in slaughterhouse!*

Impersonals — verbs which occur in a form not assigned to any specific grammatical person; usually ending with *-no, -to*:

W programie spotkania **przewidziano** prezentacje najnowszych produktów.
*In the program of the meeting **there was provided** a presentation of the newest products*

Infinitives — verbs in base form (like *to be, to have*):

Zaczal Dratewka pszczolom **pomagac**.
*Dratewka started **to help** the bees.*

One should constructions (*nalezaloby, trzeba, etc.*):

Nalezaloby zamontowac wiecej koszy na odpadki.
***One should** install more recycle bins.*

Pleonastic *it*:

Pada
[It] is raining.

Non-reflexive personal verb form used with reflexive particle *sie*:

Nie mówi sie tez "ups" i "wow", nie klamie i nie przeklina.
You don't say "whoops" and "wow", you do not lie and do not curse.

3.2   Definition of *Verb* and *Noun*

For the need of zero subject detection we introduce definitions of **verb** and **noun** based on [6] as follows:

Verb — verbs are all words having *fin, praet, bedzie, winien* part-of-speech. We exclude many words usually considered as verbs having following parts-of-speech: *impt, imps, inf, pcon, pant, pact, ppas, pred, aglt* as they cannot have subject. Aglutinates are merged with associated verbs.

Noun — definition of nouns is extended to pronouns, gerunds, numerals and some adjectives, because they frequently are subjects and share most morphosyntactic features of nouns. We also do not consider *siebie* (Eng. 'self') usually treated as pronoun as it cannot have a subject.

4   Data

For development and evaluation we used the Polish Coreference Corpus (henceforth PCC) [7]. It consists of documents 250–350 word-long. For the purpose of algorithm development we used the same subcorpus of PCC as in [6], which contains 779 document with total 22k verbs (including 13k sentences with total of 6k verbs with zero subject). The corpus was initially split into two equally sized parts: *development* and *test*. The *development* part was used to make observations about the verb properties and to verify partial results during the development process. After each change the error analysis was performed on this part of corpus. The *test* part was used to calculate the final score.

5   Algorithm

The algorithm design is based on an assumption that in the sequential end-to-end coreference resolution process we might use a system for coreferential relations recognition that makes joint decisions about coreferential relation and anaphoricity. Thus, we prefer *recall* of the zero subject detection algorithm over *precision*. In the subsequent steps the false positives might be discarded, while the false negatives will not be restored. The main part of this study is devoted to the third person verbs. Nevertheless, some specific methods for first and second person verbs are also presented in the last paragraph.

WCRFT Retagging with Guesser. The PCC corpus was tagged without a guesser and all out-of-dictionary words (proper names) were tagged as *ign*, i.e. no morphological analysis was assign. The missing tags lower the recall of used tools and rules. Because of that, we decided to retag the corpus using the WCRFT tagger [8] with a guesser module (the module try to guess the morphological tag for unknown words).

Filtering of Verbs Based on Functional Properties. First we filter verbs classified according to the subject-blocking verbs description from Section 3. We use a set of hand-written rules to determine the subject-blocking property and additionally we define arbitrarily two sets of verbs: *subject-blocking verbs* — verbs are, with high probability, subject-blocking and *reflexive subject-blocking verbs* — verbs which are highly probable to be subject-blocking when occurring with the reflexive particle *sie*, serving as black lists of base forms of verbs to discard (classify as non-coreferential). These lists were based on verbs without subject in valence frame from the Polish Valence Dictionary *Walenty* [9] and the *development* part of corpus as well as on list of pseudo-verbs used in [6], however the last list was much longer due to use in machine learning algorithm as feature describing the possibility of being subject-blocking verb, while our lists express very high confidence in the subject-blocking property of verbs.

Verb-Noun Agreement Check. In most cases we perform a validation step for subject candidate to check if it can be morphologically fitted to given verb — in Polish the main verb in clause shall determine morphology of its subject. Precisely, we check first for proper part-of-speech tag restricting adjectives to subject candidates found using relations returned by *MaltParser*. Additionally, for relative pronoun *który* we skip the proper part-of-speech requirement. Then we proceed to check grammatical case — basically we accept only nouns having nominative case as possible subjects, however due to specificity of Polish, we consider also nouns having *subst* part-of-speech tag that have accusative case, due to the fact that these two cases have often the same orthographical form, causing common tagger errors. At last we check for person, gender and number agreement.

Dependency Parser with Agreement Check. The most substantial part of the algorithm is the use of dependency parser to directly find *subject* relation between *verb* and *noun*. We use MaltParser [10], a transition-based deterministic parser using classifier predicted shift-reduce action sequence. The parser is configured to use a built-in *stackeager* parsing algorithm, designed for non-projective dependency structures. Dependency parser for Polish is trained on the Polish Dependency Bank [11]. Having every sentence parsed into a dependency tree, we find nouns that are connected to verb by a subject relation (*subj*). Subject candidates found in this step were subsequently verified as described in section Verb-Noun Agreement Check.

Annotation Relation Marker ChunkRel. Second method used to find subjects directly connected to verbs is to use *ChunkRel* [12] — a statistical tool for recognition of selected relations between noun phrases, adjective phrases and verb phrases, namely *subject*, *object* and *copula*. Analogously we find subject candidates that are directly connected to verb by the *subject* relation. However, as experiments shown, better results were obtained when subject verification was omitted. *ChunkRel* performs slightly better in indicating that verb should have subject than in indicating its subject, but yet the first is more useful for our purpose.

Contextual Subject Lookup. The statistical learning methods mentioned above were complemented with contextual subject lookup within certain range around classified verbs. Our experiments shown that the best results were achieved for a search window of 15 tokens preceding and 9 tokens succeeding the verb. In such a window we look for **noun** that is in agreement with verb, if we find some, we accept it as valid subject. Additionally, we restricted this search with punctuation in case of considering complex clause indicated by presence of other **verbs**.

Specific Subject Cases. Additionally, we consider some non-intuitive cases of subjects which occur in Polish. First we implemented recognition of constructions consisting of a neutral singular verb occurring with a numeral followed by a plural noun in genitive or dative:

W glosowaniu **wzielo** udzial **332 poslów**.
332 deputies$_{genitive}$ **took part**$_{neutral}$ *in voting*.

And second, we consider grammatical negations of *to have* in present tense taking semantic form of negation of *to be* verb. For example, "I **nie ma** nikogo, kto nas przed nimi obroni" looks like *"And (He)* **does not have** *anybody to save us from them"* but have actual semantic meaning: *"And* **there is** *nobody to save us from them"*.

First and Second Person verbs. While all considerations mentioned above were generally focused on third person verbs, we also took care of first and second person verbs manifesting different properties in the perspective of being zero subject. Excluding the usage of *MaltParser* to determine the existence of a subject we use the following properties:

Preceeding personal pronoun — In many cases the first and second person verbs occur with preceding personal pronoun, so lookup for personal pronouns is one of the ways to determine if such verb is zero-subject.

Please phrases — The first person verb **prosze** (please) occurs often in phrases like: "**Prosze** usiasc." (*Please* *sit down.*).

6   Results

First, we evaluated the influence of zero subject detection on the final coreference resolution. We tested three tools for Polish: Ruler [13], Bartek [14] and IKAR [15]. It is important to notice, that the tools cannot be compared with each other as they were developed using different data. Table 1 contains precision (P), recall (R) and their harmonic mean (F1) of coreferential links for each of the tools with different set of mentions. In the first column we show results on the corpus with all verbs annotated as zero subjects in the second we provide results on corpus annotated with verbs classified as mentions by mention detection system and in third we present results achieved on corpus with gold standard mentions. The results show the intra-tool improvement of coreference resolution — the better zero subject detection is, the higher performance is achieved. This shows, that even non-ideal zero subject detection gives better result than a naïve classification of all verbs.

In Table 2 we present results of binary classification of verbs done by *Minos* at subsequent development steps. First we used only the verb discarding rules. Next we present results for *MaltParser* with subject agreement validation and *ChunkRel* without subject agreement validation as it performs better in that configuration. Afterwards we show performance of the two parsers combined together and at last we present performance of the both parsers complemented by contextual subject lookup. The last configuration presented in this table is the one used in other evaluations presented in this paper.

Table 1. Influence of zero subject detection on coreference resolution. Results of coreference resolution on coreference-annotated subcorpus of KPWr corpus using P, R and F1 for coreferential links part of the BLANC metric.

Tool	All verbs			Mention detection			Gold standard		
	P	R	F1	P	R	F1	P	R	F1
Bartek	7.88%	52.21%	13.69%	**64.78%**	**18.59%**	**28.89%**	64.18%	27.12%	38.13%
Ruler	26.91%	24.54%	25.67%	**65.26%**	**18.21%**	**28.48%**	60.05%	26.13%	36.42%
Ikar	11.93%	43.52%	18.72%	**61.37%**	**50.16%**	**55.20%**	61.17%	59.29%	60.22%

Table 2. Minos results for different configurations on PCC (development)

Minos configuration	TP	TN	FP	FN	Precision	Recall	F1
Verbs rules	3041	938	6759	63	32.91%	98.06%	49.29%
Malt + subject agreement	2962	4872	2825	142	51.15%	95.51%	66.62%
ChunkRel	2797	5507	2190	307	56.05%	90.18%	69.13%
Malt + ChunkRel	2792	5890	1807	312	60.67%	90.02%	72.49%
Malt + ChunkRel + subject lookup	2627	6209	1005	475	72.33%	84.69%	78.02%

Table 3. Comparison of Minos and MentionDetector results on PCC (test) and KPWr

Verbs	Algorithm	PCC (test)			KPWr		
		Precision	Recall	F1	Precision	Recall	F1
All	Minos	72.33%	84.69%	78.02%	82.25%	69.55%	75.48%
	MentionDetector	71.79%	67.39%	69.60%	88.74%	53.77%	66.97%
Third	Minos	55.47%	69.49%	61.69%	70.32%	50.87%	59.03%
Person	MentionDetector	62.56%	33.62%	43.74%	78.51%	27.42%	40.65%

The final evaluation was performed on the test part of PCC and the KPWr corpus. For PCC the results for the best Minos configuration were compared with results reported in [6] for MentionDetector. Minos obtained statistically significant better performance than MentionDetector[2]. In addition we evaluated both algorithms for third person verbs only as they are more ambiguous than first and second person. The results are shown in Table 3.

7   Conclusions and Future Work

This article presented a rule based approach to zero subject detection in Polish. The evaluation on the PCC and KPWr showed that our method significantly outperforms the alternative machine learning approach presented by Kopec in [6]. We have shown that the zero subject identification process have great impact on the performance of coreference resolution systems. Finally, we implemented our algorithm as a part of Liner2 toolkit [16][3]. Concerning the fact of outperforming a machine-learning algorithm we consider applying some approach that would

---

[2] We used the McNemar test with significance level $\alpha = 0.05$.

[3] A demo is available at http://inforex.clarin-pl.eu/index.php?page=ner.

base on features extracted from *Minos* hoping to achieve further improvements. We consider also approaching the subject detection problem using deep learning methods like word embeddings.

Acknowledgments. Work financed as part of the investment in the CLARIN-PL research infrastructure funded by the Polish Ministry of Science and Higher Education. One of the authors is receiving Scholarship financed by European Union within European Social Fund.

References

1. Broda, B., Marcińczuk, M., Maziarz, M., Radziszewski, A., Wardyński, A.: KPWr: towards a free corpus of Polish. In: Calzolari, N., Choukri, K., Declerck, T., Doğan, M.U., Maegaard, B., Mariani, J., Odijk, J., Piperidis, S. (eds.) Proceedings of LREC 2012, Istanbul, Turkey. ELRA (2012)
2. Chomsky, N.: Lectures on government and binding. In: The Pisa Lectures. Foris Publications, Holland (1981)
3. Russo, L., Loáiciga, S., Gulati, A.: Improving machine translation of null subjects in italian and spanish. In: Proceedings of the Student Research Workshop at the 13th Conference of the European Chapter of the Association for Computational Linguistics, Avignon, France, pp. 81–89. Association for Computational Linguistics, April 2012
4. Rello, L., Ferraro, G., Gayo, I.: A first approach to the automatic detection of zero subjects and impersonal constructions in portuguese. Procesamiento del Lenguaje Natural **49**, 163–170 (2012)
5. Mihăilă, C., Ilisei, I., Inkpen, D.: Zero pronominal anaphora resolution for the romanian language
6. Kopeć, M.: Zero subject detection for Polish. In: Proceedings of the 14th Conference of the European Chapter of the Association for Computational Linguistics. Short Papers, Gothenburg, Sweden, vol. 2, pp. 221–225. Association for Computational Linguistics (2014)
7. Ogrodniczuk, M., Głowińska, K., Kopeć, M., Savary, A., Zawisławska, M.: Polish coreference corpus. In: Vetulani, Z. (ed.) Proceedings of the 6th Language & Technology Conference: Human Language Technologies as a Challenge for Computer Science and Linguistics, Poznań, Poland, Wydawnictwo Poznańskie, Fundacja Uniwersytetu im, pp. 494–498. Adama Mickiewicza (2013)
8. Radziszewski, A.: A tiered CRF tagger for Polish. In: Bembenik, R., Skonieczny, Ł., Rybiński, H., Kryszkiewicz, M., Niezgódka, M. (eds.) Intell. Tools for Building a Scientific Information. SCI, vol. 467, pp. 215–230. Springer, Heidelberg (2013)
9. Przepiórkowski, A., Hajnicz, E., Patejuk, A., Woliński, M., Skwarski, F., Świdziński, M.: Walenty: towards a comprehensive valence dictionary of polish. In: Chair, N.C.C., Choukri, K., Declerck, T., Loftsson, H., Maegaard, B., Mariani, J., Moreno, A., Odijk, J., Piperidis, S. (eds.) Proceedings of the Ninth International Conference on Language Resources and Evaluation (LREC 2014), Reykjavik, Iceland. European Language Resources Association (ELRA), May 2014
10. Nivre, J., Hall, J., Nilsson, J.: Maltparser: a data-driven parser-generator for dependency parsing. In: Proc. of LREC-2006, pp. 2216–2219 (2006)
11. Wróblewska, A.: Polish dependency bank. Linguistic Issues in Language Technology **7**(1) (2012)

12. Radziszewski, A., Orłowicz, P., Broda, B.: Classification of predicate-argument relations in Polish data. In: Kłopotek, M.A., Koronacki, J., Marciniak, M., Mykowiecka, A., Wierzchoń, S.T. (eds.) IIS 2013. LNCS, vol. 7912, pp. 28–38. Springer, Heidelberg (2013)
13. Ogrodniczuk, M., Kopeć, M.: Rule-based coreference resolution module for Polish. In: Proceedings of the 8th Discourse Anaphora and Anaphor Resolution Colloquium (DAARC 2011), Faro, Portugal, pp. 191–200 (2011)
14. Kopeć, M., Ogrodniczuk, M.: Creating a coreference resolution system for Polish. In: Proceedings of the Eighth International Conference on Language Resources and Evaluation, LREC 2012, Istanbul, Turkey, pp. 192–195. ELRA (2012)
15. Broda, B., Burdka, L., Maziarz, M.: IKAR: an improved kit for anaphora resolution for Polish. In: Proceedings of COLING 2012: Demonstration Papers, Mumbai, India, pp. 25–32. The COLING 2012 Organizing Committee, December 2012
16. Marcińczuk, M., Kocoń, J., Janicki, M.: Liner2 — a customizable framework for proper names recognition for Polish. In: Bembenik, R., Skonieczny, Ł., Rybiński, H., Kryszkiewicz, M., Niezgódka, M. (eds.) Intell. Tools for Building a Scientific Information. SCI, vol. 467, pp. 231–254. Springer, Heidelberg (2013)

Vocal Tract Length Normalization Features for Audio Search

Maulik C. Madhavi[1](✉), Shubham Sharma[2], and Hemant A. Patil[1]

[1] Dhirubhai Ambani Institute of Information
and Communication Technology, Gandhinagar, India
{maulik_madhavi,hemant_patil}@daiict.ac.in
[2] Indian Institute of Science, Bangalore, India
shubham@mile.ee.iisc.ernet.in
http://www.daiict.ac.in

Abstract. This paper presents speaker normalization approaches for audio search task. Conventional state-of-the-art feature set, *viz.*, Mel Frequency Cepstral Coefficients (MFCC) is known to contain speaker-specific and linguistic information implicitly. This might create problem for speaker-independent audio search task. In this paper, universal warping-based approach is used for vocal tract length normalization in audio search. In particular, features such as scale transform and warped linear prediction are used to compensate speaker variability in audio matching. The advantage of these features over conventional feature set is that they apply universal frequency warping for both the templates to be matched during audio search. The performance of Scale Transform Cepstral Coefficients (STCC) and Warped Linear Prediction Cepstral Coefficients (WLPCC) are about *3%* higher than the state-of-the-art MFCC feature sets on TIMIT database.

Keywords: Vocal tract length normalization · Audio search · Scale transform cepstral coefficients · Warped linear prediction coefficients

1 Introduction

Recently, speech-based search or retrieval technologies have gained keen attention. The simple reason could be the ease of their storage and retrieval capabilities. A speech signal is enriched with many attributes such as message conveyed from it, speaker information, emotion, age, gender, etc. Several studies have been involved in extracting such kind of information from the speech signal. National Institute Science and Technology (NIST) has started evaluation named Spoken Term Detection (STD) in *2006* which is involved in linguistic information extraction from spoken document [1]. The technology has a quite relevance with the Automatic Speech Recognition (ASR) task. However, ASR is slightly different technology where speech signal is decoded in terms of word transcription whereas STD is only interested to detect particular audio query within database [2].

© Springer International Publishing Switzerland 2015
P. Král and V. Matoušek (Eds.): TSD 2015, LNAI 9302, pp. 387–395, 2015.
DOI: 10.1007/978-3-319-24033-6_44

Inspite of being different technology all together, speech researchers have attempted STD problem via ASR system [2],[3]. For an ASR experts, the task remains merely obtaining the transcripts and assigning them confidence measure. Word-based ASR has found to be effective on well resourced languages. However, this may introduce problem in out-of-vocabulary (OOV) query such as *named entities*. To deal with this issue, researchers have come up with an subword modeling approach [4].

For low-resourced languages, it is hard to find the labeled speech corpora. Hence, speech researchers have exploited the spoken query representation rather than textual representation. Considering the fact that query is taken from speech only, the STD technology is now termed as Query-by-Example Spoken Term Detection (QbE-STD) [5], [6]. In this paper, we refer QbE-STD as *audio search*, which basically involves matching and hence, highly dependent on the type of *representation* used in the matching task. In particular, representation used in audio search should be speaker-invariant so as to perform speaker-independent audio matching. In this paper, we explore universal frequency warping in two spectral features, *viz.*, Scale Transform Cepstral Coefficients (STCC) and Warped Linear Prediction Cepstral Coefficients (WLPCC) in order to develop speaker-invariant features for audio search task.

2   Relation to Prior Work

There have been many attempts made in designing the proper representation of QbESTD. Speaker-invariant representation has been primary need for speaker-independent audio search. There have been significant attempts made in posterior-based feature representation. They are mainly supervised phonetic posteriorgram and Gaussian posteriorgram.

Posteriogram representation is found to be more robust to speech information [6],[7]. They also used Gaussian Mixture Model (GMM)- based posteriorgram representation in their studies. The K-means clustering algorithm is combined with the GMM posteriorgrams front-end to obtain more discriminant features [8]. Parallel tokenizer-based approach for QbE-STD was used in [9] to combine the evidences from different systems. Each tokenizer uses both posteriorgram of query and utterance and combines the evidences from all [9]. In addition to the posteriorgram features, researchers have exploited speaker normalization technique, *viz.*, Vocal Tract Length Normalization (VTLN) [9], [10] for audio search task. The VTLN method discussed in those studies exploit different version of Mel warped filterbank. The optimum *warping factor* is estimated using grid-search under Maximum Likelihood (ML) criteria. This requires features to be computed for all different warped Mel filterbank and then estimation for proper warping factor using state-of-the-art Lee and Rose method [11]. In this paper, we have used universal warping-based approach of VTLN for audio search [12]. In addition, we have explored two different feature sets, *viz.*, STCC and WLPCC. These features avoid exhaustive grid search and hence, less cumbersome in terms of computation yet performing well in speaker mismatched condition. The presented approach does not compare the absolute performance of audio search system presented earlier in the literature. The objective and focus of this work is to bring signal processing aspect at front end of audio search task.

3 Vocal Tract Length Normalization (VTLN)

In this paper, we have compared the performance of state-of-the-art mel frequency cepstral coefficients (MFCC) with vocal tract length (VTL) normalized features such as STCC [13] and WLPCC. For feature extraction, speech signal is divided into frames and each frame is Hamming windowed. In the following sub-Sections, we discuss and compare various feature extraction methods used in this study.

3.1 Scale Transform Cepstral Coefficients (STCC)

STCCs compensate for the differences in VTL using log-warping. Vocal tract is generally modelled as a uniform tube. For such a model, formant frequencies are inversely proportional to VTL. The spectrum of a particular speech sound of a speaker is the scaled version of the spectrum of another speaker uttering the same sound, i.e., $F_A(\omega) = F_B(\alpha_{AB}\omega)$. The scale α_{AB} is speaker-specific and is known as *warping factor*. Replacing ω by e^v, we get,

$$f_A(v) = F_A(\omega = e^v) = F_B(\alpha_{AB}e^v) = F_B(e^{(v+\ln\alpha_{AB})}) = f_B(v + \ln\alpha_{AB}). \quad (1)$$

This shows that in the log-warped domain, the spectra are shifted versions of each other with a translation factor of $\ln\alpha_{AB}$. Since Fourier transform is a shift-invariant transform and the shift factor appears in the phase part, the speaker-dependent warping factor is removed by taking the magnitude. These are used as VTL normalized features. Smoothed spectra are obtained by suppressing the pitch information by the method described in [12]. Here, mel-warping is used as it was observed that mel-warping provides better performance [14].

3.2 Warped Linear Prediction Cepstral Coefficients (WLPCC)

This feature set provides Bark scale-based frequency warping via warped linear prediction (WLP). WLP is obtained by replacing unit delays of classical LP filter by first-order all pass filters with transfer function given by [15],

$$D(z) = \frac{z^{-1} - \lambda}{1 - \lambda z^{-1}}, \quad (2)$$

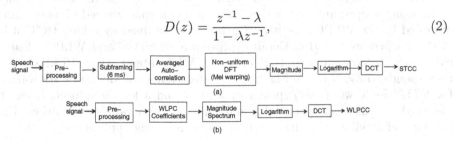

(a)

(b)

Fig. 1. Schematic block diagram for feature extraction of (a) STCC and (b) WLPCC. After, [12],[15].

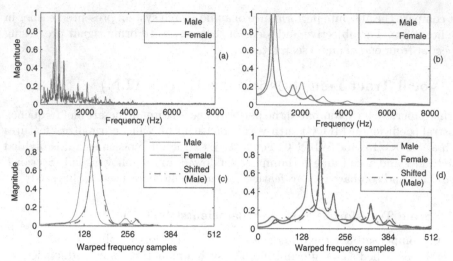

Fig. 2. Illustration of speaker normalization for vowel segment /ae/ of two different (male and female) speakers. (a) magnitude spectrum, (b) LP magnitude spectrum (c) Mel warped Fourier transform spectrum (STCC spectrum), and (d) Bark warped LP spectrum (WLPCC spectrum). In Fig. 2, (c) and (d), the spectra plotted with dotted line indicates that speaker differences are normalized.

and phase response [15],

$$\Psi(\omega) = \omega + 2\tan^{-1}\left(\frac{\lambda\sin\omega}{1 - \cos\omega}\right), \tag{3}$$

where $-1 < \lambda < 1$ is the warping factor. For $0 < \lambda < 1$, lower frequencies are compressed and higher frequencies are expanded. The reverse warping happens for $0 > \lambda > -1$. An analytical expression provides the value of λ for warping similar to Bark scale [15] depending on the sampling frequency, i.e., f_s and given by,

$$\lambda_{fs} \approx 1.0674\left(\frac{2}{\pi}\arctan\left(0.6583\frac{fs}{1000}\right)\right)^{\frac{1}{2}} - 0.1916, \tag{4}$$

WLP coefficients (WLPC) are easily obtained by Levinson-Durbin algorithm using warped autocorrelation function. Bark scale-warped LP spectrum is obtained by the WLPCs. Cepstral features are obtained by taking DCT of log of the warped spectra [16]. Detailed procedure of STCC and WLPCC feature extraction is shown in Fig. 1. Here, for TIMIT dataset $fs = 16$ kHz, which corresponds to $\lambda_{fs} = 0.575$. Fig. 2 shows effectiveness of Mel and Bark warping for VTLN for vowel (/ae/) spoken by a male and a female subject. It can be observed that these spectra overlap and hence, speaker variability reduces. That can be useful evidence in the design of audio matching application.

Table 1. Statistics of query used in this work (# Train:#Test)

Query Index	Query	Query Index	Query
1	age (3:8)	2	artists (7:6)
3	children (18:10)	4	development (9:8)
5	money (19:9)	6	organizations (7:6)
7	problem (22:13)	8	surface (3:8)
9	warm (10:5)	10	year (11:5)

4 Experimental Results

4.1 Database Used

For audio search task, we have used TIMIT corpora [17]. *10* queries are taken from training database and the details are shown in Table 1. The selection of query word is as given in [7]. TIMIT dataset consists of *3,696* training utterances and *944* test utterances. Audio search queries are taken from training set and searching is performed on testing set.

4.2 Architecture of Audio Matching System

For audio search task, the state-of-the-art dynamic programming-based Dynamic Time Warping (DTW) [18] and their variant such as segmental DTW [19] have been used prominently for audio matching. For audio search task, two-pass strategy has been employed. In the first pass, segmental DTW is performed between query word and reference utterances in order to obtain hypothetical location of query within reference utterance. The segmental DTW is an extended version of DTW. In particular, in audio search task, reference data is utterance whereas query is merely a word or a phrase. The direct application of DTW between these two signals would not make much sense as both corresponds to different length information. However, we are not aware of the duration of the query present in the actual reference data. We need to perform segmental DTW using $R = 5$. We may call this task-1 as *'localization of query'*. The task-2, the remaining unwanted reference is chopped out and conventional template matching using DTW. This phase can be called as *'scoring'* as we need to rank the audio document based on the minimum DTW distance value.Conventional Euclidean distance is used to compute the local distance between two patterns via DTW algorithm.

4.3 Results and Discussion

To evaluate the performance of audio search following evaluation measures are considered, namely, 1) $p@N$ precision at N, (i.e., number of hits in top N retrieved documents, where N corresponds to the number of actual document present in database) 2) % EER: Equal Error Rate.

Table 2. Experimental Results

Feature	p@N	%EER	Feature	p@N
MFCC	24.65	25.30	MFCC-fused	40.17
STCC	27.98	23.73	STCC-fused	44.68
WLPCC	27.13	23.25	WLPCC-fused	41.29

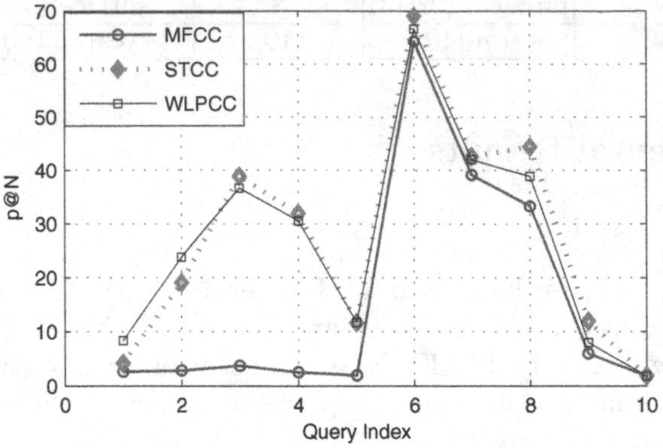

Fig. 3. Performance of audio search for each individual query.

Overall performance of the audio search system is shown in Table 2. It can be observed that the VTLN-based feature sets, namely, STCC and WLPCC performs better than MFCC alone. About *3* % absolute improvement can be observed using STCC and WLPCC features. In Table 2, the performance of each isolated query of each feature sets are mentioned as MFCC, STCC and WLPCC, respectively. In addition, distortion score from the same query are fused. This will improve the statistical confidence about the query detection task. The fused features are called as MFCC-fused, STCC-fused and WLPCC-fused, respectively. From Table 2, it can be observed that the fusing of multiple evidences indeed improve the audio search performance for all *3* feature sets. Simple averaging of distortion score is used as fused score. 5th column of Table 2 shows the performance using fused score improves. In order to investigate the performance *w. r. t.* every query, mean p@N is computed for individual query. From Fig. 3, it can be observed that for most of the query word, STCC and WLPCC performs better than MFCC. The query associated with particular query index is listed in Table 1.

4.4 Evaluation of Class Separability

In order to investigate the effectiveness of these feature sets, we conducted class separability test. For different vowel class *J*-measure is computed. Separability property of MFCC, STCC and WLPCC is compared for four vowels /iy/, /ih/, /ae/ and /ow/ taken from *13* female and *22* male speakers of *dr7* of TIMIT database. Middle *32* ms speech signal of *150* examples of each vowel is considered. Separability is measured as suggested in [20].

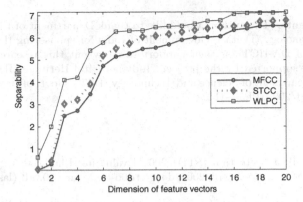

Fig. 4. Class separability of vowels for different feature sets, *viz.*, MFCC, STCC, WLPCC.

$$J = tr(S_W^{-1} S_B), \tag{5}$$

where S_B is the between groups mean square and S_W is the within groups mean square. S_B and S_W are obtained as

$$S_B = \frac{1}{I} \sum_{i=1}^{I} (M_i - M_o)(M_i - M_o)^T, \; S_W = \frac{1}{I} \sum_{i=1}^{I} R_i,$$

where M_i and R_i denote the mean feature vector and covariance matrix, respectively and $M_0 = \frac{1}{I} \sum_{i=1}^{I} M_i$, where I is the total number of phoneme classes being compared. Fig. 4 shows the separability of different feature extraction methods *w. r. t.* dimension of feature vectors. It can be observed that the separability of STCC is higher than that of MFCC and still higher for WLPCC. This along with normalization of speaker-specific spectra (as shown in Fig. 2 (c) and Fig. 2 (d)) may be the reason for better performance of STCC and WLPCC feature than MFCC for audio search task.

5   Summary and Conclusions

This paper presented audio search system using VTLN-based approach. There have been several ways to exploit VTLN in order to suppress the speaker variation for speaker-independent audio search task. This work involved universal warping-based feature extraction, namely, STCC and WLPCC. Performance of audio search system is found to be improved when these representation are used instead of conventional MFCC. Our future plan is to incorporate other VTLN aspects such as estimation of frequency warping relation for speaker pair to improve the audio search performance. In addition, we would like to explore telephone recorded speech for audio search task.

Acknowledgments. The authors would like to thank Department of Electronics and Information Technology (DeitY), Government of India for sponsoring the project and the authorities of DA-IICT for their support to carry out this research work. This work is partially supported by the project "Indian Digital Heritage (IDH) - Hampi" sponsored by Department of Science and Technology (DST), Govt. of India (Grant No: NRDMS/11/1586/2009/Phase-II).

References

1. The Spoken Term Detection (STD) 2006 Evaluation Plan (2006). http://www. itl.nist.gov/iad/mig/tests/std/2006/docs/std06-evalplan-v10.pdf (last accessed on March 25, 2015)
2. Vergyri, D., Shafran, I., Stolcke, A., Gadde, V.R.R., Akbacak, M., Roark, B., Wang, W.: The SRI/OGI 2006 spoken term detection system. In: INTERSPEECH 2007, Belgium, pp. 2393–2396 (2007)
3. Parlak, S., Saraclar, M.: Spoken term detection for turkish broadcast news. In: Proc. IEEE Int. Conf. on Acous. Speech, and Signal Process. ICASSP 2008, Las Vegas, USA, pp. 5244–5247 (2008)
4. Wallace, R., Vogt, R., Sridharan, S.: A phonetic search approach to the 2006 NIST spoken term detection evaluation. In: INTERSPEECH 2007, Belgium, pp. 2385–2388 (2007)
5. Metze, F., Anguera, X., Barnard, E., Davel, M.H., Gravier, G.: Language independent search in mediaeval's spoken web search task. Computer Speech & Language **28**(5), 1066–1082 (2014)
6. Hazen, T.J., Shen, W., White, C.M.: Query-by-example spoken term detection using phonetic posteriorgram templates. In: IEEE Workshop on Automatic Speech Recognition & Understanding, ASRU, 2009, Merano/Meran, Italy, pp. 421–426 (2009)
7. Zhang, Y., Glass, J.R.: Unsupervised spoken keyword spotting via segmental DTW on gaussian posteriorgrams. In: IEEE Workshop on Automatic Speech Recognition & Understanding, ASRU, 2009, Merano/Meran, Italy, pp. 398–403 (2009)
8. Anguera, X.: Speaker independent discriminant feature extraction for acoustic pattern-matching. In: IEEE Int. Conf. on Acoust. Speech and Signal Process., ICASSP 2012, Kyoto, Japan, pp. 485–488 (2012)
9. Wang, H., Lee, T., Leung, C., Ma, B., Li, H.: Using parallel tokenizers with DTW matrix combination for low-resource spoken term detection. In: IEEE Int. Conf. on Acoust. Speech and Signal Process., ICASSP 2013, Vancouver, BC, Canada, pp. 8545–8549 (2013)
10. Tejedor, J., Szöke, I., Fapso, M.: Novel methods for query selection and query combination in query-by-example spoken term detection. In: Proc. of 2010 Int. Workshop on Searching Spontaneous Conversational Speech. SSCS 2010, New York, NY, USA, pp. 15–20. ACM (2010)
11. Lee, L., Rose, R.C.: Speaker normalization using efficient frequency warping procedures. In: IEEE Int. Conf. on Acoust. Speech and Signal Process., ICASSP 1996, Atlanta, Georgia, USA, pp. 353–356 (1996)
12. Umesh, S., Cohen, L., Marinovic, N., Nelson, D.J.: Scale transform in speech analysis. IEEE Transactions on Speech and Audio Processing **7**(1), 40–45 (1999)
13. Umesh, S., Sanand, D.R., Praveen, G.: Speaker-invariant features for automatic speech recognition. In: IJCAI 2007, Proc. 20th Int. Joint Conf. on Artificial Intelligence, Hyderabad, India, pp. 1738–1743 (2007)

14. Sinha, R., Umesh, S.: Non-uniform scaling based speaker normalization. In: Proceedings of the IEEE International Conference on Acoustics, Speech, and Signal Processing, ICASSP 2002, May 13–17 2002, Orlando, Florida, USA, pp. 589–592 (2002)
15. Iii, J.O.S., Abel, J.S.: Bark and ERB bilinear transforms. IEEE Transactions on Speech and Audio Processing 7(6), 697–708 (1999)
16. Kim, Y., Smith, J.O.: A speech feature based on bark frequency warping-the non-uniform linear prediction (nlp) cepstrum. In: IEEE Workshop on Applications of Signal Processing to Audio and Acoustics, 1999, pp. 131–134. IEEE (1999)
17. Garofolo, J.S., Lamel, L.F., Fisher, W.M., Fiscus, J.G., Pallett, D.S., Dahlgren, N.L.: DARPA TIMIT acoustic-phonetic continous speech corpus CD-ROM (1993)
18. Sakoe, H., Chiba, S.: Dynamic programming algorithm optimization for spoken word recognition. IEEE Transactions on Acoustics Speech and Signal Processing 26(1), 43–49 (1978)
19. Park, A.S., Glass, J.R.: Unsupervised pattern discovery in speech. IEEE Transactions on Audio, Speech & Language Processing 16(1), 186–197 (2008)
20. Nicholson, S., Milner, B.P., Cox, S.J.: Evaluating feature set performance using the f-ratio and j-measures. In: Fifth European Conference on Speech Communication and Technology, EUROSPEECH 1997, Rhodes, Greece (1997)

RENA: A Named Entity Recognition System for Arabic

Ismail El bazi[✉] and Nabil Laachfoubi

Computer, Networks, Mobility and Modeling Laboratory,
FST, Hassan 1st University, Settat, Morocco
ismailelbazi@gmail.com, n.laachfoubi@hotmail.fr

Abstract. The Named Entity Recognition (NER) task aims to identify and categorize proper and important nouns in a text. This Natural Language Processing task proved to be challenging for languages with a rich morphology such as the Arabic language. In this paper, We introduce a new named entity recognizer for Arabic. This recognizer is based on Conditional Random Fields (CRF) and an optimized feature set that combines contextual, lexical, morphological and gazetteers features. Our system outperforms the state-of-the-art Arabic NER systems with a F-measure of 93.5% when applied to ANERcorp standard dataset.

Keywords: Natural Language Processing · Named Entity Recognition · Machine learning · Arabic

1 Introduction

Named Entity Recognition (NER) aims to identify and categorize proper and important nouns in a text. NER is an important preprocessing component in many Natural Language Processing applications such as Information Extraction, Machine Translation and Question Answering. The majority of research studies on the NER task were on English language where the best State-of-the-art NER system for English achieve near-human performance. For the Arabic language, NER has gained a lot of attention over the past fifteen years. Many Arabic NER systems were developed with quite good results specially in the newswire genre. In these systems, different types of features were used such as: contextual features (the surrounding words of a context window), lexical features (word length, prefix, suffix, character n-grams, punctuation), list lookup features (gazetteers, Wikipedia, DBpedia) and morphological features (Part of Speech tag, Base Phrase Chunk, Aspect, Person, Gender, Number).

In this paper, we study the influence of each features type on the performance of Arabic NER. We investigate also if the combination of these features can yield to a better state-of-the-art system.

The main contributions of this work can be summarized as follows:

- An investigation of the effectiveness of the Margin Infused Relaxed Algorithm (MIRA) for the Arabic NER;

© Springer International Publishing Switzerland 2015
P. Král and V. Matoušek (Eds.): TSD 2015, LNAI 9302, pp. 396–404, 2015.
DOI: 10.1007/978-3-319-24033-6_45

- An optimized feature set that combines contextual, lexical, morphological and gazetteers features;
- A new system that outperforms the state-of-the-art Arabic NER system on ANERcorp standard dataset .

The remainder of the paper is organized as follows: Section 2 gives background about Arabic Language and the challenges related to Arabic Named Entity Recognition. Section 3 surveys previous work on Arabic NER. Section 4 introduces our approach and the NER features used. In Section 5 the experimental setup , tools and evaluation are described, and in Section 6 the experimental results are reported. Section 7 provides final conclusions and gives directions for future work.

2 Background

2.1 The Arabic Language

The Arabic language is a Semitic language spoken in the Arab World, a region of 22 countries with a collective population of 300 million people. It is ranked the fifth most used language in the world and one of the six official languages of the United Nations.

With regards to language usage, there are three forms of the Arabic language:

1. **Classical Arabic (CA):** is the formal version of the language. It has been in usage in the Arabian Peninsula for over 1500 years. Most Arabic religious texts are written in CA;
2. **Modern Standard Arabic (MSA):** is the primary written language of the media and education. MSA is the common language of all the Arabic speakers and the most widely used form of the Arabic language. The main differences between CA and MSA are basically in style and vocabulary, but in terms of linguistic structure, MSA and CA are quite similar [1]. This is the form studied in this work;
3. **Dialectal Arabic (DA):** is the day to day spoken form of the language used in the informal communication. DA is region-specific that differs not only from one area of the Arab world to another, but also across regions in the same country.

2.2 Challenges in Arabic Named Entity Recognition

The NER task is considerably more challenging when it is targeting a morphologically rich language such as Arabic for four main reasons:

- **Absence of Capitalization:** Unlike Latin script languages, Arabic does not capitalize proper nouns. Since the use of capitalization is a helpful indicator for named entities [2], the lack of this characteristic increases the complexity of the Arabic NER task;

- **Agglutination:** The agglutinative nature of Arabic makes it possible for a Named Entity (NE) to be concatenated to different clitics. A preprocessing step of morphological analysis needs to be performed in order to recognize and categorize such entities. This peculiarity renders the Arabic NER task more challenging;
- **Optional Short Vowels:** Short vowels (diacritics) are optional in Arabic. Currently, most MSA written texts do not include diacritics, this causes a high degree of ambiguity since the same undiacritized word may refer to different words or meanings. This ambiguity can be resolved using contextual information [3];
- **Spelling Variants:** In Arabic, as for many other languages, an NE can have multiple transliterations. The lack of standardization leads to many spelling variants of the same word with the same meaning. For example, the transliteration of the Person name 'Huntington' may produce these spelling variants:

"هنتنجتون", "هنتنغتون", "هنتينجتون" or "هنتينغتون".

3 Related Work

Significant amount of work has been done in the NER task. Nadeau and Sekine [4], surveyed the literature of NER for a variety of languages and reported on the different features and techniques used. Recently Arabic NER has started to gain momentum and a lot of work has been done for this language. Similar to other languages, the proposed approaches for Arabic NER fall in three categories: handcrafted rule-based approach, Machine Learning(ML) based approach and a hybrid of both approaches.

Mesfar [5] presented a rule-based NER system for Arabic using a combination of NooJ syntactic grammars and a morphological analysis. In [6], Shaalan and Raza developed a system called NERA using a rule-based approach. it is divided into three components: gazetteers, local handcrafted grammars, and a filtering mechanism. Al-Jumaily et al. [7] introduced a rule-based Arabic NER system that can be incorporated in web applications. It uses a pattern recognition model to identify NEs by integrating patterns from different gazetteers (DBPedia, GATE and ANERGazet).

Although this approach proved to be quite precise but it has usually low coverage and it is time-consuming and labor-intensive. Recently, most of the researches focuses on ML-based methods. Numerous works have been conducted on ML-based Arabic NER by Benajiba and colleagues. Benajiba et al. [8] developed an Arabic NER system (ANERsys 1.0) based on n-grams and Maximum Entropy (ME). The system can classify four types of NEs: Person (PERS), Location (LOC), Organisation (ORG) and Miscellaneous. The authors also introduced a new corpora (ANERcorp) and gazetteers (ANERgazet). The ANERsys 1.0 had some issues with detecting long NEs. Therefore, a new version (ANERsys 2.0) has been created in [9], which adopts two-steps mechanism: first extracts the NEs boundaries then classify each of the delimited NEs. In order to enhance the accuracy of ANERsys, Benajiba and Rosso [10] changed the probabilistic model from ME to CRF. The system used the same feature set as in ANERsys 2.0. The CRF-based system achieves higher results with an overall F-measure(F1) performance of 79.21 %.

A hybrid approach combining both ML-based and Rule-based has been also used for Arabic NER. Abdallah et al. [11] developed a hybrid NER system for Arabic. The ML-based component uses J48 Decision Tree classifier, while the rule-based component is a re-implementation of the NERA system [6] using GATE framework. The system recognize three types of NEs: PERS, LOC, and ORG. Oudah and Shaalan [12] extended this hybrid NER system to identify 11 NEs. They also investigated two more ML-based approaches : SVM and Logistic Regression. Their system outperforms the state-of-the-art Arabic NER when applied to ANERcorp. A comprehensive survey of Arabic NER can be found in [13].

MIRA is an extension of the perceptron algorithm introduced by Crammer and Singer [14]. Although this algorithm was not yet used to tackle Arabic NER task till date, it was successfully applied to NER for other languages [15,16].

4 Approach

We approach the problem of NER by using supervised machine learning methods. In the literature, it has been shown that supervised typically outperform unsupervised methods for the NER task [17]. In this paper, we investigated two ML techniques: CRF and MIRA.

4.1 CRF and MIRA

CRF[18] is a probabilistic framework for building models oriented toward segmenting and labeling sequence data. It is based on undirected graphical models that combine the advantages of classification and graphical modeling. CRF has been applied successfully to Arabic NER [10].

MIRA[14] is a perceptron-like machine learning algorithm which has been employed successfully for a number of multiclass classification tasks in NLP. For k classes, MIRA maintains a matrix of k rows, one row per class. Given a new instance, it calculates the similarity-score between the instance and each of the k prototypes. Then it predicts the class which achieves the highest score.

4.2 Features

In our approach, we propose the following features:

- **Contextual Features** (CXT): The surrounding words of a context window = ±1;
- **Lexical Features** (LEX): The leading and trailing bigrams, trigrams and 4-grams characters and the stem of the word as described in [19];

- **Morphological Features** (MORPH): These features are generated by MADAMIRA tool [20]. We selected 5 morphological features to include in this work:
  - Aspect: describes the aspect of an Arabic verb. It has four possible values: Command, Imperfective, Perfective, Not applicable. However, since none of the NEs can be verbal, we use this feature as binary feature indicating if a word is marked for Aspect or not;
  - Gender: The nominal Gender. This feature has three values: Feminine, Masculine, Not applicable;
  - Person: It indicates the Person Information. The possible values are: 1st, 2nd, 3rd, Not applicable. Similar to aspect, we use it as binary feature indicating if a word is marked for Person or not;
  - Proclitic2: The conjunction proclitic. The tool produces nine values for this feature: No proclitic, Not applicable, Conjunction fa, Connective particle fa, Response conditional fa, Subordinating conjunction fa, Conjunction wa, Particle wa, Subordinating conjunction wa;
  - Voice: The verb voice. The values for this feature are: Active, Passive, Not applicable, Undefined.
- **Part-Of-Speech Tags** (POS): We use POS tags generated from MADAMIRA tool;
- **Gazetteers Features** (GAZ): A binary feature indicating the existence of the word in a gazetteer. We use three gazetteers: a location gazetteer, a person gazetteer and an organizations gazetteer.

5 Experimental Setup

5.1 Datasets

For the evaluation purposes, the standard ANERcorp dataset was used. It is a commonly used corpora and allow us to evaluate and compare our results with other existing systems. This corpora is built and tagged especially by Benajiba and colleagues for the Arabic NER task [8]. The Binajiba dataset is composed of 316 newswire articles totaling 150,286 words (11 % of the words are NEs). The distribution of different NEs is as follows:

For training, development and testing, we used a 80/10/10 split of this dataset.

Table 1. Number of NEs in ANERcorp [19]

NE	Number
Persons	689
Organizations	342
Locations	878

5.2   Tools

In this work, we used the following tools:

1. MADAMIRA [20], a system for morphological analysis and disambiguation of Arabic to generate POS tags and other morphological features;
2. CRF++[1], a CRF sequence labeling toolkit used with default parameters.
3. Miralium[2], an open source java implementation of MIRA algorithm.

5.3   Evaluation Metrics

F-measure is the standard evaluation measure of the Information Extraction and Information Retrieval systems. We chose to use the same measure to evaluate the performance of our system. To Process the results obtained by RENA, we used the standard evaluation tool provided by the CoNLL-2002 conference [21]. According to the CoNLL evaluation guidelines, a named entity is considered correct only if it is an exact match of the corresponding entity in the gold standard data file.

6   Experimental Results

Table 2 enumerates the results for the ANERcorp dataset. The best results for F1 are bolded.

In comparing lexical features with each of the other features, LEX gives better results than each feature type tested individually with both ML approaches. This result is consistent with the finding of Abdul-Hamid and Darwish in [19]. The authors introduced a simplified feature set composed mainly of lexical features and achieves good results without the use of morphological analysis which suggests that these lexical features has helped to overcome some of the orthographic and morphological complexities of Arabic language.

In addition, combining all features produces a significant improvement over the singleton features for both CRF and MIRA. We noted also that CRF provide the best results when using a combination of all features.

In comparison to results in literature [8–12], our system performs better than the state-of-the-art Arabic NER systems when applied on ANERcorp dataset as shown by Table 3. Our system outperforms the previous systems in terms of F-measure in extracting Location, Organization and Person NEs from ANERcorp with an overall F1 = 93.5 %.

Our good results compared to previous Arabic NER systems can be explained by the fact that our proposed feature set combines the lexical features introduced by Hamid and Darwish [19] which produce near state-of-the-art results with powerful features such as POS feature and morphological features. This combination create an optimized feature set that allow our CRF model to achieves state-of-the-art results.

---

[1] https://code.google.com/p/crfpp/
[2] https://code.google.com/p/miralium/

Table 2. NER results for the ANERcorp dataset

Model	Features	Location	Organization	Person	Overall
		F1	F1	F1	Average F1
CRF	CXT	80.0	69.7	53.2	67.6
	LEX	91.6	77.5	70.9	80.0
	POS	85.0	67.9	71.7	74.9
	MORPH	85.8	71.0	57.4	71.4
	GAZ	87.6	76.2	61.0	74.9
	All Features	**95.7**	**89.1**	**95.8**	**93.5**
MIRA	CXT	84.6	71.1	63.8	73.1
	LEX	88.5	76.5	67.5	77.5
	POS	86.5	69.4	70.0	75.3
	MORPH	88.6	69.4	50.8	69.6
	GAZ	88.5	78.1	60.2	75.6
	All Features	94.7	84.6	91.9	90.4

Table 3. Comparison with state-of-the-art Arabic NER systems for ANERcorp dataset.

System	Location	Organization	Person	Overall
	F1	F1	F1	Average F1
ANERsys 1.0 [8]	80.3	36.8	46.7	54.6
ANERsys 2.0 [9]	86.7	46.4	52.1	61.7
CRF-based system [10]	89.7	65.8	73.4	76.3
Abdallah et al. [11]	87.4	86.1	92.8	88.8
Oudah and Shaalan [12]	90.1	88.2	94.4	90.9
Our System	**95.7**	**89.1**	**95.8**	**93.5**

7   Conclusion and Future Work

We presented a novel NER system using CRF and an optimized feature set. We measured the impact of each features independently and in a joint combination. The experimental results indicates that employing an optimized feature set that combines contextual, lexical, morphological and gazetteers features achieves higher performance. Our system outperforms the state-of-the-art Arabic NER systems with F-measure = 93.5 % when applied to ANERcorp dataset.

To further our research we are planning to extend our NER system to identify more NEs types. We would like also to assess the impact of using different ML techniques other than CRF and MIRA on the overall performance of the system.

References

1. Ryding, K.C.: A reference grammar of modern standard Arabic. Cambridge University Press (2005)
2. Benajiba, Y., Diab, M., Rosso, P.: Arabic named entity recognition using optimized feature sets. In: Proc. of EMNLP08, pp. 284–293 (2008)
3. Benajiba, Y., Diab, M., Rosso, P.: Arabic named entity recognition: A feature-driven study. IEEE Transactions on Audio, Speech, and Language Processing **17**(5), 926–934 (2009)
4. Nadeau, D., Sekine, S.: A survey of named entity recognition and classification. Lingvisticae Investigationes **30**(1), 3–26 (2007)
5. Mesfar, S.: Named entity recognition for Arabic using syntactic grammars. In: Kedad, Z., Lammari, N., Métais, E., Meziane, F., Rezgui, Y. (eds.) NLDB 2007. LNCS, vol. 4592, pp. 305–316. Springer, Heidelberg (2007)
6. Shaalan, K., Raza, H.: NERA: Named entity recognition for arabic. Journal of the American Society for Information Science and Technology **60**(8), 1652–1663 (2009)
7. Al-Jumaily, H., Martínez, P., Martínez-Fernández, J., Van der Goot, E.: A real time named entity recognition system for arabic text mining. Language Resources and Evaluation **46**(4), 543–563 (2012)
8. Benajiba, Y., Rosso, P., BenedíRuiz, J.M.: ANERsys: an arabic named entity recognition system based on maximum entropy. In: Gelbukh, A. (ed.) CICLing 2007. LNCS, vol. 4394, pp. 143–153. Springer, Heidelberg (2007)
9. Benajiba, Y., Rosso, P.: Anersys 2.0: Conquering the ner task for the arabic language by combining the maximum entropy with pos-tag information. In: IICAI, pp. 1814–1823 (2007)
10. Benajiba, Y., Rosso, P.: Arabic named entity recognition using conditional random fields. In: Proc. of Workshop on HLT & NLP within the Arabic World, LREC, vol. 8, pp. 143–153. Citeseer (2008)
11. Abdallah, S., Shaalan, K., Shoaib, M.: Integrating rule-based system with classification for arabic named entity recognition. In: Gelbukh, A. (ed.) CICLing 2012, Part I. LNCS, vol. 7181, pp. 311–322. Springer, Heidelberg (2012)
12. Oudah, M., Shaalan, K.F.: A pipeline Arabic named entity recognition using a hybrid approach. In: COLING, pp. 2159–2176. Citeseer (2012)
13. Shaalan, K.: A survey of arabic named entity recognition and classification. Computational Linguistics **40**(2), 469–510 (2014)
14. Crammer, K., Singer, Y.: Ultraconservative online algorithms for multiclass problems. J. Mach. Learn. Res. **3**, 951–991 (2003)
15. Ganchev, K., Pereira, O., Mandel, M., Carroll, S., White, P.: Semi-automated named entity annotation. In: Proceedings of the Linguistic Annotation Workshop, 5356, Prague, Czech Republic. Association for Computational Linguistics (2007)
16. Banerjee, S., Naskar, S.K., Bandyopadhyay, S.: Bengali named entity recognition using margin infused relaxed algorithm. In: Sojka, P., Horák, A., Kopeček, I., Pala, K. (eds.) TSD 2014. LNCS, vol. 8655, pp. 125–132. Springer, Heidelberg (2014)
17. Nadeau, D., Turney, P.D., Matwin, S.: Unsupervised named-entity recognition: generating gazetteers and resolving ambiguity. In: Lamontagne, L., Marchand, M. (eds.) Canadian AI 2006. LNCS (LNAI), vol. 4013, pp. 266–277. Springer, Heidelberg (2006)

18. Lafferty, J.D., McCallum, A., Pereira, F.C.N.: Conditional random fields: probabilistic models for segmenting and labeling sequence data. In: Proceedings of the Eighteenth International Conference on Machine Learning. ICML 2001, San Francisco, CA, USA, pp. 282–289. Morgan Kaufmann Publishers Inc. (2001)
19. Abdul-Hamid, A., Darwish, K.: Simplified feature set for arabic named entity recognition. In: Proceedings of the 2010 Named Entities Workshop. NEWS 2010, Stroudsburg, PA, USA, pp. 110–115. Association for Computational Linguistics (2010)
20. Pasha, A., Al-Badrashiny, M., Diab, M., Kholy, A.E., Eskander, R., Habash, N., Pooleery, M., Rambow, O., Roth, R.: Madamira: A fast, comprehensive tool for morphological analysis and disambiguation of arabic. In Chair, N.C.C., Choukri, K., Declerck, T., Loftsson, H., Maegaard, B., Mariani, J., Moreno, A., Odijk, J., Piperidis, S. (eds.) Proceedings of the Ninth International Conference on Language Resources and Evaluation (LREC 2014), Reykjavik, Iceland. European Language Resources Association (ELRA), May 2014
21. Tjong Kim Sang, E.F.: Introduction to the conll-2002 shared task: language-independent named entity recognition. In: Proceedings of the 6th Conference on Natural Language Learning. COLING-02, Stroudsburg, PA, USA, vol. 20, pp. 1–4. Association for Computational Linguistics (2002)

Combining Evidences from Mel Cepstral and Cochlear Cepstral Features for Speaker Recognition Using Whispered Speech

Aditya Raikar$^{(\boxtimes)}$, Ami Gandhi, and Hemant A. Patil

DA-IICT, Gandhinagar, India
{aditya_raikar,hemant_patil}@daiict.ac.in, gandhi.ami18@gmail.com
http://www.daiict.ac.in, http://www.ldceahd.org

Abstract. Whisper is an alternative way of speech communication especially when a speaker does not want to reveal the information other than the target listeners. Generally, speaker-specific information is present in both excitation *source* and vocal tract *system*. However, whispered speech does not contain significant source characteristics as there is almost *no* excitation by the vocal folds, and speaker information in vocal tract system is also low as compared to the normal speech signal. Hence, it is difficult to recognize a speaker from his/her whispered speech. To address this, features based on vocal tract system characteristics such as state-of-the-art Mel Frequency Cepstral Coefficients (MFCC) and recently developed Cochlear Frequency Cepstral Coefficients (CFCC) are proposed. CHAINS (Characterizing individual speakers) whispered speech database is used for conducting experiments using GMM-UBM (Gaussian Mixture Modeling- Universal Background Modeling) approach. It was observed from the experiments that the fusion of CFCC and MFCC gives improvement in % IR (Identification Rate) and % EER (Equal Error Rate) than MFCC alone, indicating that proposed features and their score-level fusion captures *complementary* speaker-specific information.

Keywords: Whisper · MFCC · CFCC · GMM-UBM · Source features · System features · CHAINS corpus

1 Introduction

Speech is a most common and powerful way of human-to-human communication. It follows a model of transmitter and receiver, where transmitter transmits the information, i.e., speech is produced and transmitted via acoustic wave propagation and receiver receives the information i.e., speech and speaker recognition, speech perception and understanding of speech at cognitive-level, etc. Source-filter model states that human speech production system can be divided into source (which contains vocal folds that vibrates to gives pitch or fundamental frequency) and vocal tract system or filter (which gives spectral colour to our speech signal) [1].

© Springer International Publishing Switzerland 2015
P. Král and V. Matoušek (Eds.): TSD 2015, LNAI 9302, pp. 405–413, 2015.
DOI: 10.1007/978-3-319-24033-6_46

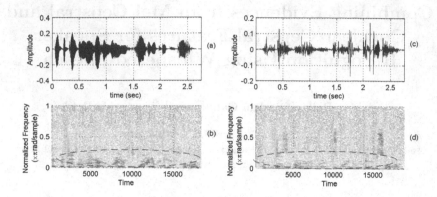

Fig. 1. Speech waveform and spectrogram of utterance "If it doesn't matter who wins why do we keep score," (a),(b) Normal speech (c),(d) Whispered speech

Every speaker has a unique speech production system which includes different vibration rate of vocal folds which gives different pitch and its dynamics, vocal tract shape and length, dialect which speaker has evolved from his/her upbringing, environment where he/she lives, which together implicitly, convey the identity of the speaker [2]. The difference in vocal tract length and shape gives different formant frequencies for different speakers [2]. An experiment was done in which excitation source and vocal tract system of a speaker was separated using speech processing toolkit (SPTK), same was done with another speaker [3]. When features were swapped, i.e, source of one speaker was synthesized to system of the other speaker, it was observed that there was no swapping of speaker identity. Hence, there is a unique interaction (which is primarily non-linear) between source and system for a given speaker. When dealing with voice-conversion problems, every possible feature is mapped which gives speaker it's identity. This means speaker information is present in both excitation source and vocal tract system (atleast dominantly) and speaker can be identified by extracting these two features [4]. Whispered speech is an alternative way of communicating specially in a situation when the speaker does not want the information to be leaked other than the target listeners. It may be because the information is very personal such as passwords, bank account balance, address, private talks, etc. Whispered speech is different than normal speech in many factors. In perception, in order to convey the information through whisper, speaker has to come near to target's ears and then pass on the information because of it's low amplitude as shown in Fig 1, less number of periodic segments, low energy as compared to normal speech [5]. In addition, pitch is almost absent in whispered speech. A narrow constriction is formed above the glottis which excites the vocal tract system without the vibration of vocal folds and hence no pitch occurs [6], [7], [8], [9]. Fig. 1 shows the absence of harmonics in whispered speech as compared to normal speech. Shift in lower formant locations, change in duration, change in spectral slope are also some of the differences observed between whispered speech and normal speech [10], [11], [12], [13]. Whispered speech is a noisy and low energy sound which degrades the speaker-specific information present in it. Hence, we need system-based features which captures speaker information maximally, in order to make an efficient speaker recognition system using whispered speech. In fact, gender can be differentiated from whispered speech, even though information is degraded by the addition of noise. A listening test was conducted in which *16* subjects were told to listen random *30* random whispered speech and tell whether it is male speech or a female speech. The result was *95.72 %*, indicating that gender information is present in whispered speech. In addition, second formant (F_2) plays an important role in recognizing gender better than the fundamental frequency [14]. This paper will be an attempt to extract speaker information from whisper and hence, speaker recognition by using MFCC and CFCC as system features whose comparison is shown in Section 4.5 .

Rest of the paper is organised as follows. Section 2 presents brief details of the system features such as CFCC, MFCC and score-level fusion of these features. Section 3 gives the decription of the database used for this system. Section 4 will deal with the listening test evaluation followed by comparison of speaker-specific information extraction by MFCC and the proposed score-level fusion of these features. Section 5 concludes the paper alongside with future research directions.

2   System Features

2.1   Cochlear Frequency Cepstral Coefficients (CFCC)

Recently, an auditory-based CFCC feature set is proposed for speaker identification task [15]. The block diagram to extract features is shown in Fig 2. The cochlear filter transform function can be defined as,

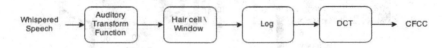

Fig. 2. Basic block diagram of extracting CFCC features [16]

$$\psi_{a,b} = \left(\frac{1}{\sqrt{a}}\right)\psi\left(\frac{t-b}{a}\right), \tag{1}$$

where

$$\psi\left(\frac{t-b}{a}\right) = \left(\frac{t-b}{a}\right)^{\alpha} exp\left[-2\pi f_1\beta\left(\frac{t-b}{a}\right)\right]\times\cos\left[2\pi f_1\beta\left(\frac{t-b}{a}\right)+\theta\right]u(t-b),$$

$$\tag{2}$$

From this, impulse response of cochlea is defined as,

$$T(a,b) = f(t) * \frac{1}{\sqrt{a}} \psi \left(\frac{t-b}{a} \right), \tag{3}$$

where $f(t)$ is the input speech signal and $a \in R^+$ is a scale or dilation variable (through which we can change the center frequency of filterbank) and $b \in R$ is a translation parameter. Value of $\alpha = 3$ and $\beta = 0.035$ is taken in $eq.(2)$ because at this value CFCC gives best performance for speaker identification [16].

As shown in Fig. 1, an auditory-based transform function is calculated using $eq.(1)$. The cochlear filterbank is designed with 28 different filters. Each cochlear filterbank response is windowed with 25 ms with $12.5\ ms$ overlap duration. Finally, log and DCT are taken to extract the CFCC features from the whispered speech. To characterize the nature of CFCC for capturing information from whispered speech, filterbank response for each filter using cochlear filter is plotted in Fig 3 which gives the comparison of whispered and normal speech in context with MFCC and CFCC features.

It can be seen from Fig. 3(e) and Fig. 3(f) that mel filterbank gives better results in case of normal speech but from Fig. 3(b) and Fig 3(c), it is clear that cochlear filters captures more information than mel filterbank. This is the main motivation of taking CFCC for extraction features from whispered speech.

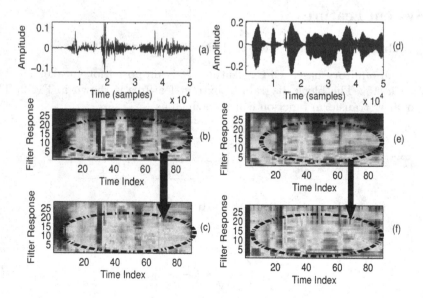

Fig. 3. Speech waveform of the utterance, "If it doesn't matter who wins why do we keep score," (a) whispered speech (d) normal speech, cochlear filterbank response of (b) whispered speech (e) normal speech, mel filterbank response of (c) whisper (f) normal.

3 Experiment Results

The following experiments will give the base that speaker information is indeed present in system features. Speaker recognition can be categorized into three methods which are recognition by listening, recognition by machine and recognition by reading spectrograms [17], [13]. Hence, experiments include listening tests to observe how well human system perform in speaker identification in whispered speech and lastly, a comparison of CFCC ,MFCC and the proposed score-level fusion of these features, to observe which feature captures speaker information the most.

3.1 Speech Database

The CHAINS corpus contains speech database of *36* different speakers, which were recorded in two sessions having a break of two months [18]. In first session, normal speech was recorded in a sound-proof room of a professional recording studio using a Neumann U87 condenser microphone. In second session, recording of whispered speech was done in a quite environment using a AKG C420 headset condenser microphone. The subject was required to utter *four* short fables, used for training and *twenty nine* sentence utterance of which *twenty* were taken from TIMIT sentences and rest *nine* were taken from CSLU speaker identification corpus for testing the system. For listening test, recording of *eleven* subjects were taken using Zoom H4n, which includes *two* normal speech sentences, *two* whispered speech sentences, *two* whispered words for each speaker.

3.2 Speaker Recognition by Listening Test of Whispered Speech

Listening test was conducted in which normal speech of *4* speakers were used for listening test. Corresponding to each speaker, *10* random whispered speeches are kept, subjects have to listen to these *10* whispers after listening to normal speech of the speaker and scale from *1-5* for each whispered speech, where 5 denotes the exact match of how close is it to that normal speech. The purpose of this experiment is to observe as to whether subjects are able to extract individual speaker information from the whispered speech or not.

Mean Opinion Score (MOS) was taken, scores were only given to whispered speech whose normal speech was taken as a reference. In addition, observations from *Table 1* shows high scores given to a whispered speech which was not of a speaker whose reference normal speech was there creating mismatch conditions and because of this scores of those whispered speech was taken whose reference normal speech was taken. *Table 1* shows MOS for each speakers. Interestingly, all subjects are known to each other as database is some of their whispers and normal speech only.

3.3 Speaker Recognition Using Spectral Features

Proposed speaker recognition system on whisper is based on an adaptive GMM-UBM based on standard CHAINS database [18]. The database contains *144* train

Table 1. Calculating Mean Opinion Score (MOS) for every speaker

MOS			
Spk 1	Spk 2	Spk 3	Spk 4
4	2.40	4.44	3.9

utterances, which are popular fablestaken from Cinderella story and *1188* test speaker utterances taken from *36* different speakers. The features are given as an input to GMM-UBM system. UBM is built with close set speaker identification approach. Every train model is adapted from UBM so as to create adaptive GMM models for each *36* speakers [19]. Models built after training are fit with with *256* Gaussian mixture components. The model is tested by giving input a test utterance against *36* different speaker models, which gives *1188* genuine trials and *41580* imposter trials. In order to calculate which speaker utterance it was, log-likelihood ratio (LLR) is calculated to measure the score of each test segment with training model. Model giving maximum score is termed as identified speaker. Finally, true and false scores are calculated using MFCC, CFCC and their score-level fusion of both to plot the Detection Error Tradeoff (DET) curve. *Table 2* shows the results for MFCC and CFCC and their score-level fusion. % Identification rate (IR) gives the efficiency of proposed feature towards identifying the correct speaker. It can be defined as,

$$\%IR = \frac{\text{Number of correctly identified speaker}}{\text{Total number of speakers}} \times 100. \tag{4}$$

% Equal Error Rate (EER) can be defined as the probability where, P_{miss} (Miss Rejection Rate) and P_{fa} (False Alarm Rate) becomes equal. Confidence interval(C.I) is calculated to show the statistical significance of proposed experiment. 95 % confidence interval can be calculated along with % IR and % EER, 95 % as (G-IR, G+IR) [20]. The value of G can be calculated as,

$$G = 1.96\sqrt{\frac{\text{IR}(100 - \text{IR})}{N}}, \tag{5}$$

where in our experiment, the value of N is *42768*. The score level fusion of features is defined as,

$$C_F = (\alpha)(C_{MFCC}) + (1 - \alpha)(C_{CFCC}), \tag{6}$$

Table 2. Speaker recognition system based on spectral features

System Features		MFCC			CFCC			CF		
Train	Test	% IR	% EER	95 % C.I.	% IR	% EER	95 % C.I.	% IR	% EER	95 % C.I.
Normal	Normal	100.0	00.51	100.0-100.0	99.66	01.20	99.63-99.68	100	00.25	100.0-100.0
Normal	Whisper	09.60	37.96	09.45-09.74	09.10	36.20	08.96-09.23	10.19	35.61	10.04-10.33
Whisper	Normal	11.03	34.76	10.88-11.18	08.00	37.54	07.86-08.13	10.27	32.91	10.12-10.41
Whisper	Whisper	99.75	0.59	99.77-99.72	99.41	1.09	99.44-99.37	99.83	0.42	99.81-99.84

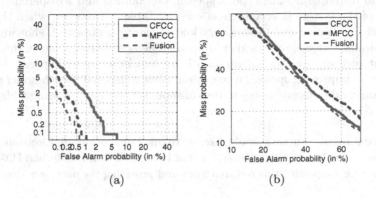

Fig. 4. DET curve for MFCC, CFCC and their score-level fusion trained and tested in (a) normal speech (left) (b) normal and whispered speech (right), respectively.

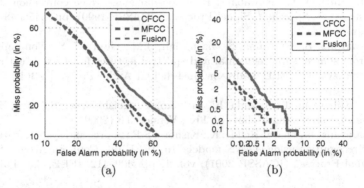

Fig. 5. DET curve for MFCC, CFCC and their score-level fusion trained and tested in (a) whispered speech and normal speech respectively (left) (b) whispered speech (right).

where C_{MFCC}, C_{CFCC} and C_F denotes scores for MFCC, CFCC and their score-level fusion, respectively. The DET plot is for *36*-dimension feature vector and at $\alpha = 0.6$, we get relatively better results. This means CFCC is able to capture some information which MFCC is not able to do so. [21]

4   Summary and Conclusions

As observed in listening test, we observed that human perception system relates well a speaker's whispered speech to it's normal speech. Speaker recognition using whispered speech was also carried out, from the Fig 4 and Fig 5, DET curve for MFCC performs better than CFCC. However, when score-level fusion of CFCC is done with MFCC, better results are obtained at majority of operating points of DET curve than MFCC alone. Speaker recognition system when trained and tested with same type of speech, i.e., normal and whispered speech, efficiency of the system is very high as seen in Table 2. However, when there is a mismatch in training and testing very low efficiency is obtained, showing that system not able to recognize speaker in mismatched conditions. Hence, this illustrates that very less speaker information is present in whispered speech. Future work includes improving speaker recognition using whispered speech under signal degradation conditions and hence improving speaker recognition using whispered speech. [22].

Acknowledgement. Authors would like to thank Department of Electronics and Information Technology (DeitY), Government of India, New Delhi, India and DA-IICT, Gandhinagar for supporting this research work and providing the necessary resources.

References

1. Abe, M., Shikano, K., Kuwabara, H.: Cross-language voice conversion. In: Int. Conf. on Acous., Speech, & Signal Process., (ICASSP-1990), pp. 345–348. IEEE, New Mexico (1990)
2. Yegnanarayana, B., Prasanna, S., Zachariah, J.M., Gupta, C.S.: Combining evidence from source, suprasegmental and spectral features for a fixed-text speaker verification system. IEEE Trans. on Speech and Audio Process. **13**(4), 575–582 (2005)
3. Imai, S., Kobayashi, T., Tokuda, K., Masuko, T., Koishida, K., Sako, S., Zen, H.: Speech signal processing toolkit (SPTK), Version 3.3 (2009)
4. Yegnanarayana, B., Sharat Reddy, K., Kishore, S.P.: Source and system features for speaker recognition using AANN models. In: IEEE Int. Conf. on Acous., Speech, and Signal Process., (ICASSP 2001), vol. 1, pp. 409–412. IEEE, Salt Lake City (2001)
5. Fan, X., Hansen, J.H.: Speaker identification within whispered speech audio streams. IEEE Trans. on Audio, Speech, and Lang. Process. **19**(5), 1408–1421 (2011)
6. Gavidia-Ceballos, L.: Analysis and modeling of speech for laryngeal pathology assessment. PhD thesis, Duke University, Durham NC, USA (1995)
7. Gavidia-Ceballos, L., Hansen, J.H.: Direct speech feature estimation using an iterative EM algorithm for vocal fold pathology detection. IEEE Trans. on Biomedical Engg. **43**(4), 373–383 (1996)
8. Meyer-Eppler, W.: Realization of prosodic features in whispered speech. The Journal of the Acoustical Society of America **29**(1), 104–106 (1957)
9. Thomas, I.: Perceived pitch of whispered vowels. The Journal of the Acoustical Society of America **46**(2B), 468–470 (1969)
10. Jovicic, S.T.: Formant feature differences between whispered and voiced sustained vowels. Acta Acustica United with Acustica **84**(4), 739–743 (1998)
11. Morris, R.W., Clements, M.A.: Reconstruction of speech from whispers. Medical Engineering & Physics **24**(7), 515–520 (2002)
12. Zhang, C., Hansen, J.H.: An entropy based feature for whisper-island detection within audio streams. In: INTERSPEECH, Brisbane, Australia, pp. 2510–2513 (2008)
13. Neustein, A., Patil, H.A.: Forensic speaker recognition. Springer (2012)

14. Childers, D.G., Wu, K.: Gender recognition from speech. Part II: Fine analysis. The Journal of the Acoustical Society of America **90**(4), 1841–1856 (1991)
15. Li, Q.: An auditory-based transfrom for audio signal processing. In: IEEE Workshop on Applications of Signal Process. to Audio and Acous., WASPAA 2009, pp. 181–184. IEEE, New York (2009)
16. Li, Q., Huang, Y.: An auditory-based feature extraction algorithm for robust speaker identification under mismatched conditions. IEEE Trans. on Audio, Speech, and Lang. Process. **19**(6), 1791–1801 (2011)
17. Bricker, P., Pruzansky, S.: Speaker recognition. In: Contemporary issues in experimental phonetics, pp. 295–326 (1976)
18. Cummins, F., Grimaldi, M., Leonard, T., Simko, J.: The CHAINS corpus: characterizing individual speakers. In: Proc. SPECOM, St. Petersburg, Russia, vol. 6, pp. 431–435 (2006)
19. Reynolds, D.A., Quatieri, T.F., Dunn, R.B.: Speaker verification using adapted gaussian mixture models. Digital Signal Processing **10**(1), 19–41 (2000)
20. Peláez-Moreno, C., Gallardo-Antolín, A., Díaz-de María, F.: Recognizing Over IP: A robust front-end for speech recognition on the world wide web. IEEE Trans. on Multimedia **3**(2), 209–218 (2001)
21. Martin, A., Doddington, G., Kamm, T., Ordowski, M., Przybocki, M.: The DET curve in assessment of detection task performance. In: Euro Conf. Speech Process. Tech., Rhodes, Greece, pp. 1895–1898 (1997)
22. Fan, X., Hansen, J.H.: Speaker identification with whispered speech based on modified LFCC parameters and feature mapping. In: IEEE Int. Conf. on Acous., Speech and Signal Process., (ICASSP 2009), pp. 4553–4556. IEEE, Taipei (2009)

Random Indexing Explained
with High Probability

Behrang QasemiZadeh[1,2][✉] and Siegfried Handschuh[1,2]

[1] Digital Libraries and Web Information Systems, University of Passau,
Passau, Germany
{behrang.qasemizadeh,siegfried.handschuh}@uni-passau.de
http://www.fim.uni-passau.de/en/digital-libraries
[2] National University of Ireland, Galway, Ireland

Abstract. Random indexing (RI) is an incremental method for constructing a vector space model (VSM) with a reduced dimensionality. Previously, the method has been justified using the mathematical
framework of Kanerva's sparse distributed memory. This justification,
although intuitively plausible, fails to provide the information that is
required to set the parameters of the method. In order to suggest criteria for the method's parameters, the RI method is revisited and described
using the principles of linear algebra and sparse random projections in
Euclidean spaces. These simple mathematics are then employed to suggest criteria for setting the method's parameters and to explain their
influence on the estimated distances in the RI-constructed VSMs. The
empirical results observed in an evaluation are reported to support the
suggested guidelines in the paper.

Keywords: Random indexing · Dimensionality reduction · Text
analytics

1 Introduction

In order to model any aspect of language, data-driven methods of natural language processing exploit patterns of co-occurrences. For example, distributional
semantic models collect patterns of co-occurrences and investigate similarities in
these patterns in order to quantify meanings. Vector spaces are mathematically
well-defined models that are often employed to serve this purpose [2].

In a vector space model (VSM), each element $\vec{s}_i$ of its standard basis—
informally, each dimension of the VSM—represents a contextual element. Given
n context elements, linguistic entities are expressed using vectors $\vec{v}$ as linear
combinations of $\vec{s}_i$ and scalars $\alpha_i \in \mathbb{R}$ such that $\vec{v} = \alpha_1 \vec{s}_1 + \cdots + \alpha_n \vec{s}_n$. The
value of α_i is acquired from the frequency of the co-occurrences of the entity that
$\vec{v}$ represents and the context element that $\vec{s}_i$ represents. Therefore, the values
assigned to the coordinates of a vector—that is, α_i—exhibit the correlation of

---

The first three pages previously appeared in [1].

© Springer International Publishing Switzerland 2015
P. Král and V. Matoušek (Eds.): TSD 2015, LNAI 9302, pp. 414–423, 2015.
DOI: 10.1007/978-3-319-24033-6_47

an entity and context elements in an n-dimensional real vector space $\mathbb{R}^n$. In this VSM, a distance function, therefore, is employed in order to discover similarities. Amongst several choices of distance metrics, the Euclidean distance is an innate choice. A VSM is endowed with the ℓ_2 norm to estimate distances between vectors, which is accordingly called a Euclidean VSM (denoted by $\mathbb{E}^n$). Salton et al.'s classic document-by-term model is, perhaps, the most familiar example of the methodology described above [3].

In distributional methods of text analysis, as the number of entities in a VSM increases, the number of context elements employed for capturing similarities between them surges. As a result, high-dimensional vectors, in which most elements are zero, represent entities. However, the proportional impact of context elements on similarities lessens when their number increases. It becomes difficult to distinguish similarities between vectors unless the values assigned to context elements are considerably different [4]. Moreover, the high dimensionality of vectors hinders the ability to compute distances with high performance. This results in setbacks known as the *curse of dimensionality*, often tackled using a *dimensionality reduction* technique.

Dimensionality reduction can be achieved using a number of methods as an auxiliary process followed by the construction of a VSM. This process improves the computational performance by reducing the number of context elements employed for the construction of a VSM. In its simple form, dimension reduction can be performed by choosing a subset of context elements using a heuristic-based *selection process*. That is, a number of context elements that account for the most discriminative information in VSM are chosen using a heuristic such as a statistical weight threshold. Alternatively, a *transformation* method can be employed. This process maps $\mathbb{R}^n$ onto $\mathbb{R}^m$, $m \ll n$, in which $\mathbb{R}^m$ is the best approximation of $\mathbb{R}^n$ in a *sense*. For example, the well-known latent semantic analysis method employs singular value decomposition (SVD) truncation, in which $\mathbb{R}^m$ gives the best approximation of the Euclidean distances in $\mathbb{R}^n$ [5].

A number of factors hamper the use of these dimension reduction methods. Firstly, a VSM at the original high dimension must be constructed. The VSM's dimension is then reduced in an independent process. Hence, the VSM at a reduced dimensionality is available for processing only after the whole sequence of these processes. Construction of the VSM at its original dimension is computationally expensive, and a delay in access to the VSM at the reduced dimension is not desirable. Secondly, reducing the dimension of vectors using the methods listed above is resource intensive. For instance, SVD truncation demands a process of the time complexity $O(n^2 m)$ and space complexity $O(n^2)$.[1] Similarly, depending on the employed heuristic, a selection process can be resource intensive too—for example, frequencies often need to be sorted by some criteria. Last but not least, these methods are *data-sensitive*: if the structure of the data being analysed changes—that is, if either the entities or context elements are updated—the dimensionality reduction process is required to be repeated and

---

[1] However, the use of incremental techniques may relax these requirements to an extent; for example, see [6].

reapplied to the whole VSM in order to reflect the updates. As a result, these methods may not be desirable in several applications, particularly when dealing with frequently updated big text-data. Random projections are mathematical tools that are employed to implement alternative dimensionality reduction techniques to alleviate the problems listed above.

In the remainder of this paper, Section 2 describes the use of random projections (RPs) in Euclidean spaces, which consequently arrives at the well-known random indexing (RI) technique. Section 3 articulates the outcome of this mathematical interpretation. To support the theoretical discussion, empirical results are reported in Section 4. Section 5 concludes this paper.

2   Random Projections in Euclidean Spaces

In Euclidean spaces, RPs are elucidated using the Johnson and Lindenstrauss lemma (JL lemma) [7]. Given an ϵ, $0 < \epsilon < 1$, the JL lemma states that for any set of p vectors in an $\mathbb{E}^n$, there exists a mapping onto an $\mathbb{E}^m$, for $m \geq m_0 = O(\frac{\log p}{\epsilon^2})$, that does not distort the distances between any pair of vectors, with high probability, by a factor more than $1 \pm \epsilon$. This mapping is given by:

$$\mathbf{M}'_{p \times m} = \mathbf{M}_{p \times n} \mathbf{R}_{n \times m}, \ m \ll p, n, \tag{1}$$

where $\mathbf{R}_{n \times m}$ is called the RP matrix, and $\mathbf{M}_{p \times n}$ and $\mathbf{M}'_{p \times m}$ denote the p vectors in $\mathbb{E}^n$ and $\mathbb{E}^m$, respectively. According to the JL lemma, if the distance between any pair of vectors $\vec{v}$ and $\vec{u}$ in $\mathbf{M}$ is given by the $d_{\mathrm{Euc}}(\vec{v}, \vec{u})$, and their distance in $\mathbf{M}'$ is given by $d'_{\mathrm{Euc}}(\mathbf{v}, \mathbf{u})$, then there exists an $\mathbf{R}$ such that $(1 - \epsilon)d'_{\mathrm{Euc}}(\mathbf{v}, \mathbf{u}) \leq d_{\mathrm{Euc}}(\mathbf{v}, \mathbf{u}) \leq (1 + \epsilon)d'_{\mathrm{Euc}}(\mathbf{v}, \mathbf{u})$.[2] Accordingly, instead of the original high-dimensional $\mathbb{E}^n$ and at the expense of a negligible amount of error ϵ, the distance between $\vec{v}$ and $\vec{u}$ can be calculated in $\mathbb{E}^m$ to reduce the computational cost of processes.

The JL lemma does not specify $\mathbf{R}$. Establishing a random matrix $\mathbf{R}$ is therefore the most important design decision when using RPs. In [7], the lemma was proved using an orthogonal projection. Subsequent studies simplified the original proof that resulted in projection techniques with enhanced computational efficiency (see [8] for references). Recently, it has been shown that a sparse $\mathbf{R}$, whose elements r_{ij} are defined as:

$$r_{ij} = \sqrt{s} \begin{cases} -1 & \text{with probability } \frac{1}{2s} \\ 0 & \text{with probability } 1 - \frac{1}{s} , \\ 1 & \text{with probability } \frac{1}{2s} \end{cases} \tag{2}$$

for $s \in \{1, 3\}$, results in a mapping that also satisfies the JL lemma [9]. Subsequent research showed that $\mathbf{R}$ can be constructed from even sparser vectors than those suggested in [9]. In [10], it is proved that in a mapping of an n-dimensional

---

[2] In addition, the lemma states that this mapping can be found in randomized polynomial time.

real vector space by a sparse $\mathbf{R}$, the JL lemma holds as long as $s = O(n)$, for example, $s = \sqrt{n}$ or even $s = \frac{n}{\log(n)}$. The sparseness of $\mathbf{R}$ consequently enhances the time and space complexity of the method by the factor $\frac{1}{s}$.

Another benefit when computing $\mathbf{M}'$ is obtained using the linearity of matrix multiplication. As stated earlier, each vector $\vec{v}_{e_i}$ in $\mathbb{E}^n$ (i.e., the ith row of $\mathbf{M}$) is given by a linear combination of the basis vectors $\vec{v}_{e_i} = w_{i1}\vec{s}_{c_1} + \cdots + w_{in}\vec{s}_{c_n}$ ($i \leq p$ and $j \leq n$). By the basic properties of the matrix multiplication, the projection of $\vec{v}_{e_i}$ in $\mathbf{M}'$ is given by $\vec{v}'_{e_i} = \vec{v}_{e_i}\mathbf{R} = w_{i1}\vec{s}_{c_1}\mathbf{R} + \cdots + w_{in}\vec{s}_{c_n}\mathbf{R}$. In turn, since by definition all the elements of $\vec{s}_{c_k}$ are zero except the kth element (i.e., 1), $\vec{v}'_{e_i}$ can be written as:

$$\vec{v}'_{e_i} = w_{i1}\vec{r}_1 + \cdots + w_{in}\vec{r}_n, \tag{3}$$

where $\vec{r}_j$ is the jth row of $\mathbf{R}$. Equation 3 means that row vectors $\mathbf{v}'_{e_i}$, thus $\mathbf{M}'$, can be computed directly without necessarily constructing the whole matrix $\mathbf{M}$. The jth row of $\mathbf{R}_{n \times m}$ represents a context element in the original VSM that is located at the jth column of $\mathbf{M}_{p \times n}$. Therefore, an entity at a reduced dimension can be computed directly by accumulating the row vectors of $\mathbf{R}$ that represent the context elements that co-occur with the entity.

The explanations above result in a two-step procedure similar to the one suggested earlier as the RI technique [11][12]: the construction of (a) *index vectors* and (b) *context vectors*. In the first step, each context element is assigned *exactly* to one *index vector*. Sahlgren [12] indicates that index vectors are high-dimensional, randomly generated vectors, in which most of the elements are set to 0 and only *a few* to 1 and -1. In the second step, the construction of *context vectors*, each target entity is assigned to a vector of which all elements are zero and that has the same dimension as the index vectors. For each occurrence of an entity (represented by $\vec{v}_{e_i}$) and a context element (represented by $\vec{r}_{c_k}$), the context vector is accumulated by the index vector (i.e., $\vec{v}_{e_i} = \vec{v}_{e_i} + \vec{r}_{c_k}$). The result is a vector space model constructed directly at reduced dimension. As can be understood, the first step of RI is equivalent to constructing the random projection matrix $\mathbf{R}$, whose elements are given by Equation 2. Each index vector is a row of the random projection matrix $\mathbf{R}$. The second step of RI deals with computing $\mathbf{M}'$. Each context vector is a row of $\mathbf{M}'$, which is computed by the iterative process justified in Equation 3.

3   The Significance of the Proposed Mathematical Justification

In contrast to previous research in which the RI's parameters were left to be decided through experiments (e.g., see [13, 14]), one can leverage the adopted mathematical framework to provide a guideline for setting the parameters of RI. In an RI-constructed VSM at reduced dimension m (i.e., $\mathbb{E}^m$), the degree of the preservation of distances in $\mathbb{E}^n$ and $\mathbb{E}^m$ is determined by the number of vectors in the model and the value of m. If the number of vectors is fixed, then the larger m is, the better the

Euclidean distances are preserved at the reduced dimension m. In other words, the probability of preserving the pairwise distances increases as m increases. Hence, m can be seen as the capacity of an RI-constructed VSM for accommodating new entities. Compared to $m = 4000$ suggested in [11] or $m = 1800$ in [12], depending on the number of entities that are modelled in an experiment, m can be set to a smaller value, such as 400.

Based on the proofs in [10], when embedding $\mathbb{E}^n$ into $\mathbb{E}^m$, the JL lemma holds as long as s in Equation 2 is $O(n)$. In text processing applications, the number of context elements (i.e., n) is often very large. When using RI, therefore, even a careful choice such as $s = \sqrt{n}$ in Equation 2 results in highly sparse index vectors. Hence, by setting only two or four non-zero elements in index vectors, distances in the RI-constructed $\mathbb{E}^m$ resemble distances in $\mathbb{E}^n$. If the dimension of index vectors (i.e., m) is fixed, then increasing the number of non-zero elements in index vectors causes additional distortions in pairwise distances. For index vectors of fixed dimensionality m, if the number of non-zero elements increases, then the probability of the orthogonality between index vectors decreases; hence, it stimulates distortions in pairwise distances (see Fig. 1)—although in some applications, distortions in pairwise distances can be beneficial.

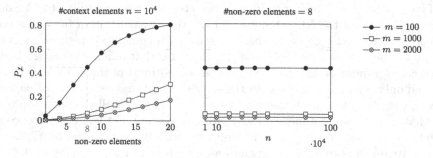

Fig. 1. Orthogonality of index vectors: the y-axis shows the proportion of non-orthogonal pairs of index vectors (denoted by $P_{\not\perp}$) for sets of index vectors of various dimensions obtained in a simulation. For index vectors of the fixed size $n = 10^4$, the left figure shows the changes of $P_{\not\perp}$ when the number of non-zero elements increases. The right figure shows $P_{\not\perp}$ when the number of non-zero elements is fixed to 8; however, the number of index vector n increases. As shown in the figure, $P_{\not\perp}$ remains constant independently of n.

4   Experimental Results

In order to show the influence of the RI's parameters on the ability of the method to preserve pairwise Euclidean distances, instead of a task-specific evaluation, an intrinsic evaluation is suggested.

In the reported experiments, a subset of Wikipedia articles chosen randomly from WaCkypedia (a 2009 dump of the English Wikipedia [15]). A document-by-term VSM at its original high dimension is first constructed from a set of 10,000 articles (shown by D). A pre-processing—that is, white-space tokenisation followed by removing non-alphabetic tokens—of documents in D results in a vocabulary of 192,117 terms. Each document in D is represented by a high-dimensional vector; each dimension represents an entry from the obtained vocabulary. Therefore, the constructed VSM using this *one-dimension-per-context-element* method has a dimensionality of $n = 192,117$.[3]

To keep the experiments a manageable size, each document d in D is randomly grouped by another nine documents from D, which consequently gives 10,000 sets of a set of ten documents. Using the constructed n-dimensional ($n = 192,117$) VSM, for each set of documents, the Euclidean distances between d and the remaining nine documents in the set are computed. Subsequently, these nine documents are sorted by their distance from d to obtain an ordered set of documents. This procedure thus results in 10,000 ordered sets of nine documents (the same steps are repeated for computing the cosine similarities).

The procedure described above is repeated, however, by calculating distances in VSMs that are constructed using the RI method. Each term in the vocabulary is assigned to an m-dimensional index vector and each document to a context vector. Context vectors are updated by accumulating index vectors to reflect the co-occurrences of documents and terms. Subsequently, the obtained context vectors are used to estimate the Euclidean distances and the cosine similarities between documents. The estimated distances are then used to create the ordered sets of documents, exactly as explained above. This process is repeated several times when the parameters of RI—that is, the dimension m and the number of non-zero elements in index vectors—are set differently.

It is expected that the relative Euclidean distances as well as the cosine similarities between documents in the RI-constructed VSMs are the same as in the original high-dimensional VSM.[4] Hence, the ordered sets of documents obtained from the estimated distances in the RI-constructed VSMs must be identical to the corresponding sets that are derived using the computed distances in the original high-dimensional VSM. For each RI-constructed VSM, therefore, the resulting ordered sets are compared with the obtained ordered sets from the original high-dimensional VSM using the Spearman's rank correlation coefficient measure (ρ). The average of ρ over the obtained sets of ordered sets of documents ($\bar{\rho}$) is reported to quantify the performance of RI with respect to its ability to preserve ℓ_2-normed distances when its parameters are set to different values: the closer $\bar{\rho}$ is to 1, the more similar the order of documents in an RI-constructed VSM and the original high-dimensional VSM.

Figure 2 shows the obtained results. Since the original VSM is high dimensional and sparse, even for $m = 1600$, two non-zero elements per index vector are sufficient to construct a VSM that resembles relative distances between vectors in the original high-dimensional vector space. In addition, because only a small number of documents are modelled (i.e., $p = 10,000$), even for $m = 100$,

---

[3] The frequencies of terms in documents are used as weights in corresponding vectors.

[4] The preservation of the cosine similarities can be verified mathematically by expressing it as the Euclidean distance when the length of vectors is normalised to unity.

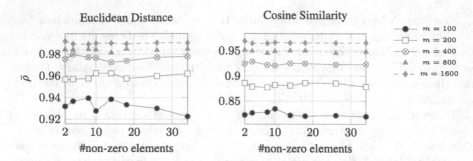

Fig. 2. Correlation between the ℓ_2-normed measures in the original high-dimensional VSM and RI-constructed VSMs: $\bar{\rho}$ shows the average of the Spearman's rank correlation between the ordered sets of documents that are obtained by computing in the original high-dimensional VSM and the RI-constructed VSMs. Results are shown for both Euclidean distances and the cosine similarities when parameters of the RI method are set to different values. The random baseline obtained in experiments is -0.002 (i.e., as expected, almost zero).

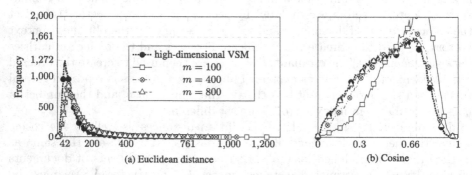

Fig. 3. A histogram of the distribution of (a) Euclidean distances and (b) cosine similarities between pairs of vectors in the original VSM of dimension 192,117 compared to the RI-constructed VSMs. For all values of m, the number of non-zero elements in index vectors is set to 2.

the estimated distances in the RI-constructed VSM show a high correlation to the distances in the original vector space (i.e., $\bar{\rho} > 0.92$ for pairwise Euclidean distances and $\bar{\rho} > 0.82$ for the cosine similarity). As expected, the generated random baseline for $\bar{\rho}$ in Figure 2 is -0.002, that is, approximately 0. For $m = 1600$, the observed pairwise distances in the RI-constructed vector space are almost identical to the original vector space, that is, $\bar{\rho} > 0.99$ for Euclidean distances and $\bar{\rho} > 0.96$ for the cosine. Figure 3 compares the distribution of distances in the original high-dimensional VSM and the RI-constructed VSMs. As expected, when m increases, these distributions become more similar to each other.

5 Discussion

Random indexing—a well-known method for the incremental construction of VSMs—is revisited and justified using the theorems proved in [10]—that is, sparse random projections in Euclidean spaces. The results from an empirical experiment are shown to explain the method's behaviour with respect to its ability to preserve pairwise Euclidean distances. Although a new method is not suggested, I would like to emphasise on several important outcomes of the description given in this paper.

Firstly, whereas the original delineation of the method did not provide a concrete guideline for setting the method's parameters, this paper embellished the previous two-step procedure with criteria for choosing the dimensionality as well as the proportion of zero and non-zero elements of index vectors.

Secondly, the proposed understanding of the RI method helps us to discern its application domain. It is shown that the employed random projections by the RI method do not preserve distances other than ℓ_2 (e.g., see [16]). Hence, it is important to note that RI-constructed VSMs can only be used for estimating similarity measures that are derived from the ℓ_2 norm—for example, the Euclidean distance and the cosine similarity. This being the case, the use of RI-constructed VSMs for estimating city block distances—such as suggested in [17]—is not justified, at least mathematically.

Thirdly, the given understanding of the method helps one to generalise the RI method to normed spaces other than ℓ_2. This generalisation can be achieved using α-stable random projections—for example, as suggested in [18,19]—and by altering Equation 2. Simply put, altering a random projection matrix $\mathbf{R}$—hence, index vectors—so that it has an α-stable distribution[5] results in new techniques similar to RI, however, for estimating distances in ℓ_α-normed spaces (e.g., see [20–22]).[6]

Last but not least, the rationale given in this paper enables one to justify several proposed variations of the RI technique mathematically. Although these methods are based on plausible intuition, similar to RI, they lack theoretical justifications. For example, based on the description given in this paper, one can identify the method proposed in [23] as a variation of RI that employs Laplacian smoothing. This idea can be generalised for coordinating other major processes that are often involved when using VSMs.

Acknowledgments. This publication has partly emanated from research supported by a research grant from Science Foundation Ireland (SFI) under Grant Number SFI/12/RC/2289.

---

[5] Note that RI uses a 2-stable projection; that is, $\mathbf{R}$ derived from Equation 2 has a standard asymptotic Gaussian distribution.

[6] In this case, new distance estimators are required.

References

1. QasemiZadeh, B.: Random indexing revisited. In: Biemann, C., Handschuh, S., Freitas, A., Meziane, F., Métais, E. (eds.) NLDB 2015. LNCS, vol. 9103, pp. 437–442. Springer, Heidelberg (2015)
2. Turney, P.D., Pantel, P.: From frequency to meaning: vector space models of semantics. Journal of Artificial Intelligence Research **37**(1), 141–188 (2010)
3. Salton, G., Wong, A., Yang, C.S.: A vector space model for automatic indexing. Communications of the ACM **18**(11), 613–620 (1975)
4. Beyer, K., Goldstein, J., Ramakrishnan, R., Shaft, U.: When is nearest neighbor meaningful? In: Beeri, C., Bruneman, P. (eds.) ICDT 1999. LNCS, vol. 1540, pp. 217–235. Springer, Heidelberg (1998)
5. Deerwester, S.C., Dumais, S.T., Landauer, T.K., Furnas, G.W., Harshman, R.A.: Indexing by latent semantic analysis. Journal of the American Society of Information Science **41**(6), 391–407 (1990)
6. Brand, M.: Fast low-rank modifications of the thin singular value decomposition. Linear Algebra and its Applications **415**(1), 20–30 (2006). Special Issue on Large Scale Linear and Nonlinear Eigenvalue Problems
7. Johnson, W., Lindenstrauss, J.: Extensions of Lipschitz mappings into a Hilbert space. In: Beals, R., Beck, A., Bellow, A., Hajian, A. (eds.) Conference on Modern Analysis and Probability (1982: Yale University). Contemporary Mathematics, vol. 26. American Mathematical Society, pp. 189–206 (1984)
8. Dasgupta, S., Gupta, A.: An elementary proof of a theorem of Johnson and Lindenstrauss. Random Structures and Algorithms **22**(1), 60–65 (2003)
9. Achlioptas, D.: Database-friendly random projections. In: Proceedings of the Twentieth ACM SIGMOD-SIGACT-SIGART Symposium on Principles of Database Systems, pp. 274–281. ACM, Santa Barbara, May 2001
10. Li, P., Hastie, T.J., Church, K.W.: Very sparse random projections. In: Proceedings of the 12th ACM SIGKDD International Conference on Knowledge Discovery and Data Mining, pp. 287–296. ACM, New York (2006)
11. Kanerva, P., Kristoferson, J., Holst, A.: Random indexing of text samples for latent semantic analysis. In: Gleitman, L.R., Josh, A.K. (eds.) Proceedings of the 22nd Annual Conference of the Cognitive Science Society, p. 1036. Erlbaum, Mahwah (2000)
12. Sahlgren, M.: An introduction to random indexing. Technical report, Swedish ICT (SICS) (2005). Retrived from https://www.sics.se/~mange/papers/RI_intro.pdf
13. Lupu, M.: On the usability of random indexing in patent retrieval. In: Hernandez, N., Jäschke, R., Croitoru, M. (eds.) ICCS 2014. LNCS, vol. 8577, pp. 202–216. Springer, Heidelberg (2014)
14. Polajnar, T., Clark, S.: Improving distributional semantic vectors through context selection and normalisation. In: Proceedings of the 14th Conference of the European Chapter of the Association for Computational Linguistics. Association for Computational Linguistics, Gothenburg, pp. 230–238, April 2014
15. Baroni, M., Bernardini, S., Ferraresi, A., Zanchetta, E.: The WaCky Wide Web: A collection of very large linguistically processed Web-crawled corpora. Language Resources and Evaluation **43**(3), 209–226 (2009)
16. Brinkman, B., Charikar, M.: On the impossibility of dimension reduction in L1. Journal of the ACM **52**(5), 766–788 (2005)

17. Lapesa, G., Evert, S.: Evaluating neighbor rank and distance measures as predictors of semantic priming. In: Proceedings of the Fourth Annual Workshop on Cognitive Modeling and Computational Linguistics. Association for Computational Linguistics, Sofia, pp. 66–74, August 2013
18. Indyk, P.: Stable distributions, pseudorandom generators, embeddings, and data stream computation. Journal of the ACM **53**(3), 307–323 (2006)
19. Li, P., Samorodnitsk, G., Hopcroft, J.: Sign Cauchy projections and chi-square kernel. In Burges, C.J.C., Bottou, L., Welling, M., Ghahramani, Z., Weinberger, K.Q. (eds.) Advances in Neural Information Processing Systems, vol. 26, pp. 2571–2579. Curran Associates, Inc. (2013)
20. Geva, S., De Vries, C.M.: TOPSIG: topology preserving document signatures. In: Berendt, B., de Vries, A., Fan, W., Macdonald, C., Ounis, I., Ruthven, I. (eds.) Proceedings of the 20th ACM International Conference on Information and Knowledge Management, pp. 333–338. ACM, Glasgow (2011)
21. Zadeh, B.Q., Handschuh, S.: Random Manhattan integer indexing: incremental L1 normed vector space construction. In: Proceedings of the 2014 Conference on Empirical Methods in Natural Language Processing (EMNLP), pp. 1713–1723. Association for Computational Linguistics, Doha (2014)
22. Zadeh, B.Q., Handschuh, S.: Random Manhattan indexing. In: 25th International Workshop on Database and Expert Systems Applications (DEXA), pp. 203–208. IEEE, Munich (2014)
23. Baroni, M., Lenci, A., Onnis, L.: ISA meets Lara: an incremental word space model for cognitively plausible simulations of semantic learning. In: Proceedings of the Workshop on Cognitive Aspects of Computational Language Acquisition, pp. 49–56. Association for Computational Linguistics, Prague, June 2007

Hungarian Grammar Writing
with LTAG and XMG

Kata Balogh[✉]

Heinrich-Heine-Universität Düsseldorf, Düsseldorf, Germany
katalin.balogh@hhu.de

Hungarian is a rather challenging language for computational linguistic applications given its flexible word order and being a discourse configurational language [1], where sentence articulation is driven by discourse-semantic functions such as *topic* and *focus*, rather that grammatical functions. Grammatical functions are morphologically marked, not by structural positions. Based on free word order and the rich morphology, mostly dependency-based parsers are developed for Hungarian (e.g. [2]). This paper presents a different approach using Lexicalized Tree-Adjoining Grammar (LTAG; [3]) and the extensible MetaGrammar (XMG; [4]). An LTAG analysis for Hungarian is challenging, not just because of the flexible word order, but also because the syntactic positions are driven by discourse semantic considerations. Despite these challenges, there are strong arguments in favor of using LTAG, especially from the computational linguistic perspective. One of the most important arguments is that TAGs are mildly context-sensitive, having the important formal properties, such that: (i) TAGs can generate crossing dependencies, (ii) parsing can be done in polynomial time and (iii) TAGs have the constant growth property (semi-linearity). LTAG is also efficient for large coverage grammar writing, as presented for English and Korean [5], French [6] or German [7]. In addition, several elegant tools are available for grammar implementation using (L)TAG, such as the XTAG tools (parser, editor, viewer) developed by the XTAG Research Group, the eXtensible MetaGrammar (XMG) by Crabbé et al [4] and the Tübingen Linguistic Parsing Architecture (TuLiPA) by Kallmeyer et al [8].

1   Core Data: Hungarian Sentence Articulation

In Hungarian sentence structure we distinguish two fields related to information structure: the postverbal and the preverbal field. The *postverbal field* by default hosts the 'argument positions' whose order is free, and the word order variations do not signal grammatical roles. In sentence (1) all six permutations of the arguments behind the verb are grammatical with the same semantic content.

(1)   Adott Pim Évának egy könyvet. / Adott Évának egy könyvet Pim. / ... etc.
      gave   Pim Eva.dat a   book.acc / gave Eva.dat a   book.acc Pim / ... etc.
      'Pim gave a book to Eva.'

© Springer International Publishing Switzerland 2015
P. Král and V. Matoušek (Eds.): TSD 2015, LNAI 9302, pp. 424–432, 2015.
DOI: 10.1007/978-3-319-24033-6_48

1.1 The Preverbal Field

The *preverbal field* hosts the so-called 'functional projections', whose order is fixed.

(2) Péter$_T$ mindenki-hez$_Q$ MARI-T$_F$ küldi.
    Peter   everyone.all   Mary.acc sends
    'It is Mary, whom Peter sends to everyone.'

Topics are sentence initial and can be iterated, while there is only a single (narrow) focus possible in the immediate preverbal position. Quantifiers stand behind the topic(s) and before the focus. Next to topic, focus and quantifiers, the preverbal field also hosts sentential negation, optative operators (*bárcsak* 'if only') and interrogative operators.

The syntactic position immediately preceding the verb is important in Hungarian for several reasons. It hosts narrow focus, sentential negation and the verbal modifier, partially in complementary distribution. In 'neutral sentences' – without narrow focus or sentential negation –, the preverbal position is occupied by the verbal modifier (VM) as in (3). When the sentence contains sentential negation (4) or a narrow focus (5) the VM stands postverbally. The verbal modifier is in complementary distribution with the narrow focus and the sentential negation, however, narrow focus and negation are not in complementary distribution, see (6).

(3) Péter meg-hívta Marit.   (4) Péter nem hívta meg Marit.
    Peter VM-invited Mary.acc   Peter neg invited VM Mary.acc
    'Peter invited Mary.'   'Peter did not invite Mary.'

(5) Péter MARIT hívta meg.   (6) Péter MARIT   nem hívta meg.
    Peter Mary.acc invited VM   Peter Mary.acc neg invited VM
    'It is Mary whom Peter invited.'   'It is Mary whom Peter did not invite.'

1.2 Verbal Modifiers

Next to the relation with the focus structure of the utterance, verbal modifiers (VM) in Hungarian are both syntactically and semantically interesting. VMs are either verbal prefixes, bare nouns or infinitives (without own VM), that are independent syntactic units and lexically selected by the verb (see [9]). As shown in the previous section, the position of the VM depends on whether narrow focus or sentential negation is present in the sentence. The type and the position of the verbal particle is related to the aspectual structure of the utterance ([10]). In sentences without narrow focus and/or sentential negation the verbal particle in the preverbal position signals perfective aspect, while standing postverbally it indicates progressive.

From the semantic point of view some more important facts can be pointed out regarding verbal modifiers. Verbal modifiers can – but not necessarily do – provide a secondary predication [9], and they can also change the argument structure of the verb. And finally, the presence of the verbal modifier can also effect the required specificity of the subject. These semantic aspects are outside of the scope of the current paper and will be implemented in the approach later.

2   LTAG and XMG in a Nutshell

Tree-Adjoining Grammar (TAG) is a tree-rewriting formalism, where the elementary structures are trees. A TAG is a set of *elementary trees* with two combinatorial operations: *substitution* and *adjunction*. The set of elementary trees is the union of a finite set of *initial trees* and *auxiliary trees*. A derivation in TAG starts with an initial tree (α) and proceeds with using either of the two operations. By *substitution* a non-terminal leaf node is replaced by an initial tree (β), while by *adjunction* an internal node is replaced by an auxiliary tree (γ).

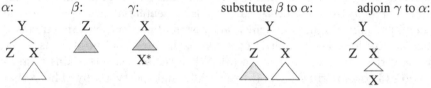

In a Lexicalized TAG (LTAG) each elementary tree contains at least one lexical element, its *lexical anchor* ($\diamond$). To increase the expressive power of the formalism, adjunction constraints are additionally introduced to restrict whether adjunction is mandatory and/or which trees can be adjoined at a given node. In particular for natural language analyses another extension of TAG is proposed, using feature structures as non-terminal nodes. Among the reasons for a Feature-based TAG (F-TAG) two important ones are generalizing agreement and case marking via underspecification. A great advantage of F-TAG with respect to grammar writing is the result of smaller grammars that are easier to maintain, as well as the possibility of modeling adjunction constraints. The shape of the elementary trees is driven by linguistic principles (see [11] and [12]), reflecting the syntactic/semantic properties of linguistic objects. Syntactic design principles determine, for example, that valency/subcategorization is expressed locally within the elementary tree of the predicate, either as a substitution node or as a footnode. In the grammar architecture *tree families* are defined, sets of tree templates representing a subcategorization frame and collecting all syntactic configurations the subcategorization frame can be realized in.

An LTAG grammar is a set of elementary trees, where most linguistic information is contained. However, the this set contains identical tree fragments, leading to multiple structure sharing. The XMG tool generates the elementary trees for a given grammar formalism, such that it factors out the redundant parts of a given tree set by identifying identical tree fragments. An additional abstraction level, the *meta-grammar*, is introduced where generalizations can be expressed. The meta-grammar is a declarative system, combining reusable tree fragments (*classes*) by conjunction and disjunction.

```
Class  ::= Name → Content
Content ::= Description | Name | Content ∧ Content | Content ∨ Content
Description ::= nᵢ → nⱼ | nᵢ →⁺ nⱼ | nᵢ →* nⱼ | nᵢ ≺ nⱼ | nᵢ ≺⁺ nⱼ | nᵢ ≺* nⱼ
             | nᵢ[f₁ : v₁, ..., fₙ : vₙ] | nᵢ(c₁ : cv₁, ..., cfₙ : cvₙ)
             | Description ∧ Description
```

The content of a class can be either a simple tree fragment or a conjunction / disjunction of two tree fragments. In the description of a tree fragment the dominance $\rightarrow$ and precedence $\prec$ relations of the nodes are given, where $\rightarrow^+$ and $\prec^+$ stand for their transitive closure and $\rightarrow^*$ and $\prec^*$ for their transitive, reflexive closure. At each node we refer to the features associated with it by $n_i[f_1 : v_1, ..., f_n : v_n]$ and each node can be marked for substitution, footnode, anchor etc. by $n_i(c_1 : cv_1, ..., cf_n : cv_n)$.

Tree fragments can be combined by conjunction and disjunction resulting in tree templates. During the combination of the tree fragments nodes get unified. There are two possible ways to specify the node equations: (1) with explicit node unification by means of node variable equations (e.g. ?VP = ?Subj.?VP;) or (2) by node polarization: annotating the nodes with colors (e.g. color = black) that declare implicitly how the nodes can be unified with other nodes. This method is based on a color matrix, according to which (i) a black node can unify with zero, one or more white nodes, producing a black node, (ii) a white node must be unified with a black node producing a black node, and (iii) a red node cannot be unified with any other node. The resulting tree fragment is a satisfying model only if it does not contain any white nodes.

3 Hungarian: Simple Sentences with XMG

In Hungarian, sentence articulation is driven by discourse semantics rather than grammatical functions. Instead of defining structural positions for grammatical functions (e.g. subject) we need to define structural positions for topic and focus (and also for quantifiers). Accordingly, the elementary trees in a given tree family need to encode the possible topic/focus structures. In the following, I will illustrate the system via the transitive tree family with two NP arguments. Both arguments can be either in postverbal position (ARG), in topic position (TOP) and in focus (FOC) position. Hence, the tree family of transitive verbs must contain the elementary trees for all of the combinations here: NP_1-ARG + NP_2-ARG, NP_1-TOP + NP_2-ARG, NP_1-FOC + NP_2-ARG, NP_1-TOP + NP_2-TOP and NP_1-TOP + NP_2-FOC. Instead of merely listing the structures, the XMG tool is used to generate the set of elementary trees, so that generalizations on positions can also be expressed.

3.1 The Nucleus and the Postverbal Field

We have three core tree fragments for the projection of the nucleus of the sentence, containing the anchor for the verbal head and the verbal modifier the verb selects for. As discussed before, we have three possible structures here, hence we need to assume three tree fragments: verb without VM and verb with VM with two different orders.

SProj1:

```
           S
           ¦
           ¦
       VP[inv=na]
           |
           V◇
```

```
class SProj1
declare ?S01 ?VP01 ?V01
{<syn>{ node ?S01 (color=black) [cat=s] ;
node ?VP01 (color=black) [cat=vp,inv=na] ;
node ?V01 (color=black,mark=anchor) [cat=v] ;
?S01 ->* ?VP01 ; ?VP01 -> ?V01 }}
```

SProj2:

```
           S
           ¦
           ¦
       VP[inv=no]
          / \
         /   \
      VM◇    V◇
```

```
class SProj2
declare ?S02 ?VP02 ?V02 ?VM02
{<syn>{ node ?S02 (color=black) [cat=s] ;
node ?VP02 (color=black) [cat=vp,inv=no] ;
node ?V02 (color=black,mark=anchor) [cat=v] ;
node ?VM02 (color=red,mark=coanchor) [cat=vm] ;
?S02 ->* ?VP02 ; ?VP02 -> ?V02 ; ?VP02 -> ?VM02 ; ?VM02 >> ?V02}}
```

SProj3:

```
           S
           ¦
           ¦
       VP[inv=yes]
          / \
         /   \
      V◇     VM◇
```

```
class SProj3
declare ?S03 ?VP03 ?V03 ?VM03
{<syn>{ node ?S03 (color=black) [cat=s] ;
node ?VP03 (color=black) [cat=vp,inv=yes] ;
node ?V03 (color=black,mark=anchor) [cat=v] ;
node ?VM03 (color=red,mark=coanchor) [cat=vm] ;
?S03 ->* ?VP03 ; ?VP03 -> ?V03 ; ?VP03 -> ?VM03 ; ?V03 >> ?VM03}}
```

The dashed lines represent the transitive closure of the dominance relation, which is necessary to make sure, that we can later combine the *SProj* fragments with the tree fragments of the topic and focus positions, where the TP and FP nodes will go between the S and the VP nodes, with dominance order S -> TP -> FP -> VP. The *inv* feature at the VP node indicates whether the verb selects a VM (*na* vs. *yes/no*), and in which order the VM and V appear in the sentence (*yes* vs. *no*).

The grammatical functions are encoded by case marking in Hungarian, hence a transitive verb requires one NP in nominative case and another one in accusative. In the postverbal field the order of the arguments is free, so in case the two NP arguments of a transitive verb both stand postverbally, we have two possible orders. The canonical argument position is given by the tree fragment below.

ArgPos:

```
          VP
         /  \
        /    \
      V◇  ≺+ NP↓[case=1]
```

```
class ArgPos
declare ?VP1 ?V1 ?NP1 ?X1
{<syn>{ node ?VP1 (color=white) [cat=vp] ;
        node ?V1 (color=white) [cat=v] ;
        node ?NP1 (color=black) [cat=np,case=?X1] ;
        ?VP1 -> ?V1; ?VP1 -> ?NP1; ?V1 >>* ?NP1 }}
```

The argument position is on the right hand side of the verbal head. The case of the NP is underspecified ([case=1]), and can be realized both by a nominative and an accusative argument:

```
          VP                          VP
         /  \                        /  \
        /    \                      /    \
      V  ≺+ NP↓[case=nom]         V  ≺+ NP↓[case=acc]
```

The precedence relation between the V node and the NP node is given by the transitive closure of precedence ($\prec^+$), that makes possible that other arguments can go in between these two nodes. Within the transitive tree family, the class `twoPostV` is defined to generate the structures, where both arguments stand in post verbal position.

$$SProj ::= SProj1 \lor SProj2 \lor SProj3$$
$$twoPostV ::= SProj \land SubjArg \land ObjArg$$

This description declares the structures with two postverbal arguments, and generates the elementary trees accordingly.

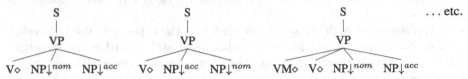

3.2   Positions in the Preverbal Field

Arguments can also be topicalized or focused, standing in designated positions in the preverbal field. In the tree fragments of the topic and focus positions the case of the argument is underspecified, and similarly to the argument position (ArgPos), can be realized as nominative or accusative.

For the topic position (TP) more structures are declared. These are necessary, since the TP can either be followed by another TP, a quantifier (QP), a focus (FP) or the VP.

TopPos1-3:

$$\text{TP}$$
$$\overset{\frown}{\text{NP}\downarrow^{[case=\boxed{1}]} \quad \prec \quad \text{TP/QP/FP}}$$

TopPos4:

$$\text{TP}$$
$$\overset{\frown}{\text{NP}\downarrow^{[case=\boxed{1}]} \quad \prec \quad \text{VP}^{[inv=no/na]}}$$

In case the TP is followed by the VP, we do not have focus in the utterance, and the *inv* feature must get the value either *na* for VP without VM, or *no* for VP with the non inverted VM-V order. The precedence relation of the NP and the TP/QP/FP/VP nodes is defined stric ($\prec$), such that no other node can interfere between them.

The focus position is always directly preceding the verb, hence it can only be followed by the VP. The *inv* feature of the VP is required to get the value *yes* for the inverted V-VM order for verbs with a VM. By this requirement we make sure that the presence of a narrow focus triggers the V-VM inversion.

FocPos:

$$\text{FP}$$
$$\overset{\frown}{\text{NP}\downarrow^{[case=\boxed{1}]} \quad \prec \quad \text{VP}^{[inv=yes/na]}}$$

In Hungarian, quantifiers also stand in a special preverbal position, after the topic(s) and before the focus. The quantifier position can be modeled similarly to the topic and focus positions; it can be followed by an FP or VP, hence we have two fragments:

QuantPos1:

$$\text{QP}$$
$$\overset{\frown}{\text{NP}\downarrow^{[case=\boxed{1}]} \quad \prec \quad \text{FP}}$$

QuantPos2:

$$\text{QP}$$
$$\overset{\frown}{\text{NP}\downarrow^{[case=\boxed{1}]} \quad \prec \quad \text{VP}}$$

3.3 Basic Sentence Structures

After defining the tree fragments of the designated positions in the pre- and postverbal fields, we can define the classes for the tree families. For an illustration I present here the tree family of transitive verbs. In non-neutral sentences one of the arguments must be in the focus position while the other argument is either postverbal or in topic position.[1]

  $NonNeutS \longrightarrow SProj \wedge ((SubjFoc \wedge (ObjArg \vee ObjTop)) \vee (ObjFoc \wedge (SubjArg \vee SubjTop)))$

In neutral sentences both arguments are either in the topic or in the postverbal position. No argument is in focus. The class *twoPostV* describes the structure with two postverbal arguments, while *topic* describes structures with one or two topical arguments.

  $NeutS \longrightarrow twoPostV \vee topic$

  $twoPostV \longrightarrow (SProj1 \vee SProj2) \wedge SubjArg \wedge ObjArg$

  $topic \longrightarrow SProj \wedge ((SubjTop \wedge (ObjArg \vee ObjTop)) \vee (ObjTop \wedge (SubjArg \vee SubjTop)))$

Using the class descriptions and the tree fragments introduced before, we can derive all possible structures in the transitive tree family, generating the elementary trees for all possible topic/focus articulations.

4 Summary and Further Work

The paper presented the analysis of basic sentence articulation in Hungarian, using the LTAG grammar formalism and the XMG tool. Despite the challenges of grammar writing for Hungarian caused by free word order and discourse configurationality, the additional descriptive level of *meta-grammar* is particularly advantageous. The paper illustrated the tree fragments and class descriptions to generate the elementary trees for the base cases in Hungarian sentence articulation, reflecting all topic/focus structures.

However, as the topic of this paper is part of a large project of grammar writing for Hungarian, further phenomena must be implemented in the system. The most important extensions to be done is implementing sentential negation and adverbials. Sentential negation has a close relation to the immediate preverbal position and as such syntactically crucial for the articulation in the preverbal field. In standard TAG analyses, negation is handled in different ways. It is either analyzed as an auxiliary tree, or being a functional element it is assumed to be part of the elementary tree of the predicate [12]. In the current system, the second option can be implemented straightforwardly. As for the adverbials, following [9] we distinguish sentential adverbials and predicate adverbials, based on what they modify: the whole statement or merely the predication. Sentential adverbials cannot enter the predication (the postverbal field), but they can mix with the topics. They can occur sentence initially, or at any position before the VP (before the main predication). Standing right before the VP, the sentential adverbial still modifies the whole statement.

---

[1] For simplicity, we skip quantifiers in the following.

4.1 Verbal Field(s) in Complex Sentences

In clausal complements we have to deal with multiple verbal fields, belonging to the matrix verb and the embedded verb. Syntactically interesting cases are sentences where arguments of the embedded verb are topicalized or focused. With this respect finite and infinite clausal complements behave differently. In finite clausal complements, the focused/topicalized argument of the embedded verb stays in its own preverbal field.

In infinitival complements the preferred position of a focused/topicalized argument of the embedded verb is in the preverbal field of the matrix verb, however, different verb groups show different behavior, as well as topic and focus behaves also differently. In Hungarian we distinguish two types of control verbs: the ones that take main stress (e.g. *fél* 'is afraid') versus the ones that avoid main stress (e.g. *akar* 'want'). These two verb groups differ in syntactic behavior with respect to the placement of the verbal modifier of the embedded infinitive.

(7) Pim (el*) fél   el-olvasni   a   levelet.
     Pim (VM) afraid VM-read.inf the book.acc
     'Pim is afraid to read the letter.'

(8) Pim el akarja (el-*)olvasni a   levelet.
     Pim VM wants (VM-)read.inf the book.acc
     'Pim wants to read the letter.'

In infinitival complements, topics preferably stand in the preverbal field of the matrix verb, however, it is not entirely out in the preverbal field of the embedded verb. Focused constituents must stand in the preverbal field of the matrix verb.

Using the XMG tool, it is relatively easy to derive the elementary trees necessary for the LTAG analysis of the pre- and postverbal fields in simple sentences. However, standard LTAG cannot handle scrambling phenomena, which makes the examples of complex sentences challenging. The problem of the standard LTAG analysis is the discontinuity of the verbal complex. Here, the VM *el-* and the embedded verb *olvasni* form a complex. By the linguistic principles behind LTAG, this dependency is represented locally: the VM is part of the elementary tree of the verb. Using standard LTAG, VM-climbing (8) is not possible, since it proposes elementary trees for sentence embedding, where sentential complements being represented as a footnode (S*), and the elementary tree of the matrix verb (*akar* 'wants') being adjoined into the elementary tree of the embedded verb. In case of VM-climbing (8), the tree of the matrix verb (*akar*) should split into more pieces when conjoined with the tree of the embedded verb.

In order to overcome this problem, a modified formalism is proposed: a TT-MCTAG (Tree-local Multicomponent TAG; [13]). In MCTAG, elementary structures are not merely trees, but sets of trees. The tree-local version of MCTAG comes with the restriction, that all trees in the set have to attach to the same elementary tree. TT-MCTAG is strongly equivalent to standard LTAG, thus by using this formalism we can overcome the problem of VM-climbing without loosing any of the attractive formal and computational properties of TAG.

References

1. Kiss, K.É. (ed.): Discourse Configurational Languages. Oxford University Press, New York/Oxford (1995)
2. Farkas, R., Vincze, V., Schmid, H.: Dependency parsing of hungarian: baseline results and challenges. In: Daelemans, W. (ed.) Proceedings of the 13th Conference of the European Chapter of the Association for Computational Linguistics, Avignon, France (2012)
3. Joshi, A.K., Schabes, Y.: Tree-adjoining grammars. In: Rozenberg, G., Salomaa, A. (eds.) Handbook of Formal Languages, pp. 69–123. Springer, Berlin (1997)
4. Crabbé, B., Duchier, D., Gardent, C., Le Roux, J., Parmentier, Y.: XMG: eXtensible MetaGrammar. Computational Linguistics **39**(3) (2013)
5. XTAG Research Group: A Lexicalized Tree Adjoining Grammar for English. Technical Report IRCS-01-03, IRCS, University of Pennsylvania (2001)
6. Crabbé, B.: Représentation informatique de grammaires d'arbres fortement lexicalisés : le cas de la grammaire d'arbres adjoints. PhD thesis, Université Nancy 2 (2005)
7. Kallmeyer, L., Lichte, T., Maier, W., Parmentier, Y., Dellert, J.: Developing a TT-MCTAG for german with an RCG-based parser. In: Calzolari, N.C.C., Choukri, K., Maegaard, B., Mariani, J., Odijk, J., Piperidis, S., Tapias, D. (eds.) LREC 2008. European Language Resources Association (ELRA), Marrakech (2008)
8. Kallmeyer, L., Lichte, T., Maier, W., Parmentier, Y., Dellert, J., Evang, K.: TuLiPA: towards a multi-formalism parsing environment for grammar engineering. In: Clark, S., King, T.H. (eds.): Coling 2008: Proceedings of the workshop on Grammar Engineering Across Frameworks, Manchester, UK, Coling 2008 Organizing Committee (2008)
9. Kiss, K.É.: The Syntax of Hungarian. Cambridge University Press (2005)
10. Kiss, K.É.: The function and the syntax of the verbal particle. In: Kiss, K.É. (ed.): Event Structure and the Left Periphery. Studies on Hungarian. Springer (2008)
11. Abeillé, A., Rambow, O.: Tree adjoning grammar: an overview. In: Abeillé, A., Rambow, O. (eds.) Tree Adjoining Grammars: Formalisms Linguistic Analyses and Processing, pp. 1–68. CSLI Publications, Stanford (2000)
12. Frank, R.: Phrase Structure Composition and Syntactic Dependencies. MIT Press, Cambridge (2002)
13. Lichte, T.: Coherent constructions as discontinuous constituents. Manuscript (2014)

Phonetic Segmentation Using KALDI and Reduced Pronunciation Detection in Causal Czech Speech

Zdenek Patc[(✉)], Petr Mizera, and Petr Pollak

Faculty of Electrical Engineering, Czech Technical University in Prague,
Prague, Czech Republic
{patczden,mizerpet,pollak}@fel.cvut.cz

Abstract. The paper describes the implementation of phonetic segmentation using the tools from KALDI toolkit. Its usage is motivated by the big development and support of topical techniques of ASR which are available in KALDI. The presented work is related to the research on pronunciation variability in casual Czech speech. For this purpose we use the automatic phonetic segmentation to analyze the particular phone boundaries, deletions, etc. We also present the tool for pronunciation detection. Both tools can be used for processing large databases as well as for an interactive work within the environment of Praat. Also the illustrative analysis of the segmentation accuracy and the design of new environment for phonetic segmentation in Praat are presented.

Keywords: Phonetic segmentation · Spontaneous speech · Pronunciation reduction · Tools · KALDI · Praat

1 Introduction

The automatic phonetic segmentation has many applications in systems using speech technology. Nevertheless its accuracy is limited in comparison to manual phonetic segmentation, on the other hand its usage is in many situations necessary. We can mention the segmentation of utterances from large speech corpora as a typical example. It is required for the Artificial Neural Network (ANN) training when manual labels are not usually available for the huge amount of speech data. As ANNs are nowadays used as the standard subpart of Automatic Speech Recognition (ASR) systems with ANN-HMM architecture as well as for the estimation of probabilistic-based speech features, the segmentation for this purpose is very important and frequently used task.

There are various approaches to obtain the phone boundaries automatically, however, the most frequently used implementations are based on the HMM-based forced alignment of trained acoustic models [1]. If needed the resulting time boundaries can be tuned by methods based on correlation of neighbouring segments or on statistical change-point detection. Concerning HMM-based phonetic segmentation, several open-source toolkits are available for building the

© Springer International Publishing Switzerland 2015
P. Král and V. Matoušek (Eds.): TSD 2015, LNAI 9302, pp. 433–441, 2015.
DOI: 10.1007/978-3-319-24033-6_49

ASR systems as well as for the realization of phonetic segmentation. The HTK toolkit [2] is well known and largely used, however, KALDI [3] toolkit introduced in 2011 starts being nowadays very popular. It is used and developed by the researchers all over the world and it contains modern training and decoding techniques which are not supported by HTK anymore.

The main purpose of this paper is to present the upgrade of our tool *Labtool* [4] for the automatic phonetic segmentation. The last version of this tool was based on the HTK tools, currently it uses KALDI tools which seem to be more efficient and which are also more suitable for the further development due to the continuous support of the new topical algorithms for ASR. In the end, the segmentation realized in this paper is more focused on the case of casual speech, which also represents challenging task in the field of ASR.

2   Automatic Phonetic Segmentation

The tools for automatic phonetic segmentation are based on various ASR toolkits and allow users to interact with them in various ways (the command-line or the graphical user interfaces). The Penn Phonetics Lab Forced Aligner tool is based on HTK with a command-line interface [5], the EasyAlign tool is also based on HTK and allows graphical user interfaces (GUI) using Praat environment [6], the SPPAS tool is based on Julius Speech Recognition Engine with GUI for interacting [7]. The comparison of these tools is presented in [8]. These tools work under all major operating system (Linux, Windows, Mac-OSX) and support various languages for the automatic phonetic segmentation. In our case, we have built Labtool [4] which allows both the command-line and the GUI interfaces when using Praat. Current features of Labtool and KALDI-based solution are described below.

2.1   KALDI-Based Segmentation

KALDI toolkit offers new interesting features for ASR, especially the acoustic modeling based on DNN and the usage of weighted finite state transducers (WFSTs) which are advantageous mainly for the purpose of Large Vocabulary Continuous Speech Recognition (LVCSR). The toolkit contains also several decoders, extensible design, and also complete recipes for particular solutions of the ASR tasks. Its availability with open-source license together with very intensive development is the main reason for its wide usage all over the world. Of course, it enables also the realization of the phonetic segmentation which is used for our purposes.

As it is well known, the phonetic segmentation is based on the forced alignment of the trained acoustic model (AM), which requires two mandatory inputs: an acoustic speech signal (in the form of the sequence of speech features) and related sequence of phones, which is more often replaced by the standard orthographic transcription and related pronunciation lexicon. The decoding algorithm in KALDI is based on WFSTs and require the HCLG graph convention [9].

To align phones in one known utterance the HCLG graph can be composed directly for the corresponding word-level orthographic transcription and lexicon containing pronunciation variants for the words included in the transcription of the processed utterance. The output of decoding is saved to the lattice format. The time information about boundaries of phones/words can be obtained by the lattice-align-phones/lattice-align-words tools saving the information in the phone-aligned/word-aligned compact lattice formats. These compact lattices can be converted using nbest-to-ctm tool to the standard ctm-format. As this format is less readable than the HTK label file (mlf) which stores both phone-level and word-level time boundaries with corresponding text symbols, conversion between ctm-format and mlf-format can be done by the simple text processing (the tool realizing this conversion is also the part of our package).

Concerning the acoustic processing of an utterance to be segmented, of course, KALDI has tools realizing the extraction of the most frequently used MFCC or PLP features, together with various normalizations (CMN, VTLN, etc.) as well as further transformations (LDA, HLDA, etc.). Finally we use our private tool *CtuCopy* in our approach, because it enables the choice of the speech features including also the noise reduction, short-time cepstral mean subtraction, usage of other filter banks, etc. CtuCopy is freely available and it has been upgraded to produce also the output in KALDI compatible format [10].

2.2   Labtool - Command Line & Praat Version

As mentioned above, we use KALDI in the *Labtool*, that is in our segmentation tool for HMM-based segmentation [4], which can be used as a command-line tool for the segmentation of huge speech corpora. Currently it supports various formats of the input signal, various speech features (KALDI- or CtuCopy-based), possible usage of noise suppression or cepstral normalization techniques, specification of proper pronunciation lexicon, etc.

Labtool is distributed at package, which can be simply extracted to any location. KALDI and further required utilities or libraries are not a part of Labtool distribution and they must be installed independently and under their license requirements.

Labtool was designed to be used also within Praat [11] which offers user-friendly environment of general speech analysis with special support for phonetic research. Concerning the phonetic segmentation, it is not a standard part of Praat because it is language dependent and the implementation would be quite complex for the tool. However, this feature can be simply added as a plugin which is also the case of our Labtool. It is controlled from Praat as well as other general scripts which are launched using control buttons added in menus at various levels.

Labtool can be installed in Praat using the script available in the distribution package. The simple installation process consists of the setting of particular paths (the main Labtool directory and KALDI location). Also the default parameters of the phonetic segmentation can be configured during the installation procedure. After restarting Praat, a new button *Align preferences* intended for the setup

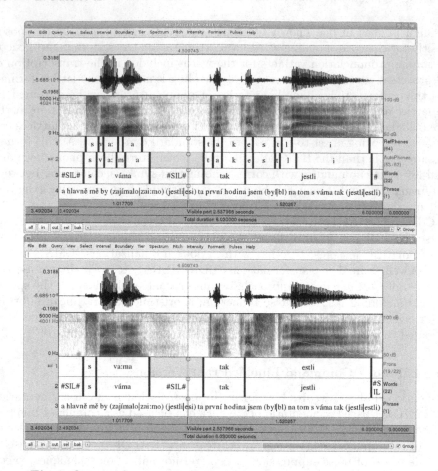

Fig. 1. Output of automatic phonetic segmentation in program Praat.

of the phonetic segmentation parameters can be seen in *File-Preferences* in the
main Praat *Objects* window. New buttons related to the phonetic segmentation
always appears in the right dynamic menu after the selection of `Sound` and
related `TextGrid` objects. The phonetic segmentation can be launched simply
by the *Align phrase* button. When the result of segmentation is available, the
output can be processed to obtain the particular word pronunciations in a more
proper format (*Get pronunciations* button). Both results of the segmentation in
the editor window can be seen in Fig. 1.

The automatically found phone boundaries are duplicated in a `RefPhones`
tier which can be manually edited. The `RefPhones` tier is modified during the
segmentation process only when it does not exist or when it is empty. The
automatic segmentation can be then repeated with different parameters and
compared to the fixed manual labels. The used phonetic alphabet can be set.
Currently world-wide used SAMPA alphabet is supported as well as our private
one for Czech language. The configuration setup and the Labtool control buttons
in the main window of Praat can be seen in Fig. 2.

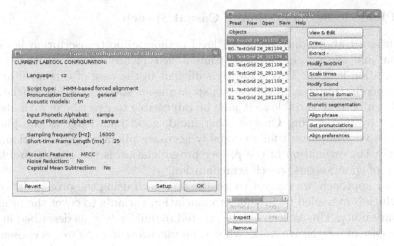

Fig. 2. Configuration of phonetic segmentation parameters and main window in program Praat.

2.3   Labtool Under MS Windows

Concerning Praat, it is an application which supports standardly both platforms Linux and Windows. The Labtool is distributed in the versions supporting both platforms. On the other hand, KALDI toolkit was designed and it is under development mainly for Linux or Unix like systems. The compilation of KALDI tools under Windows was not fully supported until recently and it can be slightly less straight forward. There are several ways how to do it, e.g. in Visual C++ or Cygwin. The compilation using Cygwin is the solution which has been finally used in our setup for Labtool usage under MS Windows platform.

3   Casual Speech Analysis

The informal spontaneous speech can contain many reductions, substitutions or insertions of phones. Furthermore, many non-standard words, word fragments, or non-speech events can occur due to mispronunciations. Also sentences are very frequently started again, so the intelligibility of the speech can be very poor. Consequently it makes the realization of the manual segmentation very hard, let alone the automatic one. Different effects of assimilation and coarticulation can be problematic, e.g. nasal(ized) vowels (not included in canonical Czech speech) can occur instead of the sequence of a nasal consonant and a standard vowel. Similar coarticulation effects can make the manual annotation also difficult, so special care about them must be paid also during the proper setup of the automatic segmentation procedure.

3.1 Phonetic Segmentation of Casual Speech

As mentioned above only the word level orthographic transcription is usually available and the proper phonetic content is typically chosen on the basis of pronunciation lexicon. This is mostly sufficient in the case of canonical or more formal speech. However, in the case of spontaneous speech, the pronunciation of a word can be very variable so it may be impossible to describe it with absolute accuracy by the lexicon. On the other hand, good knowledge of the phonetic content is very important for reasonably accurate phonetic segmentation. Consequently, the estimation of the proper pronunciation is the crucial problem in the case of spontaneous speech segmentation.

In our approach, we suggest to use the approach using the pronunciation lexicon, which is extended by further pronunciation variants to cover the pronunciation variability. This lexicon can be created manually (e.g. as described in [12]). For the case of interactive work, when pronunciation of some words chosen from the used lexicon is not satisfactory, the correct pronunciation can be specified on the level of orthographic transcription using special conventions with parentheses. For the complete automatic processing, further rules for generating other pronunciation variants for the given basic dictionary can be used. This is the procedure used standardly within the recognition of spontaneous speech [13], but for the accuracy of the phone boundaries setup, it seems to be more crucial. Typical examples considered to be frequent within the informal speech style are: variants with reductions of specific groups of phones, cross-word coarticulations or phone softenings, see [14], [15].

3.2 Detection of Reduced Pronunciation in Casual Speech

To study the rules of pronunciation reductions, a check of particular pronunciation in recorded casual corpus is reasonable [16]. As it is impossible to find them only manually in large corpora, this procedure can be simplified by the automatic segmentation followed by the pronunciation detection. It can be done on the level of the lexicon with more pronunciation variants. When the composed graph is aligned, the most probable pronunciation variant is selected, it can be stored with the related information about the utterance where it was used. Further it can be found by an query search, displayed and manually checked.

Now Labtool allows to save the pronunciations in a simple text format for the above mentioned query search as well as in the `TextGrid` format for fast usage in the environment of Praat. Particular pronunciations are saved with information about word boundaries here, so it is very easy to play them for manual checks.

4 Experimental Part

Although the main purpose of this article was to describe the updated version of the segmentation tool itself, particular results of the segmentation of selected casual speech data are presented. Realized experiments were done with

Table 1. Global results of segmentation of standard read speech.

ALIGN1CS	SPB		SPE		CPL	
	μ [ms]	σ [ms]	μ [ms]	σ [ms]	μ [ms]	σ [ms]
HTK mono1608	-3.54	17.81	-2.44	17.17	1.25	23.46
HTK mono2510	0.05	18.38	1.45	16.98	1.58	23.61
HTK mono3216	6.77	16.93	8.87	16.63	2.35	22.91
KALDI tri1608	-0.91	34.28	2.35	23.38	3.69	39.68
KALDI tri2510	-2.93	17.57	-0.71	21.04	2.53	26.74
KALDI tri3216	-5.68	17.11	-2.46	22.46	3.71	27.54

Table 2. Global results of segmentation of spontaneous speech.

NCCCz	SPB		SPE		CPL	
	μ [ms]	σ [ms]	μ [ms]	σ [ms]	μ [ms]	σ [ms]
HTK mono1608	3.95	88.02	5.28	86.96	3.72	42.12
HTK mono2510	-0.02	51.21	1.52	47.59	2.51	31.11
HTK mono3216	7.33	57.99	8.80	53.78	3.02	31.58
KALDI tri1608	-0.70	33.56	0.48	33.45	0.43	24.65
KALDI tri2510	-4.56	22.05	-3.33	24.56	1.23	25.63
KALDI tri3216	-5.60	26.64	-3.58	28.39	2.28	26.72

the manually phonetically segmented data selected from Nijmegen Corpus of Casual Czech (NCCCz) [17], [12]. We compare the results using two setups: HTK based approach (AM was trained by Expectation Maximization (EM) algorithms on clean office subset from the SPEECON database with approx. 51 hours of speech, 45 Czech monophones were modeled and target AMs consist of 32 Gaussian mixture components per each emitting state) and new KALDI based setup (context-dependent triphone AM, trained by the Viterbi-EM algorithm, 15029 number of Gaussian mixture components).

The achieved results of experiments realized on the standard read speech from SPEECON database are in Tab. 1 while the results for subset from NCCCz casual speech data are in Tab. 2. Mean values and standard deviations for criteria SPB (Shift of Phone Beginning), SPE (Shift of Phone End) and CPL (Change of Phone Length) [18] are presented.

It is possible to see that the reasonable improvement of segmentation accuracy was achieved mainly in the case of the segmentation of casual speech data. Here the impact of the new AMs was clear, especially, significantly lower standard deviations from all evaluated criteria are observable. Also the segmentation using longer frame of short-time analysis provided slightly less variable values of evaluated criteria but higher mean values were achieved and short phones were not segmented properly.

5   Conclusions

In this paper, the implementation of HMM-based phonetic segmentation using KALDI tools was presented and the upgraded version of our segmentation tool

Labtool was described. There are several new features implemented in Labtool and the detection of word pronunciation on the basis of the automatic segmentation is the most important one. It was shown within the brief experimental evaluations on spontaneous speech data that the suggested segmentation using new acoustic models yielded to more consistent results. The implementation based on KALDI toolkit enables further development. Especially, it should enable the segmentation of utterances with apriori unknown contents using efficient LVCSR system based on KALDI tools.

Acknowledgments. Research described in this paper was supported by the internal CTU grant SGS14/191/OHK3/3T/13 "Advanced Algorithms of Digital Signal Processing and their Applications".

References

1. Toledano, D.T., Hernandez, G.L.A., Grande, L.V.: Automatic Phonetic Segmentation. IEEE Transactions on Speech and Audio Processing **11**, 617–625 (2003)
2. Young, S., et al.: The HTK Book, Version 3.4.1. Cambridge (2009)
3. Povey, D., et al.: The kaldi speech recognition toolkit. In: Proc of IEEE 2011 Workshop on Automatic Speech Recognition and Understanding, ASRU 2011 (2011)
4. Pollak, P., Volin, J., Skarnitzl, R.: Phone segmentation tool with integrated pronunciation lexicon and czech phonetically labelled reference database. In: Proc. of 6th International Conference on Language Resources and Evaluation, LREC 2008, Marrakech, Morocco (2008). http://www.lrec-conf.org/proceedings/lrec2008/
5. Yuan, J., Liberman, M.: Speaker identification on the SCOTUS corpus. In: Proc. of Acoustics 2008. (2008)
6. Goldman, J.P.: Easyalign: an automatic phonetic alignment tool under praat. In: Proc. of 12th Annual Conference of the Interantional Speech Communication Association, Interspeech 2011, Firenze, Italy (2011)
7. Bigi, B.: Sppas: a tool for the phonetic segmentation of speech. In: Proc. of 8th International Conference on Language Resources and Evaluation, LREC 2012, Istanbul, Turkey, pp. 1748–1755 (2012)
8. Brognaux, S., Roekhaut, S., Drugman, T., Beaufort, R.: Automatic phone alignment. a comparison between speaker-independent models and models trained on the corpus to align. In: Isahara, H., Kanzaki, K. (eds.) JapTAL 2012. LNCS, vol. 7614, pp. 300–311. Springer, Heidelberg (2012)
9. Mohri, M., Pereira, F.C.N., Riley, M.: Speech recognition with weighted finite-state transducers. In: Handbook on Speech Processing and Speech Communication, pp. 559–584. Springer (2008)
10. Fousek, P., Mizera, P., Pollak, P.: CtuCopy feature extraction tool. http://noel.feld.cvut.cz/speechlab/
11. Boersma, P., Weenink, D.: Praat: Doing phonetics by computer (version 5.3.15) (2009). http://www.praat.org/
12. Mizera, P., Pollak, P., Kolman, A., Ernestus, M.: Impact of irregular pronunciation on phonetic segmentation of nijmegen corpus of casual czech. In: Sojka, P., Horák, A., Kopeček, I., Pala, K. (eds.) TSD 2014. LNCS, vol. 8655, pp. 499–506. Springer, Heidelberg (2014)

13. Nouza, J., Silovský, J.: Adapting lexical and language models for transcription of highly spontaneous spoken czech. In: Sojka, P., Horák, A., Kopeček, I., Pala, K. (eds.) TSD 2010. LNCS, vol. 6231, pp. 377–384. Springer, Heidelberg (2010)
14. Kolman, A., Pollak, P.: Speech reduction in czech. In: Proc. of The 14th Conference on Laboratory Phonology, LabPhone 2014, Tokyo, Japan (2014)
15. Lehr, M., Gorman, K., Shafran, I.: Discriminative pronunciation modeling for dialectal speech recognition. In: Proc. of 15th Annual Conference of the Interantional Speech Communication Association, Interspeech 2014, Singapore, pp. 1458–1462 (2014)
16. Schuppler, B., Adda-Decker, M., Morales-Cordovilla, J.A.: Pronunciation variation in read and conversational austrian german. In: Proc. of 15th Annual Conference of the Interantional Speech Communication Association, Interspeech 2014, Singapore (2014)
17. Ernestus, M., Kočková-Amortová, L., Pollák, P.: The nijmegen corpus of casual czech. In: Proc. of 9th International Conference on Language Resources and Evaluation, LREC 2014, Reykjavik, Iceland, pp. 365–370 (2014)
18. Mizera, P., Pollák, P., Kolman, A., Ernestus, M.: Accuracy of HMM-based phonetic segmentation using monophone or triphone acoustic model. In: 2013 International Conference on Applied Electronics, Applied Electronics, Pilsen, CR, pp. 45–48 (2013)

Automatic Robust Rule-Based Phonetization
of Standard Arabic

Fadi Sindran[1(⊠)], Firas Mualla[1], Katharina Bobzin[2], and Elmar Nöth[1]

[1] Friedrich-Alexander-Universität Erlangen-Nürnberg (FAU),
Lehrstuhl für Mustererkennung (Informatik 5),
Martensstraße 3 91058, Erlangen, Germany
fadi.sindran@faui51.informatik.uni-erlangen.de
http://www5.cs.fau.de
[2] Sprachenzentrum der Friedrich-Alexander-Universität Erlangen-Nürnberg (FAU),
Bismarckstr. 1 1/2 91054, Erlangen, Germany

Abstract. Phonetization is the process of encoding language sounds using phonetic symbols. It is used in many natural language processing tasks such as speech processing, speech synthesis, and computer-aided pronunciation assessment. A common phonetization approach is the use of letter-to-sound rules developed by linguists for the transcription form orthography to sound. In this paper, we address the problem of rule-based phonetization of standard Arabic. The paper contributions can be summarized as follows: 1) Discussing the transcription rules of standard Arabic which were used in literature on the phonemic and phonetic levels. 2) Important improvements of these rules were suggested and the resulting rules set was tested on large datasets. 3) We present a reliable automatic phonetic transcription of standard Arabic on five levels: phoneme, allophone, syllable, word, and sentence. An encoding which covers all sounds of standard Arabic is proposed and several pronunciation dictionaries were automatically generated. These dictionaries were manually verified yielding an accuracy of 100% with standard Arabic texts that do not contain dates, numbers, acronyms, abbreviations, and special symbols. They are available for research purposes along with the software package which performs the automatic transcription.

Keywords: Standard Arabic · Phonetic transcription · Pronunciation dictionaries · Transcription rules

1 Introduction

Letter-to-sound transcription is the conversion process from orthography to the sound system of a language. It is an essential component in state-of-the-art text to speech (TTS), computer-aided pronunciation learning (CAPL), and automatic speech recognition (ASR) systems [1]. Phonetic transcription methods can be categorized as follows [2]:

© Springer International Publishing Switzerland 2015
P. Král and V. Matoušek (Eds.): TSD 2015, LNAI 9302 pp. 442–451, 2015.
DOI: 10.1007/978-3-319-24033-6_50

- Dictionary-based methods: A lexicon with rich phonological information is utilized to perform the transcription.
- Data-oriented methods: A machine learning-based model is learnt from pronunciation lexicons, and applied then on unseen data.
- Rule-based methods: Expert linguists define letter-to-sound language-dependent rules and a lexicon of exceptions. In these approaches, the degree of difficulty is highly dependent on the extent of compatibility between the writing and sound system of the language as well as the phonetic variations of its sounds.

Arabic, spoken by more than 250 million people, is one of the six official languages of the United Nations [3]. Standard Arabic contains classical Arabic and modern standard Arabic (MSA). Classical Arabic is the language of the Holly Qur'an and books of Arabic heritage. MSA, on the other hand, is the formal language in all Arab countries [4], taught in schools and universities, used in radio and television along with local dialects, and it is the predominant language in which books and newspapers are written. Standard Arabic is known to have a clear correspondence between orthography and sound system. Therefore, in this work, we adopted a rule-based approach for the transcription.

The transcription process of Arabic text can be seen at two levels:

- Phonemic level: The transcription is performed from text to phonemes. Arabic phonemes are 28 consonants, 3 short vowels, 3 long vowels, and 2 diphthongs.
- Phonetic level: The effect of neighboring sounds on the phoneme is taken into account. The final produced sound is named allophone or actual phone. Several rules at this level are concerned with the pharyngealization and nasalization of Arabic sounds.

Computer-based research on Arabic text and speech is relatively new. Some works done in this domain considered informal languages such as Tunisian Arabic in [4] and Algiers dialect in [5]. Regarding standard Arabic phonetization, considerable contributions were made by Al-ghamdi [6, 7] and El-Imam [2]. Al-ghamdi et al., in cooperation with Arab linguists, derived transcription rules from Arabic literature books and formulated them in a manner accessible to computer scientists. They implemented these rules in an Arabic TTS system. Moreover, they presented in [7] an encoding system at the phonetic level. El-Imam, on the other hand, in his comprehensive work [2], profoundly analyzed the problems of Arabic text phonetization, letter-to-sound rules, and rule implementation including the assessment of the transcription process outcome.

In this paper, we address the following issues:

- We discuss the above-mentioned works [2], [6], and [7]. In [6], we found that more information is needed in order to apply the rules related to the Alif (IPA: ʔalif with a form as "l") correctly. Additionally, some rules on the phonetic level did not consider all the phonemes affected by phonetic variations. Regarding [7], there will be a lack of phonetic information when adopting the encoding they presented. In [2], more details are required about each rule and the priority of rule implementation.

- We proposed several modifications on the transcription rules. For instance, the ambiguities concerning the transcription of the Alif were clarified and the relevant rules were changed accordingly.
- Typically, the performance of a rule-based letter-to-sound approach is heavily dependent on the order of rules. We figured out the correct priority of several rules which is necessary to prevent any undesired overlapping.
- The texts used for training and validation cover all rules at both phonemic and phonetic levels.
- We implemented the transcription rules in a software package to accomplish the following tasks automatically and reliably:

  - Grapheme to phoneme transcription (generation of pronunciation dictionaries). Usually the linguists write these dictionaries for ASR systems manually [9].
  - Grapheme to allophone transcription.
  - Grapheme to syllable transcription.
  - Statistical analysis at the level of phonemes, allophones, and syllables.

On the phonetic level, we modified the encoding developed in [7] so that it can cover all phonetic variations of standard Arabic sounds and implemented it in the software package.

For the pronunciation dictionaries, we adopted the Arpabet coding, which is a phonemic transcription system that depends on English capital letters and digits developed by the Advanced Research Projects Agency (ARPA). Therefore, the output of the transcription can be utilized in any Arpabet-compatible system such as the TTS and ASR tools in the widely-used CMU Sphinx Toolkit.

2   Pre-transcription

In this work, we assume that the input of the transcription process is an Arabic text which is fully diacritized. Additionally, before applying the transcription rules, the following steps need to be performed:

- Cleaning the text by keeping only Arabic characters and symbols with a single sentence per line.
- Converting dates, numbers, acronyms, abbreviations, and special symbols to a proper word or sequence of words manually.
- Words of irregular spelling, i.e. words which do not follow the Arabic pronunciation rules such as "طه" (a name) and "ذٰلِك" (that) are processed separately. We collected most of these words in a separate lexicon which can be extended later to include new words.

3   Arabic Letters to Sound Rules

In our work, we adopted the rules proposed by Al-ghamdi et al. in [6] after adjusting them. For the phonemic level, these rules are listed in section 3.1 while our modifications are presented in section 3.2. Likewise, for the phonetic level, the rules of Al-ghamdi et al. are clarified in section 3.3 and discussed along with other suggestions from literature in section 3.4. In section 3.5, we describe the phonetic-level encoding.

3.1 Rules at the Phonemic Level

After [6], phonemic transcription is achieved according to the following rules (in order):

1. Convert emphatic consonants (consonants with a diacritic named Shaddah has a form as "ﹽ") to two consecutive ones as in "كَلَّمَ" (he spoke) becomes "كَلْلَمَ".

2. Delete the grapheme Alif if it is followed by two consecutive consonants, and does not come at the beginning of a phrase.

3. If the Alif is a part of the "ال" (The definite article in Arabic, IPA: ʔalluttaʕriːf) at the beginning of a phrase, it is pronounced as "ءَ" (ʔa).

4. If the Alif is a part of a verb whose third letter is diacritized with an original Dammah (IPA: dammah which is a diacritic with a form as "ﹹ"), and comes at the beginning of a phrase, we pronounce it "ءُ" (ʔu), otherwise convert the Alif to "ءِ" (ʔi) when it comes at the beginning of a phrase.

5. Convert an Alif preceded with a Fatha (diacritic has a form as "ﹷ") to a long Fatha "ﹷﹷ".

6. Convert a "و" preceded with a Dammah (diacritic has a form as "ﹹ") and followed with a Sukun (diacritic has a form as "ﹿ") to a long Dammah "ﹹﹹ".

7. Convert a "ي" preceded with a Kasrah (diacritic has a form as "ﹻ") and followed by a Sukun to a long Kasrah "ﹻﹻ".

8. Convert each two similar consecutive short vowels to a long one.

9. Convert a "ة" (taːʔ marbuːtah) to "ه" (haːʔ) when it comes at the end of a phrase.

10. Delete the shamsi "ل" (laːm). For example, the transcription of the word "النُّور" (ʔalnnuːr the light) is "ءَ نْ نُ ر" (ʔannuːr).

11. Delete the Tanwin (tanwiːn is a diacritic which has three forms: "ﹰ", "ﹲ", "ﹴ") at the end of a phrase and turn it to short vowel and "ن" anywhere else.

12. Convert "آ" (ʔalif ʔalmad) to "ءَا" (ʔaː) wherever it is found in the text.

13. Pronunciation of all types of the Hamza (the types are: "ء", "أ", "إ", "ؤ", "ئ") is "ء".

14. When three consonants come consecutively, a short vowel has to be inserted after the first one. This inserted short vowel is "ﹷ", "ﹹ", or "ﹻ" when the short vowel which precedes the first consonant is "ﹻ", "ﹹ", or "ﹷ", respectively.

3.2 Discussion and Proposed Modifications at the Phonemic Level

In rule 2, we must distinguish between the long vowel "ا" as in "ضَالِّين" (dˤaːlliːn "erratic") and "ا" (Arabic: همزة الوصل hamzat ʔalwaṣˤl) as in "فَاتِّبَاع" (following): Only in the case of Hamzat al-Wasl (the second case), the Alif must be deleted when it is followed by two consonants. In order to solve this problem:

- We made a list L1 of Arabic words which have Hamzat al-Wasl followed by an emphatic consonant with all possible prefixes. We extracted these words from the famous Arabic dictionary "Lisan Al Arab" (the Arab Tongue, Arabic: لسان العرب).
- We made a list L2 of most names which have a long vowel "ا" followed by two consonants such as "إيطَالِيَا" (Italy), and "أَلْمَانِيَا" (Germany).

The application of rule 1 and rule 2 in section 3.1 is then modified as follows:

1. Rule 1 is not applied.
2. If a word contains an Alif followed by an emphatic consonant:

   o If the word belongs to L1 e.g. "فَاتَّبَاع" (following), "وَاتِّهَام" (accusation), rule 2 is applied.
   o Otherwise e.g. "ضَالِّين" (erratic), "جَادِّين" (earnest), rule 2 is not applied.

3. If a word belongs to L2, rule 2 is not applied.
4. Rule 1 is applied.

In rule 3, we cannot distinguish whether the "ال" is a definite article or not. For example, the "ال" in the word "البَحر" (the sea) is a definite article and must be converted to "ءَل", but the "ال" in the word "التِمَاس" (request) is not a definite article. In fact, contrary to rule 3, it must be converted to "ءِل".

Moreover, rule 4 is not valid with verbs which end with "ا" or "ى" (Ɂalif maqsˤuːrah) in the past tense. The first Alif here must be converted to "ء" regardless of the diacritic at the fourth position as in the verb "ارمُوا" (throw) where the first "ا" must be always converted to "ء".

We suggest to modify rule 3 and rule 4 as follows:

- If the third letter in the word, beginning with "ال", is a shamsi letter ("ت", "ث", "د", "ذ", "ر", "ز", "س", "ش", "ص", "ض", "ط", "ظ", "ل", "ن") and is followed with Fatha or kasrah, we must convert "ال" to "ءِل". On the other hand, when it is followed with Dammah, it should be converted to "ءُل".
- Rule 4 may not be applied for verbs which end with "ا" or "ى" in the past tense. We gathered all verbs which fulfill this property in a special list. When a word belongs to this list, the Alif is converted to "ء" regardless of the diacritic at the fourth position.
- Convert "ال" in all other words to "ءَل".

Lastly, in rule 14, in the case when the first consonant is "ن" resulting after the transcription of the Tanwin, a modification needs to be made. In this case, the short vowel which must be added is "ِ" regardless of the short vowel before the first consonant. For example, the transcription of the sentence "إنَّهُ حَامِدٌ السَّمَّان" (he is Hamid the grocer) is "إنَّهُ حَامِدُن سَمَّان". Here, we add Kasrah after "ن" in the word "حَامِدُن" in spite of the fact that the short vowel before "ن" is Dammah.

3.3   Rules at the Phonetic Level

The rules at this level are used for transcription from phoneme to allophone. After [6], they are given in the order of application as follows:

1. Convert "ن" (nuːn) followed by "ب" (baː?) or "م" (miːm) to "م".

2. Convert "ن" followed by "ر" (raː?) to "ر".

3. Convert "ن" followed by "ل" (laːm) to "ل".

4. Convert "ن" to "و" (waːw) or "ي" (yaː?) with nasalization, when a word ends with "ن" and the next one starts with "و" or "ي", respectively.

5. Convert "ن" to a nasal sound of the following letter when it is followed by any letter except "و" or "ي".

6. Convert "ذ" (ðaːl) followed by "ظ" (ðˤaː?) to "ظ".

7. Convert "ت" (taː?) followed by "ط" (tˤaː?) to "ط".

8. Convert "ت" (taː?) followed by "د" (daːl) to "د".

9. Convert "د" followed by "ت" to "ت" when they come in one word.

10. Convert "د" to "ت" when "ت" comes after the word "قد" (qad).

11. Convert "ل" followed by "ر" (raː?) to "ر".

12. The "ل" and "ر" are pronounced sometimes normally (light accent, in Arabic:ترقيق "tarqiːq") and other times with pharyngealization (heavy accent, in Arabic:تفخيم "tafxiːm") depending on the context in which they occur. Further details can be checked in [6].

13. The vowels followed by "قك" or "طت" are pronounced with a heavy accent.

14. Arabic vowels preceded by emphatic consonants ("ص", "ض", "ط", "ظ", "خ", "غ", "ق"), pharyngeal "ل" or pharyngeal "ر" are pharyngealized.

15. Make short vowels shorter at the end of a word followed by another one, or when they are followed by two consonants.

16. Convert a long vowel followed by two consonants to a short one.

17. When stopping at the end of a phrase, the short vowels must be deleted.

18. Release stop sounds at the end of a phrase (qalqalah, Arabic: "قلقلة") with voiced stops "ق", "ط", "ب", "ج", "د" and aspiration with voiceless stops "ء", "ت", "ك").

3.4   Discussion of the Phonetic-Level Rules

Considering the relevant literature, one can notice that there is no complete agreement about the positions of pharyngealization and affected sounds in Arabic. Al-ghamdi et al. in [6] considered only the vowels that followed the pharyngealization sounds. This is, however, not sufficient as the two diphthongs, the sounds /ت/, /د/, /ذ/, /س/, /ض/, and the non-pharyngealized /ل/ and /ر/ must be considered as well. In fact, pharyngealization may affect some sounds before the pharyngealized sound. For example, in the word "مُستطيل" (rectangle), all the sounds /ó/, /س/, /ت/, and /ó/ before the /ط/ are pharyngealized.

Abu Salim in [9] considered only the sounds which are in the same syllable where the pharyngealization sound occurs. He ignored the sounds in the syllables which come before or after. However, these sounds can be affected as well. For instance, the sound /د/ in the word "مِنْضَدَة" (table) is pharyngealized even though it does not occur in the same syllable which contains the pharyngealized sound /ض/.

El-Imam in [2] presented an important discussion of the pharyngealization of Arabic sounds, but he did not consider the different types of syllables. We think that the phenomenon of pharyngealization in Arabic is more complex than what is described in [2], [6], and [9]. It depends on the syllables, their types, and the positions of the pharyngealized and non-pharyngealized sounds. Moreover, there are different degrees of pharyngealization. We thus believe that the study of pharyngealization in Arabic needs to be done in another research paper. In this work, we take into account only the obvious cases of pharyngealization as considered by most linguists.

3.5   Phonetic Level Encoding

The result of transcription is represented after [7] (cf. table 1) using two letters to represent the phoneme, a number to represent the emphatic status, and another number to represent the duration. We extended this encoding in order to cover all phonetic variations of Arabic phonemes. This extension is given by a label (cf. table 2) which is concatenated to the original representation.

Table 1. Encoding of transcription result after [7]

1	2	3 (Emphatic)		4 (Duration)			5 (Phonetic variations)
Two letters represent a phoneme		0	1	0	1	2	
		Non emphatic	Emphatic	Short	Shorter	Long	

Table 2. The modified label we use to represent phonetic variations

5 (Phonetic variations)				
0	1	2	3	4
The sound is a phoneme	Allophone for "ن"	Released with a schwa	Allophone for "ض" followed by "ط"	Allophone for "ت" followed by "د"
5	6	7	8	
Allophone for "د" followed by "ت"	Allophone for "ل" followed by "ر"	Allophone for "ق" followed by "ك"	Allophone for "ط" followed by "ت"	

4   Syllabication

As mentioned in section 3.4, the pharyngealization rules in Arabic are syllable-based. Moreover, syllabication is very important in Arabic TTS systems because stress depends on the type of syllables in the word [10]. Recognition of non-Arabic words as well as learning music and meters of Arabic poetry are other application areas of syllabication. The Arabic syllables can be characterized as follows:

- Every syllable must start with a consonant followed by a vowel.
- Two consecutive consonants can only occur when stopping at the end of a phrase.
- The number of vowels per word is equal to the number of syllables in this word.

There are six forms of syllables in Arabic: CV, CV:, CVC, CV:C, CVCC, and CV:CC, where C refers to a consonant, V to a short vowel, and V: to a long vowel.

In order to represent the two diphthongs, we categorized the syllables into eight types instead of six as follows: CV, CD2, CL, CVC, CD2C, CLC, CVCC, and CLCC, where L refers to a long vowel and D2 is a diphthong. The diphthong D2 can be disassembled into VC. The syllables CD2 and CD2C are thus equivalent to the syllables CVC and CVCC, respectively.

5   Training and Testing Sets

The term training in the context of this paper refers to the process of finding new rules and/or new details of existing rules. This was done manually by studying and analyzing sentences in the training corpora.

We used two corpora for training:

- **Corpus1:** Developed by King Abdul-Aziz City for Science and Technology (KACST) and contains 367 fully diacritized sentences including together 1812 words. This corpus is characterized as follows:

  - Minimum repetition of words in all sentences.
  - Every sentence contains between two and nine words.
  - Sentences can be easily pronounced.
  - The sentences were chosen in a way which minimizes the number of sentences and maximizes the phonetical information content.

  This corpus was used in a project between KACST and IBM for training a speech recognizer of Arabic words.

- **Corpus2:** The holly Qur'an without Surat al-Baqarah ("IPA: suːrat ʔal baqarah"). This corpus contains 72146 words.

The following corpora were used for testing:

- **Corpus3:** Made by Katharina Bobzin and contains 306 sentences selected from her book: "Arabisch Grundkurs" (Arabic basic course).
- **Corpus4:** A set of sentences selected from Nahj al-Balagha (Way of Eloquence) which is a book that contains sermons, letters, and sayings of Imam Ali ibn Abi Talib, written by Sharif Razi.
- **Corpus5:** Surat al-Baqarah (Arabic: سورة البقرة) which is the longest sura of the Qur'an containing 286 verses.

We wrote the above-mentioned testing corpora manually and edited them as described in section 2. Additional information and statistics about these corpora are given in table 3.

Table 3. Information and statistics about the testing corpora

Corpus	Bobzin corpus	Corpus from Nahj al-Balagha	Surat al-Baqarah corpus
Non-existent words in the first training corpus (%)	90.8356 %	96.2981 %	95.4760 %
Non-existent words in the second training corpus (%)	84.6361 %	86.5683 %	68.2140 %
All words	1340	6996	6150
Non-repeated words	742	3898	2542
Phonemes	8224	44901	38827
Allophones	8313	45470	39309
Syllables	3451	19424	16710

6   Evaluation

The transcription of a word depends on its location in the sentence. We thus used sentences to test the transcription rules, i.e. the transition between words are taken into account in the evaluation. We implemented our transcription algorithm in Matlab. As mentioned earlier, an executable version of the code is freely available for research purposes upon request. The output of the program includes:

- Phonemic dictionary of the entire text given in Arpabet.
- Phonetic dictionary represented according to the encoding proposed in section 3.5.
- Syllables transcription including syllable types.

The automatically-generated results with the three test corpora (cf. section 5) were validated manually. We checked the results at four levels: phoneme, allophone, syllable, and word. Table 4 shows the achieved accuracy at each level.

Table 4. Test results

Test corpora	Bobzin corpus	Corpus from Nahj al-Balagha	Surat al-Baqarah corpus
Phoneme accuracy (%)	99.99	100	100
Allophone accuracy (%)	99.99	100	100
Syllable accuracy (%)	99.97	100	100
Word accuracy (%)	99.93	100	100

We obtained only one single error with the word "التَّاكِسِي" (the taxi), where the long vowel Alif "ا" was deleted in the transcription. Obtaining an error in this word is expected as it is a non-Arabic word.

We could not compare our results with [6] because Al-ghamdi cited only handmade examples to test the rules. El-Imam in [2] tested the transcription rules only with words. He used two corpora for testing. The first one consists of 6000 frequently-used words in Arabic, while the second one contains 4000 popular Arabic names. His datasets are, however, not available for direct use.

7   Conclusion

We tackled the problem of rule-based automatic phonetization of Arabic text. Our method was tested on three corpora representing rich phonetical information and containing together more than 14400 words. The results show that the proposed method yields an accuracy of 100% at four levels: phoneme, allophone, syllable, and word. Complete sentences were used in the evaluation so that transition between words are considered. We generated pronunciation dictionaries from these corpora automatically and validated the correctness of the results manually. They are available for research purposes along with the software package upon request. A profound discussion of rule-based Arabic phonetization was presented and several extensions and improvements to the state-of-the-art were proposed. As future work, we consider the automatic processing of the acronyms, abbreviations, special symbols, proper names, and words with irregular spelling. In addition, more research is required to develop appropriate rules for transcription of Arabic numbers.

References

1. Mohamed, A., Elshafei, M., Al-Ghamdi, M., Al-Muhtaseb, H., Al-Najjar, A.: Arabic Phonetic Dictionaries for Speech Recognition. Journal of Information Technology Research 2(4), 67–80 (2009)
2. El-Imam, Y.A.: Phonetization of Arabic: rules and algorithms. Computer Speech Language 18(4), 339–373 (2004)
3. Hadjar, k., Ingold, R.: Arabic newspaper page segmentation. In: 7th International Conference on Document Analysis and Recognition, vol. 2, pp. 895–899 (2003)
4. Masmoudi, A., Khemakhem, M.E., Estève, Y., Belguith, L.H., Habash, N.: A corpus and phonetic dictionary for Tunisian Arabic speech recognition
5. Harrat, S., Meftouh, K., Abbas, M., Smaili, K.: Grapheme to phoneme conversion: an Arabic dialect case
6. Al-ghamdi, M., Al-Muhtasib, H., Elshafei, M.: Phonetic Rules in Arabic Script. Journal of King Saud University-Computer and Information Sciences 16(0), 85–115
7. Alghamdi, M., Mohamed El Hadj, Y.O., Alkanhal, M.: A manual system to segment and transcribe Arabic speech. In: M. KACST (AT-25-113)
8. Biadsy, F., Habash, N., Hirschberg, J.: Improving the Arabic pronunciation dictionary for phone and word recognition with linguistically-based pronunciation rules. In: Proceedings of Human Language Technologies, The Annual Conference of the North American Chapter of the Association for Computational Linguistics, NAACL, pp. 397–405 (2009)
9. Abu Salim, I.:The syllabic structure in Arabic language. In: Magazine of the Jordan Academy of Arabic, num. 33 (1987).
10. Zeki, M., Khalifa, O.O., Naji, A.W.: Development of an Arabic text-to-speech system. In: International Conference on Computer and Communication Engineering, ICCCE, Kuala Lumpur, Malaysia, (2010)

Knowledge-Based and Data-Driven Approaches for Georeferencing of Informal Documents

Daniel Ferrés[✉] and Horacio Rodríguez

TALP Research Center, Universitat Politècnica de Catalunya,
Jordi Girona 1-3, 08034 Barcelona, Spain
{dferres,horacio}@cs.upc.edu

Abstract. This paper describes Knowledge-Based and Data-Driven approaches we have followed for generic Textual Georeferencing of Informal Documents. Textual Georeferencing consists in assigning a set of geographical coordinates to formal (news, reports,..) or informal (blogs, social networks, chats, tagsets,...) texts and documents. The system presented in this paper has been designed to deal with informal documents from social sites. The paper describes four Georeferencing approaches, experiments, and results at the MediaEval 2014 Placing Task (ME2014PT) evaluation, and posterior experiments. The task consisted of predicting the most probable geographical coordinates of Flickr images and videos using its visual, audio and metadata associated features. Our approaches used only Flickr users textual metadata annotations and tagsets. The four approaches used for this task were: 1) a Geographical Knowledge-Based (GeoKB) approach that uses Toponym Disambiguation heuristics, 2) the Hiemstra Language Model (HLM), TFIDF and BM25 Information Retrieval (IR) approaches with Re-Ranking, 3) a combination of the GeoKB and the IR models with Re-Ranking (Geo-Fusion), 4) a combination of the GeoFusion with a HLM model derived from the English Wikipedia georeferenced pages. The HLM approach with Re-Ranking showed the best performance in accuracy within a margin of distance errors ranging from 10m to 1km. The GeoFusion approaches achieved the best results in accuracies from 10km to 5,000km. Both approaches achieved state-of-the-art results at ME2014PT evaluation and posterior experiments, including the best results for distance accuracies of 1000km and 5,000km in the task where only the official training dataset can be used to predict the coordinates.

Keywords: Textual Georeferencing · Toponym disambiguation · Language models · Information retrieval · Geographical gazetteers

1 Introduction

There is an increasing amount of user generated content in social platforms on the web like social networks, blogs and multimedia sharing platforms. Currently some platforms allow users to georeference (geotag) their content automatically

© Springer International Publishing Switzerland 2015
P. Král and V. Matoušek (Eds.): TSD 2015, LNAI 9302, pp. 452–460, 2015.
DOI: 10.1007/978-3-319-24033-6_51

(GPS-enabled cameras) or manually, but most of the textual and media content is not georeferenced by the users. In some applications it is important to know exactly, or at least predict with some confidence, geographical information related to these texts such as: the user location or the place where the content refers to. This paper describes Knowledge-Based and Data-Driven approaches for generic Textual Georeferencing prediction of Informal Texts. The experiments presented in this paper are focused on Flickr photos and videos geo-location prediction. The evaluation of these experiments is presented in the context of the Media Eval 2014 Placing task evaluation and posterior experiments over the same test set in order to improve our results. The MediaEval Placing Task is a multi-modal georeferencing challenge to evaluate algorithms that can predict the location coordinates of randomly selected photos and videos from Flickr [1]. The Placing Task challenge has been organized in five editions: in the 2010, 2011, 2012, 2013, and 2014 [2]. The algorithms can use visual, audio and metadata associated features to predict. Our approaches used only Flickr users textual metadata annotations and tagsets. We used four approaches for this task: 1) a Geographical Knowledge-Based (GeoKB) approach that uses Toponym Disambiguation heuristics, 2) the IR models TF-IDF, BM25 and the Hiemstra Language Model (HLM) trained with georeferenced data and posterior Re-Ranking, 3) a combination of the GeoKB top-performance heuristics and the IR models (GeoFusion), and 4) a combination of the GeoFusion with a HLM model derived from the English Wikipedia georeferenced pages.

2   Related Work

Many approaches for Textual Georeferencing of user generated content (annotations, tagsets,..) have been presented in last years [3], [4], [5], and [6]. [3] used a language model based on the tags provided by the users to predict the location of Flickr images. In addition, they used several smoothing strategies: 1) spatial neighbourhood for tags, 2) cell relevance probabilities, 3) toponym-based smoothing, and 4) spatial ambiguity-aware smoothing. [4] uses a Natural Language Processing (NLP) approach with geographical knowledge filtering: they use Google translate, OpenNLP[1], Wikipedia[2] and Geonames[3] with the use of population and higher-level categories salience. [5] uses an approach based on applying a geographical Named Entity Recognizer (Geo-NER) that uses Wikipedia and Geonames on the textual annotations. [6] presented a complex system that uses clustering (k-medoids, grid-based, and mean shift), feature selection, and language modeling algorithms to predict geographic coordinates of Flickr photos by using users tags. [7] [8] [9] [10] [11] presented text metadata based approaches at MediaEval 2014[4]. [7] [8] used probabilistic place modeling

---

[1] http://www.opennlp.com

[2] http://www.wikipedia.org

[3] http://www.geonames.org

[4] [8] and [9] approaches included also visual features-based approaches in some experiments.

and rectangular cells of size 0.01 of latitude and longitude degree as baseline. [7] used also machine tag and user modeling in some runs. [8] used some extensions that include: 1) similarity search, 2) internal grid with finer cells, and 3) spatial tag entrophy. On the other hand, [9] and [10] used document retrieval models. [9] used the BM25 and TF-IDF retrieval models (implemented by Lucene[5]) with stemming and stopwords filtering. [10] uses a language model-based document retrieval model in combination with a spatial-aware tag weighting schema and collection geo-correlation to predict test items without tags by modeling users collections. Finally, [11] presented two text-based approaches: spatial variance and graphical model framework.

We used a point-based approach (modelling individual coordinates) instead of grid-based approaches (modelling spatial regions) used in [3], [6], [8], and [7]. A main difference from the text and gazetteer based approaches systems of [4] and [5] with respect to our system is that we do not use Named Entity Recognizers and NLP processors.

3   Geographical Knowledge-Based Heuristics for Georeferencing

This approach has two phases: 1) Place Names Recognition and 2) Geographical Focus Detection. The Place Names Recognition phase uses the Geonames gazetteer for detecting the place names in the textual annotations and tagsets. The Geonames gazetteer currently contains over ten million geographical names and consists of 9 million unique features classified with a set 645 different geographical feature types[6]. We decided not to use a Named Entity Recognition and Classifier (NERC) for several reasons: i) multilingual annotations complicate the use of NLP processes such as Part-of-Speech tagging, and NERC, ii) informal documents are not suitable for most NERC systems trained in news corpora. iii) some NERC systems are not performing much better than geographical Names Recognition from gazzetteer lookup [12]. The Geonames gazetteer allows us to deal with the issues of recognizing place names in social annotations such as: multilinguality of toponyms, acronyms, lowercased place names, and word joined place names (e.g. riodejaneiro). We use the following information from each Geonames toponym entry: 1) toponym name itself, 2) country, state, and continent of the toponym, 3) feature type, 4) coordinates, and 5) population. The Place Names Disambiguation phase tries to handle the geo/non-geo ambiguity of toponyms due to the huge number of non-geographical words that could be recognized as a toponym (e.g. aurora (noun), aurora (city)) because of the usually lowercased toponym mentions in informal documents. This phase uses stopwords lists in several languages[7]

---

[5] http://lucene.apache.org

[6] The Geonames version we used at ME2014PT and posterior experiments was downloaded in 2011

[7] http://search.cpan.org/dist/Lingua-StopWords. Includes stopwords for Danish, Dutch, English, Finnish, French, German, Hungarian, Italian, Norwegian, Portuguese, Spanish, Swedish, and Russian.

(including English) and an English Dictionary obtained from the NLP tool Freeling[8] to filter out non-geographical words that could be erroneously tagged as place names. These resources had proven to be useful in previous research in Flickr metadata geolocation with the MediaEval 2010 Placing Task dataset [13].

The Geographical Focus Detection phase uses some Toponym Disambiguation heuristics [14] [15] to compute a prediction of the geographical focus of the user generated content. The "one reference per discourse" hypothesis applied to Toponym Disambiguation is assumed: one geographical place/coordinates per photo/video metadata. If there are no detected place names in the textual annotations the georeference prediction can remain unresolved or assigned the most photographed place in the world. Once detected all possible co-referents of all the place names detected by Place Names Recognition phase the following heuristics can be applied in the following order of priority: 1) Geographical Knowledge heuristics based on common-sense, 2) Population Heuristics. The Geographical Knowledge heuristics based on common-sense are similar to the Toponym Disambiguation algorithm applied by [15] to plot on a map locations mentioned in automatically transcribed news broadcasts. In our system the geo-class ambiguity between country names and city names (e.g. Brasil (city in Colombia) versus Brasil (country)) is resolved giving priority to the country names, and the geo-class ambiguity between state names and city names is resolved giving priority to city names. From the set of different places appearing in the text the following Toponym Disambiguation heuristics are applied in priority order to select the scope (focus) of the text: H1) select the most populated place that is not a state, country or continent and has its state appearing in the text[9], H2) select the most populated place that is not a state, country or continent and has its country apearing in the text, H3) otherwise select the most populated state that has its country apearing in the text. The population heuristics disambiguate between all the possible places using population information of the detected toponyms. The following rules are applied: P1) if a place exists select the most populated place that is not a country, state (administrative division type one) or a continent, P2) otherwise if a state exists, select the most populated one, P3) otherwise select the most populated country, P4) otherwise select the most populated continent. Finally once computed the toponym scope of the text, the coordinates of this toponym in the Geonames are selected as the final predicted coordinates.

4   Information Retrieval with Re-ranking for Georeferencing

Given an informal text to georeference, this approach treats this text as a an IR query and uses existing state-of-the-art IR models to retrieve a set of weighted

---

[8] http://nlp.lsi.upc.edu/freeling/

[9] An improvement with respect to our results at Media Eval 2014 Placing Task is that toponyms which have a class ambiguity of city/state or city/country do have to appear at least twice in the text to be selected by the heuristics.

coordinates relevant to the query and re-rank them with a geographical distance function. The IR models used are the TF-IDF, BM25 and Hiemstra Language Model (HLM) [16]. For each unique coordinates in the training corpus a document was created with all the textual metadata fields (title, description and user tags) content of all the photos/videos that pertain to this coordinates pair. The Terrier[10] IR software (version 4.0) was used for this process with its default settings for each IR model used[11]. The indexing process uses the default Terrier stopwords list to filter out irrelevant tokens to be indexed. A Re-Ranking process is applied after the IR process. For each query their first 1,000 retrieved documents (coordinates pairs in this case) from the IR software are used. From them we selected the subset of coordinates pairs with a score equal or greater than the two-thirds (threshold 66.66%)[12] of the top-ranked coordinates pair. Then for each geographical coordinates pair of the subset we sum its associated score (provided by the IR software) and the score of their neighbours in the subset at a threshold distance (e.g. 100km) below their Haversine distance. Then we select the one with the maximum weighted sum as the final predicted coordinates pair.

5   GeoFusion: Knowledge-Based and Data-Driven Georeferencing

In this approach the GeoKB system uses the predictions that come only from the heuristics H1, H2 and H3. When the GeoKB heuristics (applied in priority order: H1, H2, and H3) do not match then the predictions are selected from the IR approach with Re-Ranking. This approach has a variant that uses a set of 857,574 Wikipedia georeferenced pages[13] as a training set to predict when the Re-Ranking based on the training data gives an score lower than a threshold (7.0 in this case)[14]. The coordinates of the top ranked georeferenced Wikipedia page after the IR process are used as a prediction.

6   Evaluation

The MediaEval 2014 Placing Task (ME2014PT) requires that participants use systems that automatically assign geographical coordinates (latitude and longitude) to Flickr photos and videos using one or more of the following data: Flickr metadata, visual content, audio content, and social information (see [2] for more details about this evaluation). Evaluation of results is done by calculating the distance from the actual point (assigned by a Flickr user) to the predicted point

---

[10] http://terrier.org

[11] The HLM was used with $\lambda = 0.15$ and the BM25 was used with $k_1 = 1.2d$, $k_3 = 8$, and $b = 0.75d$

[12] This threshold has been chosen after tuning with the Media Eval 2011 dataset.

[13] http://de.wikipedia.org/wiki/Wikipedia:WikiProjekt_Georeferenzierung/ Hauptseite/Wikipedia-World/en

[14] This threshold was found empirically training with the MediaEval 2011 test set.

(assigned by a participant). Runs are evaluated finding how many videos were placed at least within some threshold margin of error distances (10m, 100m, 1km, 10km, 100km, 1000km, and 5000km). The ME2014PT training data consists of 5,000,000 geotagged photos and 25,000 geotagged videos, and the test data consists of 500,000 photos and 10,000 videos. This data has been extracted from the YFCC100M[15] dataset (Yahoo Flickr Creative Commons 100M). This resource has 99.3 million images and 0.7 million videos.

Our approaches were tested with two corpora for training: the ME2014 Training dataset and the YFCC100M geotagged dataset (47,959,829 geotagged items) with items that are not contained in the test set. From the ME2014PT training and the YFCC100M geotagged datasets we extracted all the unique coordinates with associated text: about 2,741,717 and 11,382,289 coordinates respectively. We did two sets of experiments:

1. Official experiments with the ME2014PT dataset and posterior experiments with gazetteer use (see the results in Table 1). Our official run1 at the benchmark was done with the HLM model and a distance threshold of 100km for Re-Ranking and it achieved the best official results in accuracies at high distances (1,000km and 5,000km). It is worth noting that in the benchmark there is not a system performing well in all distances.

Table 1. Results of Run1 at ME2014PT (use provided training dataset only) and posterior experiments (without and with gazetteers used).

	System	accuracy percentage						
		10m	100m	1km	10km	100km	1000km	5000km
Benchmark Results	CEALIST [7]	0.01	0.61	22.62	40.00	47.36	61.17	74.94
	RECOD [9]	0.55	**6.06**	21.04	37.59	46.14	61.69	76.76
	SonSens-CERTH [8]	0.50	5.85	23.02	39.92	46.87	60.11	74.80
	UQ-DKE[10]	**1.07**	4.98	19.57	**41.71**	**52.46**	63.61	77.28
	USEMP [7]	0.78	1.61	**23.48**	40.77	48.11	61.79	75.30
	ICSI/TUDelft [11]	0.24	3.15	16.65	34.70	45.58	60.67	75.03
	TALP-UPC (HLM@100km) [17]	0.29	4.12	16.54	34.34	51.06	**64.67**	**78.63**
Our Post-Evaluation Experiments	HLM@10km	0.29	4.18	17.35	**41.99**	50.97	63.38	77.91
	HLM@1km	0.30	4.65	**24.03**	41.10	49.53	62.20	75.79
	HLM@0.1km	0.46	**7.20**	22.29	38.37	46.86	60.10	74.59
	TFIDF@100km	0.29	4.21	16.84	34.32	50.15	63.52	77.69
	BM25@100km	0.29	4.24	17.01	34.63	50.60	63.88	77.93
	GeoKB	0.07	0.89	11.31	34.44	42.26	48.45	58.32
	HLM@100km+GeoKB	0.25	3.25	16.82	39.71	**53.61**	**66.78**	**80.06**
	HLM@10km+GeoKB	0.26	3.32	17.30	**43.48**	**53.47**	65.67	**79.47**
	HLM@1km+GeoKB	0.25	3.56	20.74	**42.80**	52.36	**64.76**	77.48
	HLM@0.1km+GeoKB	0.35	5.03	19.69	40.95	50.53	63.22	76.58
	TFIDF@100km+GeoKB	0.25	3.19	16.72	39.34	**53.07**	**66.10**	**79.39**
	BM25@100km+GeoKB	0.25	3.21	16.83	39.53	**53.31**	**66.30**	**79.52**
	HLM@100km+GeoKB+GeoWiki	0.25	3.25	16.82	39.72	**53.61**	66.77	80.05

---

[15] http://www.yli-corpus.org/

2. Official experiments with the use of external data and gazetters allowed and posterior experiments with the YFCC100M geotagged dataset (see the results and details of these experiments in Table 2). In these experiments our official results were not so good and achieved only the median (of all participants) in distances higher than 10km. In this case the CEALIST and USEMP [7] systems[16] got the best results.

Table 2. Overall official best results at ME2014 runs (anything allowed except crawling the exact items of the test set) and posterior experiments (training with YFCC100M geotagged).

	System	accuracy percentage						
		10m	100m	1km	10km	100km	1000km	5000km
Benchmark Results	CEALIST [7]	0.01	1.22	40.25	55.98	62.26	72.14	81.95
	RECOD [9][17]	0.59	**6.26**	21.15	37.50	46.03	61.41	75.07
	SonSens-CERTH [8]	0.50	5.85	23.02	39.92	46.87	60.11	74.80
	UQ-DKE [10]	1.08	5.05	20.23	43.68	56.03	69.08	81.14
	USEMP [7]	**2.56**	4.33	**44.14**	**61.34**	**69.10**	**78.69**	**86.52**
	ICSI/ TUDelft [11]	0.32	3.41	12.13	19.95	22.82	33.79	53.06
	TALP-UPC (HLM@100km+GeoKB) [17]	0.23	3.00	15.90	38.52	52.47	65.87	79.29
Our Post-Evaluation	training with YFCC100M geotagged photos/videos							
	HLM@100km	**20.63**	**26.64**	40.65	56.13	68.52	76.60	84.76
	BM25@100km	19.96	26.10	40.30	55.80	68.30	76.72	85.69
	TFIDF@100km	19.84	25.97	40.11	55.57	68.06	76.54	85.56
	HLM@100km+GeoKB	13.72	18.14	32.62	54.53	67.49	77.05	86.10
	BM25@100km+GeoKB	13.20	17.64	32.16	54.05	67.09	76.83	85.97
	TFIDF@100km+GeoKB	13.12	17.55	32.03	53.88	66.91	76.69	85.87

7 Conclusions

This paper presents four approaches for Georeferencing of informal user generated textual content that were used at MediaEval 2014 Placing Task (ME2014PT) and posterior experiments. The GeoFusion approaches achieved the best results in the experiments at ranges from 10 km to 5,000 km with the ME2014PT Training dataset, clearly outperforming our other approaches. The GeoFusion approaches achieved the best results at these evaluation ranges because this approach combines high precision rules based on Toponym Disambiguation heuristics and predictions that come from an IR model when these rules are not activated. When these rules are activated (144,074 cases of 510,000), they achieve accuracy percentages of 87.37% (125,878 of 144,074 items) predicting up to 100 km. By contrast, the HLM IR model trained with the ME2014PT training set with Re-Ranking achieved a 78.34% of accuracy at 100 km when evaluated over this subset (144,074 cases). The HLM approach with Re-Ranking obtained the best results in distance

---

[16] In these official experiments CEALIST and USEMP [7] systems were trained with the YFCC100M geotagged dataset.

ranges from 10m to 1 km because it captures non-geographical highly descriptive and unique keywords and place names appearing in the geographical coordinates' associated metadata that are not present in the gazetteer. The approach that uses the English Wikipedia georeferenced pages to handle difficult cases does not generally offer better performance than the original GeoFusion approach. On the other hand, the GeoFusion approaches trained with the YFCC100M only improve slightly the IR models in accuracy ranges from 1,000 km to 5,000 km. The results with the YFCC100M geotagged dataset as a training data lead to the following conclusions: 1) with YFCC100M data the accuracy of the Data-Driven approach outperforms the GeoKB approach, 2) although the YFCC100M geotagged dataset used in this study had filtered out the items appearing in the test set, some users with items in the test set could have also items in the train set, and this fact could lead the IR model to have a gain by modeling user's particular way of tagging [1]. In comparison with the results of the other participants, our IR with Re-Ranking and GeoFusion approaches achieved state-of-the-art results at ME2014PT evaluation. The HLM with Re-Ranking approach obtained the best results for accuracies at distances of 1,000 km and 5,000 km in the task where only the official training data can be used to predict. In posterior experiments using the YFCC100M geotagged dataset the IR with Re-Ranking and GeoFusion approaches outperformed the best results for accuracies from 10m to 100m with accuracy percentages of 20.63% and 26.64%. Further work should be done to assess the effects of filtering out all those users of the training set (YFCC100M) that have items that appear also in the test set.

Acknowledgments. This work has been supported by the Spanish Research Department (SKATER Project: TIN2012-38584-C06-01). TALP Research Center is recognized as a Quality Research Group (2014 SGR 1338) by AGAUR, the Research Agency of the Catalan Government.

References

1. Larson, M., Kelm, P., Rae, A., Hauff, C., Thomee, B., Trevisiol, M., Choi, J., Van Laere, O., Schockaert, S., Jones, G., Serdyukov, P., Murdock, V., Friedland, G.: The benchmark as a research catalyst: charting the progress of geo-prediction for social multimedia. In: Choi, J., Friedland, G. (eds.) Multimodal Location Estimation of Videos and Images, pp. 5–40. Springer International Publishing (2015)
2. Choi, J., Thomee, B., Friedland, G., Cao, L., Ni, K., Borth, D., Elizalde, B., Gottlieb, L., Carrano, C., Pearce, R., Poland, D.: The placing task: a large-scale geo-estimation challenge for social-media videos and images. In: Proceedings of the 3rd ACM Multimedia Workshop on Geotagging and its Applications in Multimedia, GeoMM 2014, pp. 27–31. ACM, New York (2014)
3. Serdyukov, P., Murdock, V., van Zwol, R.: Placing flickr photos on a map. In: Allan, J., Aslam, J.A., Sanderson, M., Zhai, C., Zobel, J. (eds) SIGIR, pp. 484–491 (2009)

---

[16] In this run they used both textual and visual features.

4. Kelm, P., Schmiedeke, S., Sikora, T.: Video2GPS: geotagging using collaborative systems, textual and visual features. In: Working Notes of the MediaEval 2010 Workshop, Pisa, Italy, October 24, 2010

5. Perea-Ortega, J.M., García-Cumbreras, M.A., López, L.A.U., García-Vega, M.: SINAI at placing task of mediaeval 2010. In: Working Notes of the MediaEval 2010 Workshop, Pisa, Italy, October 24, 2010

6. Laere, O.V., Schockaert, S., Dhoedt, B.: Georeferencing flickr resources based on textual meta-data. Information Sciences **238**, 52–74 (2013)

7. Popescu, A., Papadopoulos, S., Kompatsiaris, I.: USEMP at MediaEval Placing Task (2014). [18]

8. Kordopatis-Zilos, G., Orfanidis, G., Papadopoulos, S., Kompatsiaris, Y.: SocialSensor at MediaEval Placing Task (2014). [18]

9. Li, L.T., Penatti, O.A.B., Almeida, J., Chiachia, G., Calumby, R.T., Mendes-Junior, P.R., Pedronette, D.C.G., da Silva Torres, R.: Multimedia Geocoding: The RECOD 2014 Approach. [18]

10. Cao, J., Huang, Z., Yang, Y., Shen, H.T.: UQ-DKE's Participation at MediaEval 2014 Placing Task. [18]

11. Choi, J., Li, X.: The 2014 ICSI/TU Delft Location Estimation System.[18]

12. Stokes, N., Li, Y., Moffat, A., Rong, J.: An Empirical Study of the Effects of NLP Components on Geographic IR performance. International Journal of Geographical Information Science **22**(3), 247–264 (2008)

13. Ferrés, D., Rodríguez, H.: Georeferencing textual annotations and tagsets with geographical knowledge and language models. In: Actas de la SEPLN 2011, Huelva, Spain, September 2011

14. Leidner, J.L.: Toponym Resolution: a Comparison and Taxonomy of Heuristics and Methods. Ph.D. Thesis, University of Edinburgh (2007)

15. Hauptmann, A.G., Hauptmann, E.G., Olligschlaeger, A.M.: Using location information from speech recognition of television news broadcasts. In: Proceedings of the ESCA ETRW Workshop on Accessing Information in Spoken Audio, pp. 102–106. University of Cambridge, Cambridge (1999)

16. Hiemstra, D.: Using Language Models for Information Retrieval. Ph.D. thesis, Enschede (2001)

17. Ferrés, D., Rodríguez, H.: TALP-UPC at MediaEval 2014 Placing Task: Combining Geographical Knowledge Bases and Language Models for Large-Scale Textual Georeferencing

18. Larson, M.A., Ionescu, B., Anguera, X., Eskevich, M., Korshunov, P., Schedl, M., Soleymani, M., Petkos, G., Sutcliffe, R.F.E., Choi, J., Jones, G.J.F. (eds): Working Notes Proceedings of the MediaEval 2014 Workshop, Barcelona. CEUR Workshop Proceedings, Catalunya, Spain, October 16–17, vol. 1263. CEUR-WS.org (2014)

Automated Mining of Relevant N-grams in Relation to Predominant Topics of Text Documents

Jan Žižka[✉] and František Dařena

Department of Informatics, FBE, Mendel University in Brno,
Zemědělská 1, 613 00 Brno, Czech Republic
{zizka,darena}@mendelu.cz

Abstract. The article describes a method focused on the automatic analysis of large collections of short Internet textual documents, freely written in various natural languages and represented as sparse vectors, to reveal what multi-word phrases are relevant in relation to a given basic categorization. In addition, the revealed phrases serve for discovering additional different predominant topics, which are not explicitly expressed by the basic categories. The main idea is to look for n-grams where an n-gram is a collocation of n consecutive words. This leads to the problem of relevant feature selection where a feature is an n-gram that provides more information than an individual word. The feature selection is carried out by entropy minimization which returns a set of combined relevant n-grams and can be used for creating rules, decision trees, or information retrieval. The results are demonstrated for English, German, Spanish, and Russian customer reviews of hotel services publicly available on the web. The most informative output was given by 3-grams.

Keywords: Natural language · N-gram · Phrase · Relevant feature · Machine learning · Entropy · Rule generation · Sentiment analysis

1 Introduction

One of the important tasks is looking for relevant information and its generalization (knowledge) hidden in large data collections. Today, many books and articles deal with this problem because of the everyday persistent accumulation of data, not excluding textual one. Analyzing textual data written in natural languages constitutes a demanding, not easy task [4,8]. This research branch is also very attractive from the application point of view – could a machine help semantically analyze such documents similarly as human beings do? The main problem is that due to the extremely large volumes of textual data, humans cannot do it within a reasonable time period.

The following sections present a possible approach to mining knowledge, represented as informatively significant word triplets, 3-grams, from freely online written reviews coming from customers of hotel services. The mined triplets

© Springer International Publishing Switzerland 2015
P. Král and V. Matoušek (Eds.): TSD 2015, LNAI 9302, pp. 461–469, 2015.
DOI: 10.1007/978-3-319-24033-6_52

can be used for various purposes, starting from using them as three keywords for plain information retrieval up to discovering various hidden topics waiting in the data for their discovery.

2   Data and Its Preprocessing

In this research work, the data coming from publicly accessible Internet resources was used. Here, the textual data are reviews written by customers of hotel services, for example, on the web pages of *booking.com,* which helps find an accommodation for travelling people almost in every country [1]. The reviews are written on-line, after using the service, in different natural languages with no specific structure. The object of this research work was to find a method that would automatically reveal understandable multi-word phrases relevant for formulating a positive or negative opinion. In addition, those phrases were expected to express the reasons of the either positive or negative meaning in more detail – they are called here as *predominant topics.*

Unlike some previous investigations, published for example in [2,9,10], where the authors searched for significant individual words (1-grams), this work aimed at finding typical 2-, 3-, and 4-grams. From this point of view, it was necessary to eliminate meaningless (or too common) words, numerically represent the remaining words using their frequency, and then generate n-grams that would create possible relevant phrases. Due to the very low share of the individual words of a review with the common review dictionary (of the order from 10^{-2} to 10^{-3}, depending on a specific language), vectors, which represented the reviews, were very sparse. The consequence of this fact was that many very short reviews had to be eliminated. For the given n's, one-word reviews could not create n-grams for $n \geq 2$, and so like. With the increasing n, the number of usable reviews rapidly decreased. Eventually, only 2- and 3-grams were investigated; for $n \geq 4$, the set of reviews was too small and the resulting set of relevant n-grams was empty because any possible 4-gram played no role in assigning either positive or negative class. After the evaluation of experimental results, 3-grams were the most semantically contributive, therefore the rest of this article focuses on 3-grams.

To eliminate meaningless words, a very simple procedure was used: all words having only one or two characters (articles, prepositions) were removed, except some important ones like "no" (to avoid loss of phrases as "no good service"). Such an elimination would need a more careful method; it depends also on a processed language and the authors decided to avoid too demanding and time-consuming steps. A negative consequence of the elimination of those short words was additional shortening of reviews, thus even more of them could not be used for generating 2-, 3-, and more-grams.

On the other hand, a lot of meaningless separated n-grams like "the hotel in" or "was a table" disappeared, thus keeping the semantically more contributive rest. The remaining words were represented numerically using the popular method known as *tf-idf,* or *term frequency times ?inverse document frequency,* which is a numerical statistic that is intended to reflect how important a word is to a textual document in a collection and plays a role of a weighting factor [6]. All the reviews were modeled as *bag of words* [7]. A "word" was not a single word, it was a couple (2-grams) or triplet (3-grams) composed of two or three neighboring words, which created a kind of artificial multi-words, for example, a word triplet "threeneighboringwords". The candidates for relevant n-grams were generated automatically. Taking 3-grams as an example, for each review gradually $word_1word_2word_3$, $word_2word_3word_4$, ..., $word_{j-2}word_{j-1}word_j$, until $word_{m-2}word_{m-1}word_m$, where $m \geq 3$ was the number of words in a given review ($3 \leq j \leq m$). Similarly to *bag of words*, this representation can be called as *bag of n-grams* and a certain advantage here is that it keeps the original order of words to a degree given by n, losing less information than in the known case of individual words. The experiments used reviews written in four "big" languages to have enough data: English, German, Spanish, and Russian. For each language, 100,000 individual opinions were randomly selected from the whole collection of corresponding reviews. Due to the vector sparsity, lower number of selected samples gave worse or even no results for 3-grams because there were many reviews having only one or two words and the low number of remaining reviews could not provide reliable necessary probabilities.

After preparing the data for 3-grams, many reviews had to be eliminated as explained above. The relatively high number of words in the dictionaries of each tested language was mainly caused by typing errors – consequently, a word could have several incorrect variants; for example, *accommodation, accomodation, acommodation, acmodation, acomodation,* and so like. In addition, namely in negative reviews, the review writers often used interjections like *aaaargh, aaaaaaaaaaargh, nooo, nooooo,* and so on, or word forms typical for the web environment like *4you, you2, u r* (for you, you too, you are). It was also one of the several main reasons why the vectors were so sparse because even if a certain interjection was often used, the form was not always identical and just one "a" more in "aaargh/aaaargh" could generate a new word. Sometimes, a review was written in two languages, for instance, in Polish and English, and assigned to the English group of reviews. On the other hand, the frequency of such original words was not high and those terms were never included as part of a relevant n-gram; they just acted like noise, however, decreasing the accuracy. In any case, it would be useful to filter the data better to get rid of such kind of noise but because of lack of time, such non-trivial filtering was not applied – it would need to develop specific tools. The following Table 1

Table 1. The number of remaining reviews from the original randomly selected 100,000 ones and their unique word number in dictionaries for each tested language after the preprocessing phase.

language	number of reviews	dictionary size
English	67,621	19,952
German	48,515	16,197
Spanish	50,592	16,909
Russian	83,344	37,870

briefly summarizes the basic data characteristics. Note that a "word" in the dictionary is in actual fact a 3-gram. Due to the limited space, illustrative examples of the reviews cannot be demonstrated here, however, quite typical instances can be seen using [1], which was the source of the investigated textual data.

3   Searching for N-grams

An n-gram played a role as a feature from the negative or positive class assignment point of view. A certain combination of features lead to the appropriate labeling of a review. The question was "Which features (or their combinations) were relevant for positive and which for negative labeling?" There are several methods of selecting significant features, see for example [3]. In this research work, the method inspired by [5] based on Quinlan's minimization of entropy (increased information gain) was employed.

The main idea is that a heterogenous set containing instances belonging to various classes can be divided into homogenous ("pure") subsets containing instances only from one class. For a given random variable X, the entropy $H(X)$ of a quite homogenous set is zero, $H(X) = 0$. If a set contains, for example, instances of two classes 50:50, its entropy is maximal, $H(X) = 1$. The entropy minimization method supposes that all features are mutually independent.

Using a proper feature, a set with a certain entropy value can be divided between subsets having the average entropy lower than their origin. This procedure can be repeated recursively (always again for each feature) until no entropy reduction is possible. The result may be expressed as a hierarchical graph – a tree, usually called a decision tree. In its root, all instances are mixed, while the branches lead to leaves that represent subsets having zero or low entropy. Every tree node represents a question to a feature; each branch represents a rule.

The entropy is computed using the *a posteriori* probability, $p(x_i)$, of selecting randomly an instance x_i, which depends on the frequencies of instances in different classes contained in a set, where a feature numerical representation $x_i \in \mathbb{R}$. In the following Equation 1, $p(x_i)$ is the probability that a certain class member, x_i, can be randomly selected. If a set is quite homogeneous, only members of one class can be selected, thus their probability $p(x_i) = 1$ and probabilities of members of other classes are zero – and vice versa (this situation is called *no surprise*, see [5]). Because of the generalization, usually at least two class members are requested in a tree leaf. The entropy is computed using the following formula:

$$H(X) = -\sum_i p(x_i) \cdot log_2 \, p(x_i) \, . \tag{1}$$

When $p(x_i) = 0$, then, according to the L'Hôpital's rule, the indefinite expression $0 \cdot -\infty = 0$.

Typically, such a tree or rule does not include all features – only those which reduce the entropy from the labeling (classification) point of view. Each feature is tested on every tree level and one of the results is a set of relevant features. Thus, the entropy minimization method was applied to the investigated data with the goal to select those n-grams that were relevant to the review membership of either positive or negative class. The classification accuracy, measured using disjunctive testing sets containing randomly selected 50,000 reviews, was from 86.3% (English) to 74.6% (Russian).

Naturally, the n-grams having a higher frequency in a class play the decisive role but it is necessary to have in mind the fact that the same n-gram may be included in all classes; for example, the 3-gram "was not good" can be part of "I must say the *accommodation was not good*" (the *negative* class) as well as "I cannot say the *accommodation was not good*" (the *positive* class). Therefore, the combinations of n-grams (that is, features) are important – and tree branches can provide them.

To reveal as understandable knowledge as possible, the tree branches were converted into rules where a rule antecedent was represented by an n-gram and consequent by the associated positive or negative class according to the corresponding combination of n-grams on the left side of the rule. A generated rule example (for an English review):

IF *hotel_good_location* > 0 AND *no_hot_water* = 0 THEN *class* = positive,

which simply means that a hotel with a good location and hot water is classified positively, providing two relevant 3-grams for the positive class. The *tf-idf* values were mostly tested against zero, only very rarely against other numbers – this suggests a prevailing binary problem. The reason was that the vectors were very sparse, therefore even a low n-gram frequency played an important role. Additional experiments based on the binary or frequency word-representation provided the same results.

4   Overview of Results and Discussion

The results of looking for relevant n-grams (here, 3-grams) can be demonstrated with the three-word phrases for each of the four investigated languages. Each language gave tens relevant 3-grams from thousands possible (see also Table 1 above). Because of the limited space, the following tables Table 2, 3, 4, and 5 show only the first 10 most important 3-grams based on the highest rule lifts (a measure of the performance of a targeting model at predicting or classifying cases giving an enhanced response – with respect to all data samples – measured against a random choice targeting model). Looking at the n-grams, it is also possible to see the revealed sub-topics that play the most significant role. Somehow surprisingly, the generated rules often provided a kind of "negative" (complementary) knowledge, for example:

IF *very_good_breakfast* = 0 AND *owners_very_friendly* = 0
AND *room_very_good* = 0 THEN *class* = negative.

This means that if the positive features are not available, the service is negative – it is a kind of a set complement. To make this fact (actually obvious positive n-grams pointing to the negative class) visible in Table 2, 3, 4, and 5, as well as

Table 2. The first ten English positive and negative 3-grams.

English positive	English negative
good value money	could hear everything
hotel very clean	bed very hard
breakfast very good	no hot water
staff very nice	not very clean
very good value	very noisy night
close train station	no free internet
very good location	bit too small
within walking distance	not very good
very nice hotel	not good value
great value money	(not) very good breakfast

Table 3. The first ten Spanish positive and negative 3-grams.

Spanish positive / translation	Spanish negative / translation
trato personal excelente / excellent personal service	desayuno bastante pobre / rather poor breakfast
hotel bien situado / well located hotel	no dispone ascensor / no lift available
no lejos centro / not far from center	poca variedad desayuno / small breakfast choice
personal amable habitacin / friendly room staff	habitación demasiado pequeña / room too small
excelente relación calidad / excellent value	no cambiaron toallas / towels were unchanged
personal recepción amable / reception staff friendly	no pudimos dormir /we could not sleep
aire acondicionado / air-conditioning	no agua caliente / no hot water
(not) habitación daba patio / (no) courtyard overlook	no servicio habitaciones / no room service
buena relación calidad / good value	no wi fi / no wi-fi
personal hotel encantador / lovely hotel staff	falta aire acondicionado / missing air-conditioner

distinguish it from the real, original negations "not/no", the tables demonstrate it using an added term "(not)", which in this case was not part of the original 3-grams. From an unknown reason, especially the German reviews generated often such a complement. In the complementary cases, the result included one very large rule containing many negated conditions on its left side followed by short rules having only one to several left-side conditions.

Analyzing the rules, it came to light that the hotel service users shared the same opinions in relation to the service quality regardless of the used language. The work [9] demonstrates the same finding for individual words, 1-grams; however, the n-grams for n > 1 are evidently more expressive as the tables show. A small fraction of those automatically revealed 3-grams was not very useful,

Table 4. The first ten German positive and negative 3-grams.

German positive / translation	German negative / translation
freundliches hilfsbereites personal / friendly helpful staff	zimmer extrem kleine / extremely small room
(not) stark befahrenen strasse / (not) bussy street	fenster nicht öffnen / unopenable window
(not) keine klimaanlage zimmer / air-conditioned rooms	rezeption nicht besetzt / no one in reception
personal beim frühstück / staff at breakfast	frühstück nicht gut / breakfast not good
(not) zimmer bad klein / (not) small bathroom	kein warmes wasser / no hot watter
vier sterne hotel / four-star hotel	hotel schwer finden / difficult to find hotel
(not) nicht wirklich sauber / really clean	zimmer strasse laut / room at noisy street
(not) frühstück nicht gut / good breakfast	zimmer klein bad / with small bathroom
(not) keine minibar zimmer / rooms with minibars	zimmer recht klein / room quite smalli
(not) kein kühlschrank zimmer / rooms with fridges	kein wi-fi lan / no wi-fi network

Table 5. The first ten Russian positive and negative 3-grams.

Russian positive / translation	Russian negative / translation
отель удачно расположен / well located hotel	плохо говорит англииски / badly speaking English
отель находится недалеко / hotel is not far	трудно найти отель / difficult to find hotel
каждый ден меняли / changed every day	номере нет фена / no hair dryer (in) room
не далеко центра / not far from center	отсутсвие wi fi / no wi-fi
отель очен хороший / hotel very good	очен маленкая комната / too small room
болшая удобная кроват / large comfortable bed	не горячей воды / no hot water
не очен далеко / not too far	не работал кондиционер / air-condition did not work
внешний вид отеля / hotel appearance	нет русских каналов / no Russian channels
не плохой завтрак / not bad breakfast	очен скыдный завтрак / very limited breakfast
чищтый уютный номер / clean and comfortable room	не очен чисто / not very clean

for example, "not too much", "gare du nord", or "made us feel" because even if they included three words, the information context was lost anyway. When trying to mine 4- and more-grams, no results were achieved because the reviews did not share such relatively too long phrases, therefore no generalization could be made. In addition, sometimes several 3-grams contained the same three words, only their order was different. For instance, "very well located" (encoded as terms "verywelllocated") and "located very well" ("locatedverywell"). To avoid it, the data preprocessing phase should be extended.

The main goal was to automatically discover significant positive and negative review subsets represented by word triplets, which is illustrated in the tables Table 2, 3, 4, and 5. On the other hand, except the discovery what are semantically the most relevant features, the 3-grams could be also applied well to an information retrieval that might be interesting for the service provider.

One of the typical simple actions is employing key-words for searching for textual documents in databases or using a web browser for finding interesting documents. When no manually created lists of key-words are available and the number of training documents is very high, an automatic generator can be useful. For instance, in the case of the hotel service provider, n-grams obtained from customers' feedback – via generating classification rules and using those ones having higher lift – may be applied to looking for all reviews containing such words connected by the logical conjunction AND, thus retrieving information relevant to improving the service, and so like. As a side test of the obtained results, this simple review-retrieving task based on applying the keywords provided by 3-grams was tried. The accuracy of correctly retrieved reviews (containing a given 3-gram) was between 91-93%, considering the positive and negative classes. The retrieval error (7-9%) was caused by the fact that not always three neighboring words were enough – a larger context should be necessary to reach better results.

5   Conclusion

A possible automatized method of generating relevant phrases is described. Based on data/text mining procedures that benefit from tried and tested machine learning algorithms, the presented work shows how a tree/rule generator can be applied to finding 3-grams. Such n-gram based phrases are interesting and significant as the information applied to solving corresponding concrete tasks related to textual documents that have no specific structure and are written in natural languages.

The results showed that more effort should have been devoted to the data preparation phase, namely to considering various word orders as well as filtering common words bringing no specific information. Such a task is, however, dependent on a particular language and would need a specific approach – primarily for different groups of languages because some of them (for example, Russian) have more free word order than others (for instance, English); and it is the word order which plays an important role in the procedure described in the sections above. From the semantic point of view, an n-gram based phrase can be also used for discovering more particularised topics like breakfast, quiet accommodation, cleaning, staff helpfulness, Internet connection, air condition, transport, and so like. In such a case, the original base classification (good/bad) may be replaced with new classes revealed using the above described process, thus giving more information hidden in the large data collections.

Acknowledgments. This work was supported by the research grant IGA of Mendel University in Brno No. 16/2015.

References

1. booking.com: http://www.booking.com (2015)
2. Dařena, F., Žižka, J.: Text mining-based formation of dictionaries expressing opinions in natural languages. In: Proceedings of the 17th International Conference on Soft Computing Mendel 2011, June 15–17, Brno, pp. 374–381 (2011)
3. Liu, H., Motoda, H.: Computational Methods of Feature Selection. Chapman and Hall/CRC (2007)
4. Miner, G.: Practical Text Mining and Statistical Analysis for Non-structured Text Data Applications. Elsevier Inc. (2012)
5. Quinlan, J.R.: C4.5: Programs for Machine Learning. Morgan Kaufmann (1993)
6. Salton, G., Buckley, C.: Term-weighting Approaches in Automatic Text Retrieval. Information Processing & Management **24**(5), 513–523 (1988)
7. Sebastiani, F.: Machine Learning in Automated Text Categorization. ACM Computing Surveys **1**, 1–47 (2002)
8. Weiss, S.M., Indurkhya, N., Zhang, T.: Text Mining: Predictive Methods for Analyzing Unstructured Information. Springer (2005)
9. Žižka, J., Dařena, F.: Mining significant words from customer opinions written in different natural languages. In: Habernal, I., Matoušek, V. (eds.) TSD 2011. LNCS, vol. 6836, pp. 211–218. Springer, Heidelberg (2011)
10. Žižka, J., Dařena, F.: Revealing prevailing semantic contents of clusters generated from untagged freely written text documents in natural languages. In: Habernal, I., Matoušek, V. (eds.) TSD 2013. LNCS, vol. 8082, pp. 434–441. Springer, Heidelberg (2013)

Incremental Dependency Parsing and Disfluency Detection in Spoken Learner English

Russell Moore, Andrew Caines[⊠], Calbert Graham, and Paula Buttery

Automated Language Teaching and Assessment Institute,
Department of Theoretical and Applied Linguistics,
University of Cambridge, Cambridge, UK
{rjm49,apc38,crg29,pjb48v}@cam.ac.uk

Abstract. This paper investigates the suitability of state-of-the-art natural language processing (NLP) tools for parsing the spoken language of second language learners of English. The task of parsing spoken learner-language is important to the domains of automated language assessment (ALA) and computer-assisted language learning (CALL). Due to the non-canonical nature of spoken language (containing filled pauses, non-standard grammatical variations, hesitations and other disfluencies) and compounded by a lack of available training data, spoken language parsing has been a challenge for standard NLP tools. Recently the Redshift parser (Honnibal *et al.* In: *Proceedings of CoNLL* (2013)) has been shown to be successful in identifying grammatical relations and certain disfluencies in native speaker spoken language, returning unlabelled dependency accuracy of 90.5% and a disfluency F-measure of 84.1% (Honnibal & Johnson: *TACL* 2, 131-142 (2014)). We investigate how this parser handles spoken data from learners of English at various proficiency levels. Firstly, we find that Redshift's parsing accuracy on non-native speech data is comparable to Honnibal & Johnson's results, with 91.1% of dependency relations correctly identified. However, disfluency detection is markedly down, with an F-measure of just 47.8%. We attempt to explain why this should be, and investigate the effect of proficiency level on parsing accuracy. We relate our findings to the use of NLP technology for CALL and ALA applications.

Keywords: Spoken language · Learner english · Learner proficiency · Disfluency detection · Dependency parsing

1 Introduction

Most natural language processing (NLP) experiments are carried out with tools trained on (mainly written) native speaker data. Two corpora in particular, Brown and the Wall Street Journal, have been widely used for the evaluation of NLP technology [24], with the WSJ being near-ubiquitous in parser testing [5,19].

There have been various efforts to extend the range of domains on which NLP tools are trained and evaluated, including, for example, biomedical literature

© Springer International Publishing Switzerland 2015
P. Král and V. Matoušek (Eds.): TSD 2015, LNAI 9302, pp. 470–479, 2015.
DOI: 10.1007/978-3-319-24033-6_53

[21,28], Twitter posts [12,17], and spoken language [8,16]. Each new domain presents its own challenges in terms of NLP: spoken language, for instance, differs from canonical written data in multiple ways [2,3,9]. Disfluencies are an especially characteristic feature of speech, and have been the object of interest for several NLP studies, as their presence is disruptive to parsing, with knock-on effects for any applications that follow.

The convention set by Shriberg [29], and applied to a portion of the Switchboard Corpus (SWB; [14]), is to differentiate between filled pauses (FP), the 'reparandum' (RM), 'interregnum' (IM) and repair (RP) in disfluent sections of text. This annotation scheme is exemplified in Figure 1.

A flight to um Berlin I mean Munich on Tuesday
            FP   RM   IM   RP

Fig. 1. Disfluency example, annotated in the Switchboard Corpus style.

The goal then of automated disfluency detection is to identify FPs, RMs and IMs – *i.e.* here 'um Berlin I mean' – so that the resultant string is understood to actually be 'A flight to Munich on Tuesday'. Several recent papers have reported various approaches achieving upwards of four-in-five accuracy in this task [16,26,27]. For instance, Honnibal & Johnson (H&J) report how well an incremental dependency parsing model, 'Redshift' [15], was able to identify dependency relations and speech repairs in hand-annotated sections of SWB, with an unlabelled attachment score (UAS)[1] of 90.5% and 84% F-measure for disfluency detection [16].

All previous efforts in disfluency detection have targeted the transcribed speech of native speakers of English. We apply the Redshift parser to *non*-native speaker (or, 'learner') English, transcribed from business English oral examinations. The texts come from learners of different proficiency levels: we asked firstly whether Redshift would be able to handle non-native speaker data as accurately as that of native speakers, and secondly whether speaker proficiency would affect parsing accuracies.

We hypothesise that Redshift should be able to parse the language of higher proficiency learners more accurately, as this is presumed to more closely approximate the language of native speakers, on which Redshift is trained. Such an outcome would be in line with previous work showing that the RASP System [5] was better able to parse higher proficiency texts [8].

We indeed found that Redshift could analyse our non-native speaker data remarkably well, with a UAS of 91.1%. However, disfluency detection was less successful, with an F-measure of just 47.8%. We propose that this is due to learner disfluencies being more extended and less orderly than those of native speakers.

---

[1] The percentage of tokens with a correctly-identified head word (*labelled* attachment is another commonly-reported metric (LAS); this is the percentage of tokens with correctly-identified head word *and* dependency relation) [20].

As for learner proficiency, there is a general upward trend in UAS and disfluency detection as the level moves from CEFR[2] B1 'intermediate', to B2 'upper intermediate', to C1 'advanced' [11]. This finding has implications both as a diagnostic tool for learner proficiency, in the context of automated language assessment (ALA), as well as the automated provision of feedback to learners on ways to improve, of the kind required by computer-assisted language learning (CALL) systems.

2   Transition-Based Dependency Parsing

H&J's Redshift parser [16] is a transition-based dependency parser modelled on Zhang & Clark's design with a structured average perceptron for training and beam search for decoding [31]. Syntactic structure is predicted incrementally, based on a series of classification decisions as to which parsing action to take with regard to tokens on the 'stack' and in the 'buffer', two disjoint sets of word indices to the right and left of the current token.

Redshift adopts the four actions defined in Nivre's arc-eager transition system [25] – SHIFT, LEFT-ARC, RIGHT-ARC, REDUCE – and adds a novel non-monotonic EDIT transition to repair disfluencies during parsing [15]. SHIFT moves the first item of the buffer onto the stack. RIGHT-ARC does the same, adding an arc so that items 1 and 2 on the stack are connected. LEFT-ARC and REDUCE both pop the stack, with the former first adding an arc from word 1 in the buffer to word 1 on the stack. Like [1], Redshift posits a dummy ROOT token to govern the head-word of each utterance; with the ROOT token at the top of the buffer and an empty stack, the parsing of this utterance ends.

The novel EDIT transition marks the token on top of the stack as disfluent along with any rightward descendents to the start of the buffer. The stack is then popped and any dependencies to or from the deleted tokens are erased. The transition is 'non-monotonic': previously-created dependencies may be deleted and previously-popped tokens are returned to the stack (for further detail and worked examples see [15,16]).

3   Experiments

3.1   Datasets

Switchboard Corpus. The Switchboard Corpus (SWB) is a collection of transcribed two-way telephone conversations among a network of hundreds of unacquainted English speaking volunteers from across the U.S.A. [14]. The calls were computer-operated, such that no two speakers spoke together more than once, and so that no single speaker was given the same topic prompt (of which there were about 70) more than once [13]. The entire corpus contains 2320 conversations of approximately 5 minutes each, totalling about 3 million words of transcribed text.

---

[2] The 'Common European Framework of Reference for Languages': a schema for grading an individual learner's language level. For further information go to http://www.coe.int/lang-CEFR

Our focus is on a subset of SWB that was hand-annotated with both syntactic bracketing and disfluency labels as part of the Penn Treebank project [30]. The project, now discontinued, produced approximately 1.5 million words from the SWB transcripts annotated for disfluencies. But as hand-labelled syntactic bracketing is more expensive to produce, only 0.6 million tokens from the disfluency layer have corresponding syntactic annotations. Following H&J [16], we require the smaller dataset with *both* syntactic and disfluency annotations (SWB-SYN-DISF), using this as training data, which we note gives us less than half the training data used in other state-of-the-art disfluency detection systems [26,27].

We obtained SWB-SYN-DISF in CoNLL-X treebank format [6]: the same dataset that featured in [16] and which we use here to train Redshift[3]. This dataset was pre-processed in the following ways: removal of filled pauses such as 'uh' and 'um', one-token sentences and partial words, all text to lower-case, punctuation stripped, and 'you know' 'i mean' bigrams merged to single-token 'you_know' and 'i_mean'.

BULATS Corpus. Our non-native speaker data is provided by Cambridge English, University of Cambridge[4]. The corpus is a collection of transcribed recordings from their Business Language Testing Service (BULATS) speaking tests. We produced a gold-standard BULATS treebank of 5667 tokens, annotated for the same features and in the same format as SWB-SYN-DISF.

The dataset features speakers from Pakistan, India and Brazil, with Gujarati, Urdu, Sindhi, Hindi, Portuguese and Panjanbi as their first languages. The BULATS corpus was pre-processed in the same way as SWB-SYN-DISF, described above, with the removal of filled pauses, *etc.* It contains transcripts from 16 different learners and is divided into approximately 1900 tokens from each of three CEFR levels: B1, B2 and C1. We refer to these data subsets as BULATS:B1, BULATS:B2, BULATS:C1.

3.2   Procedure

System Setup. Redshift is an open-source parser so we were able to obtain the same code-base that was used in [16]. We also trained the parser in a similar manner[5]: the training data comprises the 600k token SWB-SYN-DISF dataset. The feature set, *disfl*, is an extended version of 73 templates from [32] mostly pertaining to local context as represented by twelve context tokens, plus extra features to detect 'rough copy' edits [18] and contiguous bursts of disfluency. The EDIT transition was enabled, allowing the parser to erase disfluent tokens

---

[3] Our profound thanks to Matthew Honnibal for sharing this data, and for his help in setting up Redshift.

[4] The data is currently not publicly available, though researchers may apply to Cambridge English for access to the Cambridge Learner Corpus, a collection of written essays.

[5] Note that full installation instructions for Redshift are provided at http://russellmoo.re/cs/redshift.

from the utterance. Random seed training was disabled, since we only trained the parser once (whereas H&J averaged over 20 random seeds).

We used 15 iterations to train the perceptron, and set the beam width to 32 so that the 32 best-scoring transition candidates were kept in the beam with each iteration. Redshift utilises its own part-of-speech tagger – also guided by averaged perceptron and beam search decoding – which was set to be trained in unison with the parser.

Task 1: Dependency Relations. As noted in [16], Redshift makes use of the Stanford Basic Dependencies scheme (SD), which makes strictly projective representations of grammatical relations. The dependencies indicate the binary relations between tokens within a sentence. This method of representing grammatical relations is now widely-used in the NLP community for parser evaluation and owes much to the framework of Lexical-Functional Grammar [4].

In brief, each dependency is a triple (t, h, r): t is a token in the sentence, h is a token that is the 'head' or 'governor' of t, and r is a relation label showing how t modifies h (*e.g.* *nsubj* for nominal subject, *aux* for auxiliary). It is common to represent this as a directed labelled arc joining the two tokens: $h \rightarrow_r t$.

Each token must have a single head, and the relations are constrained to form a projective tree headed by a single word, which itself is headed by a special ROOT token. To mark tokens that have been removed from the sentence by an EDIT transition, H&J added the *erased* relation to the SD set (see also [22,23]). In these cases, the token is self-governing: $t \rightarrow_{erased} t$.

The task, then, is to correctly identify any non-erased token's head: the unlabelled attachment score (UAS) is a measure of success in this respect[6].

Task 2: Disfluency Detection. We follow the evaluation defined by Charniak & Johnson [10], according to which, out of all the disfluency types, only the reparandum (RM) is the target for automatic detection. The reasoning behind this is that the filled pause (FP) and interregnum (IM) are said to be straightforwardly identified with a rule-based approach[7], whereas the RM is what needs to be *edited* out for a successful parse of the repaired string (RP). The task, then, is to successfully apply an EDIT transition to RM tokens, and results are presented as the disfluency F-measure (Disfl.F) – a computation over precision (p) and recall (r) as follows: $F = 2 \times \frac{p \times r}{p+r}$

---

[6] Following [16] we do not report the labelled attachment score (LAS) though acknowledge that this would be an interesting direction for future work.

[7] To our knowledge these rules have not been codified, presumably because it is assumed a trivial task. We assume FP detection would rely heavily on UH POS tags, which are often accurately applied – though not always, an issue briefly discussed by Caines & Buttery [8]. We assume that a bigram rule would account for the IM, or 'parenthetical', which is usually exemplified as either 'you know' or 'I mean'; but we have similar concerns here, as it is not clear that distinguishing IMs from subordinating uses of these chunks (*e.g.* 'I mean what I say'; 'you know what I mean') is as straightforward as it is presumed to be.

Table 1. Redshift parse (UAS) and disfluency (Disfl.P/R, Disfl.F) accuracies on the Switchboard and BULATS test-sets.

	Treebank	Sentences	Tokens	UAS	Disfl.P/R	Disfl.F
native	SWB-SYN-DISF:TEST	3900	45,405	90.5	n/a	84.1
	SWB-SYN-DISF:DEV	3833	45,381	90.9	92.3/76.5	83.7
non-native	BULATS:ALL	381	5667	91.1	82.6/33.6	47.8
	BULATS:B1	121	1895	88.9	85.3/31.4	45.9
	BULATS:B2	136	1879	91.2	79.2/33.2	46.8
	BULATS:C1	124	1893	93.0	83.8/37.3	51.6

4 Results

The performance of the Redshift parser in the two tasks described in section 3 is given in Table 1. **UAS** is a measure of parse accuracy, indicating the percentage of correctly-identified dependency-head relations. **Disfl.F** represents accuracy at disfluency detection, being the harmonic mean of precision **Disfl.P** and recall **Disfl.R** on this task.

Results are reported for various test-sets as follows: SWB-SYN-DISF:TEST[8] represents the averaged results from [16], with Redshift trained on the 600k token SWB-SYN-DISF dataset (90.5% UAS, 84.1% Disfl.F).

To verify Redshift's performance figures versus other parsers, H&J averaged results across twenty randomly-seeded training runs. To compare datasets, we chose to keep the parsing model constant, using a single training run with the default perceptron seeding. We benchmarked the parser on SWB-SYN-DISF:DEV and observed a near-identical score to H&J (90.9% UAS, 83.7% Disfl.F).

From our BULATS non-native speaker corpus, we have results for the CEFR level subcorpora – BULATS:B1, BULATS:B2, BULATS:C1 – as well as for the combined BULATS treebank, BULATS:ALL. For the combined data, UAS is similar to the scores obtained on the SWB corpora (91.1%) but Disfl.F is markedly reduced (47.8%). In section 5 we discuss this result, but the major factor seems to be a far lower recall rate – many disfluencies in the BULATS data are simply being missed. As for the outcome by learner proficiency, both UAS and Disf.F improve with increasing CEFR level. This indicates that the learners' speech approximates something like native speech as proficiency increases, thus both dependency relations and disfluencies are more accurately identified. We suggest that this trend makes parsing a useful diagnostic in ALA and CALL applications.

5 Discussion

Any CALL or ALA applications rely upon an accurate understanding of natural language, within which analysis of language relations in the *who did what*

---

[8] Following standard practice in the disfluency detection literature, the train/dev/test splits are those described in [10].

to whom sense are a crucial component. Spoken language is inherently disfluent and presents an acute challenge in attempts to identify these relations. As discussed here and in earlier work [16, 26, 27], a disfluent utterance may yet be appropriately parsed if the disfluent sections can be detected and removed. We have shown that the models of disfluency detection developed for native speaker data enjoy less success with learner data, at least of the kind presented here (business exam monologues). This means that our automated understanding of what the learner meant to say is impaired, with negative implications for CALL and ALA models.

We observe both a lower precision and a notably lower recall rate than is attained when parsing native speech. The implication is that in BULATS the errors are slightly harder to correct, but much harder to detect, than in SWB.

In comparison to the native speaker's FP-RM-IM-RR sequence given in Figure 1, our non-native speaker disfluencies are rarely so orderly. Filled pauses (FP) were omitted from the datasets used here, and interestingly the BULATS corpus has no clear interregna (IM) structures. 'Rough copy' edits (exact repetition, or repetition with insertion, deletion or substitution of one or more words [18]) are common in BULATS (131 of 207 disfluent sections – 63.3% – have exact repetition of at least one token between reparandum and repair) and are accommodated in Redshift with specific contextual devices that search the buffer and stack for nearby POS or word matches [16].

The remainder of the BULATS disfluencies are characterised by errors: incorrect lexical choices that are intitiated but subsequently abandoned – so-called 'false starts'. Of the 207 disfluent sections, 55 (26.6%) are speech errors of this kind. Many false starts are single token mispronunciations – *e.g.* 'health' followed by the corrective 'help', 'far' then 'fast', 'ship' then 'sitting'. These 'soundalikes' represent a challenge for disfluency detection systems that, without audio, cannot recognise them as a kind of repetition.

Many of the uncorrected reparanda in BULATS are long and not obviously related to the repair - often resembling complete grammatical structures. Some examples are given in Figure 2:

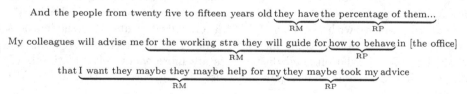

Fig. 2. Undetected or partially detected disfluency examples from the BULATS corpus.

Our future work involves further investigation of these errors and how they might automatically be recognised. We propose that these errors distinguish non-native speaker data (*e.g.* BULATS) from native speaker data (*e.g.* SWB) and that they are harder for language models to detect than rough copy, as reflected in the Disfl.F disparity.

Moreover, if disfluencies such as these can be detected more successfully in learner data, we envisage that a measure of the quantity of linguistic material edited out of the utterance will provide some measure of *distance* to canonical language, or the learner's 'target', whatever that is taken to be ('correct' standard English, as a first approximation). This, along with a measure of error correction, fits in with the ideas expressed in [8], in which increasing (normalised) parse tree probabilities from the RASP System were taken as a proxy for distance to the native speaker data on which the parser had been trained. The next step in this line of work is to develop a more appropriate model of disfluencies for non-native speaker data, that can accommodate the type of disfluency exemplified in Figure 2.

We acknowledge that at 5667 tokens in total, our BULATS dataset is dwarfed by the SWB corpus and its 45k dev/test-sets. However, preparation of a gold-standard treebank is a laborious task and has already taken many hours of work. However, new techniques – such as crowdsourcing as in [7] – are becoming available, and we intend to continue expanding the BULATS treebank, in particular with speakers of other first languages and CEFR levels. We will also collect new data for speech topics other than business, for spontaneous dialogues as well as monologues, and for tasks other than oral examinations.

Acknowledgments. This paper reports on research supported by Cambridge English, University of Cambridge. We thank Ted Briscoe, Nick Saville, Fiona Barker, Ardeshir Geranpayeh, Nahal Khabbazbashi, and Matthew Honnibal for their advice and assistance, as well as the three anonymous reviewers for their helpful feedback.

References

1. Ballesteros, M., Nivre, J.: Going to the roots of dependency parsing. Computational Linguistics **39**(1) (2013)
2. Biber, D.: Dimensions of register variation: a cross-linguistic comparison. Cambridge University Press, Cambridge (1995)
3. Brazil, D.: A grammar of speech. Oxford University Press, Oxford (1995)
4. Bresnan, J.: Lexical-Functional Syntax. Blackwell, Oxford (2001)
5. Briscoe, T., Carroll, J., Watson, R.: The second release of the RASP System. In: Proceedings of the COLING/ACL 2006 Interactive Presentations Session. Association for Computational Linguistics (2006)
6. Buchholz, S., Marsi, E.: CoNLL-X shared task on multilingual dependency parsing. In: Proceedings of the 10th Conference on Computational Natural Language Learning (CoNLL-X). Association for Computational Linguistics (2006)
7. Caines, A., Bentz, C., Graham, C., Polzehl, T., Buttery, P.: Crowdsourcing a multilingual speech corpus: recording, transcription, and natural language processing. In: Proceedings of EUROCALL 2015 (2015)
8. Caines, A., Buttery, P.: The effect of disfluencies and learner errors on the parsing of spoken learner language. In: Proceedings of the First Joint Workshop on Statistical Parsing of Morphologically Rich Languages and Syntactic Analysis of Non-Canonical Languages (2014)
9. Carter, R., McCarthy, M.: Spoken Grammar: where are we and where are we going? Applied Linguistics (in press)

10. Charniak, E., Johnson, M.: Edit detection and parsing for transcribed speech. In: Proceedings of the 2nd Meeting of the North American Chapter of the Association for Computational Linguistics (NAACL). Association for Computational Linguistics (2001)

11. Council of Europe: Common European Framework of Reference for Languages. Cambridge University Press, Cambridge (2001)

12. Foster, J., Çetinoğlu, Ö., Wagner, J., Roux, J.L., Nivre, J., Hogan, D., van Genabith, J.: From news to comment: resources and benchmarks for parsing the language of Web 2.0. In: Proceedings of the 5th International Joint Conference on Natural Language Processing (IJCNLP). Association for Computational Linguistics (2011)

13. Godfrey, J., Holliman, E.: Switchboard-1 Release 2 LDC97S62. DVD (1993)

14. Godfrey, J.J., Holliman, E.C., McDaniel, J.: SWITCHBOARD: telephone speech corpus for research and development. In: Proceedings of Acoustics, Speech, and Signal Processing (ICASSP 1992). IEEE (1992)

15. Honnibal, M., Goldberg, Y., Johnson, M.: A non-monotonic arc-eager transition system for dependency parsing. In: Proceedings of the Seventh Conference on Computational Natural Language Learning. Association for Computational Linguistics (2013)

16. Honnibal, M., Johnson, M.: Joint incremental disfluency detection and dependency parsing. Transactions of the Association for Computational Linguistics 2, 131–142 (2014)

17. Hovy, D., Plank, B., Søgaard, A.: When POS data sets don't add up: combatting sample bias. In: Proceedings of the Ninth International Conference on Language Resources and Evaluation (LREC). European Language Resources Association (2014)

18. Johnson, M., Charniak, E.: A TAG-based noisy channel model of speech repairs. In: Proceedings of the 42nd Annual Meeting of the Association for Computational Linguistics (ACL). Association for Computational Linguistics (2004)

19. Klein, D., Manning, C.D.: Accurate unlexicalized parsing. In: Proceedings of the 41st Annual Meeting of the Association for Computational Linguistics (ACL). Association for Computational Linguistics (2003)

20. Kübler, S., McDonald, R., Nivre, J.: Dependency parsing. Morgan & Claypool Publishers, Synthesis Lectures on Human Language Technologies (2009)

21. Lease, M., Charniak, E.: Parsing biomedical literature. In: Dale, R., Wong, K.-F., Su, J., Kwong, O.Y. (eds.) IJCNLP 2005. LNCS (LNAI), vol. 3651, pp. 58–69. Springer, Heidelberg (2005)

22. de Marneffe, M.C., MacCartney, B., Manning, C.D.: Generating typed dependency parses from phrase structure parses. In: Proceedings of the Fifth International Conference on Language Resources and Evaluation (LREC). European Language Resources Association (2006)

23. de Marneffe, M.C., Manning, C.D.: The Stanford typed dependencies representation. In: Proceedings of the COLING Workshop on Cross-framework and Cross-domain Parser Evaluation (2008)

24. Mikheev, A.: Text segmentation. In: Mitkov, R. (ed.) The Oxford Handbook of Computational Linguistics. Oxford University Press, Oxford (2005)

25. Nivre, J.: Algorithms for deterministic incremental dependency parsing. Computational Linguistics 34(4), 513–553 (2008)

26. Qian, X., Liu, Y.: Disfluency detection using multi-step stacked learning. In: Proceedings of the 2013 Conference of the North American Chapter of the Association for Computational Linguistics: Human Language Technologies (NAACL-HLT). Association for Computational Linguistics (2013)
27. Rasooli, M.S., Tetreault, J.: Non-monotonic parsing of Fluent umm I Mean disfluent sentences. In: Proceedings of the 14th Conference of the European Chapter of the Association for Computational Linguistics (EACL 2014). Association for Computational Linguistics (2014)
28. Rimell, L., Clark, S.: Porting a lexicalized-grammar parser to the biomedical domain. Journal of Biomedical Informatics **42**, 852–865 (2009)
29. Shriberg, E.: Preliminaries to a theory of speech disfluencies. Ph.D. thesis, University of California, Berkeley (1994)
30. Taylor, A., Marcus, M., Santorini, B.: The Penn Treebank: An Overview (2003)
31. Zhang, Y., Clark, S.: Syntactic processing using the generalized perceptron and beam search. Computational Linguistics **37**(1), 105–151 (2011)
32. Zhang, Y., Nivre, J.: Transition-based dependency parsing with rich non-local features. In: Proceedings of the 49th Annual Meeting of the Association for Computational Linguistics: Human Language Technologies (ACL-HLT). Association for Computational Linguistics (2011)

Open Source German Distant Speech Recognition: Corpus and Acoustic Model

Stephan Radeck-Arneth[1,2(✉)], Benjamin Milde[1], Arvid Lange[1,2],
Evandro Gouvêa, Stefan Radomski[1], Max Mühlhäuser[1], and Chris Biemann[1]

[1] Language Technology Group Computer Science Departement,
Technische Universität Darmstadt, Darmstadt, Germany
{stephan.radeck-arneth,milde}@cs.tu-darmstadt.de,
{radomski,max}@tk.informatik.tu-darmstadt.de,
{egouvea,arvidjl}@gmail.com
[2] Telecooperation Group Computer Science Departement,
Technische Universität Darmstadt, Darmstadt, Germany

Abstract. We present a new freely available corpus for German distant speech recognition and report speaker-independent word error rate (WER) results for two open source speech recognizers trained on this corpus. The corpus has been recorded in a controlled environment with three different microphones at a distance of one meter. It comprises 180 different speakers with a total of 36 hours of audio recordings. We show recognition results with the open source toolkit Kaldi (20.5% WER) and PocketSphinx (39.6% WER) and make a complete open source solution for German distant speech recognition possible.

Keywords: German speech recognition · Open source · Speech corpus · Distant speech recognition · Speaker-independent

1 Introduction

In this paper, we present a new open source corpus for distant microphone recordings of broadcast-like speech with sentence-level transcriptions. We evaluate the corpus with standard word error rate (WER) for different acoustic models, trained with both Kaldi[1] and PocketSphinx[2]. While similar corpora already exist for the German language (see Table 1), we placed a particular focus on open access by using a permissive CC-BY license and ensured a high quality of (1) the audio recordings, by conducting the recordings in a controlled environment with different types of microphones; and (2) the hand verified accompanying transcriptions.

Each utterance in our corpus was simultaneously recorded over different microphones. We recorded audio from a sizable number of speakers, targeting speaker independent acoustic modeling. With a dictionary size of 44.8k words and the best Kaldi model, we are able to achieve a word error rate (WER) of 20.5%.

© Springer International Publishing Switzerland 2015
P. Král and V. Matoušek (Eds.): TSD 2015, LNAI 9302, pp. 480–488, 2015.
DOI: 10.1007/978-3-319-24033-6_54

Table 1. Major available corpora containing spoken utterances for the German language.

Name	Type	Size	Recorded
SmartKom [3]	spontaneous, orthographic & prosodic Transcription	12h	synchronized directional & beamformed
Verbmobil [4]	spontaneous, orthographic transcription	appr. 180h	synchronized close-range & far-field & telephone
PhonDat [5]	spontaneous, orthographic transcription	9.5h	close-range
German Today [6]	read and spontaneous, orthographic transcription	1000h	single microphone, headset
FAU IISAH [7]	spontaneous	3.5h	synchronized close-range to far-field
Voxforge	read, orthographic transcription	appr. 55h	varying microphones, usually headsets
Alcohol Language Corpus [8]	read, spontaneous, command, orthographic transcription & control	300k words	close-range, headset
GER-TV1000h [9]	read, orthographic transcription	appr. 1000h	various microphones, broadcast data
Our speech corpus	read and semi-spontaneous, orthographic transcription	appr. 36h x 3	synchronized multiple microphones far-field & beamformed

1.1 Related Work

Table 1 shows an overview of current major German speech corpora. Between 1993 and 2000 the Verbmobil [4] project collected around 180 hours of speech for speech translation. The PhonDat [5] corpus was generated during the Verbmobil project and includes recorded dialog speech with a length of 9.5 hours. The Smartkom [3] project combines speech, gesture and mimics for a multimodal application. The data were recorded during Wizard-of-Oz experiments with a length of 4.5 minutes each. Audio was also recorded using multiple far-field microphones, but totals only 12 hours of speech data. A large German speech corpus focusing on dialect identification [6] was recorded by the "Institut für Deutsche Sprache". It contains 1000h of audio recorded in several cities in Germany, thereby ensuring a variety of different dialects. The recorded speakers were split between a group aged between 50-60 years old and a younger group between 16-20 years old. A further speech corpus is FAU IISAH [7] with 3 hours and 27 minutes of spontaneous speech.

The Voxforge corpus[1] was a first open source German speech corpus, with 55 hours of collected speech from various participants, who usually recorded the speech on their own. None of these other German corpora, except the Voxforge corpus, are available under a permissive open source license. However, Voxforge is recorded under uncontrolled conditions, and the audio recording quality is unreliable.

An introduction to distant speech recognition (DSR) is given in [10]. A key challenge is the more pronounced effects, such as noise and reverberation, that the environment has on the recorded speech. Kaldi was recently compared [11] to other open source speech recognizers, outperforming them by a large margin

---

[1] http://www.voxforge.org

in German and English automatic speech recognition (ASR) tasks. Previously, Morbini et al. [12] also compared Kaldi and PocketSphinx to commercial cloud based ASR providers.

2 Corpus

In this section, we detail the corpus recording procedure and characterize the corpus quantitatively. The goal of our corpus acquisition efforts is to compile a data collection to build speaker-independent distant speech recognition. Our target use case is distant speech recognition in business meetings as a building block for automatic transcription [13] or the creation of semantic maps [14]. We recorded our speech data as described in [15]. We employed the KisRecord[2] software, which supports concurrent recording with multiple microphones. Speakers were presented with text on a screen, one sentence at a time, and were asked to read the text aloud. While the setup is somewhat artificial in that reading differs from speaking freely, we avoid the need of transcribing the audio and thus follow a more cost-effective approach.

For the training part of the corpus, sentences were drawn randomly from three text sources: a set of 175 sentences from the German Wikipedia, a set of 567 utterances from the German section of the European Parliament transcriptions [16] and 177 short commands for command-and-control settings. Text sources were chosen because of their free licenses, allowing redistribution of derivatives without restrictions. Test and development sets were recorded at a later date, with new sentences and new speakers. These have 1028 and 1085 unique sentences from Wikipedia, European Parliament transcriptions and crawled German sentences from the Internet, distributed equally per speaker. The crawled German sentences were collected randomly with a focused crawler [17], and were only selected from sentences encountered between quotation marks, which exhibit textual content more typical of direct speech. Unlike in the training set, where multiple speakers read the same sentences, every sentence recorded in the test and development set is unique, for evaluation purposes.

The distance between speaker and microphones was chosen to be one meter, which seems realistic for a business meeting setup where microphones could e.g. be located on a meeting table. There are many more use cases, where a distance of one meter is a sensible choice, e.g. in-car speech recognition systems, smart living rooms or plenary sessions. The entire corpus is available for the following microphones: Microsoft Kinect, Yamaha PSG-01S, and Samson C01U. The most promising and interesting microphone is the Microsoft Kinect, which is in fact a small microphone array and supports speaker direction detection and beamforming, cf. [15]. We recorded the beamformed and the raw signal simultaneously, however due to driver restrictions both are single channel recordings (i.e. it is not possible to apply our own beamforming on the raw signals). We split the overall data set into training, development and test partitions in such a way that speakers or sentences do not overlap across the different sets. The audio

---

[2] http://kisrecord.sourceforge.net

Table 2. Speaker gender and age distribution

Gender	Train	Dev	Test
male	105	12	13
female	42	4	4

Age	Train	Dev	Test
41–50	2	0	1
31–40	17	17	2
21–30	108	13	10
18–20	20	1	4

Table 3. Mean and standard deviation of training set sentences read per person per text source and total audio recording times for each microphone

Corpus	Mean ± StD
Wikipedia	35 ± 12
Europarl	8 ± 3
Command	18 ± 23

Microphone	Dur. Train/Dev/Test (h)
Microsoft Kinect	31 / 2 / 2
Yamaha PSG-01S	33 / 2 / 2
Samson C01U	29 / 2 / 2

data are available in MS Wave format, one file per sentence per microphone. In addition, for each recorded sentence, an XML file is provided that contains the sentence in original and normalized form (cf. Section 3.2), and the speaker metadata, such as gender, age and region of birth.

Table 2 lists the gender and age distribution. Female speakers make up about 30% in all sets and most speakers are aged between 18 and 30 years. For analysis of dialect influence, the sentence metadata also includes the federal state (Bundesland) where the speakers spent the majority of their lives. Despite our corpus being too small for training and testing dialect-specific models, this metadata is collected to support a future training of regional acoustic models. Most speakers are students that grew up and live in Hesse.

The statistics related to the number of sentences per speaker for each of the three text sources in the training corpus is given in Table 3. Most sentences were drawn from Wikipedia. On average, utterances from Europarl are longer than sentences from Wikipedia, and speakers encountered more difficulties in reading Europarl utterances because of their length and domain specificity. Recordings were done in sessions of 20 minutes. Most speakers participated in only one session, some speakers (in the training set) took part in two sessions. Table 3 also compares the audio recording lengths of the three main microphones for each dataset. Occasional outages of audio streams occurred during the recording procedure, causing deviations in recording length for each microphone. By ensuring that the training and the dev/test portions have disjoint sets of speakers and textual material, this corpus is perfectly suited for examining approaches to speaker-independent distant speech recognition for German.

3 Experiments

In this section, we show how our corpus can be used to generate a speaker-independent model in PocketSphinx and Kaldi. We then compare both speech recognition toolkits using the same language model and pronunciation dictionary in terms of word error rate (WER) on our heldout data from unseen speakers (development and test utterances). Our Kaldi recipe has been released on Github[3] as a package of scripts, which support fully automatic generation of

---

[3] https://github.com/tudarmstadt-lt/kaldi-tuda-de

all acoustic models with automatic downloading and preparation of all needed resources, including phoneme dictionary and language model. This makes the Kaldi results easily reproducible.

3.1 Phoneme Dictionary

As our goal is to train a German speech recognizer using only freely available sources, we have not relied on e.g. PHONOLEX[4], which is a very large German pronunciation dictionary with strict licensing. We compiled our own German pronunciation dictionary using all publicly available phoneme lexicons at the Bavarian Archive for speech signals (BAS)[5] and using the MaryTTS [18] LGPL-licensed pronunciation dictionary, which has 26k entries. Some of the publicly available BAS dictionaries, like the one for the Verbmobil [4] corpus, also contain pronunciation variants which were included. The final dictionary covers 44.8k unique German words with 70k total entries, with alternate pronunciations for some of the more common words. Stress markers in the phoneme set were grouped with their unstressed equivalents in Kaldi using the `extra_questions.txt` file and were entirely removed for training with CMU Sphinx. Words from the train / development / test sets with missing pronunciation had their pronunciations automatically generated with MaryTTS. This makes the current evaluation recognition task a completely closed vocabulary one and sets the focus of the evaluation on the acoustic model (AM). Still, our pronunciation dictionary is of reasonable size for large-vocabulary speech recognition.

3.2 Language Model

We used approximately 8 million German sentences to train our 3-gram language model (LM) using Kneyser-Ney [19] smoothing. We made use of the same resources that were used to select appropriate sentences for recording read speech, but they were carefully filtered, so that sentences from the development and test speech corpus are not included in the LM. We also used 1 million crawled German sentences in quotation marks. Finally, we used MaryTTS [18] to normalize the text to a form that is close to how a reader would speak the sentence, e.g. any numbers and dates have been converted into a canonical text form and any punctuation has been discarded. The final sentence distribution of the text sources is 63.0% Wikipedia, 22.4% Europarl, and 14.6% crawled sentences. The perplexity of our LM is 101.62. We also released both the LM in ARPA format and its training corpus, consisting of eight million filtered and Mary-fied sentences.

3.3 CMU Sphinx Acoustic Model

Our Sphinx training procedure follows the tutorial scripts provided with the code. We trained a default triphone Gaussian Mixture Model - Hidden Markov

---

[4] https://www.phonetik.uni-muenchen.de/Bas/BasPHONOLEXeng.html
[5] ftp://ftp.bas.uni-muenchen.de/pub/BAS/

Model (GMM-HMM) with Cepstral Mean Normalized (CMN) features. The cepstra uses the standard 13 dimensions (energy + 12) concatenated with the delta and double delta features. The HMM has 2000 senones and 32 Gaussians per state. We further tested the influence of Linear Discriminative Analysis (LDA), Maximum Likelihood Linear Transformation (MLLT) and vocal tract length normalization (VTLN) with a warp window between 0.8 and 1.2 using a step size of 0.02. We used the newest development version of SphinxTrain (revision 12890). SphinxTrain uses Pocketsphinx for the decoding step and sphinx3-align for its forced alignment. After optimization, we ran PocketSphinx with the beamwidth set to 10^{-180} and the language weight to 20.

3.4 Kaldi Acoustic Models

We follow the typical Kaldi training recipe S5 [1,20] for Subspace Gaussian Mixture Models (SGMM) [21], using a development version of Kaldi (revision 4968). For all GMM models, our features are computed as standard 13-dimensional Cepstral Mean-Variance Normalized (CMVN) Mel-Frequency Cesptral Coefficients (MFCC) features with first and second derivatives. We also apply LDA over a central frame with $+/-$ 4 frames and project the concatenated frames to 40 dimensions, followed by Maximum Likelihood Linear Transform (MLLT) [22]. For speaker adaptation in GMM models we employ feature-space Maximum Likelihood Linear Regression (fMLLR) [23]. We also make use of discriminative training [24] using the minimum phone error rate (MPE) and boosted maximum mutual information (bMMI) [25] criteria. For deep neural network (DNN) - HMM models [26], we also use the standard training recipe with 2048 neurons and 5 hidden layers.

4   Evaluation

Table 4 shows different WER achieved on the development and test sets of our speech corpus, using the Microsoft Kinect speech recordings and different Kaldi models. Table 5 shows our results using different Sphinx models, using the

Table 4. WER across multiple Kaldi acoustic models.

Kaldi Model	WER (%)	
	dev	test
GMM	25.7	27.8
GMM+fMLLR	24.6	27.1
GMM+MPE	23.7	26.2
GMM+bMMI(0.1)	23.5	25.8
SGMM+fMLLR	19.6	21.6
SGMM+bMMI(0.1)	19.1	20.9
DNN	**18.2**	**20.5**

Table 5. WER across multiple Sphinx acoustic models.

Sphinx Model	WER (%)	
	dev	test
GMM	39.6	43.8
GMM+LDA/MLLT	40.5	44.2
GMM+VTLN	**38.3**	**39.6**

same data resources. For all models the OOV vocabulary of the development corpus was included into the pronunciation dictionary, so the results on this portion reflect the performance of the speech recognizer under a known vocabulary scenario. Adaptation methods like VTLN and fMLLR improve our WER as expected. For Kaldi, using SGMM considerably improves scores compared to purely GMM based models, with a further smaller improvement using a DNN based model. This is probably because the training corpus size is moderate ($\approx$31 hours) and SGMM models usually perform well with smaller amounts of training data [21]. Our best Kaldi model (DNN) clearly outperforms the best Sphinx model (GMM+VTLN) on the test set: 20.5% vs 39.6% WER. Such a relatively large difference in WER performance between the two systems has also been observed in [11]. Domain and training corpus have a large effect on this performance difference [12], but the results presented here do not seem to be unusual for a large vocabulary task in distant speech recognition.

5   Conclusion and Future Work

In this paper, we present a complete open source solution for German distant speech recognition, using a Microsoft Kinect microphone array and a new distant speech corpus. Distant speech recognition is a challenging task owing to the additional and more pronounced effects of the environment, like noise and reverberation. With a dictionary size of 44.8k words and by using a speaker independent Kaldi acoustic model, we are able to achieve a word error rate (WER) of 20.5%, with a comparatively modest corpus size. We have implicitly exploited the beamforming and noise filtering capabilities of a Microsoft Kinect, which we used, among other microphones for recording our speech data. The Kaldi speech recognition toolkit outperforms the Sphinx toolkit by a large margin and seems to be an overall better choice for the challenges of distant speech recognition.

Ultimately, we would like to enable open source distant speech recognition in German - the presented open source corpus and the acoustic models in this paper is a first step towards this goal. As an extension of our work, we would like to expand our data collection and speech recognition training to study the effects of overlapping and multiple speakers in one utterance, OOV words, additional non-speech audio, and spontaneous speech. This will increase the difficulty of the speech recognition task, but would make it more appealing for research on realistic scenarios. Such scenarios are not uncommon to be in the vicinity of 50% WER or more [27].

Our open source corpus is licensed under a very permissive Creative Commons license and all other resources are also freely available. Unlike other distant speech recognition recipes for German acoustic models, which make extensive use of proprietary speech resources and data with stricter licensing, our resources and acoustic models can be used without restrictions for any purpose: private, academic and even commercial. We also want to encourage the release of other German open source speech data into equally permissive licenses.

Acknowledgments. This work was partly supported by the Bundesministerium für Bildung und Forschung (BMBF), Germany within the programme "KMU-innovativ: Mensch-Technik-Interaktion für den demografischen Wandel".

References

1. Povey, D., Ghoshal, A., Boulianne, G., Burget, L., Glembek, O., Goel, N., Hannemann, M., Motlicek, P., Qian, Y., Schwarz, P., Silovsky, J., Stemmer, G., Vesely, K.: The Kaldi speech recognition toolkit. In: Proc. IEEE ASRU, pp. 1–4 (2011)
2. Huggins-Daines, D., Kumar, M., Chan, A., Black, A.W., Ravishankar, M., Rudnicky, A.I.: PocketSphinx: a free, real-time continuous speech recognition system for hand-held devices. In: Proc. ICASSP (2006)
3. Schiel, F., Steininger, S., Türk, U.: The smartkom multimodal corpus at BAS. In: Proc. LREC (2002)
4. Wahlster, W.: Verbmobil: translation of face-to-face dialogs. In: Proc. 4th Machine Translation Summit, pp. 128–135 (1993)
5. Hess, W.J., Kohler, K.J., Tillmann, H.G.: The Phondat-verbmobil speech corpus. In: Proc. EUROSPEECH (1995)
6. Brinckmann, C., Kleiner, S., Knöbl, R., Berend, N.: German today: a really extensive corpus of spoken standard german. In: Proc. LREC (2008)
7. Spiegl, W., Riedhammer, K., Steidl, S., Nöth, E.: FAU IISAH corpus - a german speech database consisting of human-machine and human-human interaction acquired by close-talking and far-distance microphones. In: Proc. LREC (2010)
8. Schiel, F., Heinrich, C., Barfüßer, S.: Alcohol language corpus: the first public corpus of alcoholized German speech. In: Proc. LREC, vol. 46(3), pp. 503–521 (2012)
9. Stadtschnitzer, M., Schwenninger, J., Stein, D., Köhler, J.: Exploiting the large-scale german broadcast corpus to boost the fraunhofer IAIS speech recognition system. In: Proc. LREC, pp. 3887–3890 (2014)
10. Woelfel, M., McDonough, J.: Distant Speech Recognition. Wiley (2009)
11. Gaida, C., Lange, P., Proba, P., Malatawy, A., Suendermann-Oeft, D.: Comparing open-source speech recognition toolkits. http://suendermann.com/su/pdf/oasis2014.pdf
12. Morbini, F., Audhkhasi, K., Sagae, K., Artstein, R., Can, D., Georgiou, P., Narayanan, S., Leuski, A., Traum, D.: Which ASR should I choose for my dialogue system? In: Proc. SIGDIAL (2013)
13. Akita, Y., Mimura, M., Kawahara, T.: Automatic transcription system for meetings of the japanese. In: Proc. INTERSPEECH, pp. 84–87 (2009)
14. Biemann, C., Böhm, K., Heyer, G., Melz, R.: Automatically building concept structures and displaying concept trails for the use in brainstorming sessions and content management systems. In: Böhme, T., Larios Rosillo, V.M., Unger, H., Unger, H. (eds.) IICS 2004. LNCS, vol. 3473, pp. 157–167. Springer, Heidelberg (2006)
15. Schnelle-Walka, D., Radeck-Arneth, S., Biemann, C., Radomski, S.: An open source corpus and recording software for distant speech recognition with the microsoft kinect. In: Proc. 11. ITG Fachtagung Sprachkommunikation (2014)
16. Koehn, P.: Europarl: A Parallel Corpus for Statistical Machine Translation. In: Proc. 10th MT Summit, Phuket, Thailand, AAMT, AAMT, pp. 79–86 (2005)

17. Remus, S.: Unsupervised relation extraction of in-domain data from focused crawls. In: Proc. Student Research Workshop of EACL, Gothenburg, Sweden, pp. 11–20 (2014)
18. Schröder, M., Trouvain, J.: The German Text-to-Speech Synthesis System MARY: A Tool for Research, Development and Teaching. IJST **6**, 365–377 (2003)
19. Kneser, R., Ney, H.: Improved backing-off for m-gram language modeling. In: Proc. ICASSP, vol. 1, pp. 181–184 (1995)
20. Ali, A., Zhang, Y., Cardinal, P., Dahak, N., Vogel, S., Glass, J.: A complete KALDI recipe for building Arabic speech recognition systems. In: Proc. IEEE SLT, pp. 525–529. Institute of Electrical and Electronics Engineers Inc. (2015)
21. Povey, D., Burget, L., Agarwal, M., Akyazi, P., Feng, K., Ghoshal, A., Glembek, O., Goel, N.K., Karafiat, M., Rastrow, A., Rose, R.C., Schwarz, P., Thomas, S.: Subspace gaussian mixture models for speech recognition. In: Proc. ICASSP, pp. 4330–4333 (2010)
22. Gales, M.J.: Semi-Tied Covariance Matrices for Hidden Markov Models. IEEE Trans. Speech and Audio Processing **7**, 272–281 (1999)
23. Gales, M.J.F.: Maximum likelihood linear transformations for HMM-based speech recognition. Computer Speech & Language **12**(2), 75–98 (1998)
24. Gales, M.: Discriminative models for speech recognition. In: 2007 Information Theory and Applications Workshop (2007)
25. Povey, D., Kanevsky, D., Kingsbury, B., Ramabhadran, B., Saon, G., Visweswariah, K.: Boosted MMI for model and feature-space discriminative training. In: Proc. ICASSP, pp. 4057–4060 (2008)
26. Dahl, G.E., Yu, D., Deng, L., Acero, A.: Context-dependent pre-trained deep neural networks for large-vocabulary speech recognition. IEEE Trans. Audio, Speech, Language Process. **20**(1), 30–42 (2012)
27. Swietojanski, P., Ghoshal, A., Renals, S.: Hybrid acoustic models for distant and multichannel large vocabulary speech recognition. In: Proc. IEEE ASRU, pp. 285–290 (2013)

First Steps in Czech Entity Linking

Michal Konkol[✉]

Department of Computer Science and Engineering, Faculty of Applied Sciences,
University of West Bohemia, Univerzitní 8, 306 14 Plzeň, Czech Republic
nlp.kiv.zcu.cz, konkol@kiv.zcu.cz

Abstract. In this paper, we present our approach for a simplified Entity Linking task in Czech, where entity mentions found in text are linked to a list of known entities. We evaluate both known and newly proposed methods for entity names similarity on a manually annotated newspaper corpus. We show that it is possible to achieve a very high accuracy in this task, which is required in many natural language processing tasks as well as in the commercial practice.

Keywords: Entity linking · Named entity · Named entity disambiguation · Czech

1 Introduction

Named entity is an (multi-word) expressions, that identifies a single object (e.g. John Doe) from a set of semantically similar objects (e.g. persons). The most often used classes of named entities are persons, organizations, locations, or dates. Named entities from these classes often hold the key information in a document, which can be exploited in many applications. There are two natural language processing tasks associated with named entities: named entity recognition and named entity disambiguation (with entity linking subtask). The named entity recognition task identifies entities in texts and classifies them. The entity linking task links the entity mentions found in text with real entities, e.g. the entity mention 'Obama' is linked with Wikipedia page of Barack Obama, the president of the USA.

The full named entity disambiguation task is not limited only to the entity linking subtask. Other subtasks are NIL detection, and NIL clustering. NIL detection tries to detect entity mentions, that are not in the knowledge base and the NIL clustering task groups these mentions in such a way, that each group refers to only a single real entity. Another related task is entity normalization, where entities are normalized to their official name, e.g. 'USA' is normalized to 'United States of America'.

In this paper, we address the task of linking entities to a list of known entities. The known entities are not associated with any data (e.g. short bios for persons), thus it is needed to link them through string similarity. This task arise very often in the research as well as in the commercial practice. We have two main goals. First, we want to propose a method, which solves this problem. Second, we need to create a new corpus in order to evaluate the proposed approach.

P. Král and V. Matoušek (Eds.): TSD 2015, LNAI 9302, pp. 489–496, 2015.
DOI: 10.1007/978-3-319-24033-6_55

2   Related Work

While the named entity recognition is well studied in Czech [1–6], the entity linking task has not been addressed yet.

The most prominent resources and systems for entity linking were introduced for the Knowledge Base Population (KBP) task of the Text Analysis Conference (TAC) [7]. The data support the full named entity disambiguation task, i.e. entity linking, NIL detection and NIL clustering. There are also data for cross-lingual entity linking [8].

Another well-known task is Web Person Search (WePS) [9]. The task is to cluster web pages returned to a person name query so that each cluster refers to a different person.

3   Corpus Creation

The corpus is based on press releases from the Czech News Agency and a list of known entities. It is important to note, that with a list of known entities we only try to solve the variety problem (multiple surface forms for one entity) and not the ambiguity problem (one surface form for multiple entities). We have chosen to use only the person names, because they have (according to our estimates) higher frequency and variety in the news domain than organizations and locations.

Each entity was assigned to the corresponding entry in the list of known entities if such an entry exists. There are situations, in which the entity can be assigned to multiple entries or we are not able to decide certainly, e.g. for entity mention 'Doe', we cannot decide if it is 'John Doe' or 'Jack Doe' from the list of know entities and even if there is only the entry 'John Doe', we cannot be certain that the document is about 'John Doe' and not some other 'Doe'. For this purpose, two types of links are defined: certain and possible. A certain link is used for entities that can be linked certainly to the list given the document, e.g. 'Johnnie' can be certainly linked to 'John Doe' only if it is obvious from the document (without external knowledge), that 'Johnnie' refers to 'John Doe'.

We have annotated 77 documents with 879 entity mentions. The list of known entities contains 21648 entries. From the 879 found entity mentions, 316 are linked to a known entity and 563 are not linked. Certain link was assigned to 253 of the linked entities and possible link to 63 entity mentions. There were 213 possible links in total, what makes the average of more than 3 possible links for the 63 entity mentions.

The certainly linked entity mentions referenced 96 distinct entries in the dictionary. Each linked entity mention is a surface form of the particular known entity. There were 38 known entities referenced by more than one surface form, so the variety is approximately 39.5%. On average the referenced known entities have 1.9 surface forms. The most surface forms (9) and links (15) were found for 'Ehud Olmert', an Israel politician and former prime minister. There are multiple documents in the corpus dealing with Israel politics. Figures 1 and 2 show the histograms of the number of surface forms and links.

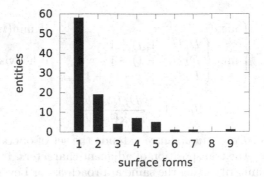

Fig. 1. Histogram of surface forms per entity.

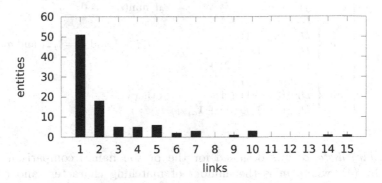

Fig. 2. Histogram of links per entity.

4   Similarity Metrics

In this section, we introduce the string similarity metrics used in our experiments. Before their introduction, we need to define our mathematical notation. We compare strings a and b. The length of the string a is denoted as $|a|$. We say that A is a set of n-grams (and their counts) contained in string a. The sum of counts of all n-grams in this set is denoted as $|A|$. The union $A \cup B$ contains all n-grams of A and B, where the count of a shared n-gram is the maximum of their original counts. The intersection $A \cap B$ contains all shared n-grams and their counts are minimums of their original counts. E.g. we have $a = $ 'aaab' and $b = $ 'aabc', the sets are $A = \{(\text{'aa'}, 2), (\text{'ab'}, 1)\}$ and $B = \{(\text{'aa'}, 1), (\text{'ab'}, 1), (\text{'bc'}, 1)\}$, $|A| = 3$ and $|B| = 3$, the union $A \cup B = \{(\text{'aa'}, 2), (\text{'ab'}, 1), (\text{'bc'}, 1)\}$ and the intersection $A \cap B = \{(\text{'aa'}, 1), (\text{'ab'}, 1)\}$.

The *Levenshtein distance* is probably the most commonly used string distance metric. It computes the minimal edit distance using three edit operations – delete, add, and substitute. The Levenshtein metric is defined as (1), where $\mathbf{1}_{a_i \neq b_i} = 1$ if $a_i \neq b_i$ and 0 otherwise. The Levenshtein distance is converted to similarity using (2).

$$D_L(i,j) = \begin{cases} \max(i,j) & \text{if } \min(i,j) = 0 \\ \min \begin{cases} D_L(i-1,j)+1 \\ D_L(i,j-1)+1 \\ D_L(i-1,j-1)+\mathbf{1}_{a_i \neq b_j} \end{cases} & \text{otherwise} \end{cases} \tag{1}$$

$$S_L = 1 - \frac{D_L(|a|,|b|)}{\max\{|a|,|b|\}} \tag{2}$$

The *Levenshtein-Damerau distance* extends the set of operations in the Levenshtein distance by the transposition of adjacent characters. It is defined by (3) and converted to similarity using the same approach as for Levenshtein distance.

$$D_{LD}(i,j) = \begin{cases} \max(i,j) & \text{if } \min(i,j) = 0 \\ \min \begin{cases} D_{LD}(i-1,j)+1 \\ D_{LD}(i,j-1)+1 \\ D_{LD}(i-1,j-1)+\mathbf{1}_{a_i \neq b_j} \\ D_{LD}(i-2,j-2)+1 \end{cases} & \text{if } i,j > 1 \text{ and } a_i = b_{j-1} \text{ and } a_{i-1} = b_j \\ \min \begin{cases} D_{LD}(i-1,j)+1 \\ D_{LD}(i,j-1)+1 \\ D_{LD}(i-1,j-1)+\mathbf{1}_{a_i \neq b_j} \end{cases} & \text{otherwise} \end{cases} \tag{3}$$

The *Jaro distance* was designed for the person names comparison and is defined by (4), where m is the number of matching characters and t is the number of transpositions. The characters are considered as matching, if they are the same and their position differs by a maximum of k characters (5). The transpositions happen if the matching characters are in different order.

$$S_J = \begin{cases} 0 & \text{if } m = 0 \\ \frac{1}{3}\left(\frac{m}{|a|} + \frac{m}{|b|} + \frac{m-t}{m}\right) & \text{otherwise} \end{cases} \tag{4}$$

$$k = \left\lfloor \frac{\max\{|a|,|b|\}}{2} - 1 \right\rfloor \tag{5}$$

The *Jaro-Winkler distance* is an improvement of the original Jaro distance. It gives higher weight to n first characters and is defined as (6). If we denote c the length of a common prefix of a and b, then $l = \max\{c,n\}$. The weight of the first characters is denoted as p, $0 \leq p \leq \frac{1}{n}$. In our experiments we use common settings $p = 0.1$ and $n = 4$. Both these metrics are named "distances", but in fact they are similarities, i.e. $0 \leq S_{J(W)} \leq 1$ and higher values are assigned to more similar strings.

$$S_{JW} = S_J + lp(1 - S_J) \tag{6}$$

The *Jaccard similarity*, *Overlap similarity*, and *Soerensen-Dice similarity* are defined by (7), (8), and (9), respectively. These similarities were not originally proposed for string similarity, but can be used for this purpose.

$$S_{Jac} = \frac{|A \cap B|}{|A \cup B|} \tag{7}$$

$$S_O = \frac{|A \cap B|}{\min\{|A|, |B|\}} \tag{8}$$

$$S_{SD} = \frac{2|A \cap B|}{|A| + |B|} \tag{9}$$

The *common prefix similarity* is simply a ration between the length of a common prefix c and the length of one of the strings. We can choose both minimal or maximal length of a and b (10), the is denoted in parentheses in the experiments.

$$S_{CP_{max}} = \frac{c}{\max\{|a|, |b|\}} \quad \text{or} \quad S_{CP_{min}} = \frac{c}{\min\{|a|, |b|\}} \tag{10}$$

The *longest common subsequence similarity* is the ratio of the length of the longest common subsequence lcs and the length of one of the strings. We can again use minimal or maximal length of a and b (11).

$$S_{LCS_{max}} = \frac{lcs}{\max\{|a|, |b|\}} \quad \text{or} \quad S_{LCS_{min}} = \frac{lcs}{\min\{|a|, |b|\}} \tag{11}$$

5 Proposed Combination

The proposed system is based on the maximum entropy classifier. We use the implementation of this algorithm from the Brainy library [10].

We use the similarities from the previous section as features, but not directly. We firstly tokenize the entities, then we align the tokens to maximize the overall similarity. For this purpose we use a (suboptimal) greedy algorithm, which seems to be sufficient for the person names. The Hungarian algorithm [11] can be used for the optimal alignment, but has higher complexity.

A missing token (one entity has more tokens than the other) is aligned to null token and the similarity is set to a constant M. Furthermore, if one of the tokens is an acronym and it can represent the other token, we set the similarity to a constant R. Both R and M are parameters of the system. Using the development data, we have set these parameters to $M = 0.5$ and $R = 0.65$.

The final similarity S is the arithmetic mean of similarities between all tokens. We use the following features for all the similarity metrics:

- Similarity
- Dissimilarity $(1 - S)$
- Intervals of length 0.1 (e.g $0.3 \leq S \leq 0.4$)
- Lesser than a threshold (0.1 step, e.g $S \leq 0.4$)
- Greater or equal than a threshold (0.1 step, e.g $S \geq 0.4$)

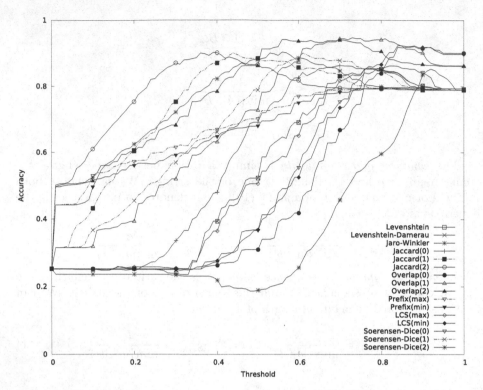

Fig. 3. Similarity metrics accuracy for various threshold settings.

6   Experiments

The experiments are based on *queries*, similarly to the KBP entity linking task. Each query contains a document with all entity annotations and one annotation (or entity mention) is chosen as the query. Each query is associated with the correct answer, i.e. the correct entry in the list of known entities or indication of unknown entity. There is a limited amount of positive examples (e.g. the entity mention matches the list entry), but very high number of negative examples (e.g. entity mention does not match the list entry). We have decided to use all positive examples and to select three times more negative examples, i.e. the positive examples forms $\frac{1}{4}$ of examples. We have tried to choose the hardest negative examples, where the entity mention is most similar to a wrong entry. The similarity was measured by the Levenshtein metric. This choice penalizes the Levenshtein metric when compared to other metrics as the negative examples the hardest for Levenshtein metric, but they may be easy for other metrics.

The first experiment was proposed to explore the data and to see the limits of similarity metrics. We compute a similarity s between the entity mention (query) and the list entry using each similarity metric and compare it with threshold t. If $s \geq t$, then we say that the mention matches the entry. Fig. 3 shows the relation

Table 1. Results for similarity metrics and their machine learning combination on the training, development, and test data.

Model	Training	Accuracy Development	Test
Levenshtein	86.37%	84.45%	83.75%
Levenshtein-Damerau	86.37%	84.45%	83.75%
Jaro-Winkler	85.66%	84.95%	83.85%
Jaccard(0)	86.96%	85.84%	85.74%
Jaccard(1)	91.07%	88.33%	89.33%
Jaccard(2)	91.07%	89.13%	86.94%
Overlap(0)	92.71%	91.03%	90.43%
Overlap(1)	95.06%	93.82%	93.52%
Overlap(2)	94.95%	92.42%	92.22%
Prefix(max)	79.20%	79.36%	79.36%
Prefix(min)	79.55%	79.66%	79.66%
LCS(max)	86.37%	84.75%	84.65%
LCS(min)	92.71%	91.72%	91.63%
Soerensen-Dice(0)	86.96%	85.54%	85.64%
Soerensen-Dice(1)	91.07%	88.33%	89.33%
Soerensen-Dice(2)	91.19%	89.43%	87.04%
ML combination	99.79%	97.21%	97.11%

between the chosen threshold and the accuracy. The values in parentheses are choices for the given metric (e.g. order of n-grams).

Our second experiments are done using a 10-fold cross-validation. For each fold, the data are divided in the ratio 80 : 10 : 10 between the training, development and test data, respectively. For each similarity metric, we estimate the optimal threshold using the training data and we apply it on the test data. For the machine learning combination of similarity metrics, we use the training data to find the optimal parameters of the maximum entropy classifier, the development data to find optimal hyperparameters of the model (e.g. the optimal compensation for missing words), and we apply the best model on the test data. The results are shown in Table 1.

We can see, that it is possible to achieve accuracy over 90% using a simple similarity metric. The highest score using similarity metric (93.52%) was achieved with Overlap similarity using bigrams. The proposed algorithm further improves the accuracy to 97.11%. These results highly surpassed our expectations.

7   Conclusion and Future Work

We have manually created a Czech corpus for a simplified entity linking task and provided the necessary statistics. The data show a rather high variety (39.5%), which can be explained by the rich morphology of Czech.

We have carried out experiments with well-known similarity metrics. The best similarity metric in our experiments was Overlap similarity with accuracy 93.52%. We also propose a classifier based combination of these similarity metrics, which achieved accuracy 97.11%.

In the future, we are going to create a new corpus for the full named entity disambiguation task.

References

1. Konkol, M., Brychcín, T., Konopík, M.: Latent semantics in named entity recognition. Expert Systems with Applications **42**(7), 3470–3479 (2015)
2. Konkol, M., Konopík, M.: Maximum entropy named entity recognition for czech language. In: Habernal, I., Matoušek, V. (eds.) TSD 2011. LNCS, vol. 6836, pp. 203–210. Springer, Heidelberg (2011)
3. Král, P.: Features for named entity recognition in Czech language. In: Proceedings of the International Conference on Knowledge Engineering and Ontology Development, KEOD 2011, Paris, France, October 26–29, pp. 437–441 (2011)
4. Konkol, M., Konopík, M.: CRF-Based czech named entity recognizer and consolidation of czech ner research. In: Habernal, I. (ed.) TSD 2013. LNCS, vol. 8082, pp. 153–160. Springer, Heidelberg (2013)
5. Konkol, M., Konopík, M.: Named entity recognition for highly inflectional languages: effects of various lemmatization and stemming approaches. In: Sojka, P., Horák, A., Kopeček, I., Pala, K. (eds.) TSD 2014. LNCS, vol. 8655, pp. 267–274. Springer, Heidelberg (2014)
6. Straková, J., Straka, M., Hajič, J.: A new state-of-the-art czech named entity recognizer. In: Habernal, I. (ed.) TSD 2013. LNCS, vol. 8082, pp. 68–75. Springer, Heidelberg (2013)
7. Simpson, H., Strassel, S., Parker, R., McNamee, P.: Wikipedia and the web of confusable entities: Experience from entity linking query creation for tac 2009 knowledge base population. In: Chair, N.C.C., Choukri, K., Maegaard, B., Mariani, J., Odijk, J., Piperidis, S., Rosner, M., Tapias, D. (eds.) Proceedings of the Seventh International Conference on Language Resources and Evaluation (LREC 2010), Valletta, Malta. European Language Resources Association (ELRA) (May 2010)
8. Ji, H., Grishman, R., Dang, H.: Overview of the TAC2011 knowledge base population track. In: TAC 2011 Proceedings Papers (2011)
9. Artiles, J., Borthwick, A., Gonzalo, J., Sekine, S., Amig, E.: Weps-3 evaluation campaign: Overview of the web people search clustering and attribute extraction tasks. In: Braschler, M., Harman, D., Pianta, E. (eds.) CLEF (Notebook Papers/LABs/Workshops) (2010)
10. Konkol, M.: Brainy: a machine learning library. In: Rutkowski, L., Korytkowski, M., Scherer, R., Tadeusiewicz, R., Zadeh, L.A., Zurada, J.M. (eds.) ICAISC 2014, Part II. LNCS, vol. 8468, pp. 490–499. Springer, Heidelberg (2014)
11. Kuhn, H.W.: The Hungarian Method for the Assignment Problem. Naval Research Logistics Quarterly **2**(1–2), 83–97 (1955)

Classification of Prosodic Phrases by Using HMMs

Zdeněk Hanzlíček[✉]

NTIS - New Technology for the Information Society, Faculty of Applied Sciences,
University of West Bohemia, Univerzitní 22, 306 14 Plzeň, Czech Republic
zhanzlic@ntis.zcu.cz
http://www.ntis.zcu.cz/en

Abstract. In this paper, we present a new approach for classification of phrase types. It is based on utilization of context-dependent hidden Markov models that consider the phonetic, prosodic and linguistic context. The classification is performed by forced-alignment for particular phrase types and selection of the type with the best alignment score. Experiments were performed on 2 large speech corpora. The classification results were successfully verified by a listening test. The speech corpora with corrected prosodemes were used in a unit selection speech synthesis framework. Another listening test confirmed that the prosody of particular phrases improved in comparison with the baseline system.

Keywords: Speech corpora · Prosodemes · Speech synthesis · Unit selection

1 Introduction

In modern speech synthesis systems [1,2], large speech corpora are utilized to learn new voices. These speech corpora usually contain several hours of speech spoken by talented speakers who are able to record such an amount of speech data in a sufficient quality. An appropriate phonetic and prosodic annotation of the particular utterances is necessary for a high quality of synthesized speech [3]. Generally, the knowledge of presence of various prosodic events in speech data and their proper description is very useful in many other applications as well.

In connection with using the large speech corpora, the automatic phonetic and prosodic annotation of speech [4,5] became an important task. This article presents a new approach for the classification of the phrase type. The research was done for the Czech language where the speech features within the last prosodic word of a phrase (corresponding to a functionally involved prosodeme[1]) are characteristic for particular types of sentences and for the phrase structure of compound/complex sentences.

Nevertheless, in the real speech data, the expectation given by the prosody models can be breached and speech features corresponding to a different type of

---

[1] Prosodemes will be explained in Section 2.

© Springer International Publishing Switzerland 2015
P. Král and V. Matoušek (Eds.): TSD 2015, LNAI 9302, pp. 497–505, 2015.
DOI: 10.1007/978-3-319-24033-6_56

phrase/prosodeme can be present. Using a speech corpus with incorrect prosodeme labels can be a source of prosody inconsistency in synthesized speech. The proposed process of prosodeme classification helps to reveal and correct such badly-labelled prosodemes. This should improve the overall quality of resulting synthetic speech.

This paper is organized as follows, Section 2 explains the prosody model used in this work. Procedure for prosodeme classification is proposed in Section 3. Section 4 describes the evaluation of performed experiments. Finally, Section 5 concludes this paper and outlines the future work.

2   Prosody Model and Prosodemes

Within this paper, the formal prosody model proposed by Romportl [6] is used. According to this model, an utterance can be divided into prosodic clauses separated by short pauses. Each prosodic clause includes one or more prosodic phrases, which contain certain continuous intonation scheme. A prosodic phrase consists of two prosodemes: null prosodeme and functionally involved prosodeme which is supposed to be related to the last prosodic word in the phrase and depends on the communication function the speaker intends the sentence to have.

For the Czech language[2], the following basic classes of functionally involved prosodemes were defined (for a more detailed prosodme categorization see [6]):

P1 – prosodemes terminating satisfactorily (typical for declarative sentences)
P2 – prosodemes terminating unsatisfactorily (typical for questions)
P3 – prosodemes non-terminating (typical for non-terminal phrases in compound/complex sentences)

Naturally, particular types of phrases do not vary solely within their last prosodic words. Some specific prosodic differences can be present throughout the whole phrase. However, those differencies are often rather content-related (e.g. emphasis on some key words) and a more complex prosody model would be required for their reasonable application. Our prosody model based on prosodemes is uncomplicated and seems to be sufficiently descriptive for the phrase type classification task [7].

Since our speech synthesis system [8] is created for neutral speech (i.e. without emphasis, expressions etc.), 3 specific prosodemes are supported[3]: P1.1, P2.2 and P3.1. A typical example of prosodemes P1.1 and P3.1 is depicted on Figure 1.

Typical speech features are related to the particular prosodeme types. For example, P1.1 is characteristic with a pitch decrease within its last syllable, a pitch increase is specific for P3.1, etc. Beside the pitch shape, spectral, duration

---

[2] A different/modified set of prosodemes can be specific for other languages.

[3] This set of prosodemes proved to be sufficient to describe basic types of phrases within the neutral speech. Other prosodemes are classified as one of those. A detailed explanation is beyond the scope of this paper.

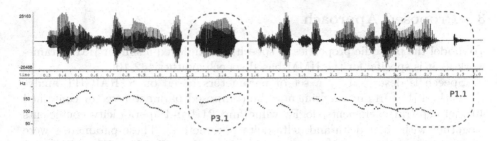

Fig. 1. Prosodemes within a declarative compound sentence **"Miloš Zeman řekl, že všechno bude jinak."** spoken by a male speaker. It is composed from 2 phrases ended by non-terminating prosodeme P3.1 and prosodeme terminating satisfactorily P1.1.

and energy features can be characteristic for particular prosodemes. However, their impact seems to be not so relevant for prosody perception or the dependence is more complex.

In real speech data, a different prosodeme than expected could be present. This could have several reasons

- The theoretical expectation is wrong since the type of utterance was incorrectly assigned.
- The utterance was read improperly. Even professional speakers have sometimes difficulties in reading many structured sentences consistently.
- The sentence is ambiguous. Depending on the meaning, specific situation or context, more prosodeme combinations are possible.

The most frequent case of prosodeme inconsistency is a compound sentence that can be split into several independent sentences. Within the compound sentence, all phrases (except the last one) should be terminated with the prosodeme P3.1. However, when the link between particular sentences is rather weak, the utterance can be split into independent sentences which are naturally terminated by the prosodeme P1.1.

Another common case is confusion of the question type. In the Czech language, wh-questions should be terminated with the prosodeme P2.3 that is similar to the P1.1[4] (i.e. the pitch is not rising), whereas yes/no questions contain the prosodeme P2.2 that is more like the P3.1 (i.e. the pitch is rising).

In any case, badly annotated speech corpora can be a source of various troubles. In speech synthesis (specifically, in unit selection method), prosodeme labels are important attributes for selecting sequence of optimal speech units for building resulting speech [9]. Using units from an inappropriate prosodeme or mixing units from different types of prosodemes can cause a decrease in the overall speech quality – prosody of synthesized speech does not correspond to the type or the structure of the sentence, some unnatural fluctuations of pitch or other prosody-related speech features occur, etc.

---

[4] In our TTS system, the prosodeme P2.3 is not differentiated from P1.1, i.e. P1.1 is used for both cases.

3   Proposed Approach

To model the prosodic properties of speech we employed a similar HMM framework as it is specific for the HMM-based speech synthesis [2,10].

Speech is described by a set of parameters based on STRAIGHT analysis method [11]. In our experiments, the composed parameter vector contained 40 mel cepstral coefficients, $\log F_0$ value and 21 band aperiodicity coefficients (together with their delta and delta-delta parameters). These parameters were modelled by a set of multi-stream context dependent HMMs. HTS toolkit[5] was employed for the model processing.

In the HMM-based speech synthesis framework, the phonetic, prosodic and linguistic context are taken into account, i.e. a speech unit (and the corresponding model) is given as a phone with its phonetic, prosodic and linguistic context information. In this manner, the language prosody is modelled implicitly – in various contexts different units/models can be used.

Within our experiments, a context-depended unit is represented by a string

$$p_1\text{-}p_2\text{+}p_3@\mathbf{P}\text{:}p_{w1}\text{-}p_{w2}@\mathbf{S}\text{:}s_{w1}|s_{h1}\text{-}s_{w2}|s_{h2}@\mathbf{W}\text{:}w_{h1}\text{-}w_{h2}/P_x$$

where all subscripted italic letters are contextual factors defined in Table 1. The other bold characters in this string template help to refer to particular factors (e.g. during model clustering).

Table 1. Contextual factors. Note: All the positions are forward and backward. Their values are limited to 5, i.e. value 5 is used for all following positions. We assume that the marginal positions are most prominent and the other positions are less relevant.

Factors		Possible values
p_1, p_2, p_3	Previous, current and next phoneme	Czech phoneme set (see e.g. [10])
p_{w1}, p_{w2}	Phone position in prosodic word (fw, bw)	
s_{w1}, s_{w2}	Syllable position in prosodic word (fw, bw)	1–5
s_{h1}, s_{h2}	Syllable position in phrase (fw, bw)	
w_{h1}, w_{h2}	Prosodic word position in phrase (fw, bw)	
P_x	Prosodeme type	P0, P1.1, P2.2, P3.1

3.1   Training Stage

The process of prosodeme correction can be roughly divided into 2 stages:

1. **Model Training** – Model parameters were estimated from speech data by using maximum likelihood criterion. 5-state left-to-right MSD-HSMM with single Gaussian output distributions were used. For a more robust model

---

[5] HMM-based Speech Synthesis System (HTS), http://hts.sp.nitech.ac.jp

parameter estimation, context clustering based on MDL (Minimum Description Length) criterion was performed. Decision trees were separately constructed for particular parameter streams and duration. In this stage, the default prosodic annotation of particular phrases is used.

2. **Prosodeme Re-Classification** – First, all utterances were divided into clauses[6]. Then, we created clause transcriptions for all considered types of functionally involved prosodemes, i.e. particular transcriptions differed only in units belonging to the last prosodic word. And finally, the best-matching transcription was selected for each clause, i.e. clauses were successively forced-aligned with particular transcriptions and the prosodeme with the best value of alignment score was selected as the correct prosodeme type for the given phrase.

After the second stage, a new corrected annotation is available. The whole process can be run iteratively with the updated annotation from the previous iteration.

Considering the main reasons for prosodeme mismatch aforementioned in Section 2, we decide to define allowed prosodeme corrections

- **P3.1 → P1.1** – corresponds to a compound/complex sentence splitted into independent phrases
- **P1.1 → P2.2 and P2.2 → P1.1** – correspond to different types of question

Without this limitation we could get some unwanted corrections, e.g. P3.1 could be classified as P2.2 or vice versa since they are very similar. Then the annotation could become less transparent. Therefore, we prefer to allow the interchangeability of some prosodemes during the unit selection process.

4   Evaluation and Results

For our experiments, we used 2 large speech corpora recorded for the purposes of speech synthesis [12]: one male and one female voice. Each corpus contained about 10,000 utterances.

4.1   Evaluation of Classification Results

The correctness of prosodeme re-classification was evaluated by a short listening test. Test contained 10 short phrases for each speaker: 5 phrases were classified as prosodeme P3.1 and 5 as P1.1. Prosodemes P2.2 were not included in the test since they are very similar to P3.1 in many cases and the listeners' decision could be often random or context-related.

The test sentences were not used for training of the HMMs. The listeners should determine the type of particular phrases. The capability of listeners to

---

[6] We used clauses instead of phrases since the division can be simply performed by the detection of pauses. As a result, only prosodemes from terminal phrases of particular clauses were considered to be functionally involved.

assign the phrase type properly was already proved in [7]. They were instructed to focus on the prosody at the end of utterances. Eight listeners participated in this test. The results presented in Table 2 showed that the HMM classification results are in agreement with the listeners' decisions in most cases (84% positive results, 5% negative and 11% indecisive) .

Table 2. Results of listening tests.

speaker	HMM classification	listeners' decision [%]		
		P1.1	P3.1	undecided
female	P1.1	**77.1**	11.4	11.4
	P3.1	5.7	**74.3**	20.0
male	P1.1	**97.1**	0.0	2.9
	P3.1	2.9	**85.7**	11.4
both	P1.1	**87.1**	5.7	7.2
	P3.1	4.3	**80.0**	15.7

4.2   Evaluation of Unit Selection Synthesis

The incorrect prosodic annotation of utterances in training data can cause a quality degradation of synthesized speech. Depending on combination of selected units, the prosodic features can be improper and unnatural. Abrupt changes in speech tempo and pitch fluctuation are common, too. Certainly, only a part of such synthesis failures is related with bad prosodeme annotation.

The selection of a suitable set of utterances is essential for the evaluation of the achieved benefit. The basic approach – evaluation of random synthetic utterances – is suitable only when a general quality improvement is expected.

In cases when the modification is related with a specific phenomenon that appears less frequently, it is better to select the suitable utterances systematically. To find out how often the performed corrections will affect the synthetic speech, a huge set of about 520,000 sentences[7] (about 1,050,000 particular phrases) were synthesized and a record of selected units for each sentence was created. By using this record, the statistics on usage of each unit from changed prosodemes were ascertained – see Figure 2 and Table 3.

According to the statistics, only about 16% and 10% utterances were affected by the changes in training corpora. The remained utterances are supposed to be unchanged[8]. Since it has no sense to compare equal sentences, we ignored those sentences and prepare a comparison listening test only with utterances that contained different units. All sentences were selected randomly, their length was

---

[7] This set of sentences is described in [13].

[8] This is only a simplified assumption which is not fully accurate, since units from changed prosodemes could be selected for other sentences, too. However for the selection of suitable sentences for a listening test, these rare cases can be ignored.

Table 3. Relative number [%] of synthetic utterances according to the number of changed units.

number of changed units	0	1	2	3–5	6–10	> 10
female speaker	89.71	3.68	1.87	3.41	1.23	0.10
male speaker	83.91	7.32	2.34	4.20	2.03	0.20

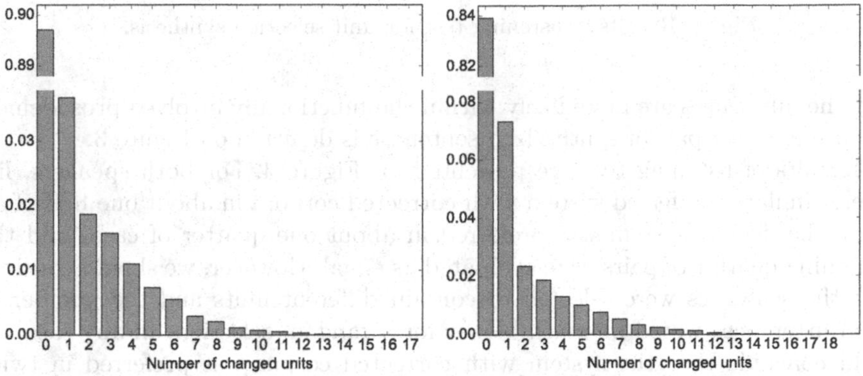

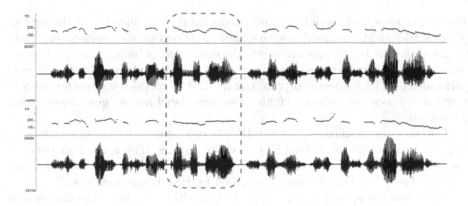

Fig. 2. Number of changed units for female and male speaker, respectively.

Fig. 3. The utterance **"Že o tom ta chůzka nebyla, jsme zjistili až na závěr."** synthesized by the male voice. It is composed from 2 phrases. Units at the end of the first phrase used by the default system are not appropriate (do not correspond to the P3.1 prosodeme). More proper units were selected by the corrected system. Besides, different units were used at the end of the utterance.eps

limited to 8–10 words. Test participants listened to 20 pairs of synthesized utterances and used 5-point scale for evaluation: A sounds significantly better than B, slightly better, equal, slightly worse, significantly worse. Moreover, they knew

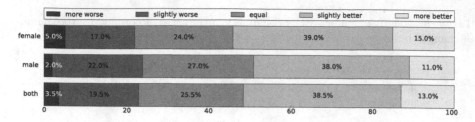

Fig. 4. Results of listening test for unit selection synthesis.

that the differences are more likely within the functionally involved prosodemes. A representative pair of synthesized sentences is depicted on Figure 3.

Results of listening test are presented on Figure 4. For both speakers, listeners similarly preferred system with corrected corpora in about one half of all cases, the default system was preferred in about one quarter of cases and the remaining quarter of pairs were evaluated as equal. However, we should consider that the sentences were selected to contain different units and the number of equal utterances would be much higher for a random selection. In any case, we could conclude that the system with corrected corpus was preferred in twice more cases than the default system.

5   Conclusion

This paper presented the initial experiments on the classification of the type of phrases. Hidden Markov models with extended contextual features were employed in this task. The basic applications are the correction of badly annotated prosodemes and the classification of phrases which type is unknown. In experiments performed on 2 large speech corpora, trained HMMs revealed various numbers of suspicious prosodemes whose default prosodeme label and its new classification do not agree.

Listening test confirmed that the HMMs can be used for the determination of proper functional prosodeme in particular phrases. The agreement between speakers and HMMs was about 84%. From the remaining cases, 12% were indecisive and only 5% were differently assigned by HMMs and speakers.

The corrected speech corpora were tested within the unit selection speech synthesis framework. For evaluation a preference listening test containing specially selected sentences was performed. In this test, listeners preferred the new TTS system above default (51% vs 22%), remaining utterances were evaluated as equal. Naturally, the obtained results substantially depend on the speech data for given speaker. For example, a smaller improvement can be achieved for a speaker whose prosody better matches the expectation of used prosodic model.

5.1   Future Work

In our future work, more experiments on prosodeme classification will be performed. We would like to include other types of prosodemes, too. For the purposes of unit selection speech synthesis, we consider to use the soft classification, i.e. particular speech units will not exclusively belong to one prosodeme type. Instead, a set of forced-alignment-score-based weights could be defined and used during the process of unit selection. We also intend to experiment with speaker-independent models. They could be employed in cases of non-professional speakers whose speech prosody is not consistent enough to train new models or the amount of speech data is low.

Acknowledgments. This research was supported by the Technology Agency of the Czech Republic, project No. TA01030476. Access to computing and storage facilities owned by parties and projects contributing to the National Grid Infrastructure MetaCentrum, provided under the programme *"Projects of Large Infrastructure for Research, Development, and Innovations"* (LM2010005), is greatly appreciated.

References

1. Hunt, A.J., Black, A.W.: Unit selection in a concatenative speech synthesis system using a large speech database. In: Proceedings of ICASSP 1996, pp. 373–376 (1996)
2. Zen, H., Tokuda, K., Black, A.W.: Statistical parametric speech synthesis. Speech Communication **51**(11), 1039–1064 (2009)
3. Ross, K., Ostendorf, M.: Prediction of abstract prosodic labels for speech synthesis. Computer Speech and Language **10**, 155–185 (1996)
4. Wightman, C., Ostendorf, M.: Automatic labeling of prosodic patterns. IEEE Transactions on Speech and Audio Processing **2**, 469–481 (1994)
5. Toledano, D., Gómez, L., Grande, L.: Automatic phonetic segmentation. IEEE Transactions on Speech and Audio Processing **11**, 617–625 (2003)
6. Romportl, J., Matoušek, J., Tihelka, D.: Advanced prosody modelling. In: Sojka, P., Kopeček, I., Pala, K. (eds.) TSD 2004. LNCS (LNAI), vol. 3206, pp. 441–447. Springer, Heidelberg (2004)
7. Hanzlíček, Z., Grůber, M.: Initial experiments on automatic correction of prosodic annotation of large speech corpora. In: Sojka, P., Horák, A., Kopeček, I., Pala, K. (eds.) TSD 2014. LNCS, vol. 8655, pp. 481–488. Springer, Heidelberg (2014)
8. Matoušek, J., Tihelka, D., Romportl, J.: Current state of Czech text-to-speech system ARTIC. In: Sojka, P., Kopeček, I., Pala, K. (eds.) TSD 2006. LNCS (LNAI), vol. 4188, pp. 439–446. Springer, Heidelberg (2006)
9. Tihelka, D., Matoušek, J.: Unit selection and its relation to symbolic prosody: a new approach. In: Proceedings of Interspeech 2006, pp. 2042–2045 (2006)
10. Hanzlíček, Z.: Czech HMM-based speech synthesis. In: Sojka, P., Horák, A., Kopeček, I., Pala, K. (eds.) TSD 2010. LNCS, vol. 6231, pp. 291–298. Springer, Heidelberg (2010)
11. Kawahara, H., Masuda-Katsuse, I., de Cheveigne, A.: Restructuring speech representations using a pitch-adaptive time-frequency smoothing and an instantaneous-frequency-based F0 extraction: Possible role of a repetitive structure in sounds. Speech Communication **27**, 187–207 (1999)
12. Matoušek, J., Tihelka, D., Romportl, J.: Building of a speech corpus optimised for unit selection TTS synthesis. In: Proceedings of LREC 2008 (2008)
13. Matoušek, J., Romportl, J.: Recording and annotation of speech corpus for Czech unit selection speech synthesis. In: Matoušek, V., Mautner, P. (eds.) TSD 2007. LNCS (LNAI), vol. 4629, pp. 326–333. Springer, Heidelberg (2007)

Adding Multilingual Terminological Resources to Parallel Corpora for Statistical Machine Translation Deteriorates System Performance: A Negative Result from Experiments in the Biomedical Domain

Johannes Hellrich[✉] and Udo Hahn

Jena University Language and Information Engineering (JULIE)
Lab Friedrich-Schiller-Universität Jena, Jena, Germany
{Johannes.Hellrich,Udo.Hahn}@uni-jena.de
http://www.julielab.de

Abstract. Unlike many other domains, biomedicine not only provides a wide range of parallel text corpora to train statistical machine translation (SMT) systems on, but also offers substantial amounts of 'parallel lexicons' in the form of multilingual terminologies. We included these lexical repositories, together with common parallel text corpora, into a MOSES-based SMT system and three commercial systems and performed experiments on four language pairs, three text genres and several corpus sizes to measure the effects of adding the lexical knowledge sources. Much to our surprise, the SMT systems additionally equipped with 'parallel lexicons' underperformed in comparison with those systems trained on parallel text corpora only. This effect could consistently be shown for all systems by BLEU scores, as well as assessments from human judges.

Keywords: Machine translation · Biomedicine · Terminologies

1 Introduction

The challenges of multi-lingualism in modern societies are manifold. Special needs for (human or machine) translation services derive, e.g., from international tourism, job mobility or business communication, while completely different demands arise from growing streams of immigrants and fugitives. Medical applications are natural targets for all sorts of translation activities, including, e.g., national public health services, international epidemia control or individual patient-doctor interactions. Hence, biomedical applications of machine translations are high on the agenda of desired outcomes of NLP and HLT.

The biomedical domain seems to be particularly suited for data-intensive NLP because of its richness of resources. On the one hand, it offers a plethora of comprehensive parallel text corpora incorporating various text genres, which are well suited for training statistical machine translation (SMT) systems. On the other hand, a substantial amount of large-scale multilingual terminologies is available, which can be considered as 'parallel lexicons' and, thus, complement the diverse

© Springer International Publishing Switzerland 2015
P. Král and V. Matoušek (Eds.): TSD 2015, LNAI 9302, pp. 506–514, 2015.
DOI: 10.1007/978-3-319-24033-6_57

parallel text corpus resources. Many of these terminologies are combined in the *Unified Medical Language System* Metathesaurus (UMLS) [1].[1] Since the integration of terminologies was already shown to increase translation quality in other, non-biomedical domains (cf. e.g., [2]) we decided to utilize the UMLS as a source for training (S)MT systems as well—prior studies had shown both negative and positive effects [3–5].

In this study, we investigate the impact of the UMLS on the translation performance of the MOSES open source SMT system [6] and complement these experiments with three commercial systems building on comparable methodological premises, namely Google's TRANSLATE,[2] Microsoft's TRANSLATOR Hub[3] and another anonymous one (we granted anonymity for all collaborating companies, if they desired so after the conduct of the experiments). We used the commercial systems' interfaces and a naïve solution for MOSES disregarding more complex configurations (as, e.g., proposed by [7,8]) to focus on the overall effect of the UMLS on different systems and for different genres, in an easy to implement and reproduceable way. Our evaluation focuses on the n-gram-based BLEU metric [9], a de-facto standard for automatically assessing SMT performance. Manual evaluations were used for some specific samples to compensate for some well-known weaknesses of BLEU and gain additional evidence for our results.

The potential of MT systems for biomedical applications has already been demonstrated by [10] who compared the performance of MOSES with Google's TRANSLATE for translating titles of scientific papers indexed in MEDLINE, without any extra efforts, e.g. the inclusion of terminological resources. After additionally varying text genres, parallel corpora, SMT systems and language pairs, we have consistent experimental evidence that SMT systems trained on parallel text corpora and additionally equipped with substantial amounts of 'parallel lexicons' in the form of multilingual terminologies underperform, most often dramatically, in comparison to systems lacking such additional lexicalized knowledge input. Hence, training SMT systems merely on parallel text corpora seems to be entirely sufficient—in the biomedical domain, at least.

2 Experimental Set-Up

2.1 Parallel Corpora and Biomedical Terminologies

Both the corpora and the terminologies were taken from the cleansed (e.g. cycle-free) source data provided for the CLEF-ER challenge[4] [11]. The corpora we used contained three different text genres: MEDLINE titles from scientific journal papers,[5] drug labels with consumer use information from the European Medicines Agency (EMEA) [12] and biomedical PATENT claims[6] from the

---

[1] http://www.nlm.nih.gov/research/umls/
[2] https://cloud.google.com/translate/docs
[3] https://hub.microsofttranslator.com/
[4] https://sites.google.com/site/mantraeu/access-content
[5] http://mbr.nlm.nih.gov/Download/
[6] Identified by their International Patent Classification code starting with 'A61K'.

European Patent Office.[7] The cleansed UMLS contains language-independent concepts, their language-specific labels (*preferred synonym*) and intralingual (within one language) as well as interlingual (crossing languages) *synonyms* for each concept. The UMLS is a hierarchical thesaurus (mainly used for information retrieval, clinical coding and accountancy) rather than a flat dictionary (as needed for MT). Thus, the entire taxonomic information was removed for our experiments, and all *synonyms* in the source language were mapped to the *preferred synonym* in the target language, for each concept (cf. also [5]), resulting in a bilingual translation dictionary without any further information. Both the UMLS and the derived dictionary contain only basic word forms and some plurals, as well as syntactically inverted forms suited for indexing (such as 'cancer, breast' instead of 'breast cancer'). Table 1 gives an overview of dictionary entries and parallel text units (i.e. titles, text fragments, claims—all roughly equivalent to sentences) in the aforementioned corpora. Vastly more translations are provided by the dictionaries for translations from English, as more English synonyms are listed in the UMLS. For tuning and evaluation 5,000 units each per language pair and corpus were set aside, the rest was used for training.

Table 1. Number of MEDLINE titles, EMEA text fragments and PATENT claims in the parallel corpora as well as translations provided by the flattened UMLS for each language pair. PATENTs are not available for all languages as indicated by dashes.

Language Pairs	MEDLINE	EMEA	PATENT	Translations into English	Translations from English
German-English	719,232	140,552	154,836	132,183	575,661
French-English	572,176	140,552	154,836	152,720	572,556
Spanish-English	247,655	140,552	—	784,972	1,456,463
Dutch-English	54,483	140,552	—	127,904	458,743

2.2 Configurations of MT Systems

We performed our experiments using a typical MOSES setup, by combining it with SRILM [13], MERT, GIZA++ [14], and the scripts of MOSES' Experiment Management System (EMS). We tested two different configurations of MOSES: a biomedical baseline, for which we trained translational and language models on parallel document corpora *as is*, and a UMLS-enhanced version, for which the translational (but not the language) model was trained on a combination of the parallel corpus and the flattened UMLS. Both configurations were tuned and evaluated with unmodified sentences from the corpora introduced in Section 2.1.

Google TRANSLATE offers no option to customize their models. So, we uploaded evaluation items to their API and calculated the BLEU score for the resulting translations with the scripts contained in the MOSES EMS. Both Microsoft's TRANSLATOR Hub and the web interface of the anonymous company's system allow to train models and include an evaluation suite used during

---

[7] http://www.epo.org/

our experiments. The anonymous system allows for the inclusion of both in-domain and background training material, the latter was applied to incorporate the terminology-derived dictionaries. The option to include glossary files was discarded on direction of the anonymous system's staff, as it is not designed for large dictionaries. In contrast, Microsoft's system offers only one way to integrate terminologies, namely by uploading dictionaries as Excel sheets, a mechanism also more apt for small dictionaries. Microsoft offers not only a training option, but also general language translation baselines which we included in our comparison.

Caveat for Patent Texts. Sentences from the PATENT corpus can be very long, both before and especially after tokenization (MOSES' tokenizer is rather aggressive in its treatment of the chemical names frequent in this subdomain). Test and tuning sentences for MOSES were thus tokenized and cropped with a script. PATENT sentences were especially problematic for the anonymous company's MT system, a working configuration could not be achieved due to lack of command over their tokenization process.

3   Evaluation

To assess the translation quality of the four systems and the effects of terminology integration we performed an automatic evaluation with BLEU, using 5,000 test sentences from the corpora introduced in Section 2.1. To better understand the negative effects of terminology integration that became evident in the course of these experiments we investigated the two MOSES configurations described in Section 2.2 in more detail, including further automatic evaluation with BLEU, manual translation quality judgments and a qualitative error analysis for some MOSES translations of MEDLINE titles.

3.1   Automatic Evaluation—System Comparison

Table 2 (next page) depicts the results for comparing MOSES and the three commercial systems from GOOGLE, MICROSOFT and the anonymous company. Training systems on texts from *any* biomedical genre notably improved translation quality over general language-trained systems. Incorporating terminological knowledge had, however, no positive effect on translation quality and, especially for MICROSOFT's system, a negative one. No system trained on in-domain texts is generally superior, each of them outperforms all the others on one of the three corpora: the anonymous system is strongest on

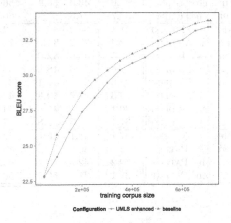

Fig. 1. BLEU scores for baseline and UMLS-enhanced MOSES systems (trained on MEDLINE, translation from English into German) over varying training corpus sizes.

EMEA, MICROSOFT's TRANSLATOR on MEDLINE and MOSES on PATENTs. Translations into English achieved higher scores than translations from English, as was to be expected due to the higher morphological complexity in the non-English languages under scrutiny.

Further automatic evaluation of different configurations of MOSES revealed that the baseline consistently outperformed the UMLS-enhanced version on all languages and corpora, even if training corpora were drastically shrunk (see Figure 1). This is puzzling as small parallel corpora alone should increase the number of *out-of-vocabulary* (OOV) words (cf. Section 3.3), the amount of which which could be reduced by terminology integration. Since language models were trained on the parallel text corpora alone, any negative influence of the terminologies via this route can be ruled out.

Table 2. BLEU scores achieved by MT systems by corpus and language pair, best system in **bold**. Multiple bold systems were not significantly different ($p \leq 0.05$) during paired bootstrap resampling [15]. Microsoft is abbreviated as MS, the anonymous system as ANON. Systems marked as 'general' use general language models not trained on texts from the corpus they are evaluated on, while those marked as 'bio' were trained on the respective parallel corpus (baseline). Systems marked as 'bio & UMLS' additionally utilized the UMLS-derived dictionary as described in Section 2. Dashes indicate missing translations (cf. Section 2.1)

From	To	Corpus	GOOGLE general	MS general	MS bio	MS bio & UMLS	MOSES bio	MOSES bio & UMLS	ANON bio	ANON bio & UMLS
de	en	EMEA	46.0	39.5	53.2	48.6	54.8	55.0	**57.7**	**57.7**
en	de	EMEA	33.7	32.7	48.0	38.0	48.4	48.1	51.0	**51.2**
en	es	EMEA	49.9	43.0	57.2	44.1	58.0	57.4	**61.5**	**61.4**
en	fr	EMEA	38.4	37.0	48.5	39.0	55.7	55.4	**58.9**	**58.7**
en	nl	EMEA	48.3	37.6	53.9	44.4	53.9	53.3	**56.7**	56.6
es	en	EMEA	49.7	46.0	60.1	52.0	61.7	61.2	**66.3**	65.9
fr	en	EMEA	47.0	43.0	56.9	56.7	58.5	58.5	**63.3**	**63.4**
nl	en	EMEA	57.7	47.6	63.5	60.7	61.6	61.6	**64.7**	**64.6**
de	en	MEDLINE	31.8	35.0	**47.4**	42.6	38.4	38.4	43.0	42.9
. en	de	MEDLINE	18.4	25.2	**39.9**	28.7	34.0	33.4	38.1	38.0
en	es	MEDLINE	39.5	45.0	**52.2**	38.3	44.0	42.6	49.4	48.8
en	fr	MEDLINE	29.0	37.6	**48.9**	36.7	43.5	42.9	47.2	47.5
en	nl	MEDLINE	30.6	33.4	**47.1**	34.7	39.1	39.0	44.5	44.5
es	en	MEDLINE	38.3	42.2	**49.8**	42.7	43.8	43.1	48.7	48.3
fr	en	MEDLINE	30.1	35.1	45.6	44.8	41.2	41.0	**45.7**	45.6
nl	en	MEDLINE	37.1	42.3	**51.8**	47.9	41.2	41.5	46.8	46.3
de	en	PATENT	37.1	38.1	55.1	52.5	**66.3**	66.0	—	—
en	de	PATENT	39.3	30.0	47.6	39.0	**60.0**	59.3	—	—
en	fr	PATENT	30.2	42.5	53.2	45.6	**57.0**	56.6	—	—
fr	en	PATENT	36.1	41.8	56.7	56.5	**59.6**	59.5	—	—

3.2 Manual Analysis

We also collected human assessments from three subjects for 'fluency' and 'accuracy' (cf. [16]), evaluating 100 MEDLINE titles translated between English and

Table 3. Averaged human assessment of fluency and adequacy of the translations provided by the baseline and the UMLS-enhanced MOSES system. Measurements were obtained from a random sample of 100 test sentences of a MEDLINE-trained system translating between English and German, trained on 709k sentences.

System	English to German		German to English	
	Fluency	Adequacy	Fluency	Adequacy
baseline	**4.13**	**4.50**	**4.18**	4.42
UMLS-enhanced	3.96	4.37	4.11	**4.43**

German (a subset of the 5,000 items used for automatic evaluation). Table 3 depicts the outcome of these judgments, again confirming the superiority of the baseline system without terminology integration. All three judges were bilingual and rated only translations in their native language. Two were native speakers of German and graduate students in the biomedical domain (judgments averaged; $\kappa = 0.23$, seemingly low, yet consistent with the literature), the third was a native speaker of English.

Upon further inspection of these 100 sentences we found lots of minimal differences affecting mainly fluency, especially regarding inflection—often both the baseline and the UMLS-enhanced MOSES configuration were equally unable to cope with German cases (e.g. *"präoperative Risiko"* instead of *"präoperatives Risiko"* for *"Preoperative risk"*). Major differences caused by including the UMLS are not always problematic for a human reader, e.g. *"Erkrankungen des Ösophagus"* ['esophageal diseases']. Yet in some cases quite bizarre translations were produced, e.g. *"pathway"* (as in 'care pathway') being translated as *"Sehbahnenverletzung"* ['optic pathway injury'] due to the UMLS entry *'Injury of optic nerve and pathways'*.

3.3 Out-of-Vocabulary Analysis

When MT systems encounter words unseen before (i.e. not contained in the parallel corpora) no model can be applied to them which, in the worst case, leads to non-translations, i.e. the unknown word from the source text appears unchanged, as the original source text item, in the target text—the OOV problem of MT. While some OOV words may be unproblematic for adequate translations (e.g. proper names) most of them will lead to low-quality translations, as is reflected in both human and automatic translations (e.g. missing n-grams for BLEU). Hence, adding a supplementary vocabulary resource, in our case the UMLS, should increase the number of words known to the MT system and thus reduce the number of OOV words.

Whereas incorporating terminological resources caused mostly negative effects for translation quality in our experiments, it significantly ($p \leq 0.01$ for paired t-test) reduced the proportion of OOV tokens and their associated types. For example a system trained on the full MEDLINE corpus for translating English to German could not translate 7.1% of the types in the test set, whereas one trained on both corpus and terminology missed only 6.5%; other corpora and languages reveal similar results.

We manually inspected the OOV types of the MEDLINE English-to-German example at the maximal corpus size to better understand the nature of the OOV errors, using the following categories (see, e.g. [17]): Missegmented words, misspelled words, valid words (missing from training material, both contained or

Table 4. Absolute frequencies and examples for five categories of OOV failure types: missegmented and misspelled words, valid English words (not) contained in the UMLS, and non-translatable words.

OOV word type	Frequency	Example
Missegmented	33	edema--which
Misspelled	2	tecnnology
Valid, not in the UMLS	43	periodontometry
Valid, in the UMLS	5	willows
Non-translatable	17	Langenstein

not contained in the English UMLS), and non-translatable words (e.g. proper names). Table 4 lists the absolute frequencies and some examples based on 100 randomly selected OOV types from the test set of the aforementioned corpus. Missegmentation is mainly caused by MOSES' tokenizer not separating on '--' (e.g. *'suicide--review'*), an issue that could easily be resolved by choosing a domain-specific tokenizer. Yet this phenomenon is, just like misspelled words, of little relevance for investigating terminology integration. Surprisingly, many of the valid words are medical in nature, e.g. the tooth mobility test *'periodontometry'*, yet neither contained in the training material, nor in the English part of the UMLS. Non-translatable words are mostly proper names and numbers which have little impact for translating between languages using the same alphabet.

4   Conclusions

We assessed the performance of the open source MOSES system and three commercial systems (Google's TRANSLATE, Microsoft's TRANSLATOR Hub and one from an anonymous company) for translating texts in the biomedical domain, both with and without adding terminological knowledge from the UMLS. Experiments were performed for four language pairs (i.e translating between English as the anchor language and French, German, Spanish as well as Dutch) and three biomedical genres (MEDLINE article titles, EMEA drug leaflets and biomedical patents).

On all genres and languages, the in-domain trained systems outperformed those for general language use. As far as the in-domain trained systems are concerned, none is a clear winner genre-wise: the anonymous system is strongest on EMEA leaflet sentences, MICROSOFT on MEDLINE titles and MOSES on PATENTs. The integration of the UMLS had no positive and often even negative effects on translation quality. While we could demonstrate a positive effect of the integration on the number of words covered by an MT system (cf. Section 3.3) the negative effects caused by the introduction of unwarranted translations seem to be stronger. In order to avoid them one might focus on OOV words only when consulting the UMLS (following [7]) or additionally cleanse the UMLS, e.g. by removing ill-suited entries (inverted index terms, cf. Section 2.1) or add frequency information from other sources.

Acknowledgments. This work was partially funded by the EU Support Action grant 296410 under the 7th EU Framework Programme within the "Intelligent Content and Semantics" programme (FP7-ICT-2011-SME-DCL). We also want to thank MICROSOFT and the anonymous company for granting us access to their SMT systems infrastructure and the support we received.

References

1. Bodenreider, O.: The Unified Medical Language System (UMLS): integrating biomedicalterminology. Nucleic Acids Research **32**(Database issue), D267–D270 (2004)
2. Arcan, M., Federmann, C., Buitelaar, P.: Experiments with term translation. In: COLING 2012 - Proceedings of the 24th International Conference on Computational Linguistics: Technical Papers. Mumbai, India, 8–15 December 2012, pp. 67–82 (2012)
3. Eck, M., Vogel, S., Waibel, A.H.: Improving statistical machine translation in the medical domain using the unified medical language system. In: Proceedings of the 20th International Conference on Computational Linguistics, COLING 2004, Geneva, Switzerland, August 23–27, 2004, pp. 792–798 (2004)
4. Jimeno Yepes, A., Névéol, A.: Effect of additional in-domain parallel corpora in biomedical statistical machine translation. In: Proceedings of the 4th International Workshop on Health Document Text Mining and Information Analysis with the Focus of Cross-Language Evaluation (Louhi 2013), February 11–12 , 2013. NICTA, Sydney (2013)
5. Pecina, P., Dušek, O., Goeuriot, L., Hajič, J., Hlaváčová, J., Jones, G.J.F., Kelly, L., Leveling, J., Mareček, D., Novák, M., Popel, M., Rosa, R., Tamchyna, A., Urešová, Z.: Adaptation of machine translation for multilingual information retrieval in the medical domain. Artificial Intelligence in Medicine **61**(3), 165–185 (2014)
6. Koehn, P., Hoang, H., Birch, A., Callison-Burch, C., Federico, M., Bertoldi, N., Cowan, B., Shen, W., Moran, C., Zens, R., Dyer, C., Bojar, O., Constantin, A., Herbst, E.: Moses: open source toolkit for statistical machine translation. In: Proceedings of the Interactive Poster and Demonstration Sessions @ ACL 2007, Prague, Czech Republic, June 25–27, 2007, pp. 177–180 (2007)
7. Daumé, H., Jagarlamudi, J.: Domain adaptation for machine translation by mining unseen words. In: ACL-HLT 2011 - Proceedings of the 49th Annual Meeting of the Association for Computational Linguistics: Human Language Technologies. Volume 2: Short Papers, Portland, OR, USA, 19–24 June, 2011, vol. 2, pp. 407–412 (2011)
8. Huang, C.C., Yen, H.C., Yang, P.C., Huang, S.T., Chang, J.S.: Using sublexical translations to handle the OOV problem in machine translation. ACM Transactions on Asian Language Information Processing (TALIP) **10**(3), #16 (2011)
9. Papineni, K., Roukos, S., Ward, T., Zhu, W.J.: Bleu: a method for automatic evaluation of machine translation. In: Proceedings of the 40th Annual Meeting of Association for Computational Linguistics, ACL 2002, Philadelphia, PA, USA, July 6–12, 2002, pp. 311–318 (2002)
10. Wu, C., Xia, F., Deléger, L., Solti, I.: Statistical machine translation for biomedical text: are we there yet? In: Proceedings of the Annual Symposium of the American Medical Informatics Association, AMIA 2011, Washington, D.C., USA, October 22–26, 2011, pp. 1290–1299 (2011)

11. Rebholz-Schuhmann, D., Clematide, S., Rinaldi, F., Kafkas, S., van Mulligen, E.M., Bui, C., Hellrich, J., Lewin, I., Milward, D., Poprat, M., Jimeno-Yepes, A., Hahn, U., Kors, J.A.: Entity recognition in parallel multi-lingual biomedical corpora: The CLEF-ER laboratory overview. In: Forner, P., Müller, H., Paredes, R., Rosso, P., Stein, B. (eds.) CLEF 2013. LNCS, vol. 8138, pp. 353–367. Springer, Heidelberg (2013)

12. Tiedemann, J.: News from opus: a collection of multilingual parallel corpora with tools and interfaces. In: Nicolov, N., Angelova, G., Mitkov, R. (eds.) RANLP 2009 - Recent Advances in Natural Language Processing, pp. 237–248. John Benjamins, Amsterdam (2009)

13. Stolcke, A.: Srlim: an extensible language modeling toolkit. In: ICSLP2002/INTERSPEECH 2002 - Proceedings of the 7th International Conference on Spoken Language Processing, Denver, CO, USA, September 16–20, 2002, pp. 901–904 (2002)

14. Och, F.J., Ney, H.: A systematic comparison of various statistical alignment models. Computational Linguistics **29**(1), 19–51 (2003)

15. Koehn, P.: Statistical significance tests for machine translation evaluation. In: EMNLP 2004 - Proceedings of the 2004 Conference on Empirical Methods in Natural Language Processing. A meeting of SIGDAT, a Special Interest Group of the ACL Held in Conjunction with ACL 2004, Barcelona, Spain, 25–26 July 2004, pp. 388–395 (2004)

16. Callison-Burch, C., Fordyce, C., Koehn, P., Monz, C., Schroeder, J.: (Meta-)evaluation of machine translation. In: Proceedings of the 2nd Workshop on Statistical Machine Translation, StatMT 2007, Prague, Czech Republic, June 23, 2007, pp. 136–158 (2007)

17. Banerjee, P., Naskar, S.K., Roturier, J., Way, A., van Genabith, J.: Domain adaptation in SMT of user-generated forum content guided by OOV word reduction: normalization and/or supplementary data? In: Proceedings of the 16th EAMT Conference, EAMT 2012, Trento, Italy, 28–30 May 2012, pp. 169–176 (2012)

Derivancze — *Derivational Analyzer of Czech*

Karel Pala and Pavel Šmerk[✉]

Faculty of Informatics, Masaryk University,
Botanická 68a, CZ-60200 Brno, Czech Republic
{pala,smerk}@fi.muni.cz

Abstract. The paper describes a new tool Derivancze, which provides an information on derivational relations between Czech words. After a summary of linguistic descriptions of Czech derivation we present a structure of our data and types of derivational relations we use. We compare our approach and results with Czech lexical network DeriNet, in particular, we discuss many differences between the two approaches. Our tool presently works with Czech data only, but the solution is general and can be used also for other languages.

Keywords: Derivational morphology · Derivational analysis · Semantics of the derivational relations

1 Introduction

Standard morphological analyzers typically provide for an input word its corresponding basic form but as a rule they do not offer (or in a limited way only) information about derivational relations between words such as, for example, in Czech *otec – otcův* (*father – father's*), *řezat – řezání* (*cut – cutting*), *učit – učitel* (*teach – teacher*), etc. This information can be very useful for text indexation in searching or in the course of the syntactic analysis of the natural language and also for other applications.

In the highly inflectional languages like Czech derivational relations (further D-relations) represent a system of both formal and semantic relations that definitely reflects cognitive structures related to what may be characterized as a language ontology. For language users derivational affixes function as formal means by which they express semantic relations necessary for using language as a vehicle of communication. The affixes denote several sorts of meanings which we will try to classify in this paper. We will deal here primarily with Czech language but presented results can be applied with the necessary modifications to all Slavonic languages, see e. g. [1] or [2].

This work was supported by the Ministry of Education of CR within the Lindat Clarin Center.

P. Král and V. Matoušek (Eds.): TSD 2015, LNAI 9302, pp. 515–523, 2015.
DOI: 10.1007/978-3-319-24033-6_58

2   Motivation

The first important reason for doing all this is a belief that D-relations and derivational nests created by them reflect basic cognitive structures existing in natural language. These cognitive structures can be partly traced down in the standard Czech grammars [3] where they can be found under the term of the onomasiological categories. The semantics of the D-relations will be in the focus of our attention in the paper.

The second good reason for paying attention to the Czech derivational morphology is a need to describe the derivational relations as formally as possible and on this ground to develop software tools allowing to handle automatically D-relations between lexemes in Czech. The obtained results can be useful for various applications such as information extraction, indexing for searching engines, textual entailment, machine translation, etc.

The third inspiring reason is to confront the traditional description of the Czech derivational morphology as it can be found in the standard Czech grammars with its formal counterpart necessary for a computer treatment.

The last reason is to present a software tool, derivational analyzer called Derivancze, and to compare it partially with an existing similar derivational tool DeriNet [4].

3   Related Work

There is a well developed theoretical description of Czech derivational morphology by Dokulil [3,5]. It has served as an excellent starting point for a further work in this area (see, for example [6,7]. Dokulil in his explanation of the D-relations intertwines both semantic aspects of the Czech word derivation and its formal aspects in an interesting but also a complicated way.

It has to be remarked that Dokulil's theory and also other derivational descriptions in standard Czech grammars adopted from it are based on the partial data containing just typical well selected examples. The situation becomes different now when we have access to almost all relevant Czech data (relatively complete lists of affixes, stems, word lists obtained from corpora) and can process them with the appropriate software tools.

In particular, Dokulil works with what he calls onomasiological categories: modifications (smaller change of the meaning within the same POS: *učitel – učitelka*; *teacher*$_{MASC}$ – *teacher*$_{FEM}$), transpositions (change of POS without change of the meaning: *dobrý – dobře*; *good – well, padat – pád*; *to fall – a fall*), mutations (with a substantial change of the meaning: *slepý – slepec*; *blind – blind man*)) and reproductions (*bác – bácnout (squab – to do squab)* which include onomatopoic derivations) — within this framework he treats most of the derivational processes in Czech. It should be remarked that Dokulil's treatment of the D-relations is rather extensive, it takes 259 pages in [3], so it is not possible to mention all the relevant points in this paper. Thus here we are trying to follow just the main and most transparent derivational processes in Czech.

We have to mention the attempts to handle Czech derivational morphology in a more formal way which have appeared recently. One of them is a tool developed by Ševčíková and Žabokrtský (2014) called DeriNet [4]. Another tool is a modified version of the Derivational Ajka developed at NLP Centre FI MU (Sedláček et al., 2005, it was not published and exists only as a computer program).

Apart from them there are two other tools. The first one is Deriv [8][1] developed at the NLP Centre FI MU and the second one is a tool called Morfio [9][2] built in ÚČNK FF UK. It has to be remarked, however, that both Deriv and Morfio are different from DeriNet and Derivancze. Particularly, Deriv is a web tool for exploring derivational relations among word forms from a morphological analyser of Czech using regular expressions. The results are linked to Czech corpora (CzTenTen, SYN2000) in order to make its manual post-editing easier. Morfio is a web interface as well allowing users to search the corpus by series of parallel queries which specify a chosen derivational model. It also analyses obtained results for the morphological productivity of affixes and estimates the completeness of the derivational model.

4 Design of Derivancze: in Constrast with DeriNet

We take advantage of the fact that there are publicly available data of the DeriNet network together with detailed description of its internals. It allows us to describe our decisions on the design of the Derivancze data by means of comparison of our approach with the approach of the DeriNet authors. The substantial differences can be drawn up in three parts.

4.1 Semantically Labelled Relations Instead of Purely Derivational Relations

The most prominent difference between the two approaches consists in our effort to classify somehow the relations between words. We work with semantically labelled relations whereas in DeriNet one finds just simple derivational relations without any explicit labels, at most they vary in their members' POS. In our view, this can be sufficient e.g. for relation adjective–adverb mentioned as a potential practical application of the network (subsection 5.1 of [4]) but for more sophisticated applications (text generation, condensation, paraphrasing or textual entailment) the more detailed information on the type of the link will be needed. This seems to be confirmed by the DErivBase [10] derivational lexicon, which is a German analog of the DeriNet, as its authors expect that for the derivationally close words also their semantic proximity will have to be captured because all existing applications assume strong correlation between derivational and sematic proximity.

---

[1] http://deb.fi.muni.cz/deriv/
[2] https://morfio.korpus.cz/

Therefore, in our data we aim at the D-relations for which a regular and transparent semantics can be found. So we are not interested in base words for *komunismus*, *rusismus* and *revmatismus*, i. e. words *komuna*, *Rus* and *revma*[3] because while from the formal point of view the derivational process is fully regular, the semantic relations inside the three pairs differ from one another. Similarly, for particular D-relations, we are not interested in word pairs which do not correspond with the semantics of the given D-relation. For example, all three contemporary Czech gramars ([3,6,7] mention *mdloba* (*faints*) as an example of a quality/property name derived by means of suffix *-oba*, but the relation to the base adjective *mdlý* (*bland*) is only formal and a "regular" *mdlost* (*blandness*) is semantically much more proper (unlike e. g. *chudoba* > *chudost* for *chudý*, similarly to English *poverty* > *poorness* for *poor*).

Moreover, in some cases we have to abandon purely formal approach and for the sake of completeness and consistency ignore the direction of the derivational process. For example, nouns describing actions or states denoted by verbs are regularly derived by means of suffix *-ní*: *pracovat–pracování* (*to work–a work / working*). But in some cases also other words can be used, e. g. *práce* in this case. From the formal point of view, *pracovat* is derived from *práce*, not conversely, but the information on direction is not interesting for real world applications: they need to give a verb and get the corresponding noun. A similar example are inhabitant names. Most of them are derived from the name of an area, but there are also many exceptions: *Vietnam–Vietnamec*, *Polsko–Polák* (*Poland*), *Rusko–Rus* (*Russia*). Clearly, the name of inhabitant is derived from the name of the area in the first case, both names are derived from some common base in the second case and the name of the area is derived from the name of the inhabitant (nation) in the third case. But from the practical point of view this is not relevant, a potential application will ask for an inhabitant name corresponding to the given area.

Even further, in some cases something similar to suppletion in inflectional morphology appears. As well as plural forms of almost all Czech words are created regularly and only a few exceptions have irregular (*přítel–přátelé*; *friend–friends*) or suppletive (*člověk–lidé*; *man–people*) forms, we can see that, for example, almost all masculine → feminine changes are expressed by a respective suffix (*učitel–učitelka* above, *dělník–dělnice*; *worker MASC–FEM*) and only a few exceptions display some irregularities (*tchán–tchyně*; *father-in-law–mother-in-law*) — or they are entirely "suppletive" (*syn–dcera*; *son–daughter*). But then, because it is hard to find any reasonable argument why the application should get an answer if it requests for a feminine form of "regular" nouns *vnuk* (*grandson*) or *medvěd* (*bear*) (in Czech *vnuč-ka* and *medvěd-ice*) and should not in case of "suppletive" nouns *syn* (*son*) or *kůň* (*horse*) (in Czech *dcera* and *kobyla*), aiming at completness and consistency, we have to admit that even word pairs like *syn–dcera* should be counted as derivational, albeit "suppletive".

On the whole, the relations in our approach are based on the formal derivational relations, but as the semantics is what matters after all, they do not fully agree with the D-relations as they are treated in the standard Czech grammars

---

[3] *communism, russism, rheumatism, commune, Russian,* and *rheuma*

— these, as we hinted above, are not suitable for use in practical applications. It has to be remarked that we are trying to follow just the main and somehow "fuzzy" tendencies as there are not any objective criteria for decision what is semantically transparent and what is disguised (and to what extent).

4.2 More than One Base Word and Semantic Equivalence

In the DeriNet network, every word is allowed to have at most one base word, but this constraint seems to be too restrictive in some cases. For instance, *virový* (*viral*) is a relational adjective derived either from *vir* or *virus* (two shapes of the same foreign word, *virus*). Another nice example offers the DeriNet network itself: the base word of *antikomunista* is *antikomunismus* (*anticommunist, anticommunism*), but the base word of *antikomunistův* is *komunistův* (*anticommunist's, communist's*) which is clearly inconsistent. But even if the authors would prefer suffixation over prefixation (or vice versa), there would be no obvious reason for such decision. In our view, much better solution is to admit that *antikomunistův* can be derived both from *antikomunista* (with suffix *-ův*) and *komunistův* (with prefix *anti-*).

From the previous explanation it immediately follows that we also need some concept of semantic equivalence, at least to be able to distinguish cases like *virový*, where the two possible base words are semantically equal, from cases like *antikomunistův*, where they are different. This equivalence is going to cover also orthographical variants (*socialismus/socializmus*, but only *socialistický*; *socialism–socialist(ic)*) and synonymic suffixes (*normalita/normálnost*, but only *normalizace* or *normální*; *normality, normalization, normal*).

4.3 Overgeneration Followed by Filtering through Language Corpora

For an initialization of the DeriNet network, only lemmata with SYN corpus [11] frequency ≥ 2 were used. In Derivancze, we prefer to acquire as many "correctly" derived pairs as possible[4] and then add frequencies from corpus or corpora. Doing this we should be able not only to obtain the same results as DeriNet, but also to make a distinction between impossible and infrequent. Moreover, we can offer this information also to the user or application to let them decide between synonymic suffixes. For example, name of the quality/property expressed by an adjective *hluchý* (*deaf*) can be both *hluchost* and *hluchota* — althought the suffix *-ost* is much more productive in general, *hluchota* is around two orders of magnitude more frequent than *hluchost*.

---

[4] We are not able to clarify what exactly means this "correctly" as it always be questionable in cases of rare word forms. It should be noted there, that contemporary Czech grammars cannot be trusted concerning statements what is possible. For example for passive verbal adjectives the grammars mention only transitive verbs or intransitive verbs with indirect object, but one can recall, e. g. *padaná jablka* (literally *fallen apples*), where the verb *padat* has no object at all. That is why we prefer to try to generate such forms for all or almost all verbs, even where they may seem "incorrect" (because not used).

5   Results

The Derivancze itself is implemented in the same way as the morphological analyzer majka [12], i. e. the data are represented as a simple list of query:response,
which is converted to a minimal finite state automaton. Derivancze does not do
any real analysis, but it only looks up all possible responses (derived forms and
D-relations) for a given query (input word). It means that all possible analyses
for all known inputs are precomputed in a compilation phase, thus the tool itself
remains very simple and fast.

The current version of data comprises the following D-relations[5]:

- k1verb, k2pas, k2proc, k2rakt, k2rpas, and k2ucel from verbs, where:
  - k1verb derives nouns describing process, action or state denoted by the
    verb (*kropit–kropení*; *sprinkle–sprinkling*),
  - k2pas and k2rpas are passive participle and past passive adjectival par-
    ticiple, i. e. two forms of adjectives which describe the patient or object
    of the action (*kropit–kropen/kropený*; *sprinkle–sprinkled*),
  - k2proc derives present active adjectival participles, i. e. adjectives describ-
    ing a subject doing the action (*kropit–kropící*; *sprinkle–sprinkling* (man)),
  - k2rakt are past active adjectival participles, i. e. adjectives describing
    subjects which have completed the action (*pokropit–pokropivší*; *sprinkle–
    who has springled st.*), and
  - k2ucel derives adjectives which describe an object used for the action
    (*kropit–kropicí*; *sprinkle–sprinkling* (machine)),
- verb → agent noun relation k1ag (*bádat–badatel*; *research–researcher*),
- adjective → name of the property relation k1prop (*rychlý–rychlost*; *fast–
  speed*),
- adjective → adverb relation k6a (*dobrý–dobře*; *good–well*),
- noun → possessive adjective relation k2pos (*otec–otcův*; *father–father's*),
- noun → relational adjective relation k2rel (*virus–virový*; *virus–viral/virus*),
  semantically perhaps the most heterogenous relation among the Derivancze
  relations,
- relations k1f, k1jmf, and k1jmr express changes in gramatical gender:
  - k1f   derives   feminines   from   general   masculines   (*doktor
    –doktorka*; *doctor*MASC *–doctor*FEM),
  - k1jmf derives feminine forms of surnames (*Novák–Nováková*), and
  - k1jmr derives family forms of surnames (*Novák–Novákovi*) — it should
    be noted that k1f and k1jmf cannot be joined because of names of
    nationalities, which can also act as surnames, but the derived forms
    differ (*Rus–Ruska* X *Rusová*, i. e. *Russian*FEM X *Mrs. Rus*),

---

[5] The names of D-relations can be seen as completely arbitrary, but in fact they are
slightly based on the morphological analyser majka tagset [13]: k1, k2 and k6 denote
that the derived word is a noun, adjective and adverb respectively.

- area or city → inhabitant name relation k1obyv (*Kanada–Kanaďan; Canada–Canadian*), formally the most heterogenous relation in Derivancze,
- noun → deminutive relation k1dem (*dům–domek; house–little house*).

The relations of the first group, verbal derivatives, are useful for tagging and syntactic analysis as the derived words somehow retain valences of the base verbs, i. e. they retain a syntactically relevant behavior (perhaps except for k2ucel), the other relations were either requested by our commercial partners (Czech search engine Seznam.cz) or useful for various kinds of text generation (e. g. [14]). The data itself are partially taken from data of morphological analyzer [15] and Czech WordNet [16,17], other relations and data were added from various sources, e. g. k1obyv is from [18], but in the most cases the Deriv tool [8] was utilized.

The Table 1 shows a distribution of the derivational pairs in Derivancze data according to the D-relation. The row var is the semantic equivalence introduced in the Section 4.2. To make a comparison with DeriNet easier, we added another two columns CzTenTen and SYN with numbers of pairs whose both members occur in respective corpus CzTenTen [19] or SYN [11] more than once (the criterion for inclusion a lemma to DeriNet was the same).

Table 1. Distribution of derivational pairs according to D-relation

Relation	# of pairs	CzTenTen	SYN
k1ag	703	588	447
k1dem	6342	5250	3342
k1f	3170	2343	1854
k1jmf	2230	2049	2114
k1jmr	2212	1786	19
k1obyv	262	241	209
k1prop	9886	7503	5975
k1verb	34781	20466	15097
k2pas	34847	11273	192
k2pos	30953	11879	6861
k2proc	15765	7040	5539
k2rakt	18106	1150	600
k2rel	20023	16782	13257
k2rpas	35017	17844	12343
k2ucel	1672	1582	1390
k6a	39065	17281	11678
var	565	406	98
total	255599	125463	81015

Obviously, the numbers of pairs occurring in corpora are rather approximative as they depend on particular lemmatization of the respective corpora (cf. [20] for CzTenTen and [21,22] for SYN). For instance, lemma of *Novákovi* in the SYN corpus is *Novák*, not *Novákovi* as in CzTenTen, thus no k1jmr pair should be found in SYN. But if the name is unknown to the morphological analysis component of the tagger, the lemma is retained equal to the word form, i. e.

Varmužovi is lemmatized as *Varmužovi*. This is the cause of the very low, but still non-zero count of k1jmr (and also k2pas and var) pairs in SYN.

Presently, Derivancze cannot be freely downloaded, but complete data is accessible through a web interface http://nlp.fi.muni.cz/projects/derivancze.

6   Conclusions and Future Work

In the paper we have presented the results of the computational analysis of basic and most regular D-relations in Czech exploiting the older unpublished derivational version of the morphological analyzer D-Ajka and re-designed it as a new derivational analyzer for Czech with the name Derivancze. Though the whole project is a "work-in-progress" and the analysis is far from complete, the number of the captured D-relations and generated derivational pairs is reasonable and covers the basic D-relations in Czech.

We have compared Derivancze with the DeriNet network: the most important differences are that Derivancze covers only semantically transparent D-relations, assigns explicit labels to them and namely prefers semantic consistency if the semantical and formal aspect of D-relations diverge from each other. It should be noted that the last seems to be novel not only for Czech derivational morphology tools and descriptions.

No evaluation has been done yet, but we plan to measure an "added value" of our data for applications which are able to exploit them.

We have also data for some other D-relations, but as they are semantically less transparent, we prefer to work on more regular relations, namely between verbs (aspectual changes, iterativity, etc.) at first. In the future we also aim to precisely describe productive derivational paradigms to be able to predict and recognize word forms derived from loanwords and other new words emerging in Czech texts.

Finally, we would like to note that the Derivancze is not only a theoretical result in Czech derivational morphology: one of its versions has been used as a concrete application in the Czech search engine Seznam.cz and it also serves as an instrument for various kinds of text generation ([14]).

References

1. Šojat, K., Srebačić, M., Tadić, M., Pavelić, T.: CroDeriV: a New Resource for Processing Croatian Morphology. In: Calzolari, N., et al. (eds.) Proceedings of LREC 2014. ELRA, Reykjavik (2014)
2. Pala, K.: Derivational Relations in Slavonic Languages. In: Proceedings of the FASSBL 2008, pp. 21–28. Croatian Language Technologies Society, Zagreb (2008)
3. Dokulil, M., et al.: Mluvnice češtiny I (Grammar of Czech I). Academia, Praha (1986)
4. Ševčíková, M., Žabokrtský, Z.: Word-Formation Network for Czech. In: Calzolari, N., et al. (eds.) Proceedings of LREC 2014. ELRA, Reykjavik (2014)
5. Dokulil, M.: Teorie odvozování slov (Theory of the Word Derivation). Academia, Praha (1962)

6. Karlík, P., et al.: Příruční mluvnice češtiny (Reference Grammar of Czech). Nakladatelství Lidové noviny, Praha (1995)
7. Čechová, M., et al.: Čeština – řeč a jazyk (Czech – Speech and Language). ISV nakladatelství, Praha (2002)
8. Hlaváčková, D., Osolsobě, K., Pala, K., Šmerk, P.: Exploring Derivational Relations in Czech with the Deriv Tool. In: NLP, Corpus Linguistics, Corpus Based Grammar Research, Bratislava, Slovakia, Tribun, pp. 152–161 (2009)
9. Cvrček, V., Vondřička, P.: Nástroj pro slovotvornou analýzu jazykového korpusu (A Tool for Word-Formation Analysis of a Language Corpus). In: Gramatika a korpus, Hradec Králové, Gaudeamus (2013)
10. Zeller, B., Padó, S., Šnajder, J.: Towards Semantic Validation of a Derivational Lexicon. In: Proceedings of COLING 2014: Technical Papers, Dublin City University and ACL, pp. 1728–1739 (2014)
11. Ústav Českého národního korpusu FF UK: Český národní korpus – SYN (Czech National Corpus – SYN), Praha (2014). http://www.korpus.cz (cited April 1, 2015)
12. Šmerk, P.: Tools for Fast Morphological Analysis Based on Finite State Automata. In: Recent Advances in Slavonic Natural Language Processing 2014, Brno, Tribun EU, pp. 147–150 (2014)
13. Jakubíček, M., Kovář, V., Šmerk, P.: Czech Morphological Tagset Revisited. In: Recent Advances in Slavonic Natural Language Processing 2011, Brno, Tribun EU, pp. 29–42 (2011)
14. Nevěřilová, Z.: Paraphrase and Textual Entailment Generation in Czech. Computación y Sistemas 18 (2014)
15. Veber, M., Sedláček, R., Pala, K., Osolsobě, K.: A Procedure for Word Derivational Processes Concerning Lexicon Extension in Highly Inflected Languages. In: Proceedings of LREC 2002, Las Palmas de Gran Canaria, pp. 1254–1259. ELRA (2002)
16. Pala, K., Hlaváčková, D.: Derivational Relations in Czech WordNet. In: Proceedings of the Workshop on Balto-Slavonic Natural Language Processing, pp. 75–81. ACL, Praha (2007)
17. Horák, A., Smrž, P.: VisDic – Wordnet Browsing and Editing Tool. In: Proceedings of GWC 2004, Brno, Czech Republic, Masaryk University, pp. 136–141 (2003)
18. Filipec, J., et al.: Slovník spisovné češtiny. Academia, Praha (1994)
19. Suchomel, V.: Recent Czech Web Corpora. In: Recent Advances in Slavonic Natural Language Processing 2012, Brno, Tribun EU, pp. 77–83 (2012)
20. Šmerk, P.: Towards Morphological Disambiguation of Czech. PhD thesis proposals, Faculty of Informatics, Masaryk University, Brno (2007) (in Czech)
21. Hajič, J.: Disambiguation of Rich Inflection (Computational Morphology of Czech). Charles Univeristy Press, Prague, Czech Republic (2004)
22. Spoustová, D., Hajič, J., Votrubec, J., Krbec, P., Květoň, P.: The best of two worlds: Cooperation of statistical and rule-based taggers for czech. In: Proceedings of the Workshop on Balto-Slavonic Natural Language Processing, Prague, pp. 67–74. ACL (2007)

Detection of Large Segmentation Errors with Score Predictive Model

Martin Matura[1](✉) and Jindřich Matoušek[1,2]

[1] Department of Cybernetics, Faculty of Applied Sciences,
University of West Bohemia, Pilsen, Czech Republic
{mate221,jmatouse}@kky.zcu.cz
[2] New Technologies for the Information Society, Faculty of Applied Sciences,
University of West Bohemia, Pilsen, Czech Republic

Abstract. This paper investigates a possibility of an utilization of regressive score predictive model (SPM) in a process of detection of large segmentation errors. SPM's scores of automatically marked boundaries between all speech segments are examined and further elaborated in an effort to discover the best threshold to distinguish between small and large errors. It was shown that the suggested detection method with a proper threshold can be used to detect all large errors for a specific type of a boundary.

Keywords: Detection of segmentation errors · Large segmentation errors · Score predictive model

1 Introduction

One of the problematic areas in concatenative speech synthesis is a segmentation of speech signal during the creation of the inventory of acoustic units [1]. Speech signal is segmented automatically but segmentation algorithms sometimes make mistakes. Basically, we divide errors to small and large where all mistakes greater than 25 milliseconds are considered to be large errors. Although there is an effort to minimize any error, they are mainly large segmentation errors that cause the biggest problems in concatenative speech synthesis. Because of errors, acoustic units have different properties (duration, spectral properties, ...) than expected which brings problems during concatenation such as a creation of speech artefacts or even an occurrence of phones or whole words that should not be in the speech [2]. Thus, if a large segmentation error occurs, it is necessary to correct it.

Before correction, however, we have to first determine that the segmentation algorithm made such a mistake. Given that the size of a large error can be very variable, it is very difficult to automatically detect these errors. However, the

This research was supported by the Technology Agency of the Czech Republic, project No. TA01011264 and by the grant of the University of West Bohemia, project No. SGS-2013-032. The access to the MetaCentrum clusters provided under the programme LM2010005 is highly appreciated.

© Springer International Publishing Switzerland 2015
P. Král and V. Matoušek (Eds.): TSD 2015, LNAI 9302, pp. 524–532, 2015.
DOI: 10.1007/978-3-319-24033-6_59

detection of large errors is an important step before their removal hence we were concerned with a proposal of a process, which would allow us to detect large segmentation errors. We use regression score predictive model (SPM) [3] which we trained using tools from LIBSVM [4].

In Section 2 we present our data and score predictive model we trained. Section 3 describes procedure of error detection. Section 4 presents conducted experiments during search for optimal score threshold and their results. Conclusions are drawn in Section 5.

2 Data and Score Predictive Model

Our data consists of 90 speech recordings (altogether 11 minutes and 9 sec of speech) and HTK master label files (MLF) [5] with information about segmentation of those recordings. Furthermore, we use files with pitchmarks' time stamps, because the segmentation of recordings was done at phone level and boundaries between phones were placed into pitchmarks [6]. Since we did not have a very high number of data we grouped all phones according to their acoustic properties into five groups fricatives (FRI), nasals + liquids (NLQ), pauses (PAU), plosives (PLO) and vowels (VOW). Our algorithm works with the boundaries between those groups thus if we talk about the boundary it is always meant the boundary between two groups.

The detection algorithm uses a regressive score predictive model (SPM). This model is trained by support vector machine method [7]. We represented each boundary in 70 speech recordings (8 minutes 49 sec) by 56 features and then we used those representations in grid parameter search for regression from [4], which uses cross-validation and finds the best parameters for training the SPM. Among features, which describe boundaries, we picked zero crossing rates, short-term energy of signal, fundamental frequency, voicedness, 12 line spectral frequencies and 12 Mel frequency cepstral coefficients as it was described in [8] in Section 3.1. Moreover, to each feature we added its dynamic description computed using the Formula (1) from [3]:

$$\Delta F(t) = \frac{\sum_{r=-M}^{M} F(t+r)*r}{\sum_{r=-M}^{M} r^2},\tag{1}$$

where F are static features, M^1 is equal to 2 and t is index of feature that is being processed. These dynamic features represent changes in static features in time, which gives us better understanding of the neighbourhood of boundaries.

After the training, we used SPM to rate automatically marked boundaries (AMB) in the remaining 20 recordings (2 minutes 20 sec). Those boundaries were marked automatically by segmentation algorithm and SPM's score should match the quality of the boundary. Furthermore, we also had manually marked boundaries (MMB) to our disposal so that we could determine what a small and

---

[1] It represents number of predecessors and successors from which we calculate the dynamics.

Table 1. Type of boundary with its error size and score value.

Type of boundary	Boundary time [s]		Error [ms]	Score of AMB
	MMB	AMB		
FRI-FRI	8.0767	8.0337	43.0	6.54
NLQ-VOW	3.5336	3.5888	55.2	8.72
FRI-FRI	6.4526	6.4465	6.1	24.48
FRI-FRI	1.0461	1.0512	5.1	43.51
FRI-FRI	4.7875	4.7825	5.0	52.19
NLQ-VOW	1.2140	1.2196	5.6	13.02
NLQ-VOW	1.7742	1.7804	6.2	37.23
NLQ-VOW	0.7170	0.7231	6.1	47.87

large mistake is. Therefore, following the rating of boundaries, we picked 2 representatives of large errors and 6 representatives of small errors to compare their scores. We assumed that the model should evaluate larger errors with a lower score and smaller errors with a higher score. This assumption was confirmed as shown in Table 1.

3   Detection Method

Previous comparison of MMB and AMB implies a possibility of distinguishing large errors from small errors using an appropriately chosen threshold score. Therefore we conducted an analysis of scores for all our tested boundaries to see its dependence on the size of the error (see Figure 1).

The figure shows that small errors tend to receive higher scores, while large errors rather get a lower score. However, there are cases where even small mistakes get a low score which is not very desirable because then our detection ratio[2] is quite small. For this reason we decided to detect large errors not only by score of AMB alone.

For each AMB we took its neighbourhood to left and to right and we considered imaginary boundaries (candidates) in places of pitchmarks. Since in the Figure 1 we saw that large mistakes do not get high score we chose the neighbourhood size dynamically according to the following formula [3], which uses information about the score,

$$d = \sqrt{\frac{\sigma_{max}}{20} * \log(\frac{100}{s})} \tag{2}$$

where σ_{max} is the maximum deviation between manual and automatic segmentation for the given type of boundary and s is score of AMB. This equation was

---

[2] Ratio between the total number of detected errors and the number of large errors that were detected.

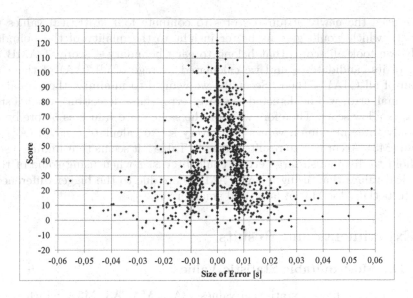

Fig. 1. Dependence of AMB score on the size of errors.

chosen because if AMB has a high score the neighbourhood will be smaller and thus we decrease the probability of corrupting a probably correct boundary.

For every candidate in the given neighbourhood we computed features, and SPM assigned them a score. The score should again represent quality of the boundary. In other words, SPM should rate candidates in the neighbourhood of small error generally with a higher score than candidates from the neighbourhood of a large error. That way we obtained to our AMB score more score representations (see Table 2) which can tell us something more about the position of AMB.

Table 2. Scores of ABM with scores of their candidates and corresponding statistical values.

Type of boundary	Score of candidates left neighbourhood		Score of AMB	Score of candidates right neighbourhood			AA	MA	A3	M3
FRI-FRI	–	–	0.29	6.54	-14.23	–	–	-2.47	0.29	-2.47	0.29
NLQ-VOW	–	10.67	12.13	8.72	1.47	–	–	8,31	9,70	10,59	10,67
FRI-FRI	18.33	3.60	32.51	24.48	13.89	45.67	–	23.08	21.41	34.22	32.51
FRI-FRI	–	–	64.33	43.51	2.74	–	–	36.86	43.51	36.86	4351
FRI-FRI	–	–	48.20	52.19	42.52	–	–	47.64	48.20	47.64	48.20
NLQ-VOW	36.88	45.93	27.92	13.02	4.22	-3.19	1.35	18.02	13.02	36.91	36.88
NLQ-VOW	–	30.73	69.03	37.23	19.22	-0.38	–	31.17	30.73	45.66	37.23
NLQ-VOW	–	–	89.52	47.87	10.16	–	–	49.19	47.87	49.19	47.87

We used the newly acquired scores to compute four statistical values (see Table 2), which would give us better insight to the quality of the boundary. Firstly, we took all scores that belong to each boundary – score of AMB and scores of its candidates – and we calculated average of all (AA) of them and median of all (MA) of them. Secondly, we wanted to minimize the probability that a small error will be detected as a large error (false detection), when a small error gets a low score. Therefore we took the same score values as before but we used only three maximum values from every set to calculate average (A3) and median (M3). According to results shown in Table 2 it seems that finding a score threshold to distinguish large errors could be easier when using some of those statistical values than using a single AMB score due to the bigger difference in score (especially for M3) of large and small errors.

4   Experiments and Results

4.1   The Most Suitable Statistic Value

At this time, we have 4 statistical values – AA, MA, A3, M3 – which we can use to identify large errors. To find out which of these values will be most suitable for identification, we performed detection with each of the value for five different score thresholds as can be seen in Table 3. Total number of all errors was 1570 where 43 of them were large errors greater than 25 milliseconds. For every statistical value we were subsequently setting score threshold to 50, 40, 30, 20 and 10 and we were evaluating the total number of detected errors and the number of large detected errors. We also added numbers for detection using only the information from the single AMB score to the last row of the table.

Table 3. Detection of large errors by using different scores thresholds and various statistical values. Errors show number of total errors detected and large errors detected (numbers in brackets) and their ratio.

	50		40		30		20		10	
	errors	ratio	errors	ratio	errors	ratio	errors	ratio	errors	ratio
AA	1049 (43)	4.10%	828 (42)	5.07%	598 (38)	6,35%	315 (36)	11.43%	73 (17)	23.29%
MA	1024 (43)	4.20%	851 (43)	5.05%	658 (39)	5.93%	382 (33)	8.64%	130 (18)	13.85%
A3	920 (42)	4.57%	718 (40)	5.57%	469 (30)	6.40%	229 (25)	10.92%	36 (9)	25.00%
M3	903 (42)	4.65%	733 (40)	5.46%	517 (32)	6.19%	250 (23)	9.20%	52 (10)	19.23%
AMB	905 (39)	4.31%	754 (37)	4.91%	577 (33)	5.72%	343 (25)	7.29%	137 (13)	9.49%

The table above shows that the best detection ratio 25.00% is obtained by using A3 and score threshold with value of 10. However, in that case we detected only 9 large errors which is only 20.93% of all large errors. In contrast, the second best result, detection with threshold 10 by using AA gave us only slightly lower detection ratio 23.29% but the number of detected errors was 17 which is almost

Table 4. Detection of large errors for fricatives and all other types of boundaries.

Type of boundary	Threshold	Total errors	Large errors	Ratio
FRI-*	17	39	12	30.76%

double. Since our priority is to find as many large errors as possible, considering a reasonable amount of false detection, we decided to use AA values (see Figure 2) for our next experiment.

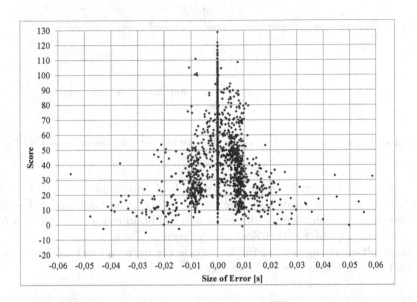

Fig. 2. Dependence of AA score on the size of errors.

4.2   Reducing Number of False Detections

By detection by using AA we achieved a better identification of large errors than if we used only the single score of AMB. Due to the distribution of small errors even in the lower score area there is still quite large number of false detections, i.e. when we label a small error as a large one. We tried to cope with this problem by dividing the data and investigating separately only one specific phone group. Until now we have tried to find a common detection threshold for all types of boundaries. Now we will focus only on the boundaries between fricatives and other groups (FRI-*), because most errors occurred right there. We set apart only errors that belong to boundaries[3] FRI-FRI, FRI-NLQ, FRI-PAU and FRI-VOW and we looked for a new score threshold to separate large errors.

---

[3] Boundary FRI-PLO is included in FRI-PAU because of the character of the signal.

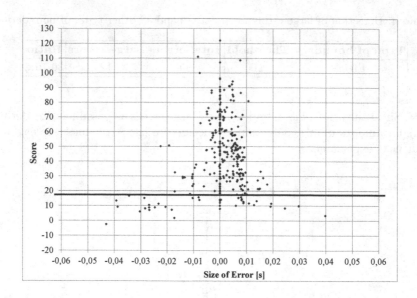

Fig. 3. Dependence of AA score on the size of errors for boundaries FRI-*.

As shown in Figure 3, we were able to find threshold value of 17 (thick black line) that can separate all large errors with a reasonable amount of false detections. By considering only limited number of boundaries our detection ratio increased from 23.29% to 30.76% as can be seen in Table 4.

This result encouraged us to further divide the data and to look for the score threshold among single specific boundaries.

4.3 Specific Boundaries

In the previous case, we worked simultaneously with all errors that belonged to FRI-FRI, FRI-NLQ, FRI-PAU and FRI-VOW. Now we take each border separately and we will try to identify individual threshold for each of them to get maximum detection ratio. We started with border FRI-NLQ where we were not able to find lower threshold so we kept the previous value of 17. Then we examined FRI-FRI and FRI-PAU where we were able to lower the threshold to 11 and detect all large errors at the same time. As the last one we processed FRI-VOW and we found out that there was no large error for this kind of boundary in our data; therefore, no detection was necessary. Table 5 summarizes results obtained for those specific boundaries.

As Table 5 shows, the detection ratio raised from the previous 30.76% to 52.17% and that is just because we analysed each border separately. Looking at the individual results it is obvious that for some types of boundaries we can get reasonable number of false detection and still detect all large errors. Given the small number of all types of boundaries it is therefore advantageous for each

Table 5. Detection of large errors for specific boundaries.

Type of boundary	Threshold	Total errors	Large errors	Ratio
FRI-FRI	11	5	3	60.00%
FRI-NLQ	17	14	7	50.00%
FRI-PAU	11	4	2	50.00%
FRI-VOW	–	–	0	–
In total	–	23	12	52.17%

type to determine its own detection threshold, which substantially increases the detection ratio.

5   Conclusion

We were investigating a method that would be able to automatically detect large errors in the automatic segmentation of speech. First, we tried to identify large errors among all boundaries. That gave us poor detection ratio (i.e. ratio between total number of detection and number of large errors detected), because of the variety of boundaries. Then we proceeded to identification of large errors in specific boundaries. There we tried to reduce the number of false detections, which is the case when we label small error as a large one.

We showed that if we work with more specific types of boundaries, we can afford to reduce the detection threshold and yet still increase the detection ratio. This phenomenon is probably caused by the way SPM models boundaries. Each type of boundary is generally differently predisposed to errors, because some boundaries are better identifiable whereas some are pretty difficult to determine even by a human expert. For this reason, range of score will show some differences among different types of boundaries. Some boundaries can generally receive a higher score and some a lower score. That is why it is then difficult to establish a common threshold simultaneously for all borders and it is preferable to approach each type as a separate unit.

In the future, we should explore the other types of boundaries and find out if they have similar behaviour. Moreover, we could conduct testing with boundaries between specific phones if we get more data.

References

1. Matoušek, J., Romportl, J.: Automatic pitch-synchronous phonetic segmentation. In: Proceedings of 9th Annual Conference of International Speech Communication Association, INTERSPEECH 2008, Brisbane, Australia, pp. 1626–1629. ISCA (2008)
2. Matoušek, J., Tihelka, D., Šmídl, L.: On the impact of annotation errors on unit-selection speech synthesis. In: Sojka, P., Horák, A., Kopeček, I., Pala, K. (eds.) TSD 2012. LNCS, vol. 7499, pp. 456–463. Springer, Heidelberg (2012)

3. Lin, C.Y., Jang, J.S.: Automatic phonetic segmentation by score predictive model for the corpora of Mandarin singing voices. IEEE Transactions on Audio, Speech, and Language Processing **15**(7), 2151–2159 (2007)
4. Chang, C.C., Lin, C.J.: LIBSVM: A library for support vector machines. ACM Transactions on Intelligent Systems and Technology **2**, 27:1–27:27 (2011). http://www.csie.ntu.edu.tw/~cjlin/libsvm
5. Young, S.J., Evermann, G., Gales, M.J.F., Hain, T., Kershaw, D., Moore, G., Odell, J., Ollason, D., Povey, D., Valtchev, V., Woodland, P.C.: HTK Book (for HTK Version 3.4). The Cambridge University, Cambridge (2006)
6. Legát, M., Matoušek, J., Tihelka, D.: On the detection of pitch marks using a robust multi-phase algorithm. Speech Communication **53**(4), 552–566 (2011)
7. Cortes, C., Vapnik, V.: Support-vector networks. Machine Learning **20**(3), 273–297 (1995)
8. Matura, M.: Phonetic Segmentation of Speech and Possibilities of its Automatic Correction. Master's thesis, The University of West Bohemia, Faculty of Applied Sciences, Pilsen (2014)

Identification of Noun-Noun Compounds in the Context of Speech-to-Speech Translation

Maria Ivanova[✉] and Eric Wehrli

University of Geneva, Language Technology Laboratory (LATL), Geneva, Switzerland
{maria.ivanova,eric.wehrli}@unige.ch
http://www.latl.unige.ch/

Abstract. Translation of noun compounds has been a challenging task in machine translation for many years. Noun compounds are used very productively and constitute a large amount of the text vocabulary. We present a work on translation of noun compounds from English to German, focusing on identifying the nominal structures in English which represent compounds. This work is part of developing a speech-to-speech translation system, and the input to the machine translation component comes from automatic speech recognition. This brings in a further difficulty to the translation of noun compounds because the input might introduce recognition errors and it does not have typical features of a written text such as punctuation and capitalization. We present a method of noun compound identification based on syntactic analysis and lexical information and discuss the challenges that we encountered during the compound identification process.

Keywords: Speech-to-speech-translation · Noun compounds · Error analysis

1 Introduction

This paper presents a work on translating noun compounds from English to German in the context of speech-to-speech translation (S2ST). The translation of compounds is a non-trivial task, causing problems to both statistical and rule-based machine translation systems [1,2]. Two major issues arise in the process: first, the identification of nominal structures which represent compounds and second, the generation of well-formed compounds in the target language. Both phases are difficult and require careful analyses beforehand. In this paper we focus on the former task.

Translation of noun compounds has been a challenging task in machine translation for many years. Noun compounds are used very productively and constitute a large amount of the text vocabulary [2]. In German and other Germanic languages (e.g., Dutch, Danish, Swedish) compounds are built by gluing different words together, possibly creating new words, which have not existed before. According to the official German spelling, the constituents of a German compound are always written together, sometimes allowing for a dash in-between,

P. Král and V. Matoušek (Eds.): TSD 2015, LNAI 9302, pp. 533–541, 2015.
DOI: 10.1007/978-3-319-24033-6_60

e.g. *Datenverarbeitung, Kaffee-Ersatz* [3]. In English, some compounds are built similarly to German by connecting two lexemes together (with or without a dash), e.g. *bookstore, brother-in-law*, other compounds consist of non-connected lexemes, e.g. *health insurance*.

In this paper we focus on the identification of noun compounds in English, and more specifically on compounds of type *noun-noun*. Not all sequences of nouns are regarded as noun compounds [2]. We need to identify the correct sequences in the source language and then to map the identified compounds to their corresponding translations in the target language. Noun compounds are reported to be the most typical compounds in English and *noun-noun* is the most frequently used type [2,4]. The identification and translation of *noun-noun* compounds is demanding enough, and for now we leave the processing of other types of compounds for future work[1].

This work is part of developing a speech-to-speech translation system, and the input to the machine translation component comes from automatic speech recognition (ASR). This introduces a further challenge to the translation of noun compounds, because the input might introduce recognition errors. Furthermore, the ASR output usually does not have typical features of a written text such as punctuation and capitalization. We present a method of noun compound identification based on syntactic analysis and lexical information and discuss the challenges that we encountered during the compound identification process.

2   Identification of *Noun-Noun* Compounds

Noun compounds as well as multi-word expressions are interesting and important for many natural language processing (NLP) applications, e.g. *machine translation, information retrieval, question answering*, etc. [2]. The first non-trivial task in handling noun compounds automatically is to identify them in the input data. To accomplish this, a clear definition of the notion of *noun compound* is needed. Up to this date, however, there have been many discussions in the linguistic literature of what exactly is a noun compound, and there is still not a stable definition for it [2,6,7].[2]

2.1   Linguistic Tests for Compound Identification

Despite the controvercies, there have been defined some linguistic tests for compoundhood [6,7]: 1) inseparability of the compound constituents; 2) a compound

---

[1] When more than two nouns are involved in the compounding process, another challenge appears - it is not always clear how to combine the constituents of the compound, i.e. its interpretation becomes ambiguous [2,5].

[2] One issue comes from the fact that sometimes compounds cannot be easily distinguished from phrases or derivations [6]. For example, *traffic lights, hair dryer, and cash machine* are considered to be compounds, but not necessarily *tomato bowl*. It might not be accepted as a single lexeme, but interpreted as a bowl that happened to contain tomatoes [7].

can be only modified as a whole (not its constituents); 3) compounds usually have different prosody (stress) than phrases; 4) the head of the compound cannot be replaced by a proform; 5) only the head of the compound is inflected. In our implementation we use the first two of these tests, as well as further restrictions, which we describe in section 4. Even though we identify compounds from speech data and potentially we can use prosodic information as a feature, in our case this feature might be ambiguous. The speech-to-speech translation project SIWIS, of which the MT component is a part [8], aims at identifying and transfering only *salient prosodic events*, i.e. particular aspects of speech such as emphasis, focus or contrast, which represent the *speaker's intention*. This means that the prosody information will not necessarily represent normal stress, which is required for linguistic test 3)[3]. With regard to linguistic tests 4) and 5), the former needs human judgement, while the latter does not always hold (e.g., *sportsman, salesperson*). Moreover, it is reported to have recently become more commonly used [9], and therefore we decided to ignore it for now.

2.2   Lexical Database

For the identification of all noun compounds in English we rely on the lexical database of the Its-2 system. For example, the two-lexeme English compound *blackbird* should be translated in German as the one-lexeme word *Amsel* [7]. The database contains also many collocations of type *noun-noun*. If any compound or collocation is present in the database, it is identified in the input text and the compound identification process ends. The second example in figure 1 illustrates how a compound from the lexical database, e.g., *health insurance*, is recognized and then the compound identification rules build another compound by combining the identified modifier *health insurance* with the head *coverage*.

The Its-2 [10] database has a rich collection of lexical items and collocations. Table 1 presents the numbers of different lexical units[4]. From the 3452 German collocations, 2402 are compounds of type *noun-noun* (e.g., *Arbeitsunfall*). Currently, there are 514 English collocations mapped to German *noun-noun* compounds[5].

---

[3] According to linguistic test 3), usually the head of a phrase is stressed, while the stress of a compound goes on the first constituent. In this way compounds can often be differentiated from phrases. In our application, however, the speaker's intention might change the normal stress in a compound.

[4] *Lexemes* are basic word forms, corresponding to dictionary entries. *Words* are inflectional forms of the lexemes. *Collocations* contain compounds and multi-word expressions.

[5] The database contains also English collocations of type *adj-noun* and *noun-prep-noun* which are mapped to German *noun-noun* compounds. Our focus in this work is only on compounds of type *noun-noun*, but later we will also rely on the database for the identification of these other types of compounds. Examples: *"war of attrition"* = *Abnutzungskrieg*, *"flag of convenience"* = *Billigflagge*.

Table 1. Lexical database.

	Lexemes	Words	Collocations
English	58 267	104 970	9826
German	43 835	450 910	3452

The lexical database contains also various lexical features for nouns, such as *gender, number, case, inflection class*, etc., which are later used in the noun compound generation component.

3   ASR Output Specifities

In S2ST systems the input to the machine translation module comes from the output of the ASR module. This brings problems, because (i) some errors of the ASR module are propagated to the MT module; (ii) typical features of written texts, such as punctuation and capitalization, are missing.

Punctuation and capitalization are helpful features for the correct grammatical analysis of written texts. Punctuation (e.g., commas, dashes, parentheses, etc.) is helpful when dealing with the structure of more complex sentences containing, for example, relative clauses, enumerations, parentheticals, etc. Lack of punctuation can lead to ambiguous and incorrect syntactic constructions with respect to the original meaning of a sentence. Here is an example which shows how the lack of punctuation can affect the analysis of noun compounds:

(1a) *Among the animals in tropical forests, antelopes and armadillos were faced with danger.*

(1b) *AMONG THE ANIMALS IN TROPICAL FORESTS ANTELOPES AND ARMADILLOS WERE FACED WITH DANGER*

The sentence in (1b) shows how the output of the ASR component looks like. In the case of English to German translation, the lack of punctuation can negatively affect the translation module by allowing it to glue all nouns together into a compound. If commas are removed from the sentences in (1a), it might not be clear whether to treat the nouns *forests* and *antelopes* as separate nouns or to merge them in a compound. In section 4 we show how our approach can solve such a problem.

4   Implementation

For the identification of noun compounds we use a syntax-based approach combined with lexical information. The syntactic analysis is performed by the Fips parser [11], which is part of the Its-2 machine translation system [10]. Its-2 is a multilingual system with a standard tranfer architecure, consisting of analysis, transfer and generation modules.[6]

The syntactic analysis gives us information about how the compound constituents are combined and this information can be helpful for the identification of correct noun compounds. We know that in both English and German the heads of most compounds are their rightmost constituents [12]. We allow for only such syntactic constructions, in which the compound constituents form a single noun phrase, with the right most noun being the head of the phrase. Figure 1 shows examples of such constructions.

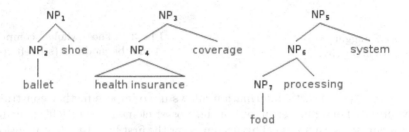

Fig. 1. Examples of noun compounds.

The above mentioned restriction can prevent us from extracting many false positives. The analyses of the sentence { *The mother gave her child biscuits.* } illustrates what can happen if we identify noun compounds relying only on part-of-speech tagging, and how syntactic information can be helpful for the correct identification. In figure 2 we can see that the nouns *child* and *biscuits* are a pair of consecutive nouns. The syntax tree[7] reveals that the two nouns belong to different syntactic constituents. This violates the requirement that the compound constituent should build a single noun phrase, and the two consecutive nouns are correctly not extracted as a noun compound.

---

[6] The choice of a rule-based machine translation system is motivated by a distinctive feature of the project, namely, the transfer of prosody, as well as by the need for grammatical information (e.g., lexical categories, constituent structures). The S2ST system should transfer from the source language to the target language important cues in speech such as *the identity of the speaker, focus, contrast or emphasis.* To transfer these prosodic events we need a link between the corresponding affected parts of the input and output languages, and we hope that a rule-based MT system can give us this link by providing more control over different linguistic structures. For example, in the context of noun compound translation, the system needs to keep track of all constituents of the noun compounds in order to assign properly the relevant prosodic event. Depending on our intermediate results and tests during the project, we are open to consider also statistical methods, if necessary.

[7] We used the phpSyntaxTree (http://ironcreek.net/phpsyntaxtree/) to visualize the syntax tree.

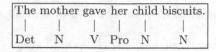

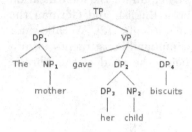

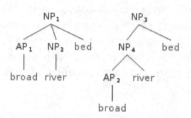

Fig. 2. POS tagging versus syntactic analysis.

Fig. 3. The whole compound should be modified (see left tree).

Moreover, the syntactic information allows us to impose further constraints. We favour only noun phrases, in which the closest element to the left of the head is a noun phrase (see figure 1). This accounts for the first linguistic test mentioned in section 2.1 We also exclude noun phrases, in which only the first element of the noun phrase is modified, but not the noun phrase as a whole. This restriction satisfies the second linguistic test mentioned in section 2.1. For example, the compound *broad river bed* should be interpreted as a broad bed for a river and not as a bed for a broad river [9]. Therefore, the correct syntactic structure for it is the one illustrated on figure 3 on the left. The syntactic structure on the right represents the semantically incorrect interpretation.[8]

The syntactic analysis approach can also help with the problem of combining consecutive nouns in a sentence when punctuation in the input is missing. In section 3 we described how the nouns *FORESTS* and *ANTELOPES* in the ASR output sentence can be incorrectly assumed to form a compound. In the syntactic tree on figure 4 these nouns belong to two different syntactic constructions and this prevents them from being put together in a noun compound.

5   Experiments and Related Work

We perform extraction of *noun-noun* compounds from English TED talks[9] data with the conditions described in section 2. The speech recognition has been performed by the UEDIN speech recognition system [13]. Our test set consisted

---

[8] Adjectival modification of the first element of a *noun-noun* compound is generally assumed to be improbable or not grammatically correct [9]. However, [9] also lists attested examples, which do not conform to this assumption (e.g. *light-rail system, instant noodle salad*). The linguistic test seems not to hold all the time, but we decided to stick to the general assumption, because the reason of the occurrence of some of the attested examples is still not clear.

[9] http://www.ted.com/

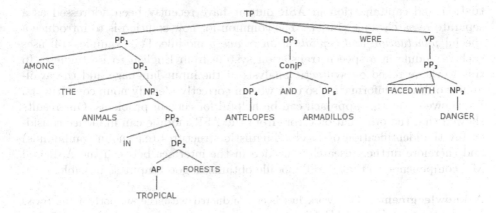

Fig. 4. Syntactic analysis: *FORESTS* and *ANTELOPES* are not combined in a compound.

Table 2. Evaluation results: identification of *noun-noun* compounds.

tst2011	Utterances	Words	WER	Compounds	**Precision**	**Recall**	**F-measure**
Talkid 1169	198	2956	14.5	39	62%	41%	49%
Talkid 1174	148	2948	5.6	60	**76%**	48%	**59%**
Talkid 1182	60	1052	7.2	19	63%	**53%**	58%
Talkid 1187	50	686	3.8	14	57%	29%	38%
Total	456	7642	7.8	132	**65%**	**43%**	**51%**

of 4 TED talks from the tst2011 set[10]. In table 2 we present our performance on the *noun-noun* compound identification.

We have manually evaluated the results and we have seen that the performance is very much influenced by propagation errors coming from ASR - in 18% of the error cases some of the constituents of the compounds were not correctly recognized, and additionally some incorrect compounds have been built based on incorrectly recognized words. In other cases, incorrect assignment of part-of-speech categories or incorrect syntactic analysis have been triggered. If words in an utterance, not necessarily the noun compound constituents, have been recognized incorrectly, they can affect the building of the whole syntactic tree in a negative way. With respect to the correctly identified noun compounds, it is interesting to know the type of identification - we have 51% of known compounds (coming from the lexical database) and 49% of compounds recognized by the compound identification rules.

[7] report very low results on the extraction of noun compounds using only part-of-speech information. Their other methods achieve good results, but they rely on the availability of multilingual data. The problems with lack of punc-

---

[10] https://wit3.fbk.eu/

tuation and capitalization in ASR output have recently been addressed as a separate research topic [14–17]. The common for now solution is to introduce a special *punctuation and capitalization recovery* module. For example, [14] uses such a module in a speech translation system from English to Portuguese. In this work we relied on syntactic analysis of the input language and the availability of lexical information so that we can correctly identify noun constituents. We showed that this approach can be helpful for certain problems. Our results showed that the propagation errors from the ASR module can influence considerably the identification of specific linguistic structures (e.g, noun compounds) and therefore further research is needed in the interface between the ASR and MT components, so that a MT module obtains as good input as possible.

Acknowledgments. This work has been conducted with the support of the Swiss NSF under grant CRSII2 141903: Spoken Interaction with Interpretation in Switzerland (SIWIS). We would like also to thank to Peter Bell from CSTR at University of Edinburgh, for providing us the speech recognized data and evaluation, and for the useful comments.

References

1. Parra Escartín, C., Peitz, S., Ney, H.: German compounds and statistical machine translation. Can they get along? In: Proceedings of the 10th Workshop on Multiword Expressions. ACL (2014)
2. Nakov, P.: On the interpretation of noun compounds: Syntax, semantics, and entailment. Natural Language Engineering **19**(3) (2013)
3. Rat für deutsche Rechtschreibung: Deutsche Rechtschreibung: Regeln und Wörterverzeichnis: Amtliche Regelung. Gunter Narr Verlag, Tübingen (2006)
4. Rackow, U., Dagan, I., Schwall, U.: Automatic translation of noun compounds. In: Proceedings of the 14th International Conference on Computational Linguistics (1992)
5. Lauer, M.: Corpus statistics meet the noun compound: some empirical results. In: Proceedings of the 33rd Annual Meeting of ACL (1995)
6. Lieber, R., Stekauer, P.: The Oxford Handbook of Compounding. Oxford Handbooks in Linguistics. OUP, Oxford (2009)
7. Ziering, P., van der Plas, L.: What good are 'nominalkomposita' for 'noun compounds': multilingual extraction and structure analysis of nominal compositions using linguistic restrictors. In: Proceedings of COLING (2014)
8. Garner, P.N., Clark, R., Goldman, J.P., Honnet, P.E., Ivanova, M., Lazaridis, A., Liang, H., Pfister, B., Ribeiro, M.S., Wehrli, E., Yamagishi, J.: Translation and prosody in Swiss languages. Nouveaux cahiers de linguistique française **31** (2014). 3rd Swiss Workshop on Prosody, Geneva, September 2014
9. Bauer, L.: When is a sequence of noun + noun a compound in english? English Language and Linguistics **2** (1998)
10. Wehrli, E., Nerima, L., Scherrer, Y.: Deep linguistic multilingual translation and bilingual dictionaries. In: Proceedings of the Fourth Workshop on Statistical Machine Translation. ACL (2009)
11. Wehrli, E.: Fips, a "deep" linguistic multilingual parser. In: Proceedings of the Workshop on Deep Linguistic Processing. ACL (2007)
12. Booij, G.: The grammar of words. An introduction to linguistic morphology. Oxford University Press (2005)

13. Bell, P., Swietojanski, P., Driesen, J., Sinclair, M., McInnes, F., Renals, S.: The UEDIN ASR systems for the IWSLT 2014 evaluation. In: IWSLT (2014)
14. Grazina, N.M.M.: Automatic Speech Translation. Universidade Técnica de Lisboa (UTL), Instituto Superior Técnico., Lisboa, Portugal (2010)
15. Batista, F., Caseiro, D., Mamede, N.J., Trancoso, I.: Recovering punctuation marks for automatic speech recognition. In: INTERSPEECH. ISCA (2007)
16. Cho, E., Niehues, J., Waibel, A.: Segmentation and punctuation prediction in speech language translation using a monolingual translation system. In: IWSLT (2012)
17. Batista, F., Caseiro, D., Mamede, N., Trancoso, I.: Recovering capitalization and punctuation marks for automatic speech recognition: Case study for portuguese broadcast news. Speech Communication **50**(10) (2008)

From Spoken Language to Ontology-Driven Dialogue Management

Dmitry Mouromtsev[1], Liubov Kovriguina[1(✉)], Yury Emelyanov[1],
Dmitry Pavlov[1], and Alexander Shipilo[2]

[1] ISST Laboratory, ITMO University, Saint-Petersburg, Russia
{d.muromtsev,yury.emelyanov,lkovriguina,vergilius}@gmail.com
http://www.ifmo.ru
[2] Saint-Petersburg State University, Saint-Petersburg, Russia
alexandershipilo@gmail.com
http://www.spbu.ru

Abstract. The paper describes the architecture of the prototype of the spoken dialogue system combining deep natural language processing with an information state dialogue manager. The system assists technical support to the customers of the digital TV provider. Raw data are sent to the natural language processing engine which performs tokenization, morphological and syntactic analysis and anaphora resolution. Multimodal Interface Language (MMIL) is used for the sentence semantic representation. A separate module of the NLP engine converts Shallow MMIL representation into Deep MMIL representation by applying transformation rules to shallow syntactic structures and generating its paraphrases. Deep MMIL representation is the input of the module generating facts for the dialogue manager. Facts are extracted using the domain ontology. A fact itself is an RDF triple containing temporal information wrapped in the move type. Dialogue manager can accept unlimited number of facts and supports mixed initiative.

Keywords: Spoken dialogue systems · Domain ontology development · Natural language processing · MMIL applications · Paraphrase generation · Information state approach

1 Introduction

The developed dialogue system is to assess customer support to the clients of digital television provider. The prototype of the system communicates with clients using chat window. The system is able to find out the problem and offer solution and give instructions how to fix the problem, otherwise it redirects the client to operator (human) or escalates the problem for another level of support.

The prototype deals with specially prepared textual data. Original training data are 150 human-human dialogues (5600 tokens) between users and technical support concerning troubles with digital TV subscription. The refinement included ellipsis recovery, discontinuity and spontaneous speech disfluency removal.

P. Král and V. Matoušek (Eds.): TSD 2015, LNAI 9302, pp. 542–550, 2015.
DOI: 10.1007/978-3-319-24033-6_61

The main goal of the project was to develop a dialogue system which will classify the problem dynamically, will be able to accept unlimited number of facts and support mixed initiative. This determined choosing information state approach for dialogue management.

2   Related Work

There is a number of plan-based dialogue systems in healthcare[1] and technology[1] which support mixed initiative and interpret the dialogue using ontologies[1][2]. We decided to develop Natural Language Processing engine performing full and deep linguistic analysis instead of using n-gram or bag-of-words models. Shallow syntax transformation and paraphrasing rules applied to shallow syntax structures to produce invariant utterance structures have been also implemented. It is a well-known fact that direct mapping of the sentence to the system command is impossible excluding some very simple cases (e.g., greeting, acknowledgement).

3   Natural Language Processing Engine

NLP engine performs full linguistic analysis of the text in Russian language including tokenization, morphological analysis and syntactic analysis. SemSyn[3] parser tags the sentence, analyzes its syntactic structure and produces a dependency tree in the output, performing syntactic ambiguity resolution. SemSyn parser performs correct morphological analysis in 95% of the cases and syntactic analysis in 85-90% of the cases. Syntactic relations, ascribed by the parser, consider semantics of the word (or the construction): the rule invocation depends on the entries' characteristics in the Tuzov semantic dictionary[4] There are about 60 relations used by the parser to resolve prepositional ambiguity, distinguish subject and object, etc. These relations can be plainly mapped to semantic roles. The XML parse tree is sent to the semantic representation module which converts it to one or more MMIL components.

4   Sentence Semantic Representation

MMIL unit, a component, obligatory includes propositional content and dialogue type. Propositional content includes events and participants of the sentence.

Dialogue types proposed in MMIL can be mapped to speech acts (see full MMIL manual in [5]).

4.1   Utterance Semantic Representation in Shallow and Deep MMIL

MMIL generation module converts XML parse tree into a number of MMIL components. Each MMIL component is referred to a single sentence or parts of the compound sentence, includes the following entities and their characteristics:

---

[1] http://www.openclinical.org/dm_homey.html

type of the dialogue act, events (entities in time dimension), events characteristics (time, aspect, mode, person, voice, polarity), participants (entities not bounded in time dimension), participants characteristics (objType, mmilId, refStatus, number, person, gender, modifier), relations between entities (a special relation "propContent" links speech act type and proposition, relations between events and participants coincide with semantic roles).

MMIL generation module uses rules dealing with sentence semantics and semantics of grammar categories of tense, aspect, modality relying on the results of A.V.Bondarko functional grammar[6].

Deep MMIL is not specified by MMIL authors. In our work Deep MMIL representation is inspired by the ideas of generative grammar, Meaning-Text theory and its further applications. Semantic roles and grammar characteristics in Deep MMIL to a large extent correspond to the relations in deep sentence structure in the tradition of generative grammar. To produce Deep MMIL component from Shallow MMIL component a number of transformations and paraphrases is performed. These modifications are aimed to produce the same deep representation for several shallow structures.

Shallow and Deep MMIL components for the input sentences "My digital channels do not work" and "My digital channels stopped working" can be seen in figure 1. Sentence 1 has the same representation for both Shallow and Deep MMIL.

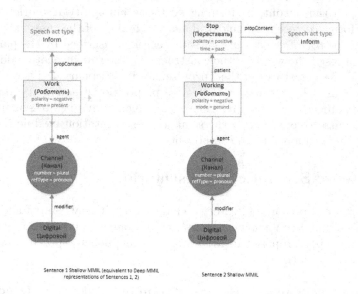

Fig. 1. Utterance representation in Shallow and Deep MMIL

5 Paraphrase Generation

Implementing paraphrase generation in a dialogue system seems necessary for several reasons. Firstly, the speaker can express the same meaning differently and the system should be able to handle it and react the same way. Secondly, paraphrases may be useful for the generation of clarification questions[7]. Moreover, it was shown that lemmatizing, synonym handling and paraphrasing improve performance of the dialogue system[8]. While developing the prototype of our dialogue system, we encountered that paraphrasing also contributes to minimization of ontology entities. There is a number of data-driven paraphrase generation techniques, which performance has already been evaluated[9]. A pivot method to generate paraphrases is to use statistical machine translation techniques when each utterance is translated into target language and then back into source[8]. In the cited paper authors used Google Translate API. In our project we used paraphrase generation method proposed by Apresyan and Cinman[10]. This method uses the notion of lexical functions introduced by I.A.Mel?cuk and A.K.Zholkovsky[11]

Mel cuk defines[11] lexical function as following: "A lexical unit f is a function that associates with a given lexical unit [= LU] L, which is the argument, or keyword, of f, a set $\{L_i\}$ of (more or less) synonymous lexical expressions (the value of f) that are selected contingent on L to manifest the meaning corresponding to f:

$$f(L) = \{L_i\}. \tag{1}$$

Below are some examples of the lexical function **OPER** (ibid.) (do), (perform) [support verb]

- OPER 1 (strike N) = to be [on]
- OPER 1 (support N) = to lend [ ]

In our study paraphrases have been manually extracted from the training corpus and matched to the set of lexical functions. SYN, OPER and ANTI turned out to be dominant lexical functions among the encountered in the corpus. The next subsection gives examples of ANTI paraphrasing rules according to the paper[10] and provides a mapping to the facts of the knowledge base using ANTI LF.

5.1 ANTI LF Paraphrase Rule

ANTI LF deals with antonyms and states that negative of the LU corresponds to the positive value of LU's antonym. Yu.D.Apresyan defines some antonyms as lexical units, for which the following statements are true: "P = !R" (ANTI2) and "stop R = begin !R" (ANTI1)[12, 18]. The following paraphrase rule is applied: X + Y =¿ ANTI1(X) + ANTI2(Y). Consider the following example: "Two hours ago internet stopped working" (" "). It is clear that internet does not work but negation is hidden inside the verb "stopped". Shallow MMIL generated for this sentence will ascribe positive polarity to this verb meanwhile the fact sent to the dialogue manager must include false value of the property corresponding to the lexeme "to work". This paraphrase rule operates with polarity values in MMIL and as list of verbs denoting action beginning and termination. The given sentence will be paraphrased into MMIL intermediate structure corresponding to the sentence "Two hours ago internet began not to work" (" .") and to the final Deep MMIL representation where "work" will have the values of polarity = 'negative' and time = 'past'.

Paraphrase generation module uses specially compiled lexical resour es, which specify the paraphrase rule, rule constraints and procedures changing Shallow MMIL structure. if all constraints are fulfilled corresponding paraphrase rule is invoked. It changes only the items of the MMIL component, no text is generated or affected. After all possible paraphrase rules have been applied Deep MMIL component is generated and sent to fact extraction module.

6   Domain Ontology

Firstly, we tried top-down approach to create domain ontology based on instructions provided for human customer support operator but in the end it proved to be unusable, because our concepts occurred rarely in real dialogues making impossible to build problem solving algorithms.

So, we switched to bottom-up approach and distinguished 6 types of problems with TV subscription and built the ontology according to empirical data.

Devices (TV set, router, set-top box, etc.), connectors (cable, switch, etc.) services and tariffs, user, etc. are represented as classes. Events are modeled as properties, like operational property that has domain of Service and range of boolean, and restrict our modelling power. The problem was that utterances always has temporal reference which is difficult to model in OWL, so temporal level was introduced to the fact structure.

6.1   Lexical Information in the Domain Ontology

Domain ontology incorporates the level of lexical semantics.

Synonyms are stored in listed as lemmas of the concept, i.e. for the boolean datatype property "insert" lemmas "insert", "stick in" and "put in" are stored. Some operations, expressed by properties like "insert", "switch on", "plug in" etc. have opposite operations. Each of these operations is stored in a separate subclass of the class defining the general name of these operations (see fig. 2^2). Such representation of concepts that denote opposite operations allows to specify complex operations using part-whole relation, e.g. rebooting,a holonym, except lemma specifying can be defined with two its meronyms: "switch on" and "switch off".

---

[2] Visualized using http://www.ontodia.org/

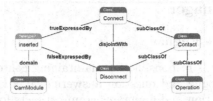

Fig. 2. Event representation in domain ontology

7   Extracting Facts for the Dialogue Manager from Deep MMIL

Fact extraction is implemented in the separate model, having Deep MMIL in the output and generally producing an RDF triple wrapped in the fact type and move type tags.

7.1   Types of Facts in the Knowledge Base of the Dialogue Manager

The developed dialogue manager supports all main concepts of the information state approach. Dialogue manager's data structure pecularity is the following: each move, except the most simplest GREET and QUIT, includes factual information embedded in the "Fact" structure. There are three types of "Fact" in the system: "SimpleFact" (stores short answers), "Agreement" (stores boolean value for user's confirmation/rejection), "PropertyFact" (stores subject-predicate-value (for datatype properties) and subject-predicate-object (for object properties) triples, each corresponding to the elementary fact) and "Alternative-Fact" (stores multiple PropertyFacts).

7.2   Fact Extraction Algorithm

This module is responsible for Deep MMIL parsing and extracting facts for the knowledge base which will be used for the dialogue manager. Basic idea of the algorithm is the following: each participant is the candidate subject (participant lemma is searched among lemmas of the domain ontology classes), each event is the candidate predicate (event lemma is searched among lemmas of the domain ontology properties)[13]. Combining move types, properties and temporal information allows to use the same ontology entities for different utterances: the only property *tv:subscribed* is stored in the ontology and corresponds to several situations, e.g. when the user 1) wants to subscribe a service, 2) has already subscribe it, 3) wants to unsubscribe the service. Extending speech act type and proposition distinction forth to the knowledge base and dialogue manager allows to reduce ontology size and keep rules more observable.

8 Dialogue Manager

8.1 Dialogue Move Engine

Dialogue move engine is responsible for maintaining context of dialogue, guiding basic flow of dialogue and question answering. It knows if user or system answered current question and maintains common facts for dialogue, as well as keeps plan for the next few actions. Dialogue move engine was created following the ideas described in GoDiS and implemented in TrindiKit with some modifications. We ve added dialogue move type TELL to provide informative messages to user and not to confuse this messages with answers to user questions. We ve also added modification to update rules to allow accommodation of incoming facts when there are no relevant questions in QUD to implement mixed initiative dialog scheme. Data structures of information state holding moves, agenda, plan, common and private beliefs were created, along with update and select rules in Java programming language, unlike traditional implementation of GoDiS in TrindiKit in Prolog and therefore they use different abstractions for rule definition and application.

8.2 Domain Binding

GoDiS defines different functions on facts: relevant (if answer fact is relevant to question fact), resolves (if answer fact resolves question fact), combine (to combine question fact with answer fact to produce resolving fact). As stated above, we have information about concepts, properties and their types in ontology and use it doing relevant, resolves and combines functions on facts: to check data types, to use concept taxonomy trees to see if answer fits the question.

8.3 Problem Solving

Dialogue move engine handles only basic part of dialogue flow. The more complex part of dialogue flow depends on things being said from both parties the more it depends on domain knowledge. The developed dialogue manager doesn't belong to command dialogue managers (like Smart Home control) or search systems or booking systems. The dialogue system should collaborate with user solving his problem, present him diagnostics options or asking to do specific task and tell if the main problem has gone. Simultaneously, user can ask questions about the data in information system, like balance, etc. The system may choose to redirect user to particular service to get additional help or query diagnostics system if there were any problems with that particular user. To implement this kind of system behaviour we chose to put the knowledge of interaction in rule system. Each time a fact comes from user, we check if any rule has fired. We used Drools Rule Engine[3] with custom domain-specific language layer.

---

[3] http://www.drools.org/

9   Evaluation, Conclusion and Future Work

Intermediate prototype evaluation performed on automatically generated Shallow MMIL components and its parts has shown the necessity of incomplete and contradictory facts handling. Future work implies massive research and development activities. NLP engine should be implemented with the parser analyzing spoken language syntax correctly. The algorithm of fact extraction needs refinement and elaboration. Ontology requires enlargement and the dialogue manager should be learnt to support partial and contradictory facts. Finally, handling large rule base is a hard task and another ways of storing and executing knowledge for problem-driven dialogue management should be elaborated.

Acknowledgments. This work was partially financially supported by the Government of the Russian Federation, Grant 074-U01.

References

1. Görz, G., Bücher, K., Ludwig, B., Schweinberger, F., Thabet, I.: Combining a lexical taxonomy with domain ontology in the erlangen dialogue system. In: Proceedings of the KI2003 Workshop on Reference Ontologies and Application Ontologies, Hamburg, Germany, September 16, 2003 (2003)
2. Maema, M.: OVR: A Novel Architecture for Voice-Based Applications. In: Thesis submitted for Master of Science (2011)
3. Boyarsky, K., Kanevsky, E.: The semantic-and-syntactic parser SEMSIN. In: International Conference on Computational Linguistics Dialog-2012 (2012)
4. Tuzov, V.: Semantic dictionary of the Russian language. http://emi.nw.ru/INDEX.html?0/Voc.html
5. Rojas-Barahona, L.M., Bazillon, T., Quignard, M., Lefevre, F.: Using MMIL for the high level semantic annotation of the french media dialogue corpus. In: Proceedings of the Ninth International Conference on Computational Semantics, pp. 375–379. Association for Computational Linguistics (2011)
6. Bondarko, A.V.: Functional grammar: a field approach, vol. 35. John Benjamins Publishing (1991)
7. Ebert, C., Lappin, S., Gregory, H., Nicolov, N.: Generating full paraphrases of fragments in a dialogue interpretation system. In: Proceedings of the Second SIGdial Workshop on Discourse and Dialogue, vol. 16, pp. 1–10. Association for Computational Linguistics (2001)
8. Gardent, C., Rojas-Barahona, L.M.: Using paraphrases and lexical semantics to improve the accuracy and the robustness of supervised models in situated dialogue systems. In: Proceedings of the 2013 Conference on Empirical Methods in Natural Language Processing, EMNLP Seattle, Seattle, Washington, USA, pp. 808–813 (2013)
9. Metzler, D., Hovy, E., Zhang, C.: An empirical evaluation of data-driven paraphrase generation techniques. In: Proceedings of the 49th Annual Meeting of the Association for Computational Linguistics: Human Language Technologies: short papers, vol. 2, pp. 546–551. Association for Computational Linguistics (2011)
10. Apresyan, Y., Zinman, L.: Perifrazirovanie na kompjutere. Vsesojuznij institut naucnoj b techniceskoj informacii **36**, 177 (1998). (in Russian)

11. Melcuk, I.: Collocations and lexical functions, vol. 31, pp. 23–53. Clarendon Press, Oxford (1998)
12. Apresyan, Y.: Lexicheskaya semantika: Sinonimiceskije sredstva jazika. Nauka, 367 (1974) (in Russian)
13. Mouromtsev, D., Kozlov, F., Kovriguina, L., Parkhimovich, O.: A combined method for e-learning ontology population based on nlp and user activity analysis, vol. 1254. CEUR-WS (2014)

Entity-Oriented Sentiment Analysis of Tweets: Results and Problems

Natalia Loukachevitch[1]([✉]) and Yuliya Rubtsova[2]

[1] Research Computing Center, Moscow State University, 119234 Moscow, Russia
louk_nat@mail.ru
[2] A.P. Ershov Institute of Informatics Systems, Siberian Branch of the Russian
Academy of Sciences, 630090 Novosibirsk, Russia
yu.rubtsova@gmail.com

Abstract. This paper summarizes the results of the reputation-oriented Twitter task, which was held as part of SentiRuEval evaluation of Russian sentiment-analysis systems. The tweets in two domains: telecom companies and banks - were included in the evaluation. The task was to determine if an author of a tweet has a positive or negative attitude to a company mentioned in the message. The main issue of this paper is to analyze the current state and problems of approaches applied by the participants.

Keywords: Sentiment analysis · Sentiment classification · Social network

1 Introduction

People often use micro blogging platform Twitter to express their opinion about the world around them. According to public sources, the number of people signed up at Twitter exceeds 500 million and this number keeps growing. So Twitter greatly facilitates the spread of information, in particular users' attitude or current news about various companies, their products and services.

Twitter differs from other social networks. The length of text messages in Twitter is limited, so one can say that automatic text analysis is carried out on the phrase or sentence level, but not on the document level. Users often write tweets via mobile phones, so there are lots of misprints like "recreation abea" instead of "recreation area" («зонаоЬдыха», instead of «зона отдыха»). The language in tweets abounds in slang, word contractions and abbreviations.

Although micro blogs are quite new, researchers pay a lot of attention to the analysis of sentiment of messages in micro blogs in total and in Twitter in particular [1–4]. Several Twitter-based sentiment-oriented evaluations were organized [5,6]. In 2012-2014 within the CLEF conference, RepLab evaluation of online reputation management systems was held [7,8]. In 2014-2015 one of tracks within Russian Sentiment Analysis Evaluation SentiRuEval was directed towards automatic reputation-oriented analysis of tweets in Russian. In this paper, we

P. Král and V. Matoušek (Eds.): TSD 2015, LNAI 9302, pp. 551–559, 2015.
DOI: 10.1007/978-3-319-24033-6_62

briefly describe the task, data and guidelines of this task, the participants' results and approaches and present our analysis of problems of current approaches.

The rest of this paper is structured as follows. In the next section, related work is given. The third section describes the task, prepared training and test collections, obtained results. The analysis of the participants' results is presented in the fourth section. Finally, section five concludes.

2   Related Work

Several evaluations were devoted to sentiment analysis of opinionated tweets. In 2013-2014, the Twitter-oriented sentiment evaluation was held within the SemEval conference. Two subtasks were given to participants: an expression-level subtask and a message-level subtask. In the message-level subtask, participants should determine whether the entire message was opinionated or not [5,6]. The task is directed to reveal, namely, an author opinion in contrast to neutral or objective information.

In 2012-2014 within the CLEF conference RepLab events devoted to the eval-uation of online reputation management systems were organized [7,8]. The tasks included the definition of the polarity for reputation classification. The goal was to decide if the tweet content has positive or negative implications for the company's reputation.

The RepLab organizers stress that the polarity for reputation is substantially different from standard sentiment analysis that should differentiate subjective from objective information. When analyzing polarity for reputation, both facts and opinions have to be considered to determine what implications a piece of information might have on the reputation of a given entity [7,8]. Tweets from four domains were provided as data sets: automotive, banking, universities and music. The training and test collections were temporally divided with at least several month intervals. The evaluation of systems was based on the overall results without subdivision to the domains.

The previous Russian-oriented evaluation of sentiment analysis systems was held within the ROMIP workshop in 2011-2013 [9]. In 2011 the participants should classify user reviews for movies, books, and digital cameras. Besides, in 2012 the tasks of sentiment classification on news quotes and sentiment-oriented retrieval of blog-posts were set. The main goal of the previously performed Russian-oriented tests was automatic sentiment analysis of texts or text frag-ments in general. This year the SentiRuEval evaluation was directed to entity-oriented sentiment analysis of user reviews and tweets in Russian. In the current paper, we present the detailed analysis of achievements and problems of partic-ipating systems in the Twitter-oriented task of SentiRuEval.

3   Twitter Entity-Oriented Task at SentiRuEval

The goal of the Twitter sentiment analysis at SentiRuEval was to find tweets influencing the reputation of a company in two domains: banks and telecom companies. Such tweets may contain sentiment-oriented opinions or positive and negative facts about the company.

Such a task is quite similar to the reputation polarity task at RepLab evaluation [7,8]. The difference is that at SentiRuEval, tweets from only two domains are taken, and the systems are evaluated for these domains separately, which gives the possibility to compare results obtained in the domains. Tweets about eight banks and seven telecom companies were taken for the evaluation. The task for participants was to define the reputation-oriented attitude of a tweet in relation to a given company: positive, negative or neutral.

In the training and test collections, fields with the list of all companies of the chosen domain were denoted. By default, the field of a company mentioned in the tweet obtained '0' (neutral attitude) value. The participants should either replace '0' with '1' (positive attitude) or '-1' (negative attitude), or leave '0', if the tweet attitude to a company mentioned in the message is neutral.

3.1 Data Collections

The datasets were collected with Streaming API Twitter[1]. The training collections had been collected from July to August 2014, whereas the test collections were collected from December 2013 to February 2014. The distribution of messages in the training and test collections according to sentiment classes is shown in Table 1. The number of tweets is not equal to a sum of neutral, positive and negative references, as users may mention more than one company in a message.

In tweets users may list companies – in this case their attitude to all companies is usually the same. They can also compare and contrast companies with each other – in this case the users' attitudes are different.

Table 1. Distribution of messages in collections according to sentiment classes

		Neutral	Positive	Negative	Number of tweets
Telecom	Training collection	2397	973	1667	5000
	Gold standard test collection	2816	413	944	3845
Banks	Training collection	3569	410	2138	5000
	Gold standard test collection	3592	350	670	4549

We noticed that sometimes users do not want to be rude and add positive emoticons to clearly negative or ironic messages. That is why simple methods based on extraction of emoticons, which are used for classification on the whole tweet level, do not work well [10,11].

Table 2. Results of the voting procedure in labeling of the tweet in test collection

Domain	The number of tweets with the same labels from at least 2 assessors	Full agreement	The final number of tweets in the test collection
Telecom	4 503 (90.06%)	2 233 (44.66%)	3 845
Banks	4 915 (98.3%)	3 818 (76.36%)	4 549

3.2 Data Annotation and Measures

To prepare the datasets, 20,000 messages were labeled including 5,000 messages in each domain for training and test collections. Each collection was labeled at least by two assessors. The gold standard test collections were labeled by three assessors. The task for assessors was to estimate the tweet reputation-oriented attitude in relation to the labeled entity. To avoid inconsistency and disputes, the voting scheme was applied to the test collections labeling. Irrelevant or unclear messages were removed from the training and test sets. The results of preparing the test collections are given in Table 2.

As the main quality measure, we used macro-average F-measure calculated as the average value between F-measure of the positive class and F-measure of the negative class ignoring the neutral class. But this does not reduce the task to the two-class prediction because erroneous labeling of neutral tweets negatively influences Fpos and Fneg. Additionally, micro-average F-measures were calculated for two sentiment classes.

3.3 Participants and Results

A total of 9 participants with 33 runs have participated in the Twitter sentiment analysis tasks. The best three results for telecom tweets are presented in the Table 3. Table 4 contains the best results for bank tweets. The baselines are based on the majority reputation-oriented category (negative one in this case).

Table 3. Top 3 results for telecom tweets according macro F

Run_id	Macro F	Micro F
Baseline	0.1823	0.337
2	**0.4882**	**0.5355**
3	0.4804	0.5094
4	0.467	0.506

Table 4. Top 3 results for telecom tweets according macro F

Run_id	Macro F	Micro F
Baseline	0.1267	0.2377
4	**0.3598**	0.343
10	0.352	0.337
2	0.3354	**0.3656**

---

[1] https://dev.twitter.com/streaming/overview

Most participants used the SVM classification method. The Participant 2 classified tweets on the basis of lemmas and syntactic links presented as triples (head word, dependent word, type of relation). The Participant 10 employed word n-grams, letter n-grams, emoticons, punctuation marks, smilies, a manual sentiment vocabulary, and automatically generated sentiment list based on calculation of pointwise mutual information (PMI) of a word occurrences in positive or negative training subsets. The Participant 3 used a rule-based approach accounting syntactic relations between sentiment words and the target entities with no machine learning applied; the performance similar to machine learning approaches was achieved. The Participant 4 applied the maximum entropy method on the basis of word n-grams, symbol n-grams, and topic modeling results.

In addition, one of the participants fulfilled independent expert labeling of telecom tweets and obtained Macro-F – 0.703, and Micro F – 0.749, which can be considered as the maximum possible performance of automated systems in this task.

In general, the best results are comparable with the results of RepLab-2012 (F-measure=0.41) [7]. Their relatively low level is due to the difficulty of the task and the limited size of the training collections.

4   Analysis of Participants' Results in Two Domains

We analyzed the obtained results from several points of view trying to explain the difference in the results level obtained in two domains, to reveal the most difficult tweets for participants, and to understand if the approaches were really entity-oriented.

4.1   Explaining the Difference in the Perfomance in Two Domains

One can see that the results in banking and telecom collections are quite different (0.36 vs. 0.488 MacroF). It can be partially explained by different baseline levels (0.1267 vs. 0.1823 MacroF), which means in this case that the number of negative tweets was much larger in the telecom domain.

For further explanations we decided to compare word distributions in test and training collections for both domains. To determine probabilities of words in each collection, we applied additive smoothing to avoid zero-probabilities (Formula 1).

$$P(w_i) = \frac{f_i + 1}{N + d}, (i = 1, ..., d) ,\qquad (1)$$

where f_i is the frequency of a term w_i in a collection, N – the number of all terms in the collection, d – the number of different terms.

Then we computed the Kullback-Leibler divergence to compare the difference of word probability distributions in the test collections in relation to the training collections (Formula 2).

$$D_{KL} = \sum_i test_i \times \ln \frac{test_i}{train_i} \quad , \qquad (2)$$

where $test_i = P(w_i)$ in the test collection, $train_i = P(w_i)$ in the training collection.

The Table 5 presents the obtained values of the KL-divergence for full collections and separately for sentiment-oriented tweets. To obtain a symmetric measure, we estimated Jensen-Shannon divergence (Formula 3) that is a symmetrized and smoothed version of the KL-divergence (Table 6).

$$D_{JS} = \frac{1}{2}(\sum_i test_i \times \ln \frac{test_i}{M}) + \frac{1}{2}(\sum_i train_i \times \ln \frac{train_i}{M}), \qquad (3)$$

where $M = \frac{1}{2}(test + train)$.

Table 5. The values of KL-divergence of word distributions in banking and telecom training and test collections

Domain	Full	Sentiment	Positive	Negative
Banks	0.465	**0.505**	0.397	**0.561**
Telecom	0.317	0.287	0.323	0.284

Table 6. The values of Jensen-Shannon - divergence of word distributions in banking and telecom training and test collections

Domain	Full	Sentiment	Positive	Negative
Banks	0.084	**0.123**	0.092	**0.139**
Telecom	0.066	0.066	0.071	0.067

From the tables we can see that the difference of word distributions in test and training collections is much larger for the banking domain, and this difference is maximal for negative tweets. In our opinion, it is due to the fact that the topics of reputation-oriented tweets greatly depend on positive or negative events with the participance of the target entities.

Our evaluation demonstrated this problem quite evidently. Our training collections in both domains were collected during July-August 2014 after Ukraine events 2013-2014. Therefore negative banking tweets in the training collection often concerned sanctions against Russian banks and their consequences. These events also concerned telecom companies because of the problems with Ukraine and Russian communication companies in Crimea, but to a lesser degree. For testing, the tweets of the December 2013-February 2014 were utilized when Ukraine events have already begun but did not influence the target entities. Certainly, in the test collections, mentions of the sanctions and Crimea events were absent.

4.2   Analyzing Difficult Tweets

The next step of our study was to analyze tweets whose classification was especially difficult for participants. With this aim, we extracted tweets that were wrongly classified by almost all participants. We obtained 71 tweets in the banking domain wrongly classified by all participants and 85 tweets in the telecom domain that were difficult for almost all participants (maximum two systems gave correct answers). We found that these tweets can be subdivided into two groups with subgroups; each subgroup requires its specific method for improving polarity classification.

The **first group** includes tweets that were misclassified because of the restricted size of the training collection, which did not contain appropriate training examples. The first subgroup of this group (**subgroup 1.1**) of the wrongly classified tweets contains evident sentiment words (such as понравиться – to like) that were absent in the training set. In this case, the general vocabulary of Russian sentiment words could help. However, the only published list of Russian sentiment words was collected automatically and does not contain polarity labels [12]. This general vocabulary should also contain sentiment slang widely used in Twitter and other social nets.

The second subgroup (**subgroup 1.2**) of the difficult tweets contains words expressing well-known positive or negative situations such as *theft* or *murder* but absent in the training collection. These words are usually considered as neutral, not-opinionated, but having positive or negative associations (so called connotations) [13]. For solving these problems, a general vocabulary of connotative words would be useful because the appearance of these words in connection with a company influences its reputation.

The third subgroup (**subgroup 1.3**) of the problematic tweets contains words and phrases describing current events, concerning the current news flow. The apperance of some events and their influence the company's reputation are very difficult to predict, their mentioning will always be absent in the training collection. In this case, the parallel analysis of the current news, revealing correlations between tweet words and general sentiment and connotation vocabulaties in news texts, can help.

The **second group** of the misclassified tweets includes tweets that are really complicated. They can include several sentiment words with different polarity orientation, mention more than one entity with different attitudes, or contain irony.

The quantitative estimation of all above-mentioned groups and subgroups showed that about 30% of difficult tweets in the banking domain could be processed on the basis of various lexical resources (a general sentiment lexicon, a lexicon of connotative words, Twitter sentiment lexicon, and etc.). For the telecom domain, the share of such tweets is about 15%. It means that integration of various vocabularies into the machine-learning framework can improve the performance of reputation-oriented automatic systems.

4.3   Understanding If Systems Were Really Entity-Oriented

At the last step of our study we checked if participants really cope with an entity-oriented task or their systems classified tweets without accounting for mentioned entities. With this aim, we extracted tweets containing more than one entity from the test collections. We extracted 58 tweets in the banking domain (15 tweets with different polarity labels), 232 tweets in the telecom domain (71 tweets with different polarity labels). Using these tweets we checked if the participating systems can classify entities from the same tweet differently. We found that only three of nine participants considered the task as entity-oriented one. Other participants always assigned the same polarity class to all entities mentioned in a tweet.

Our next question was if the entity-oriented systems achieved better results. We calculated F-measure of classification only for tweets with two or more entities and found that entity-oriented systems in both domains obtained worse results than other systems. So, the current capability of systems to solve entity-oriented tasks in tweets, meanwhile, is very limited.

5   Conclusion

In this paper we briefly presented the reputation-oriented Twitter track of SentiRuEval evaluation of sentiment-analysis systems in Russian, which was fulfilled in two domains: banking and telecommunication companies. The aim of the tweet analysis was to classify messages according to their influence on the reputation of the mentioned company. Reputation-oriented tweets may express an opinion of an author, or positive or negative facts about this company. A total of 9 teams took part in this testing, most participants applied various machine-learning approaches.

We have analyzed the results of the participants and found that the best-achieved performance in the reputation oriented-task for a specific domain is correlated with the difference between word probability distributions over training and test collections in this domain. For the reputation task, such difference can arise from current dramatic events.

Besides, at this moment, the large impact for improving results in the tweet reputation task can be based on integration of a general sentiment vocabulary and a general vocabulary of connotative words to machine-learning approaches. Also we found that the most participants solved the general task of tweet classification; entity-oriented approaches did not achieve better results.

This work is partially supported by RFBR grant 14-07-00682. All prepared collections are available for research purposes: http://goo.gl/qHeAVo.

References

1. Kouloumpis, E., Wilson, T., Moore, J.: Twitter sentiment analysis: the good the bad and the omg! In: ICWSM, vol. 11, pp. 538–541 (2011)
2. Pak, A., Paroubek, P.: Twitter as a corpus for sentiment analysis and opinion mining. In: LREC, vol. 10, pp. 1320–1326 (2010)
3. Agarwal, A., Xie, B., Vovsha, I., Rambow, O., Passonneau, R.: Sentiment analysis of twitter data. In: Proceedings of the Workshop on Languages in Social Media, pp. 30–38. Association for Computational Linguistics (2011)

4. Rubtsova, Y.V.: Development and research domain independent sentiment classifier. SPIIRAS Proceedings **5**(36), 59–77 (2014)
5. Nakov, P., Kozareva, Z., Ritter, A., Rosenthal, S., Stoyanov, V., Wilson, T.: Semeval-2013 task 2: Sentiment analysis in Twitter (2013)
6. Rosenthal, S., Nakov, P., Ritter, A., Stoyanov, V.: Semeval-2014 task 9: sentiment analysis in twitter. In: Proc. SemEval (2014)
7. Amigo, E., Corujo, A., Gonzalo, J., Meij, E., de Rijke, M.: Overview of RepLab 2012: evaluating online reputation management systems. In: CLEF 2012 Evaluation Labs and Workshop Notebook Papers (2012)
8. Amigó, E., et al.: Overview of RepLab 2013: evaluating online reputation monitoring systems. In: Forner, P., Müller, H., Paredes, R., Rosso, P., Stein, B. (eds.) CLEF 2013. LNCS, vol. 8138, pp. 333–352. Springer, Heidelberg (2013)
9. Chetviorkin, I., Loukachevitch, N.: Evaluating sentiment analysis systems in Russian. In: Proceedings of BSNLP Workshop, ACL 2013, pp. 12–16 (2013)
10. Read, J.: Using emoticons to reduce dependency in machine learning techniques for sentiment classification. In: Proceedings of the ACL Student Research Workshop, pp. 43–48. Association for Computational Linguistics (2005)
11. Hogenboom, A., Bal, D., Frasincar, F., Bal, M., de Jong, F., Kaymak, U.: Exploiting emoticons in sentiment analysis. In: Proceedings of the 28th Annual ACM Symposium on Applied Computing, pp. 703–710. ACM (2013)
12. Chetviorkin, I., Loukachevitch, N.V.: Extraction of Russian sentiment lexicon for product meta-domain. In: COLING, pp. 593–610. Citeseer (2012)
13. Feng, S., Kang, J.S., Kuznetsova, P., Choi, Y.: Connotation lexicon: a dash of sentiment beneath the surface meaning. In: ACL (1), pp. 1774–1784 (2013)

Improved Estimation of Articulatory Features Based on Acoustic Features with Temporal Context

Petr Mizera[✉] and Petr Pollak

Faculty of Electrical Engineering, Czech Technical University in Prague, K13131,
Technicka 2, 166 27 Praha 6, Prague, Czech Republic
{mizerpet,pollak}@fel.cvut.cz
www.fel.cvut.cz,noel.feld.cvut.cz/speechlab

Abstract. The paper deals with neural network-based estimation of articulatory features for Czech which are intended to be applied within automatic phonetic segmentation or automatic speech recognition. In our current approach we use the multi-layer perceptron networks to extract the articulatory features on the basis of non-linear mapping from standard acoustic features extracted from speech signal. The suitability of various acoustic features and the optimum length of temporal context at the input of used network were analysed. The temporal context is represented by a context window created from the stacked feature vectors. The optimum length of the temporal contextual information was analysed and identified for the context window in the range from 9 to 21 frames. We obtained 90.5% frame level accuracy on average across all the articulatory feature classes for mel-log filter-bank features. The highest classification rate of 95.3% was achieved for the voicing class.

Keywords: Speech recognition · Spontaneous speech · Articulatory features · Neural networks · Phonetic segmentation · Log filter-bank features

1 Introduction

Articulatory features (AFs) contain useful information about speech production and their properties are currently very interesting for applications in important areas of automatic speech recognition (ASR), e. g. spontaneous, low resource or multilingual speech recognition [1], [2], etc. AFs are used separately or in various combinations with the most common features (MFCC/PLP) for better modelling of the co-articulation, assimilation and reduction processes which are presented in spontaneous or casual speech [3]. They can also help in other technologies such as language identification [4] or speaker identification/verification systems [5] and improve the robustness of noisy speech recognition [6].

Various machine learning algorithms were used for the estimation of AFs. Among the most frequently used approaches are Artificial Neural Networks

© Springer International Publishing Switzerland 2015
P. Král and V. Matoušek (Eds.): TSD 2015, LNAI 9302, pp. 560–568, 2015.
DOI: 10.1007/978-3-319-24033-6_63

(ANN) [3], [7] and Dynamic Bayesian Networks. In addition, also other classifiers such as Hidden Markov models (HMM), k-nearest neighbour algorithm or Gaussian Mixture Model (GMM) are used [8], [9]. However, the standard Multi-Layer Perceptron (MLP) network still represents the most frequent approach used for AF estimation.

Concerning acoustic features used at the input of MLP, the Mel-frequency cepstral coefficients (MFCC) or Perceptual Linear Prediction (PLP) coefficients are commonly used. However, the suitability of cepstral based features for the state-of-the-art hybrid DNN-HMM ASR systems or recently proposed novel CNN-HMM approaches were discussed in [10], [11]. It was shown that the systems using the mel-log filter-bank features (band logarithmic energies) achieved better results in contrast to cepstral features. It was also shown that the long temporal contextual information is very important for the increase of ASR accuracy in general [12], [13].

On the basis of these results, we analysed the suitability of mel-log filter-bank features (FBANK) in the task of the MLP based AF estimation and we compared them with common cepstral features. The paper also investigates the usage of long-time context information in the MLP-based AF estimation and analyses the optimum length of contextual information in this task. At first, contextual information is created from standard MFCC/PLP/FBANK features and then also from their dynamic coefficients. Both approaches are compared in the experimental section. Another purpose of this work is to analyse the optimum setup of the number of hidden neurons in the MLP network depending on the length of the contextual window. The KALDI toolkit was used to implement this task.

2   AF Estimation

The neural network based AFs estimation is currently the most frequently used approach. Although Deep Neural Networks (DNN) are becoming very popular in the speech community nowadays and they are also used by some authors for AF estimation with very promising results [14], [15], [16], the 3-layer MLP networks are still frequently used. The whole process of AF estimation is presented within this section.

2.1   AF Classes for Czech

AFs are principally defined on the basis of particular phone generation known as the process of articulation. For the English language, several approaches of AF definition are used with slightly varying amount of classes and categories [3]. Concerning Czech, we have defined them similarly to the above mentioned English standard and we took into account phoneme categories as they are used standardly for Czech [17]. The Czech language consists of 49 phones which are defined by SAMPA standard [18] and can be categorized into the phonetic classes such as vowels, fricatives, affricates, plosives, sonorants (nasals and approximants). The brief overview of used AFs for Czech is presented in Table 1. Each class is

Table 1. AF classes for the Czech language.

AF class	Cardinality	Feature values
voicing	3	voiced, unvoiced
place_c	9	bilabial, labiodental, prealveolar, postalveolar, palatal, velar, glottal, nil
place_v	5	front, central, back, nil
manner_c	9	stop, nasals, affricates, fricatives, trills, lateral, glides, nil
manner_v	5	open, mid, close, nil
rounding	4	rounded, unrounded, nil
sonority	4	noise, sonor, nil

completed by the 'silence' value which increases class cardinality. The particular elements (phonemes) belonging to the AF categories for the Czech language are described in more detail [19].

2.2 Acoustic Features for AF Estimation

As mentioned, above the common cepstral features are standardly used for AF classification, i.e. MFCC/PLP features similarly as within GMM-HMM based ASR systems. It was presented by many authors that the long temporal contextual information is very important for increasing accuracy of ASR systems or phoneme recognition [20], [12]. The long temporal contextual information can be integrated into these systems by the TempoRAl Pattern (TRAP) based features such as DCT-TRAP, wLP-TRAPS [21] or can be commonly caught using context windows of various length. The contribution of TRAP based features in the task of AF estimation was analysed for the English language [22] and also used in our first work for the Czech language [19]. In this paper we focused on the second method where the temporal context information is created using a context window from several neighbouring short-time frames (MFCC/PLP). 90 ms was proposed as an optimum length of temporal context for English in [13] and is standardly used by other authors for AF estimation [7].

On the basis of these results, we would like to analyse the optimum length of context window in the task of Czech AF estimation. The context window is made of MFCC/PLP and their delta, delta-delta coefficients. We work with rather standard setup for MFCC/PLP features which can be summarized as follows; the length of frame 25ms, the shift of frame 10ms, Hamming window, 30 mel-scaled filter-bank in the range 64-8000 Hz for MFCC (bark filter bank for PLP), liftering of order 22 and finally 12 MFCC/PLP static coefficients with $c0$ completed by their delta and double delta coefficients. Concerning the FBANK features, 23 bands of log mel-scaled filter-bank in the range 64-8000 Hz were used.

2.3 MLP-Based AF Classification

Finally, we worked with seven AF classes and the MLP network was used for their classification. Precisely, the independent AF classifiers for each one of AF classes were created on the basis of feed-forward MLP structure consisting of the three-layers. The input layer distributes the acoustic feature vector to the hidden layer and the output layer generates posterior probabilities for corresponding AF class. The sigmoid activation function for all neurons in the hidden layer and the softmax activation function for the neurons in the output layer were selected. The size of input and output layers are related to the size of the feature vector and AF cardinality. The mini-batch Stochastic Gradient Descent training and the Newbob learning rate strategy are employed for learning of MLP networks to classify the AF classes.

3   Experiments and Discussion

Two groups of experiments were realized with the main purpose to find the optimum length of context information for AF estimation for the cases when the context window was created either from static or from differential features respectively.

3.1   Experimental Setup

The Czech SpeeCon database was used for our analyses. The corpus consists of 550 adult speakers and contains phonetically rich and balanced sentences. The database covers varied environments such as an office, entertainment, a public place or a car. The speech signals were recorded at 16 kHz sampling frequency and each utterance contains orthographic transcription. This design of the database is very suitable for creating acoustic models for speech recognition [23].

We performed all our experiments on the sub-part of this database which contained read speech from office and entertainment environment. Since the database contains only information about the orthographic transcription, the AF target for MLP learning was obtained using HMM-based phonetic forced alignment followed by the mapping of phones to articulatory features. The sub-part of the database was further divided into standard disjunctive data sets for training, testing and cross validation. The test set was manually labelled by phonetic expert on level of phoneme. More details about datasets are shown in Tab. 2.

Table 2. Data subsets used in experiments.

set	speakers	sentences	hours
training	101	3450	4.99
cross-val.	17	585	0.88
test	77	103	0.18

The initial feature vector was extracted by our CtuCopy tool [24] and then stacked with additional vectors from neighbouring context frames in a pipeline. The dimensionality of the resulting vector differed in dependence on the size of the initial vector and the context window's length. Finally, the whole feature vector was normalized using the cepstral mean and normalization (CMVN) technique and passed to the input layer of the MLP network. The implementation of the MLP-based AF classifier was done in the KALDI toolkit (mainly the *nnet* tools were used) [25]. The experiments were performed on two GPU nodes.

3.2 Results

In the course of our experiments, the accuracy of AF classification was evaluated by the standard frame level accuracy (*FAcc*) criterion. It was computed as the ratio between the number of correctly labelled short-time frames and the total number of frames.

I. The Results of MLP Size Optimization

Empirical analysis of the optimum setup of MLP size for the range from 10 to 2400 hidden layer units was performed. The dependency of the frame accuracy on the number of hidden neurons is shown in Fig. 1. Although the optimum number of hidden neurons for various MLP based AF classifiers was in the range from 400 to 800, only very little improvements were observed for more than 400 neurons. The best setups of the MLP network for training the AF classifiers which achieved the best results estimation are summarized in Tab. 3.

Table 3. The optimum setup size of MLP for the best results classification of AFs.

	Out	MFCC_0_cw_17			_PLP_0_cw_17			FBANK_0_cw_21		
	units	hids	Epoch	CV	hids	Epoch	CV	hids	Epoch	CV
Voicing	3	400	12	94.8	400	14	94.9	600	13	95.4
P_con.	9	600	12	86.3	600	13	86.5	800	14	87.6
P_vowel	5	600	14	89.2	600	14	89.4	800	16	89.9
M_con.	9	600	13	87.2	600	14	87.4	800	14	88.4
M_vowel	5	600	15	88.0	600	13	88.1	800	15	88.8
Rounding	4	500	14	89.9	500	14	90.1	600	15	90.1
Sonor	4	600	12	89.3	600	13	89.4	800	14	90.1

II. The Results of Context Length Optimization

The optimum length of the context information in AF estimation was analysed in the two similar scenarios of the experiments . Two groups of experiments focused on finding the optimum length of the context information for the case where the context window was created either from static or dynamic features.

Firstly, the experiments were focused on modelling of the context information based on the context window with static MFCC, PLP or FBANK features only. The context window size in the range from 3 to 61 frames was analysed. The results are shown in Fig. 2. The average absolute improvement of the accuracy of AF classification depending on the varying size of the context window was compared to the zero context. The results are presented in Fig. 2 and the best size

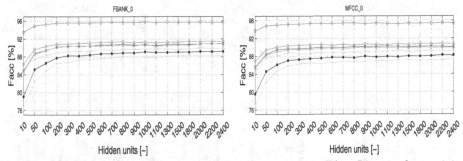

Fig. 1. The dependency of the number of hidden neurons on FAcc. Lines: red ⋄ - *voicing*, yellow * - *placed_consonant*, blue × - *placed_vowel*, black • - *manner_consonant*, cyan ▷ - *manner_vowel*, magenta ○ - *rounding*, green □ - *sonority*.

of the context window is marked by blue color of the proper bar. Secondly, the similar scenarios were applied also to differential features (dynamic and acceleration) of MFCC, PLP or FBANK. The results of experiments are presented in Fig. 3. In this case the context window size was analysed in the range from 5 to 31 frames and the average absolute improvement is also presented using a bar graph. All results presented in Fig. 2, 3 were evaluated against automatically labelled test data set.

Experiments showed that it was better to work with the static features only and to take slightly longer context instead of using differential features. In addition, this setup led to a vector with lower dimensions which sped up the process of MLP training and reduced the data requirements. Finally, the optimum length of contextual information for particular types of parametrization achieving the best results of AF classification are summarized in Table 4. The results for frame accuracy evaluated for manually labelled test set are also presented. The best results of AF classification were achieved for FBANK static features with the context of 21 frames (FBANK_0_cw_21).

Table 4. The best results estimation of AFs.

	MFCC				PLP				FBANK			
	0		0_d_a		0		0_d_a		0		0_d_a	
	cw_17		cw_9		cw_17		cw_9		cw_21		cw_9	
	test	test m.	test	test m.	test	test m.	test	test m.	test	test m.	test	test m.
Voicing	95.0	90.7	94.9	90.7	94.9	90.8	94.8	90.7	95.3	90.9	95.2	90.8
Place_con	86.9	79.5	86.0	79.2	86.2	79.3	86.0	79.0	87.9	80.2	86.5	79.4
Place_vow	89.8	82.7	89.9	82.7	89.6	82.5	89.8	82.6	90.5	83.1	90.2	82.9
Manner_con	87.8	80.1	87.4	80.0	87.2	80.1	87.3	80.1	88.7	80.5	87.9	80.2
Manner_vow	88.9	81.6	88.5	81.5	88.6	81.7	88.7	81.5	89.5	82.2	89.1	81.9
Rounding	90.3	83.2	90.2	83.0	90.2	83.1	90.4	83.3	91.0	83.9	90.8	83.5
Sonor	89.5	81.8	89.5	81.7	89.3	81.6	89.4	81.8	90.5	82.2	90.0	82.0
avg.	**89.7**	82.8	**89.5**	82.7	**89.4**	82.7	**89.5**	82.7	**90.5**	83.3	**90.0**	83.0

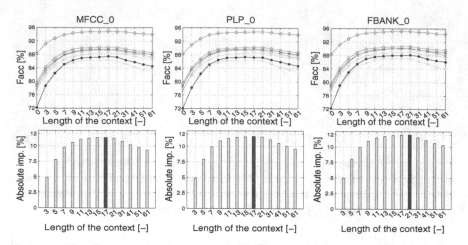

Fig. 2. The evaluation of AF estimation for lengths of context information for static features. Lines: red ◇ - *voicing*, yellow ∗ - *placed_consonant*, blue × - *placed_vowel*, black ● - *manner_consonant*, cyan ▷ - *manner_vowel*, magenta ○ - *rounding*, green □ - *sonority*.

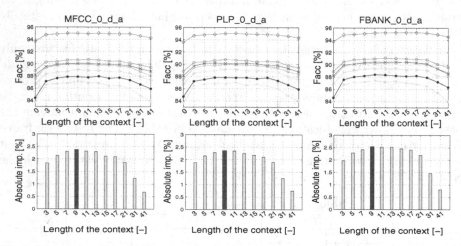

Fig. 3. The evaluation of AF estimation for lengths of context information for differential features. Lines: red ◇ - *voicing*, yellow ∗ - *placed_consonant*, blue × - *placed_vowel*, black ● - *manner_consonant*, cyan ▷ - *manner_vowel*, magenta ○ - *rounding*, green □ - *sonority*.

4   Conclusions

In this paper we presented the results of MLP based AFs classification for the Czech language using various acoustic features. We performed detailed analyses of the optimum length of contextual information. Experiments showed that the optimum length for the modelling of context information is between 150 and 210 ms for the static parameters. When differential features were used, the length decreased to 90 ms for all used features, which corresponds to conclusions in [22]. The best result of 90.5% was reached for the log mel-scaled filter-bank setting. It represented more than 2% improvement of FAcc across all AF classes in comparison to our best results obtained within previous research [19].

Acknowledgments. The research described in this paper was supported by internal CTU grant SGS14 /191/OHK3/3T/13 "Advanced Algorithms of Digital Signal Processing and their Applications".

References

1. Qian, Y., Liu, J.: Articulatory feature based multilingual mlps for low-resource speech recognition. In: INTERSPEECH. ISCA (2012)
2. Qian, Y., Xu, J., Povey, D., Liu, J.: Strategies for using mlp based features with limited target-language training data. In: 2011 IEEE Workshop on Automatic Speech Recognition and Understanding (ASRU), pp. 354–358 (December 2011)
3. Frankel, J., Magimai-Doss, M., King, S., Livescu, K., Cetin, O.: Articulatory feature classifiers trained on 2000 hours of telephone speech. In: Proceedings of Interspeech, Antwerp, Belgium (2007)
4. Carson-Berndsen, J.: Articulatory-acoustic-feature-based automatic language identification. In: ISCA - MultiLing 2006, Stellenbosch, South Africa (2006)
5. Zhang, S.X., Mak, M.: High-level speaker verification via articulatory-feature based sequence kernels and SVM. In: Proceedings of Interspeech, Brisbane, Australia (2008)
6. Kirchhoff, K.: Combining articulatory and acoustic information for speech recognition in noisy and reverberant environments. In: Proceedings of ICSLP (1998)
7. Rasipuram, R., Magimai-Doss, M.: Improving articulatory feature and phoneme recognition using multitask learning. In: Honkela, T. (ed.) ICANN 2011, Part I. LNCS, vol. 6791, pp. 299–306. Springer, Heidelberg (2011)
8. Frankel, J., Wester, M., King, S.: Articulatory feature recognition using dynamic Bayesian networks. Computer Speech & Language, 620–640 (2007)
9. Næss, A.B., Livescu, K., Prabhavalkar, R.: Articulatory feature classification using nearest neighbors. In: INTERSPEECH, Florence, Italy, ISCA, pp. 2301–2304 (2011)
10. Li, J., Yu, D., Huang, J.T., Gong, Y.: Improving wideband speech recognition using mixed-bandwidth training data in cd-dnn-hmm. In: 2012 IEEE Spoken Language Technology Workshop (SLT), pp. 131–136 (December 2012)
11. Abdel-Hamid, O., Mohamed, A.R., Jiang, H., Deng, L., Penn, G., Yu, D.: Convolutional neural networks for speech recognition. IEEE/ACM Transactions on Audio, Speech, and Language Processing 22(10), 1533–1545 (2014)
12. Morgan, N., et al.: Pushing the envelope - aside [speech recognition]. IEEE Signal Processing Magazine 22(5), 81–88 (2005)
13. Pinto, J.P., Prasanna, S.R.M., Yegnanarayana, B., Hermansky, H.: Significance of contextual information in phoneme recognition. Idiap-RR Idiap-RR-28-2007, IDIAP (2007)
14. Yu, D., Siniscalchi, S., Deng, L., Lee, C.: Boosting attribute and phone estimation accuracies with deep neural networks for detection-based speech recognition. In: Acoustics, Speech and Signal Processing (ICASSP), pp. 4169–4172 (2012)

15. Abdel-hamid, O., Rahman Mohamed, A., Jiang, H., Penn, G.: Applying convolutional neural networks concepts to hybrid NN-HMM model for speech recognition. In: Acoustics, Speech and Signal Processing (ICASSP), pp. 4277–4280 (2012)
16. Hinton, G., et al.: Deep neural networks for acoustic modeling in speech recognition. Signal Processing Magazine (2012)
17. Volín, J.: Phonetic and phonology. In: Cvrček, V., et al. (eds.) Grammar of Contemporary Czech. Karolinum (2013) 35–64; In Czech language: Mluvnice současné češtiny
18. Wells, J.C., Batusek, R., Matousek, J., Hanzl, V.: Czech SAMPA Home Page (2003). http://www.phon.ucl.ac.uk/home/sampa/czech-uni.htm
19. Mizera, P., Pollak, P.: Robust neural network-based estimation of articulatory features for czech. Neural Network World **24**(5), 463–478 (2014)
20. Schwarz, P.: Phoneme recognition based on long temporal context, PhD Thesis (2009)
21. Park, J., Diehl, F., Gales, M.J.F., Tomalin, M., Woodland, P.C.: Efficient generation and use of mlp features for arabic speech recognition (2008)
22. Kirchhoff, K.: Robust speech recognition using articulatory information. PhD thesis, Der. Technischen Fakultaet der Universitaet Bielefeld (June 1999)
23. Pollak, P., Cernocky, J.: Czech SPEECON adult database. Czech Technical University in Prague & Brno University of Technology, Technical report (April 2004)
24. Fousek, P., Mizera, P., Pollak, P.: Ctucopy feature extraction tool. http://noel.feld.cvut.cz/speechlab/
25. Povey, D., Ghoshal, A., et al.: The Kaldi speech recognition toolkit. In: Proc. of ASRU 2011, IEEE 2011 Workshop on Automatic Speech Recognition and Understanding (2011)

Do Important Words in Bag-of-Words Model of Text Relatedness Help?

Aminul Islam[✉], Evangelos Milios, and Vlado Kešelj

Faculty of Computer Science, Dalhousie University, Halifax, Canada
{islam,eem,vlado}@cs.dal.ca

Abstract. We address the question of how Bag-of-Words (BoW) models of text relatedness can be improved by using important words in the text-pair instead of all the words. To find important words in a text, we use a new approach based on word relatedness. We use two text relatedness methods: Latent Semantic Analysis (LSA) and Google Trigram Model (GTM) on five different datasets where words in the text-pair are sorted based on importance. We compare the use of a small number of important words against the use of all the words in the texts, and we find that both LSA and GTM achieve better results on four of the data sets and the same result on the fifth dataset.

Keywords: Text relatedness · Word relatedness · Bag-of-Words · GTM · LSA

1 Introduction

Text relatedness[1] methods have many applications in natural language processing (NLP) and related areas such as text or document clustering [1], text summarization [2], word sense disambiguation [3], information retrieval [4], image retrieval [5], text categorization [6], and database schema matching [7]. In general, most works on text relatedness can be abstracted as a function of word relatedness [8]. This reflects the direct importance of words chosen in determining text relatedness. Thus, the question arises: Does taking into account all or some of the words in texts impact performance of text relatedness? This question is important especially for BoW models where a text (such as a sentence or a document) is represented as an unordered collection of words, disregarding grammar and even word order. It is obvious that removing stop words provides better relatedness scores. The question is whether removing some unimportant words in addition to the stop words provides better text relatedness scores. To the best of our knowledge, no text relatedness method has tried to address this issue. The reasons for using BoW models are threefold. First, most BoW models do not require extra NLP tools and resources. Second, most BoW models are time-efficient as they do not use extra NLP tools and resources. Third, the

---

[1] We use 'relatedness' and 'similarity' interchangeably in this paper, though 'similarity' is a special case or a subset of 'relatedness'.

© Springer International Publishing Switzerland 2015
P. Král and V. Matoušek (Eds.): TSD 2015, LNAI 9302, pp. 569–577, 2015.
DOI: 10.1007/978-3-319-24033-6_64

performance accuracy of some BoW models is comparable to the methods that use substantial NLP tools and resources (e.g., [9,10]) as shown in this study. In a classical BoW text relatedness method, a pair of unordered texts is fed into a method and a relatedness score is returned for the pair (Figure 1 without dotted box). We propose the application of the text relatedness method to a small(er) set of important words instead of the set of all words, and our hypothesis is that the quality of the results (in terms of relatedness score) is at least as good in the former case as in the latter (Figure 1).

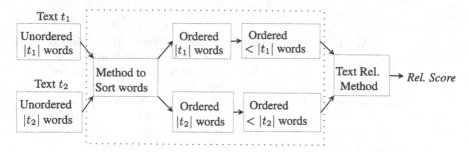

Fig. 1. High-level overview of the proposed approach based on sorted words

The rest of this paper is organized as follows: A brief overview of the related work is presented in Section 2. An unsupervised approach to find important words is described in Section 3. Two BoW methods of text relatedness are briefly described in Section 4. A brief description of five evaluation datasets and the experimental results is in Section 5. We address some contributions and future related work in Conclusion.

2   Related Work

To the best of our knowledge, the use of important words against all the words has not been compared in different BoW methods to text relatedness. Existing work on determining text relatedness is broadly categorized into three major groups: corpus-based (e.g., [11]), knowledge-based (e.g., [12]) and hybrid measure (e.g., [13]). A detailed survey of text relatedness can be found in [14]. In general, BoW methods fall into corpus-based category. In [11] a corpus-based semantic text similarity (STS) measure is proposed as a function of string similarity, word similarity, and common-word-order similarity. For determining word similarity, they focused on corpus-based measures [15,16]. In [12] a word-to-word relatedness measure is used based on the construction of semantic links between individual words from WordNet. In [13] a hybrid measure using three dictionaries (i.e., WordNet, OntoNotes, and Wiktionary) along with the Brown corpus is proposed.

Most methods in SemEval-2012 [17] and SemEval-2013 [18] shared task on semantic textual similarity are hybrid where different tools and resources were used. The UKP system [9], the top-ranked system for SemEval STS 2012 task [17], uses a trained log-linear regression model and combines multiple text similarity methods based on lexical-semantic resources. The UMBC_EBIQUITY-CORE system [10], the top-ranked system for SemEval STS 2013 core task [18], uses a term alignment algorithm augmented with penalty terms. It also uses a word similarity feature that combines LSA word similarity and WordNet. The tools and resources used by this system are monolingual corpora, WordNet, KB Similarity, lemmatizer, POS tagger, and time and date resolution [18].

3   Finding Important Words Using Word Relatedness

We use the assumption that a word in a text is more 'important' if it is more representative as well as more discriminative compared to the rest of the words in the text. A word in a text is more representative if it gives higher relatedness scores with the rest of the words in the text. This notion can be captured by the mean relatedness scores of a word with the rest of the words in the text. Again, a higher standard deviation of the relatedness scores of a word with the rest of the words in the text is an indication of how discriminative a word is compared to the other words. Combining these two notions indicates that the word that maximizes the mean (μ) and the standard deviation (δ) of the relatedness scores of itself with the rest of the words in the text is more important than others. Thus, we call this approach 'Mean+STD' (henceforth, $\mu+\delta$).

$$M_1 = \begin{array}{c|cccccc} & w_1 & w_2 & w_3 & \cdots & w_{b-1} & w_b \\ \hline w_1 & w_{11} & w_{12} & w_{13} & \cdots & w_{1(b-1)} & w_{1b} \\ w_2 & w_{21} & w_{22} & w_{23} & \cdots & w_{2(b-1)} & w_{2b} \\ w_3 & w_{31} & w_{32} & w_{33} & \cdots & w_{3(b-1)} & w_{3b} \\ \cdots & \cdots & \cdots & \cdots & \cdots & & \\ w_{b-1} & w_{(b-1)1} & w_{(b-1)2} & w_{(b-1)3} & \cdots & w_{(b-1)(b-1)} & w_{(b-1)b} \\ w_b & w_{b1} & w_{b2} & w_{b3} & \cdots & w_{b(b-1)} & w_{bb} \end{array}$$

Fig. 2. Word relatedness symmetric matrix

For each word against the rest of the words in a text, T, consisting of b words[2], we compute $\frac{b^2-b}{2}$ word-pair relatedness values. Using the special character of symmetric matrices can save time and storage during the word-pair relatedness calculation. Generally, a square matrix, M_1, of order b requires storage for b^2 elements. If M_1 is a symmetric matrix then it can be stored in about less than half the space, $\frac{b^2-b}{2}$ elements (shown as grey cells in Figure 2). Only the upper (or lower) triangular elements of M_1 excluding the main diagonal need to be explicitly stored. The implicit elements of M_1 can be retrieved interchanging row and column numbers. An efficient data structure for storing a symmetric matrix is a simple linear array. If the upper triangular elements of M_1 excluding the main diagonal are retained, the linear array, B, is organized as:

---

[2] After removing punctuation and stop words.

$B = \{w_{12}, w_{13}, \cdots, w_{1b}, w_{23}, \cdots, w_{2b}, \cdots, w_{(b-1)b}\}$. The indexing rule to retrieve the element w_{ij} from B is: $w_{ij} \leftarrow B[(i-1)(b-1) - (i-1)i/2 + j - 1]$

To find important words using this assumption and data structure, we use two algorithms based on the word relatedness method proposed by [19]. We use word relatedness method proposed by [19], though other word relatedness methods could be used instead. The arithmetic means of each of b rows (excluding the diagonal cells shown in Figure 2) are computed using Algorithm 1. The initial text, T, relatedness scores of $\frac{b^2-b}{2}$ word-pair, and arithmetic means of each of b rows generated by Algorithm 1 are used by Algorithm 2 to compute the summations of means and standard deviations of each of b rows and then to sort the b words by this score in descending order.

Algorithm 1. Computing the arithmetic means of each of b rows excluding the diagonal cells

Require: A linear *array*,
$\quad B = \{w_{12}, w_{13}, \cdots,$
$\quad w_{1b}, w_{23}, \cdots, w_{2b}, \cdots, w_{(b-1)b}\}$
Ensure: A linear *array*,
$\quad C = \{\mu_1, \mu_2, \cdots, \mu_b\}$

$\quad\quad\quad\quad\quad\quad\quad\quad\quad\quad\quad\quad \triangleright$

$|C| = b$ and $\mu_i \in C$ is the mean of the relatedness scores of w_i with the rest $b-1$ words

1: $i \leftarrow 1$
2: **while** $i \leq b$ **do**
3: $\quad j \leftarrow 1$, $sum \leftarrow 0$
4: $\quad$ **while** $j \leq b$ **do**
5: $\quad\quad$ **if** $j > i$ **then**
6: $\quad\quad\quad w_{ij} \leftarrow B[(i-1)(b-1) - (i-1)i/2 + j - 1]$
7: $\quad\quad$ **else**
8: $\quad\quad\quad$ **if** $i > j$ **then**
9: $\quad\quad\quad\quad w_{ij} \leftarrow B[(j-1)(b-1) - (j-1)j/2 + i - 1]$
10: $\quad\quad\quad$ **end if**
11: $\quad\quad$ **end if**
12: $\quad\quad sum \leftarrow sum + w_{ij}$
13: $\quad\quad$ increment j
14: $\quad$ **end while**
15: $\quad C[i] \leftarrow \frac{sum}{b-1}$
16: $\quad$ increment i
17: **end while**

Algorithm 2. Sorting based on the summation of mean and standard deviation of each of b rows

Require: $B = \{w_{12}, w_{13}, \cdots,$
$\quad w_{1b}, w_{23}, \cdots, w_{2b}, \cdots, w_{(b-1)b}\}$,
$\quad C = \{\mu_1, \mu_2, \cdots, \mu_b\}$ and
$\quad T = \{w_1, \cdots, w_b\}$
Ensure: $T' = \{w_1', \cdots, w_b'\}$

1: $i \leftarrow 1$
2: **while** $i \leq b$ **do**
3: $\quad j \leftarrow 1$, $sum \leftarrow 0$
4: $\quad$ **while** $j \leq b$ **do**
5: $\quad\quad$ **if** $j > i$ **then**
6: $\quad\quad\quad w_{ij} \leftarrow B[(i-1)(b-1) - (i-1)i/2 + j - 1]$
7: $\quad\quad\quad sum \leftarrow sum + (w_{ij} - \mu_i)^2$
8: $\quad\quad$ **else**
9: $\quad\quad\quad$ **if** $i > j$ **then**
10: $\quad\quad\quad\quad w_{ij} \leftarrow B[(j-1)(b-1) - (j-1)j/2 + i - 1]$
11: $\quad\quad\quad\quad sum \leftarrow sum + (w_{ij} - \mu_i)^2$
12: $\quad\quad\quad$ **end if**
13: $\quad\quad$ **end if**
14: $\quad\quad$ increment j
15: $\quad$ **end while**
16: $\quad \sigma_i \leftarrow \sqrt{\frac{sum}{b-1}}$
17: $\quad D[i] \leftarrow \mu_i + \sigma_i$
18: $\quad$ increment i
19: **end while**
20: $T' \leftarrow$ Sort T by D in desc. order

4 Methods Used

This section presents the two methods to text-relatedness that are used in this study: the LSA [20] and the GTM [19].

4.1 Latent Semantic Analysis

LSA is a method for extracting and representing the similarity of meaning of words and texts by statistical computations applied to a large corpus of text. The underlying idea is that the aggregation of all the word contexts with the presence and absence of a given word provides a set of mutual constraints that generates a representation of the similarity of meaning of words and set of words to each other [20]. LSA is based on Singular Value Decomposition (SVD), a mathematical matrix decomposition technique, to condense a very large matrix of word-by-context data into a much smaller, but still high-typically 50-500 dimensional semantic space. A semantic space in LSA is a mathematical representation of a large corpus of text and every term, text, or novel combination of terms has a high dimensional vector representation. In our experiment section in this paper, we have used the general knowledge space (college level) available on the LSA Web site (http://lsa.colorado.edu/). These spaces use a variety of texts, novels, newspaper articles, and other information, from the TASA (Touchstone Applied Science Associates, Inc.) corpus.

4.2 The Google Trigram Model

GTM uses Google n-grams [21] to compute word relatedness of representative words from each text to ultimately compute text-pair relatedness as described in [19] and available on the GTM Web site (http://ares.research.cs.dal.ca/gtm/). In word-pair relatedness, the frequencies of the Google trigrams that start and end with the given pair and the unigram frequencies of the pair are taken into account. The text relatedness method first separates shared words between texts being compared before computing the word relatedness. The count of the separated words and the relatedness scores of representative words between the texts are then normalized using the texts' lengths.

The GTM constructs a $(|t_1| - \delta) \times (|t_2| - \delta)$ 'relatedness matrix', M_2, using the word relatedness method described in [19], where t_1, t_2 are the two texts with $|t_1| \leq |t_2|$, the number of shared words between texts t_1 and t_2 is denoted by δ, and w_{ij} is the relatedness between word i of t_1 and word j of t_2. In order to determine the relatedness between t_1 and t_2, the following strategy is applied:

$$sim(t_1, t_2) = \frac{(\delta + \sum_{i=1}^{|t_1|-\delta} \mu(A_i)) \cdot (|t_1| + |t_2|)}{2 \cdot |t_1| \cdot |t_2|} \quad (1)$$

where A_i is a set of elements from row i of matrix M_2 such that each element in the set is greater than or equal to the summation of the mean and the standard deviation of that row, and $\mu(A_i)$ is the mean of A_i.

5   Evaluation Datasets and Results

There are very few publicly available collections of long text pairs with their ground truth of text relatedness. Four out of the five datasets used are from the SemEval-2012 [17] and SemEval-2013 [18] shared task on semantic textual similarity. For all of the datasets, preprocessing (i.e., punctuation and stop words removal) is done. Maximum (Max.), minimum (Min.), average (Avg.) number of words, and standard deviation (STD) from the average before and after the preprocessing for all the datasets are shown in Table 1.

Table 1. Statistics of Datasets used

| Dataset | # of pairs | Number of Words | | | | | | | |
| | | After Stop word removal | | | | Before Stop word removal | | | |
		Max.	Min.	Avg.	STD	Max.	Min.	Avg.	STD
ABC1225	1225	59	23	39.32	8.42	126	45	80.2	17.5
OnWN2012	750	22	1	4.75	3.07	55	4	15.13	6.67
SMTeuroparl2012	459	12	1	6.80	3.48	37	2	21.40	9.40
SMT2013	750	57	1	16.24	7.46	188	2	52.79	23.24
HDL2013	750	16	2	5.93	2.44	38	8	14.42	3.65

5.1   ABC1225

This dataset [22] contains 1225 pairs of documents evaluated by human judges. To create this dataset, 50 documents were selected from the Australian Broadcasting Corporation's news mail service, which provides text emails of headline stories (henceforth, ABC1225) and 83 human subjects were used. In [22] it was also produced the 'inter-rater' correlation (0.605).

Using GTM for text relatedness, $\mu+\delta$ method for sorting words (henceforth, GTM with $\mu+\delta$) and 23 sorted words from this dataset, we achieve Pearson correlation $r = 0.588$ with human ratings, while using all the 59 words we achieve Pearson's $r = 0.519$ (Figure 3) and the difference is statistically significant at the 0.05 level[3]. Similarly, using LSA for text relatedness and $\mu+\delta$ for sorting words (henceforth, LSA with $\mu+\delta$), we achieve Pearson's $r = 0.543$ for 33 sorted words while for all the words we achieve Pearson's $r = 0.531$ and the difference is statistically significant at the 0.05 level.

5.2   OnWN2012

This dataset from SemEval-2012 [17] contains 750 pairs of glosses from OntoNotes 4.0 and WordNet 3.1 senses (henceforth, OnWN2012). The similarity between the sense pairs was generated using simple word overlap. On this dataset, using

---

[3] To compare between two dependent (overlapping) correlations, we use the procedure mentioned in [23]. In the rest of the paper, the difference between two Pearson's r is not statistically significant at the 0.05 level, unless otherwise specified.

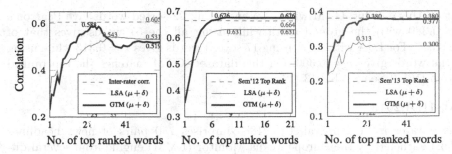

Fig. 3. Corr. with human ratings for ABC1225   **Fig. 4.** Corr. with ground truth for OnWN2012   **Fig. 5.** Corr. with ground truth for SMT2013

9 sorted words for both GTM with $\mu+\delta$ and LSA with $\mu+\delta$ gives the highest Pearson's $r = 0.676$ and $r = 0.631$, respectively with the ground truth while using all the 22 words also gives the same correlation (Figure 4). GTM also marginally outperforms the top ranked system [9] of SemEval-2012 [17].

5.3 SMTeuroparl2012

SemEval-2012 [17] selected 459 unique human evaluated French to English test pairs (henceforth, SMTeuroparl2012) from the 2008 ACL Workshops on Statistical Machine Translation (WMT). In Figure 6, it is shown that on this dataset, GTM with $\mu+\delta$ and using 9 sorted words give Pearson's $r = 0.505$ with human ratings while using all the words gives that of $r = 0.503$. For LSA with $\mu+\delta$, using only 3 sorted words gives Pearson's $r = 0.309$ while using all the words gives $r = 0.287$ and the difference is statistically significant at the 0.05 level.

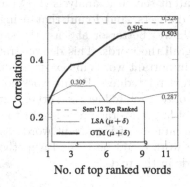
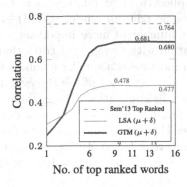

Fig. 6. Corr. with human ratings for SMTeuroparl2012   **Fig. 7.** Corr. with human ratings for HDL2013

5.4 SMT2013

This dataset from SemEval-2013 [18] comprises 750 pairs of sentences used in machine translation evaluation (henceforth, SMT2013). In Figure 5, it is shown that on this dataset, GTM with $\mu+\delta$ and using 22 sorted words give Pearson's $r = 0.380$ with the ground truth while using all the 57 words gives that of $r = 0.377$. For LSA with $\mu+\delta$, using 17 sorted words gives $r = 0.313$ while using all the words gives $r = 0.300$. On this dataset, GTM matches the top ranked system [10] of SemEval-2013 [18].

5.5   HDL2013

This dataset from SemEval-2013 [18] comprises 750 pairs of news headlines (HDL) gathered by the Europe Media Monitor (EMM) engine from several different news sources (henceforth, HDL2013). In Figure 7, it is shown that on this dataset, GTM with $\mu+\delta$ and using 13 sorted words give Pearson's $r = 0.681$ with the ground truth while using all the 16 words gives that of $r = 0.680$. For LSA with $\mu+\delta$, using 9 sorted words gives Pearson's $r = 0.478$ while using all the words gives $r = 0.477$.

Both LSA and GTM achieve better correlation with the ground truth using a smaller number of sorted words than using all the words on all the datasets, except on OnWN2012 where the same correlation is achieved. On the dataset with the highest average text length (i.e., ABC1225), both methods achieve statistically significant (at the 0.05 level) better correlation with the ground truth using a smaller number of sorted words than using all the words. We do not formally evaluate the '$\mu+\delta$' approach of finding important words, though it is shown that the approach improves the text relatedness result in comparison with the classical BoW models.

6   Conclusion

We applied two text relatedness methods—latent semantic analysis (LSA) and the Google trigram model (GTM)—to five datasets, and found that using a smaller number of more important words achieves better or at least the same correlation with the ground truth than using all the words. This demonstrates that in most cases a smaller number of more important words in text-pair could improve the accuracy of the BoW models of text relatedness. Future work could consider the application of other BoW models of text relatedness and other methods of finding important words on datatsets of larger average text length to see whether there is any correlation between the number of important words used and the average text length. Another future work could be to find a methodological way to select a threshold on the words in the text.

References

1. Liu, T., Liu, S., Chen, Z., Ma, W.Y.: An evaluation on feature selection for text clustering. In: Fawcett, T., Mishra, N. (eds.) Proc. 20th Intl. Conf. on Machine Learning, pp. 488–495 (2003)
2. Erkan, G., Radev, D.: Lexrank: Graph-based lexical centrality as salience in text summarization. Journal of Artificial Intelligence Research **22**, 457–479 (2004)
3. Schutze, H.: Automatic word sense discrimination. Computational Linguistics **24**(1), 97–124 (1998)

4. Park, E., Ra, D., Jang, M.: Techniques for improving web retrieval effectiveness. Information Processing and Management **41**(5), 1207–1223 (2005)
5. Coelho, T., Calado, P., Souza, L., Ribeiro-Neto, B., Muntz, R.: Image retrieval using multiple evidence ranking. IEEE Tran. on Knowledge and Data Engineering **16**(4), 408–417 (2004)
6. Ko, Y., Park, J., Seo, J.: Improving text categorization using the importance of sentences. Information Processing and Management **40**, 65–79 (2004)
7. Islam, A., Inkpen, D., Kiringa, I.: Applications of corpus-based semantic similarity and word segmentation to database schema matching. The VLDB Journal **17**(5), 1293–1320 (2008)
8. Ho, C., Murad, M.A.A., Kadir, R.A., Doraisamy, S.C.: Word sense disambiguation-based sentence similarity. In: Proc. of COLING 2010, pp. 418–426 (2010)
9. Bär, D., Biemann, C., Gurevych, I., Zesch, T.: UKP: Computing semantic textual similarity by combining multiple content similarity measures. In: SemEval 2012, pp. 435–440 (2012)
10. Han, L., Kashyap, A.L., Finin, T., Mayfield, J., Weese, J.: UMBC-EBIQUITY-CORE: semantic textual similarity systems. In: Proceedings of the Second Joint Conference on Lexical and Computational Semantics. Association for Computational Linguistics (June 2013)
11. Islam, A., Inkpen, D.: Semantic text similarity using corpus-based word similarity and string similarity. ACM Trans. Knowl. Discov. Data **2**, 10:1–10:25 (2008)
12. Tsatsaronis, G., Varlamis, I., Vazirgiannis, M.: Text relatedness based on a word thesaurus. J. Artif. Int. Res. **37**(1), 1–40 (2010)
13. Guo, W., Diab, M.T.: Modeling sentences in the latent space. In: ACL (1), The Association for Computer Linguistics, pp. 864–872 (2012)
14. Gomaa, W.H., Fahmy, A.A.: A survey of text similarity approaches. International Journal of Computer Applications **68**(13), 13–18 (2013)
15. Islam, A., Inkpen, D.: Second order co-occurrence PMI for determining the semantic similarity of words. In: Proceedings of the International Conference on Language Resources and Evaluation, Genoa, Italy, pp. 1033–1038 (May 2006)
16. Islam, A., Milios, E., Keselj, V.: Comparing word relatedness measures based on Google *n*-grams. In: Proceedings of COLING 2012: Posters, Mumbai, India, The COLING 2012 Organizing Committee, pp. 495–506 (December 2012)
17. Agirre, E., Cer, D., Diab, M., Gonzalez-Agirre, A.: *sem 2012 task 6: A pilot on semantic textual similarity, June 7–8, pp. 385–393 (2012)
18. Agirre, E., Cer, D., Diab, M., Gonzalez-Agirre, A., Guo, W.: *SEM 2013 shared task: Semantic textual similarity, pp. 32–43 (June 2013)
19. Islam, A., Milios, E., Kešelj, V.: Text similarity using Google tri-grams. In: Kosseim, L., Inkpen, D. (eds.) Canadian AI 2012. LNCS, vol. 7310, pp. 312–317. Springer, Heidelberg (2012)
20. Landauer, T., Foltz, P., Laham, D.: Introduction to latent semantic analysis. Discourse Processes **25**(2–3), 259–284 (1998)
21. Brants, T., Franz, A.: Web 1T 5-gram corpus version 1.1. Technical report, Google Research (2006)
22. Lee, M.D., Pincombe, B., Welsh, M.: An empirical evaluation of models of text document similarity. In: Proc. of the 27th Annual Conf. of the Cog. Sci. Society, pp. 1254–1259 (2005)
23. Zou, G.Y.: Toward using confidence intervals to compare correlations. Psychological Methods **12**(4), 399–413 (2007)

Increased Recall in Annotation Variance Detection in Treebanks

Pablo Faria[(⊠)]

University of Campinas, Language Studies Institute, Campinas, Brazil
pablofaria@gmail.com

Abstract. Automatic inconsistency detection in parsed corpora is significantly helpful for building more and larger corpora of annotated texts. Inconsistencies are inevitable and originate from variance in annotation caused by different factors as, for instance, the lack of attention or the absence of clear annotation guidelines. In this paper, some results involving the automatic detection of annotation variance in parsed corpora are presented. In particular, it is shown that a generalization procedure substantially increases the recall of the variant detection algorithm proposed in [1].

Keywords: Treebank · Inconsistency detection · Quality control

1 Introduction

Variation in annotation (and, thus, potential inconsistencies) in parsed corpora tends to be more frequent than the intuitions of annotators would suggest, as indicated by a number of studies on the detection of annotation inconsistencies carried out in recent years [1–6, amongothers]. However, although being a potential problem both for information extraction from the corpus and for parser training, the actual impact of inconsistencies is difficult evaluate fully. Nevertheless, inconsistencies are something to be avoided since these may not only impact the extraction of information from the corpus – demanding different queries to extract the same kind of information (to the extent that the relevant inconsistencies are predictable) – but also because it may have a negative effect on the performance of parsers trained on the annotated portion of the corpus.

A corpus can be inconsistent in two different ways: by having violations of clear and strict annotation guidelines (e.g., applying a wrong label) and/or by lacking guidelines for certain kinds of expressions which leads to variation in annotation (see [7]). Both are inconsistencies, but the former are true errors. Since going through a (growing) corpus again and again in order to find inconsistencies is painful and quite inefficient, developing automatic methods to do so is a welcome contribution to the field of corpus linguistics.

This research is funded by Sao Paulo Research Foundation – FAPESP – through grants no. 13/18090-6 and 14/17172-1.

© Springer International Publishing Switzerland 2015
P. Král and V. Matoušek (Eds.): TSD 2015, LNAI 9302, pp. 578–586, 2015.
DOI: 10.1007/978-3-319-24033-6_65

In this paper, an updated and extended version of the alternative algorithm in [1] is presented along with results of its application to the Tycho Brahe Corpus (TBC, [8]). The TBC is a parsed corpus of historical texts in Portuguese, but the algorithm discussed here may be applied to any text annotated in the Penn Treebank ([9]) style. The updated version of the algorithm presented here is able to detect one more type of variation (missed by the previous one) and also covers a significantly larger portion of the corpus as a consequence of a generalization procedure applied to the detected variants. The paper is organized as follows: **Section 2** introduces the algorithm in [1]. In **Section 3** the new generalization procedure is discussed. Experimental results are reported on **Section 4** followed by a discussion. Related work is presented in **Section 5** and the paper ends with some concluding remarks in **Section 6**.

2   Alternative Approach

The algorithm proposed by Faria ([1]) is an alternative implementation of the algorithm for variance detection proposed in Dickinson & Meurers' ([2]) study. Their algorithm searchs for *variation nuclei*, that is, sequences of tokens (words and punctuation) occurring two or more times in a corpus which were at least once analyzed as a constituent (i.e., there exists a label that dominates all and only the tokens in the sequence) and that exhibit two or more distinct labels (including a "non-constituent" pseudo-label NIL assigned for non-constituent occurrences of a given nucleus). After extracting the variation nuclei, the algorithm generates "variation n-grams", that is, it filters out variation nuclei for which there is no similar context (i.e., surrounding tokens) of occurrence. This second step is necessary for their algorithm to obtain higher precision although at the cost of lowering its recall.

Leaving the second step aside, it turns out that the criterion for finding variation nuclei based on root labels of partial trees may produce the opposite of the desired results. First, two instances of a nucleus may have the same label but inconsistent internal structures as the left pair in Figure 1 exemplifies.[1] Although being structurally distinct, they will be considered as equivalents since they have the same root label, PP.

On the other hand, although not forming a constituent of its own, a particular instance of a nucleus may still be syntactically consistent with another instance. Take, for example, the right pair of trees in Figure 1. There are two instances of the nucleus "d@ @os homens", both consistent with each other on the structure assigned. The only difference is that the NP node in the second tree contains one more element which turns out to be irrelevant in this case: with respect to those three words both trees are equivalent. Nonetheless, Dickinson & Meurers' method will take them as distinct variants, with the second being labelled as NIL.

As an alternative criterion, [1] uses what the author calls "dominance chains" (DC), that is, the set of sequences of nodes from the root label to each terminal

---

[1] The symbol "@" in these data is used in TBC for contracted words that undergone splitting.

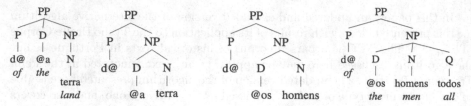

Fig. 1. Examples of inconsistency involving constituents of the same label (left pair) and consistency between a constituent and a "non-constituent" instance of "d@ @os homes" (right pair).

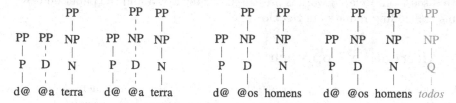

Fig. 2. Dominance chain sets give the desired results.

symbol in a (partial) tree. Variants are distinguished on the basis of their sets of DCs which gives us the desired opposite results for the trees in Figure 1, as Figure 2 allows us to infer. A *local implementation* of the original method and the alternative algorithm were compared based on their results for the current publicized version of the TBC. It consists of 16 parsed texts which comprise (after some clean-up) a total of 707,685 tokens distributed over 34,265 sentences. A partial corpus with 1,000 sentences was used to compare both methods.

Results were manually checked to determine whether a variation nucleus was a genuine case of variation in annotation or just a false positive, that is, variants consistent with each other. Absolute measures of recall are not obtainable given that one would have to go sentence by sentence to list the inconsistencies in the corpus. Therefore, measures of recall are relative, that is, the union set of the true variants found by both algorithms is taken as the goal. That said, the original method found 606 variation nuclei whereas the alternative found 202. From these, 188 are nuclei found by both methods. Thus, 418 nuclei were exclusively found by the original whereas the alternative found 14 exclusive nuclei. It turns out, however, that all 418 nuclei were verified to be false positives, that is, consistent structures taken as variants because of non-constituent occurrences of the nuclei involved. On the other hand, the 14 nuclei found by the alternative algorithm were true variants.[2]

---

[2] Being variants here does not imply being true errors of annotation. Some variants derive from semantic ambiguities of the words involved, being thus legitimate, while others derive from the permissiveness of the guidelines which in some cases do not establish a particular analysis.

Consequently, for the purposes of a relative comparison, the 202 nuclei found by the alternative can be taken as the "target" for the original method which, thus, shows a precision of 31.02% and relative recall of 93.06%. This result demonstrates that the precision of the method can be substantially improved with only a modification in the criterion for distinguishing variants. In fact, even without resort to variation n-grams, the alternative algorithm shows for the partial corpus an "error detection precision" – that is, the proportion of variation nuclei that turned out to involve true errors in annotation – of 61.29% for $n \geq 2$, not a bad result given that the recall is also improved.[3] If we consider non-erroneous cases that are still useful to highlight certain aspects of the annotation system which are possibly inconsistent[4], precision goes to 69.35% ($n \geq 2$).

3 Generalization of Variants

The algorithm in [1] introduced in the previous section was further developed here in order to increase its recall. Given that variation nuclei are initially discovered by means of multiple occurrences of specific token sequences, we are at first limited by the number and variety of sequences that repeat in a corpus. Therefore, it is likely that some or even many other instances of the same *type* of variation are lost simply because the sequences of tokens involved do not repeat or do repeat but with the same incorrect annotation. We could use part-of-speech (POS) tags instead of words to overcome this problem, but it turns out that this strategy overgeneralizes significantly, since non-constituent repetitions of sequences of POS tags abound. In a sense, POS tags are in general too abstract, while tokens may be too specific (see a related discussion in [6]).

The solution found is in between these two options. Once a variation nucleus is found on the basis of token sequences, the "types" of its variants are used to match instances of the same template for different sequences of words. "Type of variant" here means simply a variant abstracted away from its terminal nodes, as shown in Figure 3. For each variant, the algorithm searches for compatible (partial) trees in the corpus. This generalization is less radical because the POS tags are constrained by the DCs. Consequently, although the method's recall is still bound by the number and variety of repeating sequences of words, it nonetheless maximizes the number of instances found for each type of variant.

3.1 Some Minor Changes

Some minor changes and adjustments were made to the algorithm with some important but smaller effects on its output. Now the preprocessing procedure

---

[3] Nuclei of size 1 are very likely cases of ambiguity caused by variation in part-of-speech tags.

[4] As one example, in the TBC, interrogative pronouns head phrases whose labels start with a "W" (like WNP, WPP, etc.). But this decision is inconsistent with the general option of starting with the syntactic category and then adding dash tags to represent more specific properties. Thus, it would be more consistent in this case to have something like a -WH dash tag added to the label, since that would prevent the algorithm supposing that WNP and NP are distinct categories.

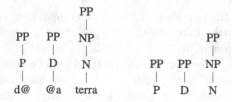

Fig. 3. A variant of the nucleus "d@ @a terra" and its type.

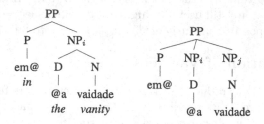

Fig. 4. Two formerly non-distinguishable variants of the nucleus "em@ @a vaidade".

removes punctuation marks from the corpus, since they can be fixed automatically by other means.[5] This change is expected to increase the chance of finding larger nuclei and also occurrences of nuclei that would otherwise not be exact repetitions because of intervening punctuation marks. Another change was in the way dominance chains are represented. Now it is possible to distinguish between distinct sister nodes with the same label. One consequence of this change is that the particular type of variation shown in Figure 4, which was missed by the first version of the algorithm, is now accounted for.

4   Experiments and Discussion

In the experiments described in this section, the same version of TBC used in [1] is also used. In this version, the total number of parsed sentences is 34,265 from which 32,458 (94.73%) remained after the clean up procedure, which also removes functional labels, null elements (e.g., "*pro*") and punctuation. Two sets of experiments are presented, each followed by the relevant discussion. The first set evaluates only the impact of removing punctuation from the corpus, under the hypothesis that more and larger variation nuclei may be found if punctuation is removed. The second set – with punctuation removed – evaluates the effect of the generalization procedure. It is expected that many more instances of variants will show up, specially for those variants that stand for correct annotations. On the other hand, generalization is expected to make the number of nuclei decrease,

---

[5] At least for corpora following the guidelines of the Penn Treebank which dictates that punctuation must be located as high as possible in a tree.

Set	Max n	Mean n	Nuclei	Variants	Instances	Mean σ	Covering
With punct.	3	1.26	168	406	4,694	10.20	96.11%
Without punct.	6	1.29	168	405	4,693	10.20	96.11%
Generalization	6	1.84 (3.90)	44 (2,473)	172	6,357	23.65	98.21%

Fig. 5. Results for the partial corpus (for $n \geq 1$)

Set	Max n	Mean n	Nuclei	Variants	Instances	Mean σ	Covering
With punct.	20	2.58	2,500	7,517	58,027	6.53	63.27%
Without punct.	18	2.52	2,421	6,740	49,714	6.55	60.46%
Generalization	18	3.33 (4.08)	698 (67,989)	3,291	144,376	38.27	86.09%

Fig. 6. Results for the whole corpus (for $n \geq 2$)

since many of them actually belong to the same "type" of variation. Tables 5 and 6 summarize the results for the partial and the whole corpus.

In the tables presented, columns *Max n* and *Mean n* show, respectively, the size (in number of tokens) of the largest nuclei found and mean size of all nuclei found. The column *Nuclei* lists the number of nuclei found. *Variants* lists the total number of variants found and *Instances* the total number of instances (i.e., actual subtrees in the corpus) for the set of variants (with possible overlaps). The column *Mean σ* lists the mean of the standard deviation in the number of instances per variant for each nuclei. Finally, the column *covering* is a measure of the proportion of corpus sentences involved in the detected variation as an indirect measure of recall.

4.1 Effects of the Minor Changes

Removing punctuation seems to have a positive effect for the partial corpus, although not dramatic, but for the full corpus we see an overall decrease in the measures, contrary to expectation. Although a careful inspection is needed before drawing any conclusions, a possible explanation is that many inconsistencies involve only punctuation. Thus, if that is the case and since they can be fixed automatically and easily, it is a good idea to ignore them and focus on more relevant cases. Regarding the representation of variants, the modification was expected not to lead to a significant increase, since the type of variation it captures (see Figure 4) is unlikely to happen very often.

4.2 Effects of Generalization

As expected, the generalization procedure leads to a substantial increase in the number of nuclei (the number inside parentheses) and instances found. The number of variants now relates to "types of variants" as discussed above and that explains it being smaller than those of the other experiments. Finally, the outcome of this procedure becomes clear in the measures *mean σ* and *covering*.

The algorithm covers much more data (for the whole corpus) and the significant increase in the mean σ suggests that a tentative classification of variants as either correct or incorrect may be possible. Since correct variants are expected to recur while incorrect ones tend to be rare, we may more confidently classify them based on their number of instances, as in [3].

5   Related Work

Accurately comparing different detection methods is not yet an easy task for reasons that go from the accessibility to the corpora used to the access to the actual algorithms. Nevertheless, a raw comparison may be drawn between the present algorithm and the ones in [2], [3] and [4,5], all sharing the same basic strategy of variance detection based on repetitions of sequences of tokens. Over a sample of 100 variation nuclei taken from the results for the *whole corpus*, the present algorithm shows a precision of 77% in detecting true errors – 84% if we consider unclear cases and cases involving inconsistencies in the annotation guidelines.

In [2], after applying heuristics for classification, the authors report a precision of up to 78.9% in the detection of true errors, from a random sample of 100 "variation n-grams" out of 6,277 obtained from the Wall Street Journal (WSJ) corpus [10]. Using another corpus, precision in subsequent work ([11]) is reported to be 80% (by the same kind of random sampling). In both cases, recall is sacrificed (in the process of generating n-grams) in order to obtain higher precision. Also using the WSJ corpus, [3] reports a precision of 71.9% in a sample with the first 100 rules induced for annotation correction. Their algorithm induces a Synchronous Tree Substitution Grammar from a "pseudo-parallel corpus" of partial trees. They measured the precision of each rule and 70 achieved 100% precision, that is, they hit only incorrect instances.

Finally, [4,5] propose a Tree Adjoining Grammar (TAG) based approach in which trees are compared on the basis of their derivation trees, obtained by the decomposition of the corpus into elementary trees. Their algorithm also group inconsistencies by type and generalize them in order to increase recall. They report that, of the first 10 different types of derivation tree inconsistencies found, all 10 appear to be real cases of annotation inconsistency. They used the Penn Arabic Treebank (ATB) [12] and a subset of the English treebank newswire section of the Ontonotes 4.0 release [13].

6   Final Remarks

As expected, the modifications to the algorithm had positive effects on the overall goal of improving the recall of the algorithm proposed in [1]. Another benefit is that the generalization procedure makes revision faster, since it provides all the related variants and instances at once for each type of variation. Thus, an automated script for revision can be specified with all relevant contexts at hand. Of course, additional qualitative analyses are still necessary in order to better assess the impact of these modifications. For example, it is important to analyze how the generalization procedure affects the relative number of instances for correct and incorrect annotations. It will be also important to evaluate the algorithm on corpora of different languages, on a variety of genres, and on different sets of labels and POS tags, since all these factors may affect the results.

It is worth noting that studies like the present one also contribute to the field of natural language parsing. Parsers not only benefit from more consistent corpora, but also (and crucially) benefit from more consistent annotation systems. In that regard, one issue that still lacks a good solution is how to deal with dash tag, coindexing, and empty category inconsistencies in phrase structure treebanks. The present algorithm as well as the ones considered in Section 5 all lack a good treatment of these sorts of inconsistency, since they involve more abstract properties of syntactic trees. This is, thus, an important issue for future work.

Acknowledgments. Thanks to Sao Paulo Research Foundation – FAPESP – for funding this research. Also thanks to the University of Pennsylvania and its Department of Linguistics where part of this research has been conducted. Finally, thanks to the reviewers for their useful comments and suggestions.

References

1. Faria, P.: Using dominance chains to detect annotation variants in parsed corpora. In: 2014 IEEE 10th International Conference on e-Science (e-Science), vol. 2, pp. 25–32. IEEE (2014)
2. Dickinson, M., Meurers, W.D.: Detecting inconsistencies in treebanks. In: Proceedings of TLT, vol. 3, pp. 45–56 (2003)
3. Kato, Y., Matsubara, S.: Correcting errors in a treebank based on synchronous tree substitution grammar. In: Proceedings of the ACL 2010 Conference Short Papers, ACLShort 2010, pp. 74–79. Association for Computational Linguistics, Stroudsburg (2010)
4. Kulick, S., Bies, A., Mott, J.: Using derivation trees for treebank error detection. In: ACL (Short Papers), pp. 693–698 (2011)
5. Kulick, S., Bies, A., Mott, J.: Further developments in treebank error detection using derivation trees. In: Chair, N.C.C., Choukri, K., Declerck, T., Doğan, M.U., Maegaard, B., Mariani, J., Odijk, J., Piperidis, S. (eds.) Proceedings of the Eight International Conference on Language Resources and Evaluation (LREC 2012), European Language Resources Association (ELRA), Istanbul, May 2012
6. Krasnowska, K., Przepiórkowski, A.: Detecting syntactic errors in dependency treebanks for morphosyntactically rich languages. In: Kłopotek, M.A., Koronacki, J., Marciniak, M., Mykowiecka, A., Wierzchoń, S.T. (eds.) IIS 2013. LNCS, vol. 7912, pp. 69–79. Springer, Heidelberg (2013)
7. Blaheta, D.: Handling noisy training and testing data. In: Proceedings of the ACL 2002 Conference on Empirical Methods in Natural Language Processing, EMNLP 2002, vol. 10, pp. 111–116. Association for Computational Linguistics, Stroudsburg (2002)
8. Galves, C., Faria, P.: Tycho brahe parsed corpus of historical portuguese (2010). http://goo.gl/cu4N6w

 9. Marcus, M.P., Marcinkiewicz, M.A., Santorini, B.: Building a large annotated corpus of english: the penn treebank. Comput. Linguist. **19**(2), 313–330 (1993)
10. Taylor, A., Marcus, M., Santorini, B.: The penn treebank: An overview. In: Abeillé, A. (eds.) Treebanks: Building and Using Syntactically Annotated Corpora, pp. 5–22. Kluwer Academic Publishers (2003)
11. Dickinson, M., Meurers, W.D.: Detecting errors in discontinuous structural annotation. In: Proceedings of the 43rd Annual Meeting on Association for Computational Linguistics, ACL 2005, pp. 322–329. Association for Computational Linguistics, Stroudsburg (2005)
12. Maamouri, M., Bies, A., Kulick, S., Krouna, S., Gaddeche, F., Zaghouani, W.: Arabic treebank part 3–v3.2. Linguistic Data Consortium LDC2010T08 (2010)
13. Weischedel, R., Palmer, M., Marcus, M., Hovy, E., Pradhan, S., Ramshaw, L., Xue, N., Taylor, A., Kaufman, J., Franchini, M., El-Bachouti, M., Belvin, R., Houston, A.: Ontonotes 4.0. Linguistic Data Consortium LDC2011T03 (2011)

Using Sociolinguistic Inspired Features for Gender Classification of Web Authors

Vasiliki Simaki[✉], Christina Aravantinou,
Iosif Mporas, and Vasileios Megalooikonomou

Multidimentional Data Analysis and Knowledge Management Laboratory,
Department of Computer Engineering and Informatics, University of Patras,
26500-Rion, Greece
{simaki,aravantino,vassilis}@ceid.upatras.gr, imporas@upatras.gr

Abstract. In this article we present a methodology for classification of text from web authors, using sociolinguistic inspired text features. The proposed methodology uses a baseline text mining based feature set, which is combined with text features that quantify results from theoretical and sociolinguistic studies. Two combination approaches were evaluated and the evaluation results indicated a significant improvement in both combination cases. For the best performing combination approach the accuracy was 84.36%, in terms of percentage of correctly classified web posts.

Keywords: Text classification algorithms · Sociolinguistics · Gender identification

1 Introduction

The expansion of text-based social media is impressive and the need of classifying the provided information into sub categories is an important task. This categorization can be made in terms of topic, genre, author, gender, age, etc. according to the informational need and the purpose of the users. This is implemented by identifying differential features characterizing the demanded purpose. Every social media user leaves his digital fingerprints on the web, not only by declaring personal information, but unconsciously through his writing style. One of the most important issues on this field is the identification of the user's gender and the classification of documents according to this specification. It is a challenging task, given that in the typical case the gender is identified without taking into account the personal information the user provides, but estimated only using the content of his/her texts.

Gender classification is an important field of text mining with many commercial applications. The knowledge of the user's gender is important to companies in order to promote a product or a service, if it is preferable mostly by women or men. Market analysis and advertising professionals are interested in which product or service is more talked or liked between the two groups, and

© Springer International Publishing Switzerland 2015
P. Král and V. Matoušek (Eds.): TSD 2015, LNAI 9302, pp. 587–594, 2015.
DOI: 10.1007/978-3-319-24033-6_66

should be addressed to women or men. Gender classification is also considerable in e-government services and social science studies. Useful conclusions can be extracted about the different trends among women and men, different topics of interests, political views, social concerns, world theories, and many other issues. Since it is quite difficult for social scientists to manually go through large volumes of data, computer-based solutions supported by the recent advances in natural language processing and machine learning have techniques been proposed. In parallel to computer-based solutions sociolinguists have offered essential knowledge in support of the task of gender identification from written language.

Sociolinguistics is the specific scientific domain of linguistics which studies the influence of social factors into the written and spoken language. Factors as gender, age, education, etc., delimitate the linguistic diversity and variation, the linguistic choices that people and social groups make in everyday life. The differences between men and women's language can be detected in their texts, due to the separate linguistic choices they make. These choices can be identified in all levels of linguistic analysis (from the phonetic to the pragmatic one) and they may be conscious or not, differentiating the speaker's attitude from the standard language in a given communicative occasion [1].

In our study, an interdisciplinary methodology for the detection of the author's gender is proposed, based on features derived from two different disciplines, the gender linguistic variation and the gender classification. These two kinds of features are fused in order to achieve higher accuracy and prove that linguistics and text mining, when combined, can contribute to better gender identification results.

The remainder of this paper is organized as follows. In Section 2 we present the background work in the field of gender identification, after theoretical, empirical and computational studies. Section 3 describes our methodology and in Section 4 the experimental part of our work is presented. Finally, in Section 5 we conclude this work.

2   Related Work

Several studies related to author's gender discrimination have been reported in the literature, both based on computer-based methods (text mining) and theoretical models (sociolinguistic). The first ones concentrate on efficient computational algorithms while the latter ones on social cues expressed through linguistic expressions on written text.

As considers text mining based approaches, they typically consider author's gender identification as a text classification issue [2,3]. Koppel et al. [4] propose text classification methods to extract the author's gender from formal texts, using features such as n-grams and function words that are more usual in authorship attribution. This research combines stylometric and text classification techniques, in order to extract the author's gender. Argamon et al. [5] have applied factor analysis for gender and age classification in texts mined from the blogosphere. Ansari et al. [6] have used frequency counting of tokens, tf-idf and POS-tags to find the gender of blog authors. In Burger et al. [7] a study on gender recognition of texts from Twitter was presented, where the content of the tweet combined with the username and other information related to the user we used. Many recent studies around gender classification deal with social media and they propose methods that identify the gender [8–10] and in some cases the age of the web users [11]. Most of the reported approaches implement their experiments, taking into account features, such as gender-polarized words, POS tags and sentence length, in order to obtain best classification results. In Sarawgi et al. [12] a comparative study of gender attribution, without taking into account the topic or the genre of the selected text is presented. Holgrem and Shyu [13] applied machine learning techniques using a feature vector containing word counts, in order to detect the author's gender of Facebook statuses. In Rangel and Rosso [14] a set of stylistic features to extract the gender and age of authors using a large set of documents from the social web written in Spanish was presented. Marquardt et al. [15] evaluated the appropriateness of several feature sets for age and gender classification in social media.

Except the text-mining approaches, sociolinguistic studies offer valuable information about the gender characterization of a text. The basic concept of sociolinguistics, and more specifically the gender linguistic variation, is perceived as a socially different but linguistically equal way to say the same thing [8]. A general opinion about the women's language is that women tend to make a more conservative use of language by using more standard types than men [16]. Women use non-normative forms only when they adapt socially prestigious changes, local linguistic elements, communicative indirection, and under specific communicative situations [17,18]. Under standard conditions, they have a smaller vocabulary than men, using a narrower range of different lexical types. Compared to men discourse, women tend to use more complex syntactic structures by forming many explanatory secondary phrases in the period. The use of "empty" adjectives which have the sense of admiration and/or approval is also frequent in women's language, as well the use of questions in place of statements [19–21]. Moreover, specific lexical choices that women do unlike men (use of norm types, avoid bad words, etc.), researchers observe their effort in many cases to decline the illocutionary force of their utterances. This phenomenon is achieved by using palliative forms like tag-questions, interrogative intonation instead of affirmations, extension of requests and hedges of uncertainty. As considers women's language, they use different politeness, agreement and disagreement strategies than men and more sentimental expressions, indirect requests and hypercorrected grammar types [22,23]. Men on the other hand, tend to use more bad words, slang types and coarse language. They insert in their vocabulary non-norm forms and neologisms. In Alami et al. [24] study of the lexical density in male and female discourse and comparison of the relationship to the discourse length is performed. Eckert [25] merged existing and traditional theories, in order to create patterns about the gender-specific variation, and analyzed the meaning and the social context around a given linguistic attitude. In recent studies [26–28] researchers discuss the social factor and the stylistic information in different communicative situation in order to explain the specific linguistic choice of speakers.

3   Gender Classification Methodology

Most of the previous studies in the field of gender identification are based either in theoretical analysis and empirical findings or in computational approaches. The first kind of research, conducted by expert sociolinguists, can reveal frequent but also rare differential characteristics after empirical studies. These studies confirm existing theories and they create new rules. However, theoretical studies are time consuming, since working with large and different data collections is tedious, especially when need to verify rare discriminative rules which will probably appear only in large volumes of text data. On the other hand, computational approaches based on data mining algorithms can perform efficient and fast process of large data collections; however, the results are frequently biased to the specifications of the dataset used. Moreover, infrequent discriminative rules either will not appear in the evaluation text or they will be considered by the algorithm as outliers rather than newly discovered patterns.

The objective of the present approach for author gender identification is to exploit existing knowledge from the sociolinguistics domain in order to enhance the performance of the dominating text mining solutions. Thus, we combine sociolinguistic characteristics and data-driven features for gender classification. Specifically, a number of well-known and widely used in text mining methods features for text, author and gender classification are used to build a baseline feature vector [29]. This feature vector is combined with features inspired from sociolinguistic studies in order to enhance the gender discriminative ability of a classification engine. The sociolinguistic characteristics of gender variation may be summarized as: 'syntactic complexity', 'use of adjectives', 'sentence length', 'different politeness and agreement/disagreement strategies', 'tag questions', 'slang types', 'bad words', 'sentimental language', 'lexical density', 'interrogative intonation' and 'vocabulary richness'.

The baseline (BASE) feature set and the features inspired from sociolinguistics (SLING) are presented in Table 1. The baseline feature vector has length equal to 24 and the sociolinguistic-inspired list of features has length equal to 11.

For the combination of the baseline (BASE) and sociolinguistic-inspired (SLING) features we relied on two fusion approaches. In the first approach (early combination), the SLING features are appended to the BASE vector and the concatenated feature vector is processed by a classification algorithm. In the second approach (late combination), the data-driven (BASE) and the knowledge-based (SLING) vectors are separately processed by classification engines and the results are fused by a second-stage classifier. In both early and late fusion scenarios both data-based (from data mining) and sociolinguistic-inspired knowledge is utilized in the classification procedure.

Table 1. The BASE and SLING features used in author's gender classification.

BASE features	SLING features
# of characters per web post	normalized # of the sentence verbs
normalized # of alphabetic characters	normalized # of adjectives per comment
normalized # of upper case characters	normalized # of the text's words
# of occurrence of each alphabetic character	# of standard polite, agreement / disagreement phrases
normalized # of digit characters	# of tag question phrases
normalized # of tab ('\t') characters	# of slang types
normalized # of space characters	# of bad words
normalized # of special characters ("@", "#", "$", "%", "&", "*", "~", "^", "_", "=", "+", ">", "<", "[", "]", "{", "}", "|", "\", "/")	normalized # of sentimentally polarized words of the comment, according to SentiWordNet[30]
total # of words	normalized # of the document's content words
normalized # of words with length less than 4 characters	normalized # of the question marks to the total # of the document's punctuation
# of punctuation symbols (".", ",", "!", "?", ":", ";", "'", "\"")	normalized # of different words per comment
average word length	
# of lines	
average # of characters per sentence	
# of sentences	
normalized # of unique words	
# of paragraphs	
average # of words per sentence	
# of "hapax legomena"	
# of "hapax dislegomena"	
normalized # of characters per word	
# of function words	
average # of sentences per paragraph	
average # of characters per paragraph	

4   Experimental Setup and Results

The text mining based and sociolinguistic-inspired combination methodology described in Section 3 was evaluated using a dataset collection of users' comments on web. Our dataset consists of user comments in English about various topics extracted from forums and web sites. It contains comments from different sources, covering various thematic areas both from gender-preferential sites and forums, like fashion (typically preferred by women) or cars (typically preferred by men) and neutral web sources (like news, health etc). The size of the corpus is 326,736 words. The number of the characters is equal to 1,643,547. The gender division between men and women is 42% and 58% respectively.

For the classification stage, we relied on several dissimilar machine learning algorithms, which have extensively been reported in the literature. In particular, we used a multilayer perceptron neural network (MLP) and support vector machines (SVMs), using radial basis kernel (RBF) and polynomial kernel (poly). Furthermore, we employed Adaboost.M1, which is a boosting algorithm combined with decision trees (AdaBoost) and a bagging algorithm using decision trees (Bagging). Finally, we used three decision tree algorithms, namely the random tree (RandTree), the random forest (RandForest) and the fast decision tree learner (RepTree). All classifiers were implemented using WEKA toolkit [31]. In order to avoid overlap between training and test subsets a 10-fold cross validation evaluation protocol was followed. The performance results in terms of percentages of correctly classified web posts are tabulated in Table 2. The best performance per setup is indicated in bold.

Table 2. Gender classification results using different combination setups and algorithms.

	BASE	SLING	BASE+SLING (early fusion)	BASE+SLING (late fusion)
MLP	82.31	66.87	82.51	**84.36**
SVM(rbf)	67.49	50.00	68.31	83.13
SVM(poly)	82.72	63.17	**84.16**	82.92
Bagging	82.72	69.35	83.54	82.30
Boosting	82.10	69.14	82.51	81.07
RepTree	**82.92**	67.08	80.86	81.48
RandForest	82.72	**69.34**	82.72	79.84
RandTree	79.84	66.05	81.07	75.51

As can be seen in Table 2, the use of SLING features improves gender classification accuracy by almost 1,5% comparing to the best BASE alone. Specifically, the best BASE performance was 82.92% using the RepTree classifier, while the overall best performance was 84.36%, which was achieved with the late combination approach and the MLP classification algorithm. The SLING approach standalone does not offer competitive performance comparing to the BASE setup, however in both fusion setups there is an increase of performance which shows the importance of the sociolinguistic-inspired features. As considers the evaluated classification algorithms for the case of early combination where the fusion feature vector is of length equal to 24+11=35, the SVM algorithm outperforms all others, probably to the fact that it does not suffer from the curse of dimensionality. In the late fusion case, where the fusion vector consists of the probability of being male/female from BASE and SLING (i.e. 2+2=4 length) the MLP classifier performs better than SVM.

5   Conclusions

The exploitation of the existing knowledge extracted from theoretical and sociolinguistic studies and the transformation of this qualitative information to quantitative metrics can improve text-based gender classification accuracy. The use of sociolinguistic-inspired text features is not essential only for combination with typical text mining features, as demonstrated in this article, but can also be used to fine-tune computational algorithms by supporting the training of statistical-based models through definition initialization values and restriction of range of values of free parameters which will protect from models biased to specific data.

References

1. Archakis, A., Kondyli, M.: Introduction to sociolinguistic issues. Nisos(in Greek), Athens (2004)
2. Cheng, N., Chandramouli, R., Subbalakshmi, K.P.: Author gender identification from text. The International Journal of Digital Forensics & Incident. Response **8**(1), 78–88 (2011)
3. Soler, J., Wanner, L.: How to Use Less Features and Reach Better Performance in Author Gender Identification. Proceedings of LREC 2014 (2014)
4. Koppel, M., Argamon, S., Shimoni, A.R.: Automatically categorizing written texts by author gender. Literary and Linguistic Computing **17**(4), 401–412 (2002)
5. Argamon, S., Koppel, M., Pennebaker, J.W., Schler, J.: Mining the Blogosphere: Age, gender and the varieties of self-expression. First Monday 12(9) (September 2007)   DOI=Http://www.uic.edu/htbin/cgiwrap/bin/ojs/index.php/fm/article/view/2003
6. Ansari, Y.Z., Azad, S.A., Akhtar, H.: Gender Classification of Blog Authors. In: International Journal of Sustainable Development and Green Economic. Volume 2. (2013) ISSN no.:2315–4721
7. Burger, J.D., Henderson, J., Kim, G., Zarrella, G.: Discriminating gender on Twitter. In: Proceedings of the Conference on Empirical Methods in Natural Language Processing (EMNLP '11), Stroudsburg, PA, USA, Association for Computational Linguistics (2011) 1301–1309
8. Kobayashi, D., Matsumura, N., Ishizuka, M.: Automatic Estimation of Bloggers' Gender. In: Proceedings of International Conference on Weblogs and Social Media, Boulder: Omnipress (2007)
9. Zhang, C., Zhang, P.: Predicting gender from blog posts. Technical report, University of Massachusetts Amherst, USA (2010)
10. Mukherjee, A., Liu, B.: Improving gender classification of blog authors. In: Proceedings of the 2010 conference on Empirical Methods in natural Language Processing (EMNLP'10). (2010) 207–217 DOI=http:/www.aclweb.org/anthology/D10-1021
11. Peersman, C., Daelemans, W., Van Vaerenbergh, L.: Predicting Age and Gender in Online Social Networks. In: Proceedings of the 3rd International Workshop on Search and Mining User-Generated Contents (SMUC'11), Glasgow, UK (2011) 37–44

12. Sarawgi, R., Gajulapalli, K., Choi, Y.: Gender Attribution: Tracing Stylometric Evidence Beyond Topic and Genre. In: Proceedings of the Fifteenth Conference on Computational Natural Language Learning (Portland, USA, 9 - 24 June, 2011), Stroudsburg, PA, USA, Association for Computational Linguistics (2011) 78–86

13. Holmgren, J., Shyu, E.: Gender classification of facebook posts. (2013)

14. Rangel, F., Rosso, P.: Use of Language and Author Profiling: Identification of Gender and Age. In: Proceedings of the Tenth International Workshop on Natural Language Processing and Cognitive Science, Marseille, France (October 2013)

15. Marquardt, J., Farnadi, G., Vasudevan, G., Moens, M.F., Davalos, S., Teredesai, A., De Cock, M.: Age and Gender Identification in Social Media. Proceedings of CLEF 2014 Evaluation Labs (2014)

16. Gordon, E.: Sex, speech, and stereotypes: Why women use prestige speech forms more than men. Language in society **26**(1), 47–63 (1997)

17. Cameron, D.: Gender, Language, and Discourse: A Review Essay. Signs: Journal of women, Culture and Society 23(4) (1998) 945–973

18. Cameron, D.: Language, Gender, and Sexuality: Current Issues and New Directions. Applied linguistics 26(4) (2005) 482–502 DOI=10.1093/applin/ami027

19. Bucholtz, M.: You da man: Narrating the racial other in the production of white masculinity. Journal of sociolinguistics **3**(4), 443–460 (1999)

20. Bucholtz, M., Liang, A.C., Sutton, L.A.: Reinventing identities: The gendered self in discourse. Oxford University Press, New York (1999)

21. Fishman, P.M.: Interaction: The work women do. In: Language, Gender and Society, Rowley, Mass.: Newbury House (1983) 89–102

22. Lakoff, R.: Talking Power: The Politics of Language. Basic Books, New York (1990)

23. Lakoff, R.: Language and Women's Place. Harper and Row, New York (1975)

24. Alami, M., Sabbah, M., Iranmanesh, M.: Male-Female Discourse Difference in Terms of Lexical Density. Research Journal Of Applied Sciences, Engineering and Technology. **5**, 5365–5369 (2013)

25. Eckert, P.: Three waves of variation study: The emergence of meaning in the study of sociolinguistic variation. Annual review of Anthropology. **41**, 87–100 (2012)

26. Moore, E., Podesva, R.: Style, indexicality, and the social meaning of tag questions. In: Language in Society. Volume 38, Cambridge Univ Press (2009) 447–485

27. Bucholtz, M.: From 'Sex Differences' to Gender Variation in Sociolinguistics. In: University of Pennsylvania Working Papers in Linguistics (Papers from NWAV 30). Volume 8, University of Pennsylvania, Department of Linguistics (2002) 33–45

28. Bucholtz, M.: Theories of Discourse as Theories of Gender: Discourse Analysis in Language and Gender Studies. In: The Handbook of Language and Gender, Oxford Blackwell (2003) 43–68

29. Zheng, R., Li, J., Chen, H., Huang, Z.: A framework for authorship identification of online messages: Writing-style features and classification techniques. Journal of the American Society for Information Science and Technology **57**(3), 378–393 (2006)

30. Esuli, A., Sebastiani, F.: Sentiwordnet: A publicly available lexical resource for opinion mining. Proceedings of Language Resources and Evaluation (LREC). **6**, 417–422 (2006)

31. Witten, I.H., Frank, E.: Data Mining: Practical machine learning tools and techniques, (2nd edn. Elsevier, Morgan-Kaufman Series of Data Management Systems), San Francisco (2005)

Modified Group Delay Based Features for Asthma and HIE Infant Cries Classification

Anshu Chittora$^{(\boxtimes)}$ and Hemant A. Patil

Dhirubhai Ambani Institute of Information and Communication Technology,
Gandhinagar, Gujarat, India
{anshu_chittora,hemant_patil}@daiict.ac.in
http://www.daiict.ac.in

Abstract. Modified group delay features have shown promising results for automatic speech recognition (ASR) task. In this paper, features are derived from the modified group delay function. These features are then used to classify asthma and hypoxy ischemic encephalopathy (HIE) infant cries. Our experimental results show that the performance of the proposed features is better than the state-of-the-art feature set, *i.e.*, Mel frequency cepstral coefficients (MFCC). Best classification accuracy is achieved with the proposed features is *90.38* % as opposed to *84.92* % obtained with MFCC, when applied to a SVM classifier with radial basis function kernel. The proposed feature set performs much better for classification of asthma infant cries. However, for HIE both features perform equally well. The class separability distance of the group delay feature is higher than the MFCC feature (for most of the features Bhattacharya class separation distance is higher by *0.2* units compared to MFCC), which also confirms that the proposed features are better than MFCC.

Keywords: Group delay · Fourier transform · Support Vector Machine (SVM) classifier · Magnitude spectrum · Phase spectrum

1 Introduction

Diagnosis of pathologies in infants is a difficult task. It is due to the inability of an infant to convey the symptoms in the form of speech and inability of the caretakers to understand the symptoms of diseases. Cry and facial expressions are the only means to communicate their needs. It has been found that cry carries acoustic features which vary with the reasons of crying, e.g., cry acoustics for pain, discomfort and hunger are different. Researchers have worked on the analysis of cry using pitch, harmonics, short time spectral features, formants and energy. Classification of various pathological cries has also been reported by some researchers for classification of normal and deaf infant cries, normal and cardiac disorders, normal and asphyxia. All have reported significant differences in cry patterns of normal and pathological cries [1-5]. In this paper, we are proposing a classification of asthma and Hypoxy Ischemic Encephalopathy (HIE) infant cries using modified group delay-based features. In our work, we are assuming

© Springer International Publishing Switzerland 2015
P. Král and V. Matoušek (Eds.): TSD 2015, LNAI 9302, pp. 595–602, 2015.
DOI: 10.1007/978-3-319-24033-6_67

that the classification of asthma and HIE cries from the normal infant cries is possible and hence the goal here is to investigate about discrimination ability of proposed feature set for different pathologies.

Asthma is a chronic inflammatory disease of the airways in which airways become blocked or narrowed. Almost five percent of infants (less than one year after the birth) suffer from asthma [6]. Diagnosis of asthma in early ages especially in infancy is very difficult. HIE is a condition in which brain does not receive enough oxygen. Because of oxygen deficiency, brain cells may begin dying, resulting in brain damage due to severe oxygen deficiency after the birth [7]. HIE is a leading cause of deaths or severe impairments such as delays in physical and mental development among infants.

Mel frequency cepstral coefficients (MFCC) feature set has been used as the state-of-the-art feature by many researchers for infant cry classification [3]. In this paper, features are derived from the modified group delay function from the infant cry signal. Modified group delay has been used widely for speech recognition task [8]. It has also been used for speaker verification, formant analysis, and signal reconstruction from minimum phase systems [9],[10]. Performance of the proposed features is compared with the state-of-the-art MFCC feature set and is found to be better for proposed problem under study.

The rest of the paper is organized as follows: Section 2 presents details of the modified group delay whereas method for feature extraction is detailed in Section 3. Section 4 briefs the details of Support Vector Machine (SVM) classifier. In Section 5, experimental results are discussed. Conclusions and future work are reported in Section 6.

2   Modified Group Delay Function

Modified group delay function has been proved effective in semi-automatic segmentation of speech and features derived from the modified group delay function have been applied for language identification, speech recognition and speaker identification [8]-[9]. Normally, the information in the speech signal is represented by the short-time Fourier transform (STFT). In that method, the magnitude of the FT is considered for feature extraction and phase of FT is ignored. The significance of phase in speech recognition has been shown by the researchers. It has been proved that in presence of significant phase distortions, the recognition of speech by human ear is very poor. The information in the FT phase can be extracted by the negative derivative of the phase, $i.e.$, group delay function, is given by

$$\tau(\omega) = -\frac{d(\theta(\omega))}{d(\omega)} \tag{1}$$

where $\theta(\omega)$ is the unwrapped phase function. The group delay function can be computed from the FT magnitude function as given below [9]:

$$\tau_x(\omega) = \frac{X_R(\omega)Y_R(\omega) + X_I(\omega)Y_I(\omega)}{|X(\omega)|^2} \tag{2}$$

where X(ω) and Y(ω are the FT of the signal x(n) and nx(n) , respectively. The subscripts R and I represents the real and imaginary parts of the FT, respectively. The group delay function requires that the signal be minimum phase signal. The group delay function becomes spiky on/ near the unit circle in z-domain [8]. To remove the source information, the denominator term $|X(\omega)|$ can be replaced by its cepstrally smoothed envelope. The cepstrally smoothed envelope is given by $S_c(\omega)$. The modified group delay is represented as [9]:

$$\tau_x(\omega) = \frac{X_R(\omega) * Y_R(\omega) + X_I(\omega) * Y_I(\omega)}{S_c(\omega)^2} \tag{3}$$

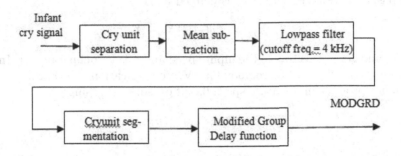

Fig. 1. Block diagram for extraction of features from modified group delay function.

To reduce the dynamic range of modified group delay spectrum and peaks at the formant locations, two parameters named, α and γ were introduced in [9]. The new modified group delay is defined as

$$\tau_c(\omega) = -\frac{\tau(\omega)}{|\tau(\omega)|}(|\tau(\omega)|)^\alpha \tag{4}$$

where $\tau(\omega) = -\frac{d(\theta(\omega))}{d(\omega)}$ and the parameters range from $0 < \alpha < 1, 0 < \gamma < 1$.

3   Feature Extraction

Features are extracted from the proposed method as shown in the Fig. 1. Initially, the cry recording from an infant is divided in several cry units. From the cry unit, mean is subtracted. The cry signal is filtered with a 4th order Butterworth lowpass filter to remove high frequency noise. Since most of the signal information is contained in signal below *4 kHz*, the filter cutoff frequency is taken as *4 kHz*. The cry signal is segmented into non-overlapping frames of *10 ms* duration. For each of the frame, modified group delay as mentioned in eq. (4) is computed. In our experiments, we have kept $\gamma = 1$ and $\alpha = 0.1$. For the same frame, MFCC are also calculated. One feature vector is extracted for a cry unit. Mean is taken over all the frames of a cry unit (same is done for MFCC feature as well). The modified group delay is taken as feature vector for the proposed work.

4 Support Vector Machine (SVM) Classifier

Support vector machine (SVM) classifier is a popular pattern classifier used for classification and regression. SVM does not take the assumption of distribution of features being Gaussian. Especially in cases, where the data size is very small, one cannot work with Gaussian approximation. SVM is a discriminative approach, where the boundaries are learned froms the training data. SVM transforms the data to a high-dimensional feature space using Covers theorem of separation [11]. Data which is nonlinear in the original space becomes linear in the high-dimensional space using kernel functions. Hence, it is possible to get a linear hyperplane in the high-dimensional space which can form a decision boundary. Radial basis function (RBF) kernel is defined as [12]:

$$K(X_i, X_j) = exp(-\gamma(|X_i - X_j|^2)) \qquad (5)$$

where X_i and X_j are vectors in the input space and γ is kernel parameter. In our work, we have used LIBSVM toolbox for SVM classification [13]. K corresponds to the inner product in a feature space based on some mapping given by

$$K(x, y) = < \phi(x), \phi(y) > \qquad (6)$$

5 Experimental Results

5.1 Database Used

In this paper, infant cry data was collected from the NICU units of King George Hospital (K.G.H), Visakhapatnam and Child Clinic, Visakhapatnam, India. The duration of recorded infant cry varied from *30 s* to *2* min. Data was collected with a Cenix digital voice recorder at a sampling rate of *12 kHz* and *12*-bits quantization. The infant cries were recorded during vaccination and while infants were crying due to inconvenience during medical examination, wet diaper or hunger (details of data collection is available in [14]). The number of cries with asthma pathology are *7* infant cries and of infants suffering from HIE are *13*. One cry per infant was recorded. These infant cries are then divided into cry units for further analysis. A cry unit is defined as the cry sound produced during one expiratory and inspiratory cycle. The cry samples were divided into cry units manually. In asthma database, there are total *182* cry units and in HIE dataset, there are *205* cry units. Per infant number of cry units is not same, it depends on the cry duration.

5.2 Experimental Results

For each of the cry unit, features are extracted as explained in Section 3. These features are then applied to a SVM classifier for training and testing. In our experiments, we have considered RBF kernel for SVM classifier. Classification accuracy (in %) is defined as the ratio of total number of cry units correctly classified to total number of cry units given for testing, multiplied by *100*. In our experiments 7-fold cross-validation is performed to validate the results since the sample size is very small. In one fold one of asthma infants cry and two HIE infants cries are considered in testing and remaining infants are taken for training the system.

Table 1 shows the classification accuracy (in %) for the asthma and HIE cry samples classification with both MODGRD and MFCC feature sets using SVM classifier with RBF kernel. It can be observed that the proposed features are performing better in classification of pathologies. Table 2 shows the classification accuracy of the same experiment with different values of gamma (γ) parameter in RBF kernel function. It is observed that reducing to *0.125* improves the performance of proposed features whereas it decreases the classification accuracy of MFCC features set.

Table 1. Classification accuracy (in %) with MFCC and proposed MODGRD features applied to SVM classifier with RBF kernel with $\gamma = 0.25$ for 7-fold cross-validation.

Feature	Feature Size	Classification Accuracy (in %)
MODGRD	1x256	90.38
MFCC	1x12	84.92

Table 2. Classification accuracy (in %) with MFCC and proposed MODGRD features with radial basis function (RBF) kernel applied to SVM classifier for 7-fold cross-validation.

Value of γ	MFCC	MODGRD
0.25	84.92	90.38
0.125	87.42	90.82

Table 3. Confusion matrix with MFCC feature set using $\gamma = 0.25$ with RBF kernel function for 1-fold experiment.

	Asthma	HIE
Asthma	27	7
HIE	0	18

Confusion matrices are shown in Table 3 and Table 4 to compare the performances of the two features in classification of asthma and HIE infant cries. For the same experimental setup confusion matrices are shown. These results are reported when infant *4* of asthma and infant *1* and *2* of HIE are considered for testing. Most of the time in our 7-fold cross validation experiment, asthma cry units are classified as HIE. However, MODGRD features captures the differences in two cry patterns and can classify the two pathologies almost correctly. MFCC and MODGRD are able to classify HIE correctly. However, MODGRD is also capable of classifying asthma samples correctly.

Table 4. Confusion matrix with MODGRD feature set using $\gamma = 0.25$ with RBF kernel function for 1-fold experiment.

	Asthma	HIE
Asthma	34	0
HIE	0	18

To compare the performances of the proposed features and MFCC, their class separability distances are calculated using Bhattacharya bound [15]. The Bhattacharya distance is a measure of similarity of two discrete or continuous probability distributions. It is used to measure separability of classes in classification. Bhattachraya distance between two classes is given by [15]

$$\tau(\omega) = -\frac{d(\theta(\omega))}{d(\omega)} \tag{7}$$

where p and q are the two classes, σ_p and σ_q are the variances of p and q classes, respectively. μ_p and μ_q are the means of classes p and q, respectively. For better visibility of the curves, only first 12 coefficients of the feature vectors are considered in the plot. Fig. 2 shows that the class separation of the proposed features is higher than the MFCC feature set.

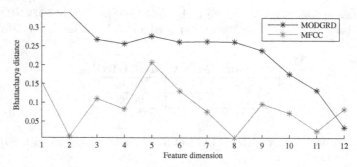

Fig. 2. Class separability of MFCC and proposed features.

Fig. 3 shows the signal waveform and their respective modified group delay functions for the three cases, normal, HIE and asthma infant cries. It can be observed that the in case of normal and HIE infant cries, we get higher number of peaks of group delay function. However, in asthma infant cries, we get well separated peaks in the group delay function. These peaks in the group delay functions correspond to formants of the infants. Hence, there may be changes in the formants and their harmonics of the infants corresponding to a particular pathology.

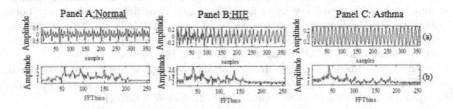

Fig. 3. Modified group delay function for normal (Panel A), HIE (Panel B) and asthma (Panel C) infant cries. In all subfigures, figure (a) corresponds to infant cry signal and figure (b) corresponds to its group delay function.

6   Summary and Conclusions

In this paper, features derived from the group delay are proposed for the classification of asthma and HIE infant cries. It has been observed that the modified group delay-based cepstral features outperform the MFCC feature set which has been the state-of-the-art method in infant cry classification. Both the features are capable of capturing the information in HIE cry samples. However, classification of asthma pathology is better in the proposed features. The improved performance of the proposed features may be due to its property that it captures the phase information of the signal, while MFCC is derived from the magnitude spectrum of the FT, ignoring the phase spectrum completely. This difference in two feature sets suggests that in asthma patients, phase variations are very frequent. These fast variations in phase of the signal may be introduced by the blocked airways. It has also been observed that the selection of kernel function is also important in the classification work. In this work, we have assumed that classification of normal and pathological cry samples is possible with some features and after this classification, we can use the proposed method for classification of asthma and HIE diseases in infants. As a part of our future work, we would like to come up with a feature set which can classify normal cries from pathological (asthma and HIE) infant cries.

Acknowledgments. Authors would like to thank Department of Electronics and Information Technology (DeitY), Government of India, New Delhi, India and authorities of DA-IICT for carrying out this research work.

References

1. Xie, Q., Ward, R.K., Laszlo, C.A.: Determining normal infants level-of-distress from cry sounds. In: IEEE Canadian Conf. on Electrical and Computer Engineering, pp. 1094–1096 (1993)
2. Patil, H.A.: Cry baby: using spectrographic analysis to assess neonatal health from an infants cry. In: Neustein, A. (ed.) Advances in Speech Recognition, Mobile Environments, Call Centers and Clinics, pp. 323–348. Springer (2010)

3. Reyes-Galaviz, O.F., Cano-Ortiz, S.D., Reyes-Garcia, C.A.: Evolutionary-neural system to classify infant cry units for pathologies identification in recently born babies. In: 7th Mexican Int. Conf. on Artificial Intell. (MICAI), pp. 330–335 (2008)

4. Patil, H.A.: Infant identification from their cry. In: 7th Int. Conf. on Advances in Pattern Recognition, ICAPR, vol. 4, pp. 107–110. ISI, Kolkata (2009)

5. Hariharan, M., Sindhu, R., Yaccob, S.: Normal and hypoacoustic infant cry signal classification using time frequency analysis and general regression neural network. Comput Methods Programs in Biomed. **108**(2), 559–569 (2012)

6. Asthma   details.   http://www.desimd.com/health-education/paediatrics/ understanding-asthma-in-infants (last accessed on March 30, 2015)

7. HIE details. http://en.wikipedia.org/wiki/ (last accessed on March 30, 2015)

8. Hegde, R.M., Murthy, H.A., Gadde, V.R.R.: Significance of modified group delay feature in speech recognition. IEEE Trans. on Audio, Speech and Lang. Process. **15**(1), 190–202 (2007)

9. Murthy, H.A., Gadde, V.: The modified group delay function and its application to phoneme recognition. In: IEEE Int. Conf. on Acoustics, Speech, and Signal Processing (ICASSP), vol. 1, pp. 68–71 (2003)

10. Yegnanarayana, B., Saikia, D., Krishnan, T.: Significance of group delay functions in signal reconstruction from spectral magnitude or phase. IEEE Trans. on Acoustics, Speech and Signal Processing **32**(3), 610–623 (1984)

11. Boser, B.E., Guyon, I., Vapnik, V.: A training algorithm for optimal margin classifiers. In: Proc. of th Annual Workshop on Computational Learning Theory, pp. 144–152. ACM Press (1992)

12. SVM kernel details. http://en.wikipedia.org/wiki/Polynomial_kernel (last accessed on March 30, 2015)

13. Chang, C.-C., Lin, C.-J.: LIBSVM A library for support vector machines. ACM Transactions on Intelligent Systems and Technology **2**, 27:1–27:27 (2011). http:// www.csie.ntu.edu.tw/~cjlin/libsvm (last accessed on August 25, 2014)

14. Buddha, N., Patil, H.A.: Corpora for analysis of infant cry. In: Int. Conf. on Speech Databases and Assessments, Oriental COCOSDA, Hanoi, Vietnam, December 4–6, 2007

15. Bhattacharyya, A.: On a measure of divergence between two statistical populations defined by their probability distributions. Bulletin of the Calcutta Mathematical Society **35**, 99–109 (1943)

Self-Enrichment of Normalized LMF Dictionaries Through Syntactic-Behaviors-to-Meanings Links

Imen Elleuch[1]([✉]), Bilel Gargouri[1], and Abdelmajid Ben Hamadou[2]

[1] FSEGS, B.P. 1088, 3018 Sfax, Tunisia
{imen.elleuch,bilel.gargouri}@fsegs.rnu.tn
[2] ISIMS, B.P. 242, 3021 Sakiet-Ezzit Sfax, Tunisia
abdelmajid.benhamadou@isimsf.rnu.tn

Abstract. This paper reports on the status of an ongoing work to enrich the syntactic extension of normalized LMF dictionaries. It proposes an approach to find out the syntactic behaviors associated with the lexical entries and to link them to the corresponding meanings of these entries. The used corpus is constructed from texts associated with each meaning in the dictionaries, such as definitions and contexts. These texts are largely available and semantically controlled because of their association to the meanings, which fosters the establishment of efficient links. The proposed approach has been implemented and tested on an available Arabic normalized dictionary. The experiment concerned about 9,800 verbal entries. Both the identification of syntactic behaviors and the establishment of syntactic-behaviors-to-meanings links have been evaluated.

Keywords: Self-enrichment · LMF dictionaries · Syntactic behaviors · Meanings

1 Introduction

Syntactic lexicons are obviously an extremely valuable basic element for Natural Language Processing (NLP) systems. As demonstrated by [1], an exhaustive and detailed syntactic lexicon improves the performance of syntactic parsers. Also, as has been repeatedly argued in [2,3], a syntactic lexicon is the core component resource for information extraction or for machine translation systems. Syntactic lexicons contain a large amount of knowledge such as the sub-categorization frames that provide the syntactic behaviors of the entries. These syntactic behaviors specify the number and type of the arguments used. The automatic acquisition of sub-categorization frames has been an active research area since the mid-90s [4,5].

Because of their importance, several syntactic lexicons have emerged for different languages. For French, we can mention the lexicon-grammar [6] that contains a detailed list of lexical unit complementation. Another example is the Lefff (Lexicon of French Inflected Forms) [7] that is a morphological and syntactic lexicon that includes partial sub-categorization knowledge constructed using an automatic corpus analysis and a fusion data from different resources.

© Springer International Publishing Switzerland 2015
P. Král and V. Matoušek (Eds.): TSD 2015, LNAI 9302, pp. 603–610, 2015.
DOI: 10.1007/978-3-319-24033-6_68

For English, we can enumerate the Comlex Syntax [8], which is a handmade moderately broad coverage lexicon including detailed knowledge about the syntactic characteristics of each lexical item. The VerbNet [9] is another lexicon for the English language that is based on Levins [10] verb classification.

For Arabic, there are few examples of such syntactic lexicons. One of them is the Arabic syntactic lexicon [11], which represents a syntactic lexicon compliant to the LMF standard. The Arabic VerbNet [12] is an Arabic version of English VerbNet classifying verbs into classes sharing the same syntactic and semantic proprieties based on the Levins [10] verbs classification.

The main criticisms that we can level at the existing syntactic lexicons concern their models and their contents. Indeed, each lexicon is modeled with respect to a particular use without providing solutions for the extension and re-use. The represented knowledge is not well detailed, notably the link between syntactic and semantic knowledge. In particular, we note the absence of correspondence between each meaning of a lexical entry and the syntactic.

In this context, the Lexical Markup Framework (LMF) standard [13] provides effective solutions to the modeling problems. Thus, several works have started to build lexicons compliant to this standard including syntactic lexicons such as the "syntactic lexicon for Arabic verbs" [11] and the LG-LMF [14]. However, the problem of enrichment persists. Indeed, these lexicons describe the syntactic behaviors related to the lexical entries but not to their meanings because they use external dictionaries as well as corpus or other linguistic resources to identify the syntactic behavior.

In this paper, we propose an approach to find out the syntactic behaviors associated with the lexical entries in LMF dictionaries and to link them to the corresponding meanings. This approach is based on the analysis of a corpus constructed from the textual contents associated with each meaning in LMF dictionaries such as definitions and contexts. These contents are largely available and semantically controlled because of their association to the meanings, which fosters the establishment of efficient links.

The remainder of this paper will be devoted primarily to the presentation of the proposed approach to the self-enrichment of LMF normalized dictionaries with syntactic behavior related to meaning of lexical entries. Thereafter, the experimentation carried out on an available normalized Arabic dictionary will be described. Finally, related works and their comparison with our study will be given. Some future works will be announced in the conclusion.

2   Related Works

In this section, we will present some works related to syntactic lexicons. We will define the Arabic syntactic lexicon [11] and the Arabic VerbNet [12] syntactic lexicon for the Arabic language.

The Arabic syntactic lexicon [11] is a lexical resource dedicated to the representation of syntactic properties of Arabic verbs using the LMF standard. This lexicon was built semi automatically using the editor Lexus. The enrichment approach of the lexicon with syntactic behavior of Arabic verbs consists of three steps: the manually specification of the sub-categorization frames accepted by verbs in Arabic language. Then the enrichment of the lexicon with the syntactic behaviors identified in the previous step by using the Lexus editor. And finally, the edition of Arabic verb lemmas and the affectation of one or many sub-categorization frames to each entered verb. The lexicon contains 2,500 verb lemmas with an average of 2.7 sub-categorization frames per verb. This is a small syntactic lexicon of Arabic verbs including only 17 sub-categorization frames that do not represent the specificities of Arabic language. Also, the sub-categorization frames were affected to verbs without consider meanings.

The Arabic VerbNet [12] is a class-based lexicon for Arabic verbs based on Levin's classification [10]. The same building procedure used in the English Verb-Net was re-used to construct the VerbNet for the Arabic language. To describe verbs, both syntactic and semantic features were considered. Indeed, all the verbs sharing syntactic and semantic properties were grouped together into the same class. Each frame includes the root, the derived forms the present participle and the past participle of the verb, the thematic roles of arguments, the sub-categorization and the syntactic and semantic description of each verb. Some Arabic verb classes were subdivided into subclasses. The Arabic VerbNet lexicon contains 291 verb classes and 7,937 verbs represented by 1,202 frames. These frames take into account, together with the syntactic features, the thematic role of arguments of the syntactic behaviors.

Even though all the approaches presented in the above studies on the Arabic language suggest some interesting ideas, each one of them includes some shortcomings. Indeed, the syntactic lexicon of [11] is a very small lexicon representing only the syntactic aspects of very few Arabic verbs while, the Arabic VerbNet does not represent the native features of Arabic verbs because its a simple translation of the classes used in the English VerbNet.

3   Proposed Approach

In this section we will present an overview concerning the LMF syntactic model and its specificities and the syntactic proposed approach to enrich LMF dictionaries with the syntactic behaviors of lexical units using corpora.

3.1   Basis Concepts

In an LMF normalized dictionary, each meaning of a lexical entry is represented by a class named Sense. Each Sense can be attached to the Definition class that represents a narrative description of a sense. It is displayed for human users to facilitate their understanding of the meaning of a Lexical Entry and is not meant to be treated by computer programs. A Sense instance can have zero to many definitions. Contrary to the Definition class, which is dedicated to human users, the Context class selected from the Machine Readable Dictionary (MRD) extension supports electronic machine readable dictionary access for both human use and machine processing. Nonetheless, the Context class represents a text string that provides authentic context for the use of the word form managed by the lemma of lexical entry. It is in a zero to many aggregate associations with the Sense class.

Based on the specificities of the Context LMF class used to describe the uses of the meanings related to the lexical entries by a simple sentence on the one hand, and also with the aim to find out the syntactic behaviors associated to the meanings of the lexical entries in LMF normalized dictionaries on other hand, we propose to analyze this textual content in order to identify the syntactic behavior related to each meaning.

Therefore, the analysis of the textual content of Context in LMF dictionaries represents the main idea used in the proposed approach to identify syntactic behaviors connected to the meanings of the lexical entries in LMF dictionaries.

3.2   Steps of the Approach

The proposed approach for the identification of the syntactic behaviors associated with the meanings of the lexical entries in LMF normalized dictionaries consists of five steps as shown in Figure 1.

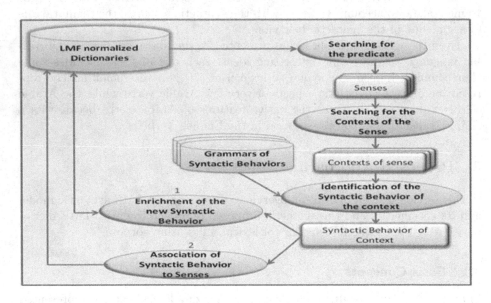

Fig. 1. Proposed approach

To describe each step of the proposed approach, we take the "leave" verb from the Wiktionary [1] represented as an LMF lexical entry. As shown in Figure 2

---

[1] http://en.wiktionary.org/wiki/leave

```
▼<LexicalEntry id="14">                              ▼<Context>
   <feat att="partOfSpeech" val="verb"/>                <feat att="text" val="I left the country and I left my wife"/>
   ▼<Lemma>                                          </Context>
      <feat att="writtenForm" val="leave"/>          ▼<Context>
   </Lemma>                                              <feat att="text" val="We leave tomorrow"/>
   ▼<Sense id="14P1">                                 </Context>
      ▼<Context>                                      ▼<Definition>
         <feat att="text" val="I left my car at home and took a bus to work"/>   <feat att="text" val="To depart, to separate from"/>
      </Context>                                         </Definition>
      ▼<Context>                                      </Sense>
         <feat att="text" val="The lightning left her dazzled for several minutes"/>  ▼<Sense id="14P3">
      </Context>                                      ▼<Context>
      ▼<Definition>                                      <feat att="text" val="When my father died, he left me the house"/>
         <feat att="text" val="To have a consequence or remnant"/>   </Context>
      </Definition>                                   ▼<Definition>
   </Sense>                                              <feat att="text" val="To transfer something"/>
   ▼<Sense id="14P2">                                  </Definition>
      ▼<Context>                                      </Sense>
         <feat att="text" val="I left him to his reflections"/>   <SyntacticBehaviour id="L4C1" subcategorizationFrames="SVC1C2"/>
      </Context>                                      <SyntacticBehaviour id="L4C2" subcategorizationFrames="SVC"/>
                                                    </LexicalEntry>
```

Fig. 2. "leave" verb in LMF dictionary

given bellow, this verb has three senses described by Contexts and Definitions and two syntactic behaviors.

Searching for the Predicate. A predicate is a well-known and extensively studied concept in linguistics. It can be a verb, a noun, an adjective or an adverb. The proposed approach can be applied to any predicate. Therefore, the first step aims to search for the treated predicate and thereafter the attached meaning represented by the Sense LMF class.

Illustration:
For the example of Fig. 2, the first step identifies the predicate having "14" identifier that corresponds to the "leave" verb. This predicate has three senses identified respectively by "14P1", "14P2" and "14P3" identifiers.

Searching for the Contexts of the Senses. Each Sense is described by Context classes. These Contexts are largely available and semantically controlled because of their association with the meanings and their representation as simple sentences without complex structure. Indeed, when aiming to attach syntactic behaviors to Senses, it is sufficient to treat these efficient Contexts. For this reason, this step intends to search for one Sense all attached Contexts.

Illustration:
The application of the second step of the proposed approach on the "leave" predicate identifies two contexts for the first sense "14P3", three contexts for the sense "14P2", and one context for sense "14P3".

Identification of the Syntactic Behavior of the Context. All Contexts searched previously will be treated in order to identify their corresponding syntactic behaviors. The identification of the syntactic behavior is performed by

Grammars. These later must be constructed for each syntactic behavior of the treated language and they should be able to differentiate both essential and optional supplement in simple sentence. Thus, this step needs the Contexts of Sense and the Grammars of syntactic behaviors in input.

Illustration:
For the first context "I left my car at home and took a bus to work" of the first sense "l4P1", Grammars of syntactic behaviors is able to segment this sentence in order to identify: "I": the Subject, "left": an inflected form of the "leave" treated predicate, "my car": First Object, "at home": Second Object, "and": conjunction and "took a bus to work": other sentence doesnt contain the treated "leave" predicate. So, this second sentence will not be segmented. Grammars applied to the first sentence of the context recognize the syntactic behavior SVC1C2 (Subject Verb First Complement Second Complement). The same treatment applies to other contexts identify: the same SVC1C2 syntactic behavior for the second context of the first sense, the SVC1C2 (Subject Verb First Complement, Second Complement), SVC and SV (Subject Verb) syntactic behaviors for the three contexts of the second sense and the SVC1C2 for the context of the third sense.

Enrichment of the New Syntactic Behavior. Each predicate in the LMF normalized dictionaries has a list of syntactic behaviors. But when we apply all grammars of syntactic behaviors, new syntactic behaviors can appear. The new syntactic behavior must be added to the list of syntactic behaviors of the treated predicate in the LMF dictionaries.

Illustration:
In the LMF dictionary, "leave" verb predicate has two syntactic behaviors: SVC and SVC1C2 whereas, the treatment of contexts in the last step finds the SV new syntactic behavior that doesnt appear with the "leave" verb. Indeed, this step aims to enrich the LMF dictionary with adding a new SyntacticBehaviour LMF class instance:

<SyntacticBehaviour id="l4C3" subcategorizationFrames="SV">

Association of Syntactic Behaviors to Senses. When the syntactic behaviors of senses are detected, all that remains to do is to associate these syntactic behaviors with the Senses in the LMF dictionary. Thus, the purpose of this step is the association of the current Sense with the relevant syntactic behavior.

Illustration:
The sense "l4P1" has the SVC1C2 syntactic behavior. The sense "l4P2" has the SVC, SVC1C2 and SV syntactic behaviors. The sense "l4P3" has the SVC1C2 syntactic behavior. The performed association adds for each SyntacticBihaviour the identifier of related senses:

<SyntacticBehaviour id="l4C1" **senses="l4P1 l4P2 l4P3"** subcategorizationFrames="SVC1C2">
<SyntacticBehaviour id="l4C2" **senses= "l4P2"** subcategorizationFrames="SVC">
<SyntacticBehaviour id="l4C3" **senses= "l4P2"** subcategorizationFrames="SV">

4   Experiment and Results

4.1   Experiment

In order to evaluate our proposed approach, we carried out an experiment using the El-Madar Arabic LMF dictionary [15]. This dictionary contains about 37,000 lexical entries: 10,800 verbs, 3800 roots and 22,400 nouns. These lexical entries incorporate morphological knowledge such as the part-of-speech, some inflected forms, some derived forms etc. It includes also semantic features like the synonymy relationships that can relate one sense with other senses of different lexical entries. On the other hand, regarding syntactic knowledge, El-Madar dictionary includes 155 syntactic behaviors of Arabic verbs and 5,000 verbs are linked to those syntactic behaviors. The experiment concerned only the verbal predicates.

4.2   Results

El-Madar dictionary [14] contains up to now 10,800 verbs. However, the experiment that we have performed treats only 9,800 verbs because there are actually 1,000 verbs that are not yet enriched by their appropriate Senses. The application of the proposed approach to these verbs generates 30,300 affectations that are added between the syntactic behaviors and meanings. A human expert has evaluated a sample of 1,550 affectations selected with a proportion of 10 affectations per kind of syntactic behavior. The expert has identified 246 incorrect affectations and 137 missing affectations. So, for these 1,550 affectations we obtain 90.15% rate for the Recall and 83.60% rate for the Precision.

The error rate detected is due to the fact that the sentences representing the Contexts are complex and the developed Grammars for the analysis don't treats certain complexes linguistic phenomena or the Contexts given by the author of the dictionary do not exhibit the appropriates syntactic behaviors of the verb.

5   Conclusion and Perspectives

This paper reports on an ongoing approach which aims to enrich an LMF normalized dictionary with the syntactic behaviors related to the meanings of predicates. It is a self-enrichment because the approach uses the textual content named Context to identify the syntactic behaviors of each meaning of a lexical entry. This Context is a largely available text in the LMF normalized dictionary and it is semantically controlled because of its association with the meanings. This association fosters the establishment of efficient links between the syntactic behaviors of treated predicate and their meanings. An experiment was carried out on an available El-Madar Arabic LMF dictionary. In this experiment, we have been able to link syntactic behaviors with the meanings of 9,800 verbs.

In the future, we intend to enrich an LMF normalized dictionary with Context from external corpora. Thus, the constructed grammars must be able more elaborated in order to take into account complexes sentences with linguistic phenomena like metaphor. Moreover, we intend to propose an approach to enrich semantic predicates.

References

1. Carroll, J., Fang, A.C.: The automatic acquisition of verb subcategorisations and their impact on the performance of an hpsg parser. In: Proceedings of the 1st International Conference on Natural Language Processing. Springer (2004)
2. Jijkoun, V., De Rijke, M., Mur, J.: Information extraction for question answering: Improving recall through syntactic patterns. In: Proceedings of the 20th International Conference on Computational Linguistics, p. 1284. Association for Computational Linguistics (2004)
3. Surdeanu, M., Harabagiu, S., Williams, J., Aarseth, P.: Using predicate-argument structures for information extraction. In: Proceedings of the 41st Annual Meeting on Association for Computational Linguistics, vol. 1, pp. 8–15. Association for Computational Linguistics (2003)
4. Manning, C.D.: Automatic acquisition of a large sub-categorization dictionary from corpora. In: Proceedings of the 31st Annual Meeting on Association for Computational Linguistics, pp. 235–242. Association for Computational Linguistics (1993)
5. Briscoe, T., Carroll, J.: Automatic extraction of sub-categorization from corpora. In: Proceedings of the Fifth ACL Conference on Applied Natural Language Processing, pp. 356–363. Association for Computational Linguistics (1997)
6. Gross, M.: Méthodes en syntaxe: régime des constructions complétives, Hermann, vol. 1365 (1975)
7. Sagot, B.: The lefff, a freely available and large-coverage morphological and syntactic lexicon for french. In: Proceedings of the 7th International Conference on Language Resources and Evaluation (LREC 2010) (2010)
8. Grishman, R., Macleod, C., Meyers, A.: Comlex syntax: Building a computational lexicon. In: Proceedings of the 15th Conference on Computational linguistics, vol. 1, pp. 268–272. Association for Computational Linguistics (1994)
9. Kipper, K., Korhonen, A., Ryant, N., Palmer, M.: A large-scale classification of english verbs. Language Resources and Evaluation 42(1), 21–40 (2008)
10. Levin, B.: English verb classes and alternations: A preliminary investigation. University of Chicago press (1993)
11. Loukil, N., Haddar, K., Hamadou, A.B.: A syntactic lexicon for arabic verbs. In: Proceedings of the 7th International Conference on Language Resources and Evaluation (LREC 2010) (2010)
12. Mousser, J.: A large coverage verb taxonomy for arabic. In: Proceedings of the 7th International Conference on Language Resources and Evaluation (LREC 2010) (2010)
13. Francopoulo, G., George, M.: Language resource management-lexical markup framework (lmf). Technical report, Technical Report ISO/TC 37/SC 4 (2008)
14. Laporte, E., Tolone, E., Constant, M.: Conversion of lexicon-grammar tables to lmf. application to french. LMF: Lexical Markup Framework, pp. 157–187 (2013)
15. Khemakhem, A., Gargouri, B., Haddar, K., Ben Hamadou, A.: Lmf for arabic. chapter in the book LMF: Lexical Markup Framework, pp. 83–96 (2013)

Author Index

Printed in the United States
By Bookmasters